Plant Essential Oil with Biological Activity II

Plant Essential Oil with Biological Activity II

Editors

Hazem Salaheldin Elshafie
Ippolito Camele
Adriano Sofo

MDPI

Basel • Beijing • Wuhan • Barcelona • Belgrade • Novi Sad • Cluj • Manchester

Editors
Hazem Salaheldin Elshafie
University of Basilicata
Potenza
Italy

Ippolito Camele
University of Basilicata
Potenza
Italy

Adriano Sofo
University of Basilicata
Matera
Italy

Editorial Office
MDPI
St. Alban-Anlage 66
4052 Basel, Switzerland

This is a reprint of articles from the Special Issue published online in the open access journal *Plants* (ISSN 2223-7747) (available at: https://www.mdpi.com/journal/plants/special_issues/Essential_Oil_2).

For citation purposes, cite each article independently as indicated on the article page online and as indicated below:

Lastname, A.A.; Lastname, B.B. Article Title. *Journal Name* **Year**, *Volume Number*, Page Range.

ISBN 978-3-7258-0379-8 (Hbk)
ISBN 978-3-7258-0380-4 (PDF)
doi.org/10.3390/books978-3-7258-0380-4

Contents

About the Editors

Hazem Salaheldin Elshafie

Hazem Salaheldin Elshafie is a Researcher at the University of Basilicata, School of Agricultural, Forestry, Food and Environmental Sciences (SAFE), Via dell'Ateneo Lucano 10, 85100 Potenza (PZ), Italy. He received qualification as an Assistant Professor in "Microbiology" from the Supreme Council of Egyptian Universities, Cairo, Egypt, in 2019. He also received qualification as a second-level University Professor in "Plant Pathology" from the National Scientific Qualification, Ministry of Education, Ministry of University and Research, in 2021. He obtained a PhD in 2012 from Basilicata University, Italy; a Master of Science in 2009 from the Mediterranean Agronomic Institute of Bari, Italy; and a Bachelor of Science in 2004 from the Faculty of Science, Zagazig University, Egypt. Hazem Salaheldin Elshafie is an Associate Editor of *Frontiers in Microbiology* and a Topic Editor of *Plants*, MDPI.

Ippolito Camele

In 1983, Ippolito Camele graduated from the University of Naples "Federico II" with a degree (Laurea) in agricultural sciences. He spent two years as a graduate student at the Institute of Plant Pathology at the University of Naples under the guidance of Prof. A. Ragozzino, carrying out studies in Plant virology. He was a researcher (May 1988 – March 2002) at the Institute of Plant and Forest Pathology, University of Basilicata. Since the 1st of March 2002, he has been acting as an associate professor at the same institution. He had worked in "Dipartimento per le Agro-Biotecnologie dell'ENEA di Roma (Casaccia)" in the field of molecular virology. He served as a visiting scientist at the University of Bologna, where he largely focused on the research of the molecular detection and characterization of phytoplasmas. His general scientific interests are as follows: plant disease; natural products; molecular diagnosis; bioactive substances; microbiology; and biological control. Since 1993, Prof. Camele has taught various plant pathology courses at the University of Basilicata and is responsible for several research projects and international collaborations. Ippolito Camele has coauthored more than 280 publications in national and international journals. Currently, Prof. Camele serves as an Associate Editor in the *Frontiers in Microbiology* journal and an Editorial Board Member of the MDPI journal *Plants*. In 2017, he obtained the academic title of "Full Professor" in Plant Pathology.

Adriano Sofo

Adriano Sofo is an Associate Professor of Agricultural Chemistry and Plant Biology at the University of Basilicata. He spent three years (1999–2002) at the University of Basilicata, Italy, with a Doctorate in Crop Productivity. From 2000 to 2001, he also was a Researcher at the National Agency for New Technologies, Energy and Sustainable Economic Development (ENEA), Italy. For postdoctoral training, in 2002, he worked at the Institute of Molecular Biology and Biotechnology, Heraklion, Greece, within a Marie Curie Fellowship. He then trained as a Postdoctoral Researcher at the University of Basilicata, where he also worked as an Assistant Professor in Agricultural Chemistry. In 2015, he was awarded a Fulbright Research Scholar grant to spend at the University of California, Davis. In 2017, he received a fellowship award from the OECD Co-operative Research Programme at the University of Waikato, New Zealand. In 2019, he was a visiting professor at Kindai University, Nara, Japan, under a JSPS Research Scholar Grant. In 2021, he benefited from a DAAD Research Stay at the University of Bremen, Germany. In 2022, he was awarded a Visiting Faculty Program Fellowship at the Weizmann Institute of Science, Israel. In 2023, he was a visiting professor at the University of California, Davis. Since 2022, he has been a member of the EGU's Biodiversity Task Force. He has been a National Geographic Explorer since 2023.

plants

MDPI

Editorial

Plant Essential Oil with Biological Activity (II)

Hazem S. Elshafie and Ippolito Camele *

School of Agricultural, Forestry, Food and Environmental Sciences, University of Basilicata, Via dell'Ateneo Lucano 10, 85100 Potenza, Italy; hazem.elshafie@unibas.it
* Correspondence: ippolito.camele@unibas.it; Tel.: +39-0971-205544; Fax: +39-0971-205503

Abstract: Essential oils (EOs) are concentrated hydrophobic liquids that originate from plants and contain different bioactive chemicals and volatile substances. Several plant essential oils (PEOs) are obtained from a variety of medicinal plants and have been utilized in folk medicine and traditional pharmacopoeia. They have a long history of usage as antibacterial medicines to treat various human, animal, and plant diseases. The extraction of essential oils frequently involves fractional distillation with a variety of organic solvents. EOs can be used successfully in the food and cosmetics industries in addition to their traditional use as antimicrobial agents. This Special Issue covers various significant PEOs and their individual chemical constituents and biological-pharmaceutical functions. Further information focused on the chemical characterizations, modes of action, and biopharmaceutical properties of PEOs. This Special Issue includes seventeen research papers from different geographical zones.

Keywords: aromatic plants; biochemical characterization; plant disease; pharmaceutical properties; sustainability

Citation: Elshafie, H.S.; Camele, I. Plant Essential Oil with Biological Activity (II). *Plants* **2023**, *12*, 3616. https://doi.org/10.3390/plants12203616

Received: 12 October 2023
Accepted: 17 October 2023
Published: 19 October 2023

1. Introduction

Several scientific topics related to the biological activity of different plant essential oils (PEOs) have been published in this Special Issue. With regard to the seventeen papers that make up the second volume of this Special Issue, "Plant Essential Oil with Biological Activity II," they cover several points either from chemical characterization point of view or even many biopharmaceutical properties and medical applications. The research in this volume included important essential oils (EOs) from different countries such as: *Diplosthephium juniperinum*, *Hedyosmum strigosum*, and *Dacryodes peruviana* (Ecuador); *Psidium guajava* (India); *Cupressus sempervirens* (Slovakia); *Artemisia rutifolia* EO (Russia); and *Origanum vulgare* (Italy), etc., as discussed below in detail.

2. An Overview of the Most Important Research

2.1. Southern America

In this Special Issue, three research papers about three important essential oils from Ecuador were published. In particular, a research paper was carried out by Salinas et al. [1] to biochemically characterize the EO extracted from *Diplosthephium juniperinum* in Ecuador. The results of this research showed moderate inhibitory effects regarding the acetylcholinesterase and butyrylcholinesterase enzymes and also low antioxidant activities, whereas another paper was carried out by Cartuche et al. [2] for studying the biological activity profiling of *Hedyosmum strigosum* EO, an aromatic native shrub from southern Ecuador. The results of this research demonstrated that the main compounds of this EO were thymol, -phellandrene, thymol acetate, and linalool, accounting for more than 51% of the EO composition. In addition, *H. strigosum* EO showed strong antioxidant and antimicrobial activities and moderate acetylcholinesterase inhibitory effects. Another study was carried out by Espinoza et al. [3], who investigated the in vivo anti-inflammatory efficacy of the copal (*Dacryodes peruviana*) EO native species from Ecuador. The results showed a moisturizing effect and an alleviation of several events occurring during the inflammatory process after topical treatment with the EO, such as a decline in skin edema, a reduction in leukocytic infiltrate, and a decrease in inflammatory cytokines, and hence they concluded that this EO could be an attractive treatment for skin inflammation.

2.2. Middle East

Several studies have been also carried out on some important EOs from the Middle East region, especially Egypt and Saudia Arabia, as follows: (i) Eos of *Jatropha intigrimma*, *J. roseae*, and *J. gossypifolia* (Egypt), which was carried out by Gamal El-Din et al. [4]; (ii) EO of *Devrra triradiata* (Saudi Arabia), carried out by Guetat et al. [5]; (iii) EO of *Acacia nilotica* (Egypt), carried out by El Gendy et al. [6]; (iv) EO of *Thyme vulgaris* (Egypt) carried out by Abd-Ellatif et al. [7]; (v) EOs of six different cultivars of *Citrus reticulata* (Egypt), carried out by Fahmy et al. [8]; and (vi) volatile EOs extracted from aerial parts of male and female ecospecies of *Ochradenus arabicus* (Saudi Arabia), carried out by Abd-ElGawad et al. [9].

In addition, this Special Issue also contained a study carried out by Khan et al. [10] on the potential antimicrobial and anticancer properties of seed extracts from *Citrullus colocynthis* (Saudia Arabia) by using different organic solvents, such as methanol, hexane, and chloroform.

This Special Issue also included important research on the comparative metabolic study of *Tamarindus indica* (Egypt) from various organs (bark, leaves, seeds, and fruits) and evaluated their anti-inflammatory effects. This study was carried out by Aly et al. [11] and concluded that the tested extracts from various organs of *T. indica* showed considerable anti-inflammatory and wound-healing activities.

2.3. Western Africa

One research paper has been published also in this volume regarding an EO from Northwest Africa. The study was carried out by Benali et al. [12], who studied the chemical profiling and biological properties of EOs extracted from *Lavandula stoechas* collected from three Moroccan sites. The chemical GC/MS profiles of the three studied EOs indicated that their biosyntheses varied depending on the site of growth. The studied EOs have also explicated promising antibacterial activities against Gram-positive and Gram-negative bacteria such as *Bacillus subtilis* and *Pseudomonas aeruginosa*. These important biological characteristics of *L. stoechas* EOs proved that this plant is a valuable source of naturally occurring bioactive chemicals with therapeutic effects.

2.4. Eastern Asia

Another study in this Special Issue focused on the chemical composition, antimicrobial effect, and anticancer activity of an EO extracted from *Psidium guajava* (India). This research was carried out by Alam et al. [13]. The GC–MS revealed that this EO was composed mainly from limonene and caryophyllene and concluded that this EO has promising antimicrobial and anticancer activities and could be a useful source for developing a natural therapeutic agent for oral infections and oral cancer.

2.5. Eastern Europe

The second volume of this Special Issue also included some important research papers from Europe. Among them, Galovičová et al. [14] evaluated the antioxidant, antibiofilm, antimicrobial (*in situ* and *in vitro*), insecticidal, and antiproliferative activities of *Cupressus sempervirens* EO (Slovakia). They concluded that *C. sempervirens* could be a suitable natural alternative as a biocontrol agent against different types of microorganisms, as well as suitable for controlling biofilms and harmful agricultural pests.

Dylenova et al. [15] studied the chemical diversity of *Artemisia rutifolia* EO (Russia) and evaluated its antimicrobial and antiradical activities. Their results showed that this EO can be classified into Tajik and Buryat-Mongol chemotypes and has strong antimicrobial activity against Gram-positive bacteria and fungi and high antiradical activity. The authors of this study concluded that the EO from *A. rutifolia* in the Russian flora indicates the prospects of the species as a raw material for the pharmaceutical and cosmetic industry.

2.6. Western Europe

Zinno et al. [16] studied the chemical composition and biological activities of two *Origanum vulgare* genotypes widely cultivated in Sicily (Italy). Their results demonstrated that these studied EOs have high antimicrobial activities, both in vitro and in a food matrix challenge test. These results suggested their potential use as biocontrol agents against a wide spectrum of foodborne pathogens.

3. Conclusions

As a result, the studies in this Special Issue demonstrated that several examined PEOs had positive potential for use in many bio-pharmacological applications, as well as in the agriculture and food sectors. Many studies in this issue emphasized the potential uses of a variety of PEOs in the agri-food industry, where they exhibit promising antimicrobial activities against a wide range of food deterioration microorganisms and prolong shelf-lives of processed food. Additionally, the investigated EOs and their primary components have been employed successfully as potential natural substitutes for synthetic drugs against a number of phytopathogens. Numerous researchers have studied the mechanisms of biological ability and have linked this potentiality to a distinctive chemical makeup, which is mainly composed of terpenoids and phenolic chemicals.

Funding: This research received no external funding.

Informed Consent Statement: Not applicable.

Acknowledgments: Personally, we would like to thank a lot and express our deep gratitude to all authors who contributed and participated in this Special Issue. We are so grateful to the editorial team staff of this journal for their kind collaboration in managing all stages of this Special Issue.

Conflicts of Interest: The authors declare no conflict of interest.

References

1. Salinas, M.; Calva, J.; Cartuche, L.; Valarezo, E.; Armijos, C. Chemical Composition, Enantiomeric Distribution and Anticholinesterase and Antioxidant Activity of the Essential Oil of *Diplosthephium juniperinum*. *Plants* **2022**, *11*, 1188. [CrossRef] [PubMed]
2. Espinoza, L.C.; Valarezo, E.; Fábrega, M.J.; Rodríguez-Lagunas, M.J.; Sosa, L.; Calpena, A.C.; Mallandrich, M. Characterization and In Vivo Anti-Inflammatory Efficacy of Copal (*Dacryodes peruviana* (Loes.) H.J. Lam) Essential Oil. *Plants* **2022**, *11*, 3104. [CrossRef] [PubMed]
3. Cartuche, L.; Calva, J.; Valarezo, E.; Chuchuca, N.; Morocho, V. Chemical and Biological Activity Profiling of *Hedyosmum strigosum* Todzia Essential Oil, an Aromatic Native Shrub from Southern Ecuador. *Plants* **2022**, *11*, 2832. [CrossRef] [PubMed]
4. Gamal El-Din, M.I.; Youssef, F.S.; Altyar, A.E.; Ashour, M.L. GC/MS Analyses of the Essential Oils Obtained from Different *Jatropha* Species, Their Discrimination Using Chemometric Analysis and Assessment of Their Antibacterial and Anti-Biofilm Activities. *Plants* **2022**, *11*, 1268. [CrossRef] [PubMed]
5. Guetat, A.; Abdelwahab, A.T.; Yahia, Y.; Rhimi, W.; Alzahrani, A.K.; Boulila, A.; Cafarchia, C.; Boussaid, M. *Deverra triradiata* Hochst. ex Boiss. from the Northern Region of Saudi Arabia: Essential Oil Profiling, Plant Extracts and Biological Activities. *Plants* **2022**, *11*, 1543. [CrossRef] [PubMed]
6. El Gendy, A.E.-N.G.; Essa, A.F.; El-Rashedy, A.A.; Elgamal, A.M.; Khalaf, D.D.; Hassan, E.M.; Abd-ElGawad, A.M.; Elgorban, A.M.; Zaghloul, N.S.; Alamery, S.F.; et al. Antiviral Potentialities of Chemical Characterized Essential Oils of *Acacia nilotica* Bark and Fruits against Hepatitis A and Herpes Simplex Viruses: In Vitro, In Silico, and Molecular Dynamics Studies. *Plants* **2022**, *11*, 2889. [CrossRef] [PubMed]
7. Abd-Ellatif, S.; Ibrahim, A.A.; Safhi, F.A.; Abdel Razik, E.S.; Kabeil, S.S.A.; Aloufi, S.; Alyamani, A.A.; Basuoni, M.M.; ALshamrani, S.M.; Elshafie, H.S. Green Synthesized of *Thymus vulgaris* Chitosan Nanoparticles Induce Relative WRKY-Genes Expression in *Solanum lycopersicum* against *Fusarium solani*, the Causal Agent of Root Rot Disease. *Plants* **2022**, *11*, 3129. [CrossRef] [PubMed]
8. Fahmy, N.M.; Elhady, S.S.; Bannan, D.F.; Malatani, R.T.; Gad, H.A. *Citrus reticulata* Leaves Essential Oil as an Antiaging Agent: A Comparative Study between Different Cultivars and Correlation with Their Chemical Compositions. *Plants* **2022**, *11*, 3335. [CrossRef] [PubMed]

9. Abd-ElGawad, A.M.; Assaeed, A.M.; El Gendy, A.E.-N.G.; Dar, B.A.; Elshamy, A.I. Volatile Oils Discrepancy between Male and Female *Ochradenus arabicus* and Their Allelopathic Activity on *Dactyloctenium aegyptium*. *Plants* **2023**, *12*, 110. [CrossRef] [PubMed]
10. Khan, M.; Khan, M.; Al-hamoud, K.; Adil, S.F.; Shaik, M.R.; Alkhathlan, H.Z. Diversity of *Citrullus colocynthis* (L.) Schrad Seeds Extracts: Detailed Chemical Profiling and Evaluation of Their Medicinal Properties. *Plants* **2023**, *12*, 567. [CrossRef] [PubMed]
11. Aly, S.H.; El-Hassab, M.A.; Elhady, S.S.; Gad, H.A. Comparative Metabolic Study of *Tamarindus indica* L.'s Various Organs Based on GC/MS Analysis, In Silico and In Vitro Anti-Inflammatory and Wound Healing Activities. *Plants* **2023**, *12*, 87. [CrossRef] [PubMed]
12. Benali, T.; Lemhadri, A.; Harboul, K.; Chtibi, H.; Khabbach, A.; Jadouali, S.M.; Quesada-Romero, L.; Louahlia, S.; Hammani, K.; Ghaleb, A.; et al. Chemical Profiling and Biological Properties of Essential Oils of *Lavandula stoechas* L. Collected from Three Moroccan Sites: In Vitro and In Silico Investigations. *Plants* **2023**, *12*, 1413. [CrossRef] [PubMed]
13. Alam, A.; Jawaid, T.; Alsanad, S.M.; Kamal, M.; Balaha, M.F. Composition, Antibacterial Efficacy, and Anticancer Activity of Essential Oil Extracted from *Psidium guajava* (L.) Leaves. *Plants* **2023**, *12*, 246. [CrossRef]
14. Galovičová, L.; Čmiková, N.; Schwarzová, M.; Vukic, M.D.; Vukovic, N.L.; Kowalczewski, P.Ł.; Bakay, L.; Kluz, M.I.; Puchalski, C.; Obradovic, A.D.; et al. Biological Activity of *Cupressus sempervirens* Essential Oil. *Plants* **2023**, *12*, 1097. [CrossRef] [PubMed]
15. Dylenova, E.P.; Zhigzhitzhapova, S.V.; Emelyanova, E.A.; Tykheev, Z.A.; Chimitov, D.G.; Goncharova, D.B.; Taraskin, V.V. Chemical Diversity of *Artemisia rutifolia* Essential Oil, Antimicrobial and Antiradical Activity. *Plants* **2023**, *12*, 1289. [CrossRef]
16. Zinno, P.; Guantario, B.; Lombardi, G.; Ranaldi, G.; Finamore, A.; Allegra, S.; Mammano, M.M.; Fascella, G.; Raffo, A.; Roselli, M. Chemical Composition and Biological Activities of Essential Oils from *Origanum vulgare* Genotypes Belonging to the Carvacrol and Thymol Chemotypes. *Plants* **2023**, *12*, 1344. [CrossRef]

plants

MDPI

Article

Chemical Composition, Enantiomeric Distribution and Anticholinesterase and Antioxidant Activity of the Essential Oil of *Diplosthephium juniperinum*

Melissa Salinas [1,2], James Calva [2], Luis Cartuche [2], Eduardo Valarezo [2] and Chabaco Armijos [2,*]

1 Maestría en Química Aplicada, Universidad Técnica Particular de Loja, San Cayetano s/n, Loja 1101608, Ecuador; masalinas4@utpl.edu.ec
2 Departamento de Química y Ciencias Exactas, Universidad Técnica Particular de Loja, Loja 1101608, Ecuador; jwcalva@utpl.edu.ec (J.C.); lecartuche@utpl.edu.ec (L.C.); bevalarezo@utpl.edu.ec (E.V.)
* Correspondence: cparmijos@utpl.edu.ec

Citation: Salinas, M.; Calva, J.; Cartuche, L.; Valarezo, E.; Armijos, C. Chemical Composition, Enantiomeric Distribution and Anticholinesterase and Antioxidant Activity of the Essential Oil of *Diplosthephium juniperinum*. *Plants* **2022**, *11*, 1188. https://doi.org/10.3390/plants11091188

Academic Editors: Hazem Salaheldin Elshafie, Ippolito Camele and Adriano Sofo

Received: 7 April 2022
Accepted: 21 April 2022
Published: 28 April 2022

Publisher's Note: MDPI stays neutral with regard to jurisdictional claims in published maps and institutional affiliations.

Abstract: The aim of this study was to extract and identify the chemical compounds of *Diplosthephium juniperinum* essential oil (EO) from Ecuador and to assess its anticholinesterase and antioxidant properties. The EO chemical composition was determined by GC–MS. A total of 74 constituents of EO were identified, representing 97.27% in DB-5ms and 96.06% in HP-INNOWax of the total EO. The major constituents (>4.50%) identified were: α-pinene (21.52, 22.04%), geranyl acetate (10.54, 7.78%), silphiper-fol-5-ene (8.67, 7.38%), α-copaene (8.26, 8.18%), 7-*epi*-silphiperfol-5-ene (4.93, 5.95%), and germacrene D (4.91, 6.00%). Enantioselective analysis of the volatile fraction of *D. juniperinum* showed: (+)-α-pinene as a pure enantiomer and 5 pairs of enantiomeric compounds. Among them, (−)-β-Pinene and (−)-Germacrene D presented a high enantiomeric excess of 93.23 and 84.62%, respectively, while (−)-α-Thujene, (−)-Sabinene and (S)-4-Terpineol with a lower enantiomeric excess of 56.34, 47.84 and 43.11%, respectively. A moderate inhibitory effect was observed for Acetylcholinesterase (AChE) and Butyrylcholinesterase (BuChE) enzymes with IC_{50} values of 67.20 ± 7.10 and 89.00 ± 9.90 μg/mL, respectively. A lower antioxidant potential was observed for the EO measured through DPPH and ABTS radical scavenging assays with SC_{50} values of 127.03 and >1000 μg/mL, respectively. To the best of our knowledge, this is the first report of the chemical composition, enantiomeric distribution and, anticholinesterase and antioxidant potential of the EO of *D. juniperinum*. As future perspective, further in-vivo studies could be conducted to confirm the anticholinesterase potential of the EO.

Keywords: *D. juniperinum*; essential oil; GC/MS; GC/FID; enantiomers; AChE; BuChE; antioxidant activity

1. Introduction

Asteraceae family is the largest group of vascular plants in the world and is composed mainly of flowering plants (angiosperms). Also called Compositae, the Asteraceae family comprises approximately 32,205 species belonging to 1911 plant genera [1], and grouped into 13 subfamilies [2,3]. Many species of this family are mainly herbaceous plants, however it can include trees, shrubs and sub-shrubs to vines [4]. Asteraceae occurs on all continents except Antarctica. On a global scale, the diversity of Asteraceae reported, is distributed as follows: South America (6316 species), Asia (6016 species), North America (5404 species), Africa (4631 species), Europe (2283 species), Oceania (1444 species), and the Pacific Islands (174 species) [5].

Despite its large number of species, a small number of them have been used for human and animal consumption as weeds (*Bidens*, *Cirsium*, *Hypochaeris* and *Sonchus* genera) [6,7], for its toxic and insecticidal properties, in gardening, for ornamental use (*Aster*, *Bellis*, *Cosmos*, *Chrysanthemum*, *Gazania* and *Gerbera* genera), in the food industry as oil plants

(*Helianthus annus* and *Carthamus tinctorius*), in the pharmaceutical (secondary metabolites with important biological activities) [8]. Important medicinal plants such as *Matricaria chamomilla*, *Artemisia absinthium* and *Tussilago farfara* belongs to this family [9]. Numerous members of the Asteraceae family are important as aromatic plants, from which essential oil (EO) can be extracted. These EOs are used in alternative and traditional medicine, and as ingredients for pharmaceutical and cosmetics industries. The Asteraceae EOs have a broad spectrum of bioactivity biological owing to the presence of active chemical compounds [10]. However, in Asteraceae family, as well as, in Boraginaceae and Fabaceae families, have been reported a series of chemical compound of alkaloid nature that are toxic for livestock and humans. These natural compounds named pyrrolizidine alkaloids are natural toxins occurring in Asteraceae family, extracted mainly with organic solvents from the plant material [11], in contrast to EOs that are a mixture of volatile compounds of terpene nature, that have a low or minimum toxicity with some exceptions, as safrole which is a natural compound present in EOs from *Piper* genus [12] or extracted commonly from *Sassafras* genus [13].

Of the 18,500 species of vascular plants registered in Ecuador, orchids are the most diverse, with 4200 species, followed by Asteraceae with 918 species, which belong to 217 genera, 7 of them endemic [14]. In addition, in Ecuador this family is recognized for the number of endemic species, with 370 specimens, located in second after orchids. Endemic Asteraceae are mainly shrubs (195 species) and herbs (97 species). The Ecuadorian Andes are the center of diversity and endemism in this family, although there are species in the Amazon, Coast and Galapagos (the four natural regions of Ecuador). Of the endemic species found in Ecuador, 32 are exclusive to Galapagos [15]. Asteraceae species exhibit a wide altitude spectrum from near sea level to 5000 m of altitude. In Ecuador the diversity of this family increases from 2000 to 3000 m a.s.l, registering a maximum between 2900 to 3000 m a.s.l. [9].

Diplostephium is a genus of trees, shrubs, and subshrubs that are part of the flora of the upper limit of the Andean forests, paramos, jalcas and punas in the neotropical mountains. Currently this genus is composed of 111 accepted species names [16], distributed from Costa Rica to Chile in high elevation cloud forests (2500–3000 m), puna habitats (3800–4200 m) and paramos (3000–4500 m) [17]. In Latin America, 63 species have been reported for Colombia, 39 for Peru, 26 for Ecuador, 10 for Venezuela, three for Chile and one for Bolivia [18]. *Diplostephium juniperinum* Cuatrec (Kunth), known as "monte de baño" (bath grass) is an endemic shrub of Ecuador, distributed in the Andean regions between 2000 to 3400 m a.s.l., especially in the Andean provinces of Azuay and Loja [14] and is used by indigenous Saraguro (Loja, province) in postpartum herbal bath [19]. This species has been found only in Ecuador, its natural habitat is subtropical or tropical moist montane forests. The *D. juniperinum* plant is a 0.8 m tall shrub which topped is round. This species has branches densely compacted, bracts green with reddish purple tinge, disk flowers dull yellow, ray florets white and tipped with pale lavender below [20].

Ecuador is considered a megadiverse country because has many species per unit surface area. Currently, this country occupies the sixth position worldwide as a biodiversity hotspot [21]. However, the fact that there are few studies of its aromatic plant species, especially of the aromatic species of the Asteraceae family, and that study of the *D. juniperinum* EO having not been previously reported in the literature have stimulated our interest in investigating the EO extracted from this species. For that reason, the aim of this research was to determine the chemical composition, and enantiomeric distribution of the EO of *D. juniperinum*, as well as, to assess its antioxidant and anticholinesterase properties and thus, contribute to the phytochemical characterization of *Diplostephium* species in Ecuador. In addition, the search for new natural products or compounds with biological interest is of relevance for the pharmaceutical and cosmetic industry nowadays.

2. Results

2.1. Physical Properties

Through hydrodistillation from fresh aerial parts of *D. juniperinum*, a pale-yellow EO was obtained, with a low extraction yield of 0.12 ± 0.01% (*w/w*), a relative density of 0.79 ± 0.02 gr/mL, refractive index $[n^{20}]$ 1.48 ± 0.01 and a specific rotation $[\alpha]_D{}^{20} = -34.18 \pm 0.01°$.

2.2. Chemical Composition

A total of 74 constituents were identified, representing 97.27% in DB-5ms and 96.06% in HP-INNOWax of the total EO composition. The main constituents (>4.50%) identified were: α-pinene (21.52, 22.04%) (a), geranyl acetate (10.54, 7.78%) (b), silphiperfol-5-ene (8.67, 7.38%) (c), α-copaene (8.26, 8.18%) (d), 7-epi-silphiperfol-5-ene (4.93, 5.95%) (e), and germacrene D (4.91, 6.00%) (f) (Figure 1 and Table 1).

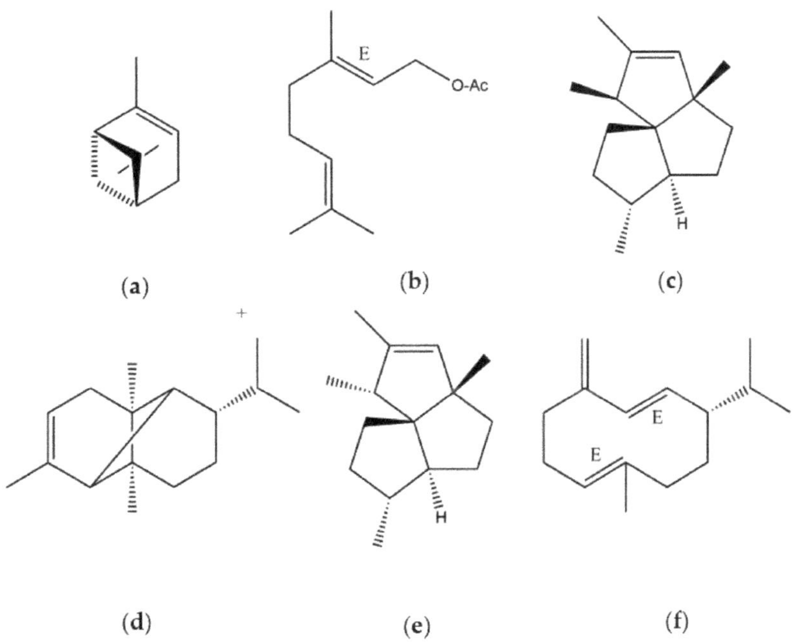

Figure 1. Principal compounds of essential oil of *D. juniperinum:* (**a**) α-pinene; (**b**) geranyl acetate; (**c**) silphiperfol-5-ene; (**d**) α-copaene; (**e**) 7-*epi*-silphiperfol-5-ene; (**f**) germacrene D.

Table 1. Chemical composition of *D. juniperinum* essential oil.

N°	Compound	DB-5ms					HP-INNOWax				
		LRI [a]	LRI [b]	Ref.	%	SD	LRI [a]	LRI [b]	Ref.	%	SD
1	α-Thujene	919	924	[22]	0.17	0.04	-	-	-	-	-
2	α-Pinene	926	932	[22]	21.52	3.76	1062	1066	[23]	22.04	4.06
3	Camphene	939	946	[22]	0.54	0.05	1083	1084	[23]	0.59	0.06
4	Thuja-2,4(10)-diene	943	953	[22]	0.20	0.05	1124	1122	[24]	0.29	0.03
5	Sabinene	964	969	[22]	1.74	0.22	1120	1120	[25]	1.83	0.13
6	β-Pinene	969	974	[22]	3.54	0.42	1109	1108	[25]	3.68	0.52
7	Myrcene	986	988	[22]	0.68	0.14	-	-	-	-	-
8	α-Phellandrene	1004	1002	[22]	0.75	0.07	1161	1163	[25]	1.38	0.16
9	α-Terpinene	1013	1014	[22]	0.34	0.16	1176	1185	[26]	0.28	0.01
10	ρ-Cymene	1020	1020	[22]	0.39	0.48	1273	1270	[25]	1.39	0.17
11	Limonene	1024	1024	[22]	1.97	0.54	1198	1199	[25]	2.07	0.13
12	1,8 Cineole	1028	1026	[22]	0.77	0.18	1208	1206	[25]	0.95	0.01
13	Z-β-Ocimene	-	-	-	-	-	1236	1236	[25]	0.11	0.02
14	E-β-Ocimene	1043	1044	[22]	2.46	0.26	1254	1253	[23]	2.72	0.33
15	ϒ-Terpinene	1052	1054	[22]	0.39	0.11	1244	1244	[25]	0.46	0.01
16	Terpinolene	1079	1086	[22]	0.23	0.14	1284	1290	[27]	0.17	0.01
17	n-Nonanal	1106	1100	[22]	0.22	0.04	-	-	-	-	-

Table 1. *Cont.*

N°	Compound	DB-5ms LRI [a]	DB-5ms LRI [b]	DB-5ms Ref.	DB-5ms %	DB-5ms SD	HP-INNOWax LRI [a]	HP-INNOWax LRI [b]	HP-INNOWax Ref.	HP-INNOWax %	HP-INNOWax SD
18	α-Campholenal	1123	1122	[22]	0.34	0.09	-	-	-	-	-
19	*trans*-Pinocarveol	1135	1135	[22]	0.31	0.10	-	-	-	-	-
20	Borneol	1167	1165	[22]	0.31	0.16	-	-	-	-	-
21	Terpinen-4-ol	1176	1174	[22]	0.75	0.16	1609	1600	[27]	0.66	0.01
22	Myrtenal	1189	1195	[22]	0.24	0.07	1623	1631	[24]	0.13	0.01
23	Thymol, methyl ether	1227	1232	[22]	0.28	0.08	-	-	-	-	-
24	Silphiperfol-5-ene	1331	1326	[22]	8.67	0.40	1464	1495	[27]	7.38	2.45
25	α-Cubebene	1338	1348	[22]	0.15	0.08	1449	1460	[26]	0.23	0.01
26	7-*epi*-Silphiperfol-5-ene	1345	1345	[22]	4.93	0.30	1437	1424	[27]	5.95	0.99
27	Cyclosativene	1354	1369	[22]	0.20	0.00	-	-	-	-	-
28	α-Copaene	1365	1374	[22]	8.26	0.55	1479	1483	[23]	8.18	1.40
29	β-Cubebene	1377	1387	[22]	0.99	0.01	1528	1531	[23]	0.34	0.01
30	Geranyl acetate	1383	1379	[22]	10.54	0.55	1766	1761	[25]	7.78	1.07
31	α-Gurjunene	1392	1409	[22]	1.90	0.16	1514	1520	[23]	2.24	0.12
32	Pinocarvone	-	-	-	-	-	1563	1559	[28]	0.17	0.04
33	β-Gurjunene	-	-	-	-	-	1574	1559	[28]	0.58	0.67
34	*E*-Caryophyllene	1414	1419	[22]	2.09	0.05	1580	1586	[23]	1.07	1.33
35	β-Copaene	1425	1430	[22]	0.16	0.06	-	-	-	-	-
36	Aromadendrene	-	-	-	-	-	1615	1613	[25]	0.17	0.06
37	*cis*-Muurola-3,5-diene	1444	1448	[22]	0.27	0.14	-	-	-	-	-
38	α-Humulene	1449	1452	[22]	1.22	0.14	1652	1657	[23]	1.09	0.11
39	allo-Aromadendrene	1443	1458	[22]	0.19	0.06	1626	1633	[23]	0.37	0.25
40	*cis*-Cadina-1(6),4-diene	1469	1461	[22]	0.43	0.10	-	-	-	-	-
41	*cis*-Muurola-4(14),5-diene	1474	1465	[22]	1.40	0.16	1646	1648	[29]	0.29	0.02
42	*cis*-Verbenol	-	-	-	-	-	1661	1663	[30]	0.22	0.09
43	Germacrene D	1477	1484	[22]	4.91	0.90	1691	1697	[25]	6.00	0.92
44	β-Selinene	1482	1489	[22]	0.96	0.06	1698	1702	[29]	1.04	0.02
45	ϒ-Muurolene	1485	1478	[22]	0.51	0.25	1687	1681	[26]	0.11	0.02
46	Viridiflorene	-	-	-	-	-	1679	1686	[26]	0.27	0.03
47	α-Selinene	-	-	-	-	-	1704	-	-	0.65	0.18
48	α-Muurolene	-	-	-	-	-	1711	1717	[23]	1.69	0.11
49	Bicyclogermacrene	1490	1500	[22]	1.11	0.67	1716	1723	[25]	1.05	0.13
50	α-Amorphene	1497	1483	[22]	1.49	0.40	1674	1679	[31]	1.40	0.15
51	Germacrene A	1500	1508	[22]	0.37	0.21	-	-	-	-	-
52	Silphiperfolan-6-α-ol	1503	1507	[22]	0.17	0.04	-	-	-	-	-
53	*trans*-Muurola-4(14),5-diene	1498	1493	[22]	0.68	0.01	-	-	-	-	-
54	δ-Cadinene	1517	1522	[22]	3.96	1.05	1745	1750	[29]	4.00	1.04
55	*trans*-Cadina1,4-diene	1528	1533	[22]	0.15	0.04	-	-	-	-	-
56	α-Cadinene	1532	1537	[22]	0.13	0.03	-	-	-	-	-
57	Germacrene B	1549	1559	[22]	0.23	0.09	1809	1814	[25]	0.16	0.06
58	*trans*-Calamenene	-	-	-	-	-	1821	1821	[29]	0.40	0.13
59	*epi*-Cubebol	-	-	-	-	-	1889	1899	[32]	0.18	0.04
60	α-Calacorene	-	-	-	-	-	1904	1894	[26]	0.13	0.01
61	Palustrol	1563	1567	[22]	0.17	0.04	1918	1915	[23]	0.15	0.02
62	Germacrene D-4-ol	1572	1574	[22]	0.40	0.16	2055	2044	[23]	0.32	0.05
63	1,5-Epoxy-salvial(4)14-ene	-	-	-	-	-	1930	1912	[32]	0.12	0.03
64	Caryophyllene oxide	1575	1582	[22]	0.59	0.09	1971	1967	[23]	0.37	0.07
65	Ledol	1597	1602	[22]	0.45	0.23	2026	2017	[23]	0.60	0.05
66	Cubenol	-	-	-	-	-	2031	2023	[26]	0.33	0.07
67	Silphiperfol-6-en-5-one	1613	1624	[22]	0.29	0.21	2100	2131	[24]	0.39	0.16
68	1-*epi*-Cubenol	1623	1627	[22]	0.16	0.05	2065	2048	[29]	0.23	0.15
69	Spathulenol	-	-	-	-	-	2155	2144	[27]	0.36	0.02
70	*epi*-α-Cadinol	1639	1638	[22]	0.15	0.12	2179	2167	[23]	0.19	0.05
71	*epi*-α-Muurolol	1642	1640	[22]	0.17	0.10	2195	2196	[27]	0.29	0.07
72	α-Muurolol (=Torreyol)	1645	1644	[22]	0.25	0.15	2185	2178	[29]	0.17	0.01
73	*epi*-α-Bisabolol	-	-	-	-	-	2210	2218	[25]	0.16	0.05
74	α-Cadinol	1653	1652	[22]	0.43	0.05	2244	2255	[27]	0.50	0.15
	Monoterpene hydrocarbons (%)				34.53					35.61	
	Oxygenated monoterpenes (%)				13.21					9.92	
	Sesquiterpene hydrocarbons (%)				45.34					44.80	
	Oxygenated sesquiterpenes (%)				3.24					4.17	
	Other compounds (%)				0.95					1.39	
	Total (%)				97.27					96.06	

LRI [a], Linear retention index calculated; LRI [b], Linear retention index from Reference; Ref, References; % Percentage and SD Standard Deviation, both values were conveyed as means of three determinations.

Sesquiterpene (SH) and monoterpene hydrocarbons (MH) predominated in the chemical composition of the *D. juniperinum* EO. The percentages of SH were 45.34% and 44.80% in DB-5ms andHP-INNOWax, respectively. MH in DB-5ms column represented 34.53% and in HP-INNOWax 35.61%. Oxygenated monoterpenes represented 13.21% (DB-5ms) and 9.92% (HP-INNOWax), followed by oxygenated sesquiterpenes with 3.24% (DB-5ms) and 4.17% (HP-INNOWax) and finally, other compounds with 0.95% in DB-5ms and 1.39% in HP-INNOWax.

2.3. Enantiomeric Composition

Enantioselective analysis of the volatile fraction of *D. juniperinum* showed: (+)-α-pinene as pure enantiomer and 5 pairs of enantiomeric compounds, among them; (−)-β-Pinene and (−)-Germacrene D reported a high enantiomeric excess of 93.23 and 84.62%, respectively, while (−)-α-Thujene, (−)-Sabinene and (S)-4-Terpineol with a lower enantiomeric excess of 56.34, 47.84 and 43.11%, respectively (Table 2).

Table 2. Enantioselective analysis of *D. juniperinum* essential oil.

Enantiomeric Compounds	LRI [a]	Enantiomeric Distribution (%)	*ee* (%) ± SD
(+)-α-Thujene	921	21.83	56.34 ± 0.12
(−)-α-Thujene	924	78.17	
(+)-α-pinene	930	100	100 ± 0.01
(+)-β-Pinene	958	3.39	93.23 ± 1.92
(−)-β-Pinene	966	96.61	
(+)-Sabinene	984	26.08	47.84 ± 2.12
(−)-Sabinene	995	73.92	
(+)-4-Terpineol	1279	71.56	43.11 ± 0.98
(−)-4-Terpineol	1288	28.44	
(+)-Germacrene D	1468	7.69	84.62 ± 0.13
(−)-Germacrene D	1474	92.31	

LRI [a], Linear retention index calculated; *ee* (%) ± SD, percentage of excess enantiomeric ± standard deviation values were conveyed as means of three determinations.

2.4. Anticholinesterase Activity

In this study, we evaluated for the first time the anti-cholinesterase activity of *D. juniperinum* EO by measuring the rate of reaction. Results showed a moderate inhibition effect with IC_{50} values of 67.20 ± 7.10 and 89.00 ± 9.90 µg/mL against AChE and BuChE, respectively. Donepezil hydrochloride was used as a positive control and their value of IC_{50} is presented in Table 3.

Table 3. AChE and BuChE inhibition of *D. juniperinum* essential oil.

Sample	AChE	BuChE
	IC_{50} (µg/mL) ± SD	
D. juniperinum	67.20 ± 7.10	89.00 ± 9.90
Donepezil	0.04 ± 0.01	3.60 ± 0.20

IC_{50}, Half maximal inhibition concentration expressed as µg/mL.

2.5. Antioxidant Activity

The results obtained for DPPH and ABTS radical scavenging of the *D. juniperinum* EO as presented in Table 4, and expressed as the concentration of the EO that scavenge or decrease the concentration of the radical at 50% (SC_{50}). Trolox was used as a positive control.

Table 4. Antioxidant activity of *D. juniperinum* essential oil.

Sample	ABTS	DPPH
	SC_{50} (µg/mL—µM *) ± SD	
D. juniperinum	127.03 ± 0.58	>1000
Trolox *	23.27 ± 1.05	29.99 ± 1.04

SC_{50}, Half scavenging capacity expressed as µg/mL—µM *.

3. Discussion

Average yields of EO were calculated based on the fresh plant material of the aerial parts of *D. juniperinum* was similar with *D. antioquense* EO with 0.16% and much higher that the reported in *D. rosmarinifolius* with a very low yield 0.0045% [33].

D. juniperinum species does not present previous chemical studies of the volatile fraction, however, the EOs obtained from plants of the same genus such as *D. antioquense* and *D. rosmarinifolius* collected in Colombia were analyzed by GC/MS and GC/FID determining to β-copaene (17.78%), (*Z*)-para-mentha-2,8-dien-1-ol (14.29%), β-pinene (13.75%), δ-cadinene (11.42%) and *E*-caryophyllene (6.54%) as majority constituents of *D. antioquense*, while in *D. rosmarinifolius* EO were found to *E*-caryophyllene (16.07%), 1*R*-α-pinene (13.79%), δ-cadinene (8.54%), limonene (8.23%), α-caryophyllene (8.15%) and γ-terpineol (8.11%) [33].

Two monoterpenes identified as main constituents of EO of *D. juniperinum*, α-pinene and geranyl acetate were isolated and reported with biological activities in other studies; α-pinene exhibits antinociceptive [34], anti-inflammatory [35–37], antidepressant [38] and antioxidant properties [39], while, geranyl acetate has shown significant anti-Candida potential [40] and antinociceptive properties [41].The rare sesquiterpenes silphiperfol-5-ene and 7-*epi*-silphiperfol-5-ene were found in *Pteronia* genus of the Asteraceae family [27], but there are no reports of their isolation or biological activity.

Germacrene D, is one of the main components identified in the *D. juniperinum* EO, and the enantiomer (−)-germacrene D was found with an *e.e.* of 92.31 %. Biologically, this sesquiterpene exerted promising results, potentially influence in the attraction and oviposition of females of the species *Heliothis virescens* [42]. Chiral compounds have great importance for the identification of adulterations due to EOs have different proportions of each enantiomer [43] and this enantiomeric characterization is also important in the olfactory profile [44].

Natural acetylcholinesterase inhibitors, such as galantamine, are usually used in the pharmacological industry as a drug to treat Alzheimer's disease [45], the search for future AChE and BuChE inhibitors guarantee the alleviation of symptoms related to the aforementioned disease and the reduction of mortality rates [46]. Several studies on the anticholinesterase activity of EOs and almost none on their main components showed that EOs are complex mixtures and their final activities are due to the combined effects of the all components [47], therefore, the inhibitory activity of the EO is probably the result of a complex interaction of its chemical components, producing synergistic or antagonistic inhibitory responses [48].

Anti-cholinesterase effect of EO from *Diplostephium* genus, has not been reported, the monoterpenes are the kind of compounds predominant in them. As mentioned by Aazza and collaborators [49] the α-pinene, limonene and sabinene, are responsible for the anticholinesterase effect. Additionally, (+)-α-Pinene as reported by Miyazawa and Yamafuji [50], presented an IC_{50} of 0.40 mM against acetylcholinesterase and, this compound was identified in the EO of *D. juniperinum* at a concentration of 21% and enantiomerically pure, which could explain the moderate effect observed for this EO against AChE and BuChE enzymes. Therefore, it is important to know the main constituents of the EOs, their proportion and chiral composition because they are the ones that give their biological potential.

A literature review on the *Diplostephium* genus indicates that few studies have been conducted on its species, one of them is on the ethanolic extract of *D. phylicoides*, which shows a high antioxidant activity (IC_{50} = 13.80 μg/mL) attributed to the presence of flavonoids in its composition [51].

In other study, α-pinene reported a lower antioxidant effect, with an IC_{50} of 12.57 ± 0.18 mg /mL [52]. Similar results for the EO of *D. juniperinum* for ABTS and DPPH assays with an SC_{50} of ca. 120 μg/mL and >1000 μg/mL were observed. The importance of knowing the antioxidant properties of EO is due to their implication in counteracting the harmful effects on biological entities by free radicals or reactive oxygen species [53].

Several studies have demonstrated that extracts of Astareaceae species have a positive impact on human health, thanks to their anti-inflammatory, antimicrobial and antioxidant, and antimicrobial [54]. Recently, species of the Asteraceae family have been considered as a sustainable planning tool in cities for their phytoremediation properties as air pollutant removal, soil protection, shaping landscapes, etc. [55]. Further studies can be conducted to validate the anticholinesterase effect in in vivo studies, however, the low yield obtained for this species could difficult such approximation. In order to obtain a better amount needed for in vivo assays, oil extraction optimization studies could be carried out, including the study of intrinsic and extrinsic parameters related to the species, such as plant age, phenological stage, soil type, amount of shade and season of the year when the species is harvested [56]. This further research could complement the current one.

4. Materials and Methods

4.1. Materials

Methanol and dichloromethane from analytical HPLC grade, anhydrous sodium sulphate, 2,2′-azinobis-3-ethylbenzothiazoline-6-sulfonic acid (ABTS), 2,2-diphenyl-1-picrylhy drazyl (DPPH), Butyrylcholinesterase from equinum serum, Acetylcholinesterase from *Electrophorus electricus*, phosphate buffered saline, Ellman's reagent (5,5′-dithiobis(2-nitrobenzoic acid), Acetylthiocoline iodide, donepezil hydrochloride, were purchased from Sigma-Aldrich (San Luis, MO, USA).The standard aliphatic hydrocarbons were purchased from ChemService (West Chester, PA, USA). Helium was purchased from INDURA (Guayaquil, Ecuador). All chemicals were of analytical grade and used without further purifications.

4.2. Plant Material

Leaves, stems and flowers of *D. juniperinum* were collected in October 2020 in Las Antenas sector, at the border between Saraguro and San Lucas, Loja province, at an altitude of 3210 m a.s.l. and located at 9,593,252 N, 696,030 E coordinates. The plant material collected under permit MAE-DBN-2016-048 granted by the Ministry of Environment of Ecuador (MAE), was identified and classified by José Miguel Andrade, botanist at UTPL. A specimen sample was deposited at the Herbarium of the Universidad Técnica Particular de Loja (HUTPL) with voucher code PPN-as-057.

4.3. Distillation of the Essential Oil

The EO from fresh aerial parts of *D. juniperinum* was extracted by steam hydrodistillation in a Clevenger-type apparatus for approximately 3 h. Three distillations were carried out with 1300, 1320 and 1410 g of fresh plant material, respectively. After obtaining the EO it was separated from the aqueous phase and dried with anhydrous sodium sulfate, filtered and stored in an amber sealed vial at −4 °C, until its analytical and biological assays. The procedure was performed three times [57].

4.4. Physical Properties of Essential Oil

The relative density, refractive index and optical rotation of the EO of *D. juniperinum* were determined in triplicate at 20 °C. The relative density was determined according to the AFNOR NF T 75-11 method (equivalent to ISO 279: 1998, using a pycnometer of 1 mL capacity and an analytical balance (Mettler AC 100), the refractive index according to AFNOR method NF 75-112 (ISO 280:1998) in a refractometer model ABBE (BOECO, Hamburg, Germany). The specific optical rotation was determined with the ISO 592-1998 standard method in an automatic polarimeter (Hanon P-810) [23].

4.5. Chemical Characterization of Essential Oil

4.5.1. Sample Preparation of EO

Quantitative and qualitative characterization of EO from *D. juniperinum* required sample preparation of the volatile fractions. Ten µL of EO was diluted in 990 µL in dichloromethane (CH_2Cl_2) obtaining a 1:100 v/v solution. The samples were used in the chemical analyses described below [25].

4.5.2. Qualitative and Quantitative Analysis

Qualitative identification was performed using the analytical technique of Gas Chromatography coupled to Mass Spectrometry (GC/MS). One μL of each sample was injected in duplicate in split mode (40:1) at 20 °C into an Agilent Technologies model 6890N gas chromatograph (GC) with an autoinjector model 7683 and a mass spectrometer model 5973 INERT (Santa Clara, CA, USA). The GC equipment operates in electron-ionization mode at 70 eV, with helium as carrier gas (1.00 mL/min in constant flow), the GC oven operated with temperature ramp from 60 °C to 250 °C with a gradient of 3 °C/min and the ion source at 250 °C. Additionally, the capillary columns DB-5ms (5%-phenyl-methyl polysiloxane, 30 m × 0.25 mm i.d., 0.25 μm film thickness;) and HP-INNOWax, (polyethylene glycol, 30 m × 0.25 mm i.d., 0.25 μm film thickness both purchased from J & W Scientific, Folsom, CA, USA, were used. The procedure was performed for triplicate.

The identification of the aromatic compounds was performed by comparison of the mass spectra and the linear retention index (LRI) with those reported in literature. The LRI was determined experimentally according to Van Den Dool and Krats [58], for which it was necessary to inject a homologous series of C_9 to C_{24} alkanes in the same conditions of the EO.

Quantitative analysis of the EO of *D. juniperinum* was performed using a gas chromatography coupled to a flame ionization detector (GC/FID). The previously prepared samples were injected under the same analytical conditions as the qualitative GC/MS method, and the chromatography columns were the same. The percentage of aromatic compounds was determined by comparing the GC peaks with the total area of the identified peaks [59]. A calibration curve was built for each column as previously described by Gilardoni et al. [60], using isopropyl caproate (0.6, 1.8, 4.3, 8.3, 16.8, and 34.3 mg of isopropyl caproate in 10 mL of cyclohexane) and n-nonane (7 mg) as calibration standard and internal standard respectively. The LOD (0.4 μg/mL) and LOQ (1.2 μg/mL) were stablished. Both calibration curves generated a correlation coefficient of 0.995.

4.5.3. Enantioselective Analysis of Essential Oil

Enantiomeric compounds present in the EO of *D. juniperinum* were determined by GC/MS on a capillary column with 2,3-diethyl-6-tert-butyldimethylsilyl-β-cyclodextrin stationary phase. The injection conditions used were the same in GC/MS. In addition, enantiomerically pure standards were injected under the same conditions to determine the elution order of the EO enantiomers [61].

4.6. AChE and BuChE Inhibition Spectrophotometric Analysis

Cholinesterase (ChEs) inhibition of EO was determined for the enzymes (i) acetylcholinesterase (AChE) and (ii) butyrylcholinesterase (BuChE). The procedure was followed as described by Ellman et al. [62] and Calva et al. [57]. Phosphate buffered saline (pH = 7.4), DTNB (5,5′-dithiobis-(2-nitrobenzoic acid) ion (1.5 mM) a reagent that reacts with thiocholine to give the yellow coloration and the EO sample in DMSO (1% v/v) were prepared. The reaction of DTNB is monitored by measuring its absorption at 412 nm. AChE, from Electrophorus electricus (Sigma-Aldrich, C3389, St. Louis, MO, USA) and BuChE, from horse serum, (Sigma-Aldrich, SRE020, St. Louis, MO, USA) are dissolved in PBS (pH = 7.4) at 24 mU/mL. Preincubation was carried out for 10 min and acetylcholine iodide (1.5 mM) is added to initiate the reaction. The reaction is monitored for 30 min at 30 °C in a PherastarFS detection system (BMG Labtech). Inhibitory concentration (IC_{50}) values were calculated in the online package GNUPLOT (www.ic50.tk, www.gnuplot.info) (accessed on 1 March 2022). Measurements were performed by triplicate. The reference drug inhibitor of ChEs was Donepezil, for AChE and BuChE with an IC_{50} value of 100 nM and 8500 nM, respectively. False positives are not excluded for high concentrations (>100 ug/mL) of amine or aldehyde compounds [59].

4.7. Antioxidant Spectrophotometric Analysis

4.7.1. DPPH Assay

The DPPH radical scavenging assay was developed according to the metodologhy proposed by Thaipong et al. [63] with slight modifications, using 2,2-diphenyl-1-picrylhydryl free radical (DPPH-). A working solution was prepared dissolving 24 mg of DPPH in 100 mL methanol and was stabilized in an EPOCH 2 microplate reader (BIOTEK, Winooski, VT, USA) at 515 nm until an absorbance of 1.1 ± 0.01 was reached. The antiradical reaction between EO and free radical was performed at different concentrations of EO (1, 0.5 and 0.25 mg/mL). In a 96-microwell plate, 270 µL of DPPH adjusted working solution and 30 µL of EO sample was placed. The reaction was monitored at 515 nm for 60 min at room temperature. Trolox and methanol were used as positive control and blank control, respectively. The results were expressed as SC_{50} (scavenging concentration of the radical at 50%) and calculated according to the corresponding curve fitting of data with GraphPadPrism v.8.0.1. Measurements were performed in triplicate.

4.7.2. ABTS Assay

The antioxidant power measured against $ABTS^{\bullet+}$ cation (2,2′-azinobis-3-ethylbenzothiazoline-6-sulfonic acid) was determined as reported by Arnao et al. [64] and Thaipong et al. [63] with slight modifications as described. Briefly, the assay started with the preparation of a stock solution of the radical by reacting equal volumes of ABTS (7.4 µM) and potassium persulfate (2.6 µM) for 12 h under stirring. The standard solution was prepared by dissolution in methanol to an absorbance of 1.1 ± 0.02 measured at 734 nm in an EPOCH 2 microplate reader (BIOTEK, Winooski, VT, USA). The antiradical reaction was evaluated over a time of 1 h in the dark at room temperature by plating 270 µL of ABTS working adjusted solution and 30 µL of EO from *D. juniperinum* at different concentrations (1, 0.5 and 0.25 mg/mL). Trolox and methanol were used as positive control and blank control, respectively. The results were expressed as SC_{50} (scavenging concentration of the radical at 50%) and calculated according to the corresponding curve fitting of data with GraphPadPrism v.8.0.1. Measurements were performed in triplicate

5. Conclusions

The fresh aerial parts of *D. juniperinum* afforded, an essential oil in quite a low yield (0.12% by weight). The EO obtained was composed exclusively of sesquiterpenes and monoterpenes hydrocarbons, whose major constituents were α-pinene (about 22%) and geranyl acetate (about 10%). The enantioselective analysis showed (+)-α-pinene as a pure enantiomer and 5 pairs of enantiomeric compounds. The EO also manifested a moderate inhibition activity against AChE and BuChE and a lower antioxidant potential was observed for the EO measured through DPPH and ABTS radical scavenging assays. As future perspective, further in-vivo studies could be conducted to confirm the anticholinesterase potential of the EO. In addition, this genus that reported bioactive compounds, could be of interest for the development of new applications such as in the food industry, as enrichment of the food matrix to enhance their beneficial properties and also the substitution of synthetic antioxidants.

Author Contributions: Conceptualization, C.A.; methodology, M.S.; investigation, M.S., J.C. and L.C.; writing—original draft preparation, M.S.; writing—review and editing, J.C., L.C. and E.V.; supervision, C.A. All authors have read and agreed to the published version of the manuscript.

Funding: This research received no external funding.

Institutional Review Board Statement: Not applicable.

Informed Consent Statement: Not applicable.

Data Availability Statement: Not applicable.

Acknowledgments: We are grateful to the Universidad Técnica Particular de Loja (UTPL) for supporting this investigation and open access publication.

Conflicts of Interest: The authors declare no conflict of interest.

References

1. The Plant List. Compositae. Available online: http://www.theplantlist.org/ (accessed on 2 August 2021).
2. Stevens, P.F. Angiosperm Phylogeny Website Version 14. Available online: http://www.mobot.org/MOBOT/research/APweb/ (accessed on 24 February 2020).
3. Naim, D.M.; Mahboob, S. Molecular identification of herbal species belonging to genus *Piper* within family Piperaceae from northern Peninsular Malaysia. *J. King Saud. Univ. Sci.* **2020**, *32*, 1417–1426. [CrossRef]
4. Valarezo, E.; Aguilera-Sarmiento, R.; Meneses, M.A.; Morocho, V. Study of Essential Oils from Leaves of Asteraceae Family Species *Ageratina dendroides* and *Gynoxys verrucosa*. *J. Essent. Oil Bear. Plants* **2021**, *24*, 400–407. [CrossRef]
5. Panero, J.L.; Crozier, B.S. Macroevolutionary dynamics in the early diversification of Asteraceae. *Mol. Phylogenet. Evol.* **2016**, *99*, 116–132. [CrossRef]
6. Del Vitto, L.A.; Petenatti, E.M. Asteraceae of economic and environmental importance: First part. Morphological and taxonomic synopsis, environmental importance and plants of industrial value. *Multequina* **2009**, *18*, 87–115.
7. Oliveira Amorim, V.; Bautista Pousada, H. Asteraceae da Ecorregião Raso da Catarina, Bahia, Brasil. *Rodriguésia* **2016**, *67*, 785–794. [CrossRef]
8. Encyclopedia Britannica. Asteraceae: Plant Family. Available online: https://www.britannica.com/ (accessed on 1 December 2021).
9. Rivero-Guerra, A.O. Diversity of endemic species of Asteraceae (Compositae) in the flora of Ecuador. *Collect. Bot.* **2020**, *39*, e001. [CrossRef]
10. Abad, M.J.; Bedoya, L.M.; Bermejo, P. Essential Oils from the Asteraceae Family Active against Multidrug-Resistant Bacteria. In *Fighting Multidrug Resistance with Herbal Extracts, Essential Oils and Their Components*; Rai, M.K., Kon, K.V., Eds.; Academic Press: San Diego, CA, USA, 2013; Chapter 14; pp. 205–221. [CrossRef]
11. Hama, J.R.; Strobel, B.W. Occurrence of pyrrolizidine alkaloids in ragwort plants, soils and surface waters at the field scale in grassland. *Sci. Total Environ.* **2021**, *755*, 142822. [CrossRef] [PubMed]
12. Gupta, M.P.; Arias, T.D.; Williams, N.H.; Bos, R.; Tattje, D.H.E. Safrole, the main component of the essential oil from *Piper auritum* of Panama. *J. Nat. Prod.* **1985**, *48*, 330. [CrossRef]
13. Kemprai, P.; Protim Mahanta, B.; Sut, D.; Barman, R.; Banik, D.; Lal, M.; Proteem Saikia, S.; Haldar, S. Review on safrole: Identity shift of the 'candy shop' aroma to a carcinogen and deforester. *Flavour Frag. J.* **2020**, *35*, 5–23. [CrossRef]
14. Jørgesen, P.M.; León-Yáñez, S. Catalogue of the Vascular Plants of Ecuador. Available online: http://legacy.tropicos.org/ProjectAdvSearch.aspx?projectid=2 (accessed on 7 November 2021).
15. León-Yánez, S.; Valencia, R.; Pitmam, N.; Endara, L.; Ulloa Ulloa, C.; Navarrete, H. Libro Rojo de Plantas Endémicas del Ecuador. Available online: https://bioweb.bio/floraweb/librorojo/home (accessed on 1 December 2021).
16. The Plant List. Diplostephium. Available online: http://www.theplantlist.org/ (accessed on 12 August 2021).
17. Vargas, O.; Madriñán, S. Clave para la identificación de las especies del género *Diplostephium* (Asteraceae, Astereae) en Colombia. *Rev. Real Acad. Cienc. Exactas Fis. Nat.* **2006**, *30*, 489–494.
18. Vargas, O.M.; Madriñán, S. Preliminary Phylogeny of *Diplostephium* (Asteraceae): Speciation Rate and Character Evolution. *Lundellia* **2012**, *2012*, 1–15. [CrossRef]
19. Andrade, J.M.; Lucero-Mosquera, H.; Armijos, C. Ethnobotany of indigenous Saraguros: Medicinal plants used by community healers "Hampiyachakkuna" in the San Lucas Parish, Southern Ecuador. *Biomed. Res. Int.* **2017**, *2017*, 1–20. [CrossRef] [PubMed]
20. Vargas Oscar, M. A Nomenclator of *Diplostephium* (Asteraceae: Astereae): A List of Species with Their Synonyms and Distribution by Country. *Lundellia* **2011**, *2011*, 32–51. [CrossRef]
21. Mestanza-Ramón, C.; Henkanaththegedara, S.M.; Vásconez Duchicela, P.; Vargas Tierras, Y.; Sánchez Capa, M.; Constante Mejía, D.; Jimenez Gutierrez, M.; Charco Guamán, M.; Mestanza Ramón, P. In-Situ and Ex-Situ Biodiversity Conservation in Ecuador: A Review of Policies, Actions and Challenges. *Diversity* **2020**, *12*, 315. [CrossRef]
22. Adams, R.P. *Identification of Essential Oil Components by Gas, Chromatography/Mass Spectrometry*, 4th ed.; Allured Publishing Corporation: Carol Stream, IL, USA, 2007; ISBN 10-1932633219.
23. Salinas, M.; Bec, N.; Calva, J.; Ramírez, J.; Andrade, J.M.; Larroque, C.; Vidari, G.; Armijos, C. Chemical composition and anticholinesterase activity of the essential oil from the Ecuadorian plant *Salvia pichinchensis* Benth. *Rec. Nat. Prod.* **2020**, *14*, 276–285. [CrossRef]
24. Demirci, B.; Başer, K.; Aytaç, Z.; Khan, S.I.; Jacob, M.R.; Tabanca, N. Comparative study of three *Achillea* essential oils from eastern part of Turkey and their biological activities. *Rec. Nat. Prod.* **2018**, *12*, 195–200. [CrossRef]
25. Armijos, C.; Matailo, A.; Bec, N.; Salinas, M.; Aguilar, G.; Solano, N.; Calva, J.; Ludeña, C.; Larroque, C.; Vidari, G. Chemical composition and selective BuChE inhibitory activity of the essential oils from aromatic plants used to prepare the traditional Ecuadorian beverage *horchata lojana*. *J. Ethnopharmacol.* **2020**, *263*, 113–162. [CrossRef]

26. Paolini, J.; Muselli, A.; Bernardini, A.F.; Bighelli, A.; Casanova, J.; Costa, J. Thymol derivatives from essential oil of *Doronicum corsicum* L. *Flavour Fragr. J.* **2007**, *22*, 479–487. [CrossRef]
27. Viljoen, A.M.; Kamatou, G.P.; Coovadia, Z.H.; Özek, T.; Başer, K.H.C. Rare sesquiterpenes from South African *Pteronia* species. *S. Afr. J. Bot.* **2010**, *76*, 146–152. [CrossRef]
28. Capetanos, C.; Saroglou, V.; Marin, P.D.; Simić, A.; Skaltsa, H.D. Essential oil analysis of two endemic *Eryngium* species from Serbia. *J. Serb. Chem. Soc.* **2007**, *72*, 961–965. [CrossRef]
29. Montalván, M.; Peñafiel, M.A.; Ramírez, J.; Cumbicus, N.; Bec, N.; Larroque, C.; Bicchi, C.; Gilardoni, G. Chemical composition, enantiomeric distribution, and sensory evaluation of the essential oils distilled from the Ecuadorian species *Myrcianthes myrsinoides* (Kunth) Grifo and *Myrcia mollis* (Kunth) dc. (Myrtaceae). *Plants* **2019**, *8*, 511. [CrossRef] [PubMed]
30. Başer, K.H.C.; Demirci, B.; Kirimer, N.E.; Satil, F.; Tümen, G. The essential oils of *Thymus migricus* and *T. fedtschenkoi* var. handelii from Turkey. *Flavour Fragr. J.* **2002**, *17*, 41–45. [CrossRef]
31. Maggio, A.; Bruno, M.; Guarino, R.; Senatore, F.; Ilardi, V. Contribution to a Taxonomic Revision of the *Sicilian Helichrysum* Taxa by PCA Analysis of Their Essential-Oil Compositions. *Chem. Biodivers.* **2016**, *13*, 151–159. [CrossRef] [PubMed]
32. Özek, G.; Bedir, E.; Tabanca, N.; Ali, A.; Khan, I.A.; Duran, A.; Baser, K.; Özek, T. Isolation of eudesmane type sesquiterpene ketone from *Prangos heyniae* H. Duman & MF Watson essential oil and mosquitocidal activity of the essential oils. *Open Chem.* **2018**, *16*, 453–467. [CrossRef]
33. Carrillo-Hormaza, L.; Mora, C.; Alvarez, R.; Alzate, F.; Osorio, E. Chemical composition and antibacterial activity against *Enterobacter cloacae* of essential oils from Asteraceae species growing in the Páramos of Colombia. *Ind. Crops. Prod.* **2015**, *77*, 108–115. [CrossRef]
34. Him, A.; Ozbek, H.; Turel, I.; Oner, A.C. Antinociceptive activity of alpha-pinene and fenchone. *Pharmacologyonline* **2008**, *3*, 363–369.
35. Kim, D.S.; Lee, H.J.; Jeon, Y.D.; Han, Y.H.; Kee, J.Y.; Kim, H.J.; Shin, H.J.; Kang, J.; Lee, B.S.; Kim, S.J.; et al. Alpha-Pinene Exhibits Anti-Inflammatory Activity Through the Suppression of MAPKs and the NF-κB Pathway in Mouse Peritoneal Macrophages. *Am. J. Chin. Med.* **2015**, *43*, 731–742. [CrossRef] [PubMed]
36. Martin, S.; Padilla, E.; Ocete, M.A.; Galvez, J.; Jiménez, J.; Zarzuelo, A. Anti-inflammatory activity of the essential oil of *Bupleurum fruticescens*. *Planta Med.* **1993**, *59*, 533–536. [CrossRef] [PubMed]
37. Zhou, J.Y.; Tang, F.D.; Mao, G.G.; Bian, R.L. Effect of alpha-pinene on nuclear translocation of NF-kappa B in THP-1 cells. *Acta Pharmacol. Sin.* **2004**, *25*, 480–484. [PubMed]
38. Ahmad, A.; Husain, A.; Mujeeb, M.; Khan, S.A.; Najmi, A.K.; Siddique, N.A.; Damanhouri, Z.A.; Anwar, F. A review on therapeutic potential of *Nigella sativa*: A miracle herb. *Asian Pac. J. Trop. Biomed.* **2013**, *3*, 337–352. [CrossRef]
39. Singh, H.P.; Batish, D.R.; Kaur, S.; Arora, K.; Kohli, R.K. alpha-Pinene inhibits growth and induces oxidative stress in roots. *Ann. Bot.* **2006**, *98*, 1261–1269. [CrossRef] [PubMed]
40. Zore, G.B.; Thakre, A.D.; Rathod, V.; Karuppayil, S.M. Evaluation of anti-Candida potential of geranium oil constituents against clinical isolates of *Candida albicans* differentially sensitive to fluconazole: Inhibition of growth, dimorphism and sensitization. *Mycoses* **2011**, *54*, 99–109. [CrossRef] [PubMed]
41. Quintans-Júnior, L.; Moreira, J.C.F.; Pasquali, M.A.B.; Rabie, S.M.S.; Pires, A.S.; Schröder, R.; Rabelo, T.K.; Santos, J.P.A.; Lima, P.S.S.; Cavalcanti, S.C.H.; et al. Antinociceptive activity and redox profile of the monoterpenes (+)-camphene, p-cymene, and geranyl acetate in experimental models. *ISRN Toxicol.* **2013**, *2013*, 1–11. [CrossRef] [PubMed]
42. Mozuraitis, R.; Stranden, M.; Ramirez, M.I.; Borg-Karlson, A.K.; Mustaparta, H. (-)-Germacrene D increases attraction and oviposition by the tobacco budworm moth *Heliothis virescens*. *Chem. Senses* **2002**, *27*, 505–510. [CrossRef] [PubMed]
43. Lis-Balcnin, M.; Ochocka, R.J.; Deans, S.G.; Asztemborska, M.; Hart, S. Differences in bioactivity between the enantiomers of α-pinene. *J. Essent. Oil Res.* **1999**, *11*, 393–397. [CrossRef]
44. Brenna, E.; Fuganti, C.; Serra, S. Enantioselective perception of chiral odorants. *Tetrahedron Asymmetry* **2003**, *14*, 1–42. [CrossRef]
45. Thomsen, T.; Kewitz, H. Selective inhibition of human acetylcholinesterase by galanthamine in vitro and in vivo. *Life Sci.* **1990**, *46*, 1553–1558. [CrossRef]
46. Blanco-Silvente, L.; Castells, X.; Saez, M.; Barceló, M.A.; Garre-Olmo, J.; Vilalta-Franch, J.; Capellà, D. Discontinuation, Efficacy, and Safety of Cholinesterase Inhibitors for Alzheimer's Disease: A Meta-Analysis and Meta-Regression of 43 Randomized Clinical Trials Enrolling 16 106 Patients. *Int. J. Neuropsychopharmacol.* **2017**, *20*, 519–528. [CrossRef] [PubMed]
47. Orhan, I.; Kartal, M.; Şener, B. Activity of Essential oils and individual components against Acetyland butyrylcholinesterase. *Z. Naturforsch* **2008**, *63*, 547–553. [CrossRef]
48. Leporini, M.; Bonesi, M.; Loizzo, M.R.; Passalacqua, N.G.; Tundis, R. The essential oil of *Salvia rosmarinus* Spenn. from Italy as a Source of Health-Promoting Compounds: Chemical Profile and Antioxidant and Cholinesterase Inhibitory Activity. *Plants* **2020**, *9*, 798. [CrossRef]
49. Aazza, S.; Lyoussi, B.; Miguel, M.G. Antioxidant and antiacetylcholinesterase activities of some commercial essential oils and their major compounds. *Molecules* **2011**, *16*, 7672–7690. [CrossRef] [PubMed]
50. Miyazawa, M.; Yamafuji, C. Inhibition of acetylcholinesterase activity by bicyclic monoterpenoids. *J. Agric. Food Chem.* **2005**, *53*, 1765–1768. [CrossRef] [PubMed]
51. Rodriguez-Aguirre, O.E.; Torrenegra-Guerrero, R.D.T. Flavonoids, Terpenes and the Anti-Oxidant Activity of *Diplostephium phylicoides* (HBK) Wedd. *Indian J. Sci. Technol.* **2018**, *11*, 1–7. [CrossRef]

52. Wang, C.Y.; Chen, Y.W.; Hou, C.Y. Antioxidant and antibacterial activity of seven predominant terpenoids. *Int. J. Food Prop.* **2019**, *22*, 230–238. [CrossRef]
53. Pham-Huy, L.A.; He, H.; Pham-Huy, C. Free radicals, antioxidants in disease and health. *Int. J. Biomed. Sci.* **2008**, *4*, 89–96.
54. Rolnik, A.; Olas, B. The Plants of the Asteraceae Family as Agents in the Protection of Human Health. *Int. J. Mol. Sci.* **2021**, *22*, 3009. [CrossRef] [PubMed]
55. Nikolić, M.; Stevović, S. Family Asteraceae as a sustainable planning tool in phytoremediation and its relevance in urban areas. *Urban For. Urban Green.* **2015**, *14*, 782–789. [CrossRef]
56. Oliveira, M.; Brugnera, D.; Cardoso, M.; Guimarães, L.; Piccoli, R. Rendimento, composição química e atividade antilisterial deóleos essenciais de espécies de *Cymbopogon*. *Rev. Bras. Plantas Med.* **2011**, *13*, 8–16. [CrossRef]
57. Calva, J.; Bec, N.; Gilardoni, G.; Larroque, C.; Cartuche, L.; Bicchi, C.; Montesinos, J.V. Acorenone B: AChE and BChE Inhibitor as a Major Compound of the Essential Oil Distilled from the Ecuadorian Species *Niphogeton dissecta* (Benth.) J.F. Macbr. *Pharmaceuticals* **2017**, *10*, 84. [CrossRef] [PubMed]
58. Van Den Dool, H.; Kratz, P.D. A generalization of the retention index system including linear temperature programmed gas—Liquid partition chromatography. *J. Chromatogr.* **1963**, *11*, 463–471. [CrossRef]
59. Villalta, G.; Salinas, M.; Calva, J.; Bec, N.; Larroque, C.; Vidari, G.; Armijos, C. Selective BuChE Inhibitory Activity, Chemical Composition, and Enantiomeric Content of the Essential Oil from *Salvia leucantha* Cav. Collected in Ecuador. *Plants* **2021**, *10*, 1169. [CrossRef] [PubMed]
60. Gilardoni, G.; Matute, Y.; Ramírez, J. Chemical and Enantioselective Analysis of the Leaf Essential Oil from *Piper coruscans* Kunth (Piperaceae), a Costal and Amazonian Native Species of Ecuador. *Plants* **2020**, *9*, 791. [CrossRef] [PubMed]
61. Calva, J.; Cartuche, L.; González, S.; Montesinos, J.V.; Morocho, V. Chemical composition, enantiomeric analysis and anti-cholinesterase activity of *Lepechinia betonicifolia* essential oil from Ecuador. *Pharm. Biol.* **2022**, *60*, 206–211. [CrossRef]
62. Ellman, G.L.; Courtney, K.D.; Andres, V.; Featherstone, R.M. A new and rapid colorimetric determination of acetylcholinesterase activity. *Biochem. Pharmacol.* **1961**, *7*, 88–95. [CrossRef]
63. Thaipong, K.; Boonprakob, U.; Crosby, K.; Cisneros-Zevallos, L.; Hawkins Byrne, D. Comparison of ABTS, DPPH, FRAP, and ORAC assays for estimating antioxidant activity from guava fruit extracts. *J. Food Compos. Anal.* **2006**, *19*, 669–675. [CrossRef]
64. Arnao, M.B.; Cano, A.; Acosta, M. The hydrophilic and lipophilic contribution to total antioxidant activity. *Food Chem.* **2001**, *73*, 239–244. [CrossRef]

plants

MDPI

Article

GC/MS Analyses of the Essential Oils Obtained from Different *Jatropha* Species, Their Discrimination Using Chemometric Analysis and Assessment of Their Antibacterial and Anti-Biofilm Activities

Mariam I. Gamal El-Din [1,†], Fadia S. Youssef [1,†], Ahmed E. Altyar [2] and Mohamed L. Ashour [1,3,*]

1 Department of Pharmacognosy, Faculty of Pharmacy, Ain-Shams University, Abbasia, Cairo 11566, Egypt; dr_mariam_gamal_eldin@pharma.asu.edu.eg (M.I.G.E.-D.); fadiayoussef@pharma.asu.edu.eg (F.S.Y.)
2 Department of Pharmacy Practice, Faculty of Pharmacy, King Abdulaziz University, P.O. Box 80260, Jeddah 21589, Saudi Arabia; aealtyar@kau.edu.sa
3 Department of Pharmaceutical Sciences, Pharmacy Program, Batterjee Medical College, P.O. Box 6231, Jeddah 21442, Saudi Arabia
* Correspondence: ashour@pharma.asu.edu.eg
† These authors contributed equally to this work.

Citation: Gamal El-Din, M.I.; Youssef, F.S.; Altyar, A.E.; Ashour, M.L. GC/MS Analyses of the Essential Oils Obtained from Different *Jatropha* Species, Their Discrimination Using Chemometric Analysis and Assessment of Their Antibacterial and Anti-Biofilm Activities. *Plants* **2022**, *11*, 1268. https://doi.org/10.3390/plants11091268

Academic Editors: Hazem Salaheldin Elshafie, Ippolito Camele and Adriano Sofo

Received: 15 April 2022
Accepted: 5 May 2022
Published: 9 May 2022

Publisher's Note: MDPI stays neutral with regard to jurisdictional claims in published maps and institutional affiliations.

Abstract: The essential oils of *Jatropha intigrimma*, *J. roseae* and *J. gossypifolia* (Euphorbiaceae) were analyzed employing GC/MS (Gas Chromatography coupled with Mass Spectrometry) analyses. A total of 95 volatile constituents were identified from *J. intigrimma*, *J. gossypifolia* and *J. roseae* essential oils, accounting for 91.61, 90.12, and 86.24%, respectively. Chemometric analysis using principal component analysis (PCA) based on the obtained GC data revealed the formation of three discriminant clusters due to the placement of the three *Jatropha* species in three different quadrants, highlighting the dissimilarity between them. Heneicosane, phytol, nonacosane, silphiperfol-6-ene, copaborneol, hexatriacontane, octadecamethyl-cyclononasiloxane, 9,12,15-Octadecatrienoic acid, methyl ester and methyl linoleate constitute the key markers for their differentiation. *In vitro* antibacterial activities of the essential oils were investigated at doses of 10 mg/mL against the Gram-negative anaerobe *Escherichia coli* using the agar well diffusion method and broth microdilution test. *J. gossypifolia* essential oil showed the most potent antimicrobial activity, demonstrating the largest inhibition zone (11.90 mm) and the least minimum inhibitory concentration (2.50 mg/mL), followed by the essential oil of *J. intigrimma*. The essential oils were evaluated for their anti-adhesion properties against the Gram-negative *E. coli* biofilm using a modified method of biofilm inhibition spectrophotometric assay. *J. intigrimma* essential oil showed the most potent biofilm inhibitory activity, demonstrating the least minimum biofilm inhibitory concentration (MBIC) of 31.25 µg/mL. *In silico* molecular docking performed within the active center of *E. coli* adhesion protein FimH showed that heneicosane, followed by cubebol and methyl linoleate, displayed the best fitting score. Thus, it can be concluded that the essential oils of *J. gossypifolia* and *J. intigrimma* leaves represent promising sources for antibacterial drugs with antibiofilm potential.

Keywords: antibacterial; antibiofilm; chemometrics; essential oils; euphorbiaceae; GC/MS; *Jatropha*; molecular docking; sustainability of natural resources; drug discovery

1. Introduction

Essential oils are natural, volatile components with a complex nature, possessing mostly fragrant odors manufactured by the plants as secondary metabolites. They are commonly prepared either by hydro-distillation or steam distillation; meanwhile, they are highly popular due to their observable biological potential, which is highly attributed to their different classes of compounds, particularly terpenoids. They are popular for their antimicrobial, antiviral, anti-inflammatory, analgesic, spasmolytic, anticancer, anti-aging and anesthetic activities, in addition to their wide consumption in the preservation of foods [1–5].

Microorganisms, a severe hazard attacking human beings, are characterized by the formation of an architectural colony inside an extracellular matrix of polymeric substances termed a biofilm. Bacterial biofilms are highly pathogenic and can trigger nosocomial infections [6,7]. It is worth highlighting that the National Institutes of Health (NIH) declared that 65% of microbial and 80% of chronic infections are accompanied by biofilm formation, which in turn occurs through many steps [4]. These steps comprise the attachment of the bacteria with living or non-living surfaces that are consequently followed by the production of a micro-colony that in turn forms three-dimensional structures and ends up, after maturation, with detachment [8]. Bacterial biofilms are highly contributed to the pronounced resistance of bacteria toward both antibiotics as well as the human immune system and thus prohibition of biofilm formation is highly adopted as a successful strategy combating microbial infections and antibiotic resistance [9]. Hence, searching for effective anti-biofilm agents, particularly from natural origin, has become mandatory worldwide.

Jatropha is a genus of flowering plants belonging to the Euphorbiaceae that includes approximately 175 succulent plants, shrubs and trees. *Jatropha* is an extensively strong and economical plant genus that natively grows in tropical and subtropical regions propagating on wasteland. Different species of the genus have been reported for their antimicrobial activities, as well as their richness in diterpenes [10]. *J. intigrimma*, *J. roseae* and *J. gossypifolia* are three species that belong to the genus *Jatropha*. The essential oils of *J. intigrimma* and *J. gossypifolia* leaves were previously analyzed for their chemical composition. In addition, *J. intigrimma* and *J. gossypifolia* leaf oils were reported to possess strong antimicrobial activities *versus Bacillus cereus* and *Staphylococcus aureus* for the former and against *Escherichia coli*, *Enterococcus faecium*, and *Staphylococcus aureus* for the latter [11,12].

Thus, herein a comparative study was performed for the first time on the leaves of *J. intigrimma*, *J. roseae* and *J. gossypifolia* essential oils that were qualitatively and quantitatively examined for their volatile constituents employing GC/MS (Gas Chromatography coupled with Mass Spectrometry analyses. The volatile oil yield, the major volatile constituents and their percent in each of the examined oils were estimated. This was consequently followed by their discrimination using chemometric analysis to easily detect the differences among the different species, which undoubtedly reflects the variation in their biological behavior. The different essential oils were investigated for their antibacterial activities against the Gram-negative anaerobe *Escherichia coli*. In addition, their inhibitory activities against *E. coli* biofilm formation were assessed for the first time, as this represents the major cause of gastroenteritis, urinary tract infections and neonatal meningitis. The major compounds identified in the bioactive essential oil were further subjected to *in silico* studies to confirm the obtained results. Thus, herein we aimed to find new antimicrobial agents of natural origin with anti-biofilm potential that could be incorporated in pharmaceutical dosage form applied topically to eliminate microbial infection.

2. Results

2.1. Chemical Composition of the Essential Oils of J. intigrimma, J. gossypifolia and J. roseae Leaves

A comparative investigation of the volatile constituents of the three *Jatropha* species, namely *J. intigrimma* Jacq., *J. gossypifolia* L. and *J. roseae* Radcl.-Sm., was conducted for the first time in the present study. The chemical compositions of the essential oils of the fresh leaves of the three species were qualitatively and quantitatively investigated by GC-MS (Figure 1) and compared with the previous results obtained by investigating the essential oils of *J. intigrimma* and *J. gossypifolia* grown in Nigeria. The yields of *J. intigrimma*, *J. gossypifolia* and *J. roseae* essential oils were estimated as 0.31 ± 0.11, 0.21 ± 0.09, and $0.19 \pm 0.11\%$ (v/w), respectively. A total of 95 volatile constituents were identified from the GC/MS analyses of *J. intigrimma*, *J. gossypifolia* and *J. roseae* essential oils, accounting for 91.61, 90.12, and 86.24% of their total oil content, respectively. A list of the identified volatile constituents, the percentage of each volatile component, their experimental retention indices and the literature retention indices, in an order of increasing retention indices (RIs) on the Rtx-5MS column, are summarized in Table 1.

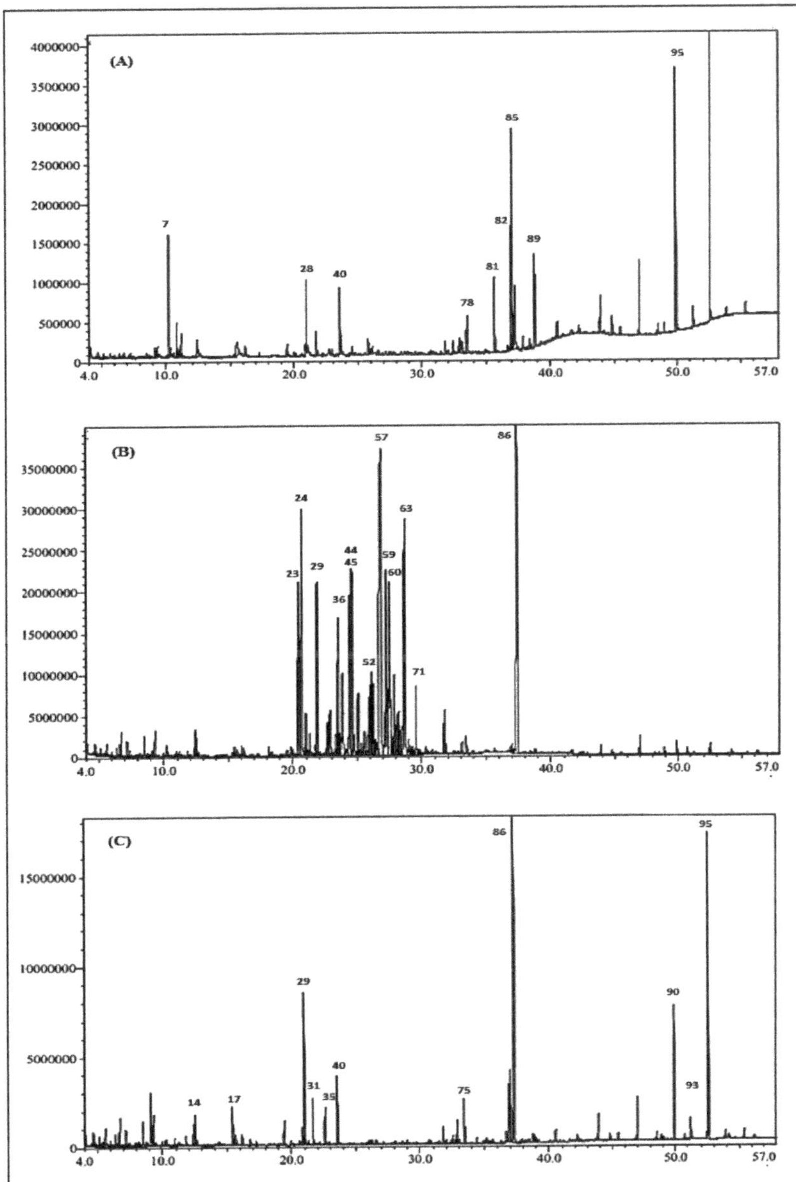

Figure 1. GC-chromatograms of the essential oils obtained from (**A**): *J. intigrimma,* (**B**): *J. gossypifolia* and (**C**): *J. roseae* leaves using the Rtx-5MS column.

Table 1. Essential oil compositions obtained from *J. intigrimma*, *J. gossypifolia* and *J. roseae* leaves using the Rtx-5MS column.

No.	Compounds [a]	*RI*		Composition (%)			Identification
		Measured [b]	Reported [c]	*J. intigrimma*	*J. gossypifolia*	*J. roseae*	
1.	*n*-Nonane	880	900	-	0.20	0.78	MS, RI
2.	α-Pinene	915	915	-	0.08	-	MS, RI
3.	2-Methyl-nonane	947	951	-	0.05	0.15	MS, RI
4.	β-Pinene	987	982	-	0.30	-	MS, RI
5.	*trans*-2-(2-Pentenyl)furan	991	985	-	-	0.25	MS, RI
6.	*p*-Cymene	1014	1017	-	0.03	-	MS, RI
7.	D-Limonene	1018	1018	5.35	0.23	-	MS, RI
8.	(E)-β-Ocimene	1038	1038	1.90	0.09	-	MS, RI
9.	2,5-dimethyl-Nonane	1045	1042	-	0.04	0.09	MS, RI
10.	γ-Terpinene	1048	1048	1.76	0.07	-	MS, RI
11.	*p*-Cymenene	1069	1069	-	0.12	0.49	MS, RI
12.	α-Terpinolene	1088	1088	0.85	0.23	0.85	MS, RI
13.	β-Linalool	1090	1090	1.23	0.34	-	MS, RI
14.	Isophorone	1094	1094	-	0.24	1.23	MS, RI
15.	Nonanal	1098	1102	-	0.05	-	MS, RI
16.	1-Nonanol	1162	1159	-	0.03	0.09	MS, RI
17.	Methyl salicylate	1185	1187	-	-	1.87	MS, RI
18.	Safranal	1189	1189	1.90	-	-	MS, RI
19.	Decanal	1197	1195	-	0.10	0.48	MS, RI
20.	Cumaldehyde	1231	1230	-	-	0.42	MS, RI
21.	Carvacrol	1297	1298	-	0.01	0.08	MS, RI
22.	4-Vinylguaiacole	1311	1311	-	-	0.38	MS, RI
23.	α-Longipinene	1324	1327	0.64	-	-	MS, RI
24.	α-Cubebene	1340	1344	-	0.19	-	MS, RI
25.	Isosativene	1358	1359	-	4.08	-	MS, RI
26.	α-Copaene	1369	1369	-	5.87	-	MS, RI
27.	Longicyclene	1374	1374	-	-	0.74	MS, RI
28.	β-Bourbonene	1376	1376	3.14	0.17	-	MS, RI
29.	Silphiperfol-6-ene	1379	1380	-	-	6.90	MS, RI
30.	*Z*-β-Caryophyllene	1404	1407	1.55	-	-	MS, RI
31.	α-Guaiene	1404	1409	-	-	2.00	MS, RI
32.	*E*-β-Caryophyllene	1409	1409	-	2.97	0.06	MS, RI
33.	α-Ionone	1417	1421	-	-	0.07	MS, RI
34.	β-Ionone epoxide	1437	1430	-	0.14	-	MS, RI
35.	Neryl-acetone	1440	1445	-	0.48	1.65	MS, RI

Table 1. *Cont.*

No.	Compounds [a]	RI		Composition (%)			Identification
		Measured [b]	Reported [c]	*J. intigrimma*	*J. gossypifolia*	*J. roseae*	
36.	Humulene	1445	1445	-	0.60	-	MS, RI
37.	Alloaromadendrene	1453	1453	-	0.80	-	MS, RI
38.	Cadina-1(6),4-diene	1465	1469	-	0.21	-	MS, RI
39.	Germacrene D	1474	1474	-	2.09	-	MS, RI
40.	*β*-Ionone	1477	1478	3.53	0.39	3.09	MS, RI
41.	Cubebol	1488	1484	-	2.17	-	MS, RI
42.	*α*-Muurolene	1491	1491	-	0.37	-	MS, RI
43.	*β*-Himachalene	1500	1500	-	0.21	-	MS, RI
44.	*δ*-Cadinene	1507	1507	-	3.55	-	MS, RI
45.	*δ*-Guaijene	1524	1526	-	4.02	-	MS, RI
46.	Cubenol	1534	1538	-	0.91	-	MS, RI
47.	*α*-Calacorene	1544	1541	-	2.8	-	MS, RI
48.	*β*-Caryophyllene oxide	1556	1556	-	1.25	-	MS, RI
49.	4,8,12-Trimethyltrideca- 1,3,7,11-tetraene	1565	1565	0.84	0.40	-	MS, RI
50.	Globulol	1569	1568	-	0.95	-	MS, RI
51.	Spathulenol	1569	1569	-	3.63	-	MS, RI
52.	Pseudoionone	1575	1581	-	-	0.33	MS, RI
53.	Caryophyllene oxide	1578	1578	-	-	0.11	MS, RI
54.	Guaiol	1581	1584	-	-	0.19	MS, RI
55.	Humulene epoxide	1592	1592	-	0.33	-	MS, RI
56.	Davanone	1594	1592	-	-	0.28	MS, RI
57.	Copaborneol	1597	1593	-	15.7	-	MS, RI
58.	*1-epi*-Cubenol	1619	1617	tr.	-	tr.	MS, RI
59.	Muurola-4,10(14)-dien-1β-ol	1624	1630	-	4.62	-	MS, RI
60.	Caryophylla-4(12),8(13)-dien-5*α*-ol	1634	1640	-	4.82	-	MS, RI
61.	*α*-Cadinol	1650	1660	-	1.42	-	MS, RI
62.	Bulnesol	1664	1666	-	1.45	-	MS, RI
63.	Eudesma-4(15),7-dien-1*β*-ol	1680	1686	-	7.01	-	MS, RI
64.	Ylangenol	1698	1693	-	0.24	-	MS, RI
65.	(2E,6E)-Farnesol	1699	1695	-	-	0.22	MS, RI
66.	14-Hydroxy-*α*-humulene	1721	1718	-	0.13	0.07	MS, RI
67.	Z-ligustilide	1732	1741	-	0.13	-	MS, RI
68.	Benzyl benzoate	1758	1750	-	0.27	-	MS, RI
69.	3-Octadecene	1778	1784	-	-	0.21	MS, RI
70.	Tetradecanoic acid, 1-methylethyl ester	1805	1812	-	-	0.13	MS, RI

Table 1. *Cont.*

No.	Compounds [a]	RI		Composition (%)			Identification
		Measured [b]	Reported [c]	*J. intigrimma*	*J. gossypifolia*	*J. roseae*	
71.	Farnesyl acetate	1816	1818	-	1.06	-	MS, RI
72.	Hexahydrofarnesyl acetone	1825	1827	-	0.56	0.74	MS, RI
73.	Eudesmol acetate	1830	1830	-	tr.	-	MS, RI
74.	*n*-Hexadecan-1-ol	1861	1854	-	-	0.39	MS, RI
75.	7,10-Hexadecadienoic acid, methyl ester	1875	1894	1.50	-	1.36	MS, RI
76.	Palmitoleic acid, methyl ester	1886	1886	1.20	-	-	MS, RI
77.	Farnesyl acetone	1903	1897	-	tr.	-	MS, RI
78.	Hexadecanoic acid methyl ester	1906	1906	3.14	0.29	0.69	MS, RI
79.	1-Hexadecanol, acetate	1986	1978	-	-	0.28	MS, RI
80.	Octadecanal	1998	1999	-	0.08	0.31	MS, RI
81.	Geranyl linalool	2009	2002	3.36	-	-	MS, RI
82.	Sclareolide	2066	2065	-	0.01	0.85	MS, RI
83.	Methyl linoleate	2077	2076	5.65	-	-	MS, RI
84.	9,12-decadienoic acid, methyl ester	2077	2075	-	-	2.47	MS, RI
85.	9,12,15-Octadecatrienoic acid, methyl ester	2085	2085	10.77	-	3.18	MS, RI
86.	Phytol	2096	2096	3.85	10.33	15.25	MS, RI
87.	Verrucarol	2132	2025	-	-	0.32	MS, RI
88.	Sandaracopimarinal	2157	2185	0.93	-	-	MS, RI
89.	Octadecamethyl-cyclononasiloxane	2198	2200	8.42	0.48	-	MS, RI
90.	Heneicosane	2276	2109	-	-	12.67	MS, RI
91.	Octacosane	2764	2800	-	0.16	-	MS, RI
92.	Squalene	2797	2790	0.50	0.03	0.13	MS, RI
93.	Nonacosane	2856	2900	-	-	5.87	MS, RI
94.	Tetrateracontane	3113	3028	4.30	0.54	4.02	MS, RI
95.	Hexatriacontane	3209	3597	28.44	-	14.50	MS, RI
	Monoterpene hydrocarbons			9.01	0.50	-	
	Oxygen containing monoterpene			3.13	0.58	1.36	
	Sesquiterpene hydrocarbons			0.64	24.55	8.96	
	Oxygen containing sesquiterpene			3.53	50.31	4.61	
	Fatty acid esters			22.26	-	7.83	
	Others			53.04	14.18	63.48	
	Total identified components			91.61	90. 12	86.24	

[a] Arrangement of the compounds based on their elution on RTX-5MS column. [b] Kovats index determined experimentally on RTX-5MS column relative to a standard mixture of C8–C30 n-alkanes. [c] Published Kovats retention indices. Identification was based on comparison of the compounds mass spectral data (MS) and Kovats retention indices (RI) with those of NIST Mass Spectral Library (2011), Wiley Registry of Mass Spectral Data 8th edition and literature.

Twenty-two volatile constituents were identified in the oil of *J. intigrimma* leaves, in which fatty acid esters represented the most prevailing class, constituting 22.26% of the oil constituents. In *J. intigrimma* oil, 9,12,15-octadecatrienoic acid methyl ester (10.77%), methyl linoleate (5.65%), and hexadecanoic acid methyl ester (3.14) were the most abundant fatty acid esters identified. Furthermore, hexatriacontane, octadecamethyl cyclononasiloxane, D-limonene, phytol and β-ionone were present in *J. intigrimma* oil in considerable quantities, representing 28.44, 8.42, 5.35, 3.85 and 3.53%, respectively.

Concerning *J. gossypifolia* oil, 61 volatile components were identified in which copaborneol, phytol and eudesma-4(15), 7-dien-1β-ol constituted the major volatile constituents of the oil, representing 15.70, 10.33 and 7.01% of the oil content, respectively. It is worth mentioning that sesquiterpene hydrocarbons and oxygenated sesquiterpenes are the predominant classes in *J. gossypifolia* oil, accounting for 74.86% of the oil content. Isosativene (4.08%), α-copaene (5.87%), spathulenol (3.63%), muurola-4,10(14)-dien-1β-ol (4.62%) and caryophylla-4(12),8(13)-dien-5α-ol (4.82%) were the most abundant sesquiterpenes present.

Meanwhile, 44 constituents were identified in *J. roseae* oil, where phytol, hexatriacontane and heneicosane constituted the major volatile constituents, representing 15.25, 14.50 and 12.67% of *J. roseae* oil, respectively. It is noteworthy that *J. roseae* oil is rich in diterpenes and higher alkanes, accounting for 63.48 % of the oil. However, different sesquiterpene hydrocarbons, oxygenated sesquiterpenes and fatty acids esters can also be observed in *J. roseae* oil viz. silphiperfol-6-ene (6.90%), β-ionone (3.09%), α-guaiene (2.00%), 7,10-hexadecadienoic acid, methyl ester (1.36%), 9,12-octadecadienoic acid, methyl ester (2.47%) and 9,12,15-octadecatrienoic acid, methyl ester (3.18%). A scheme representing the major compounds present in the three *Jatropha* species is presented in Figure 2.

2.2. Discrimination of the Three Jatropha Species Using GC Data Coupled with Chemometrics

Chemometric analysis was adopted using an unsupervised pattern recognition technique represented by principal component analysis (PCA) based on the obtained GC data. Chemometric analysis constitutes an advanced approach for the better discrimination of closely related species, relying upon data gathered from different chromatographic and spectroscopic techniques. Principal component analysis (PCA) was initially performed to categorize data and to correlate between the examined samples and the used variables [13]. PCA based upon the number as well as the relative peak area of volatile constituents obtained from GC spectra for different *Jatropha* species, illustrated in Figure 3, revealed the formation of three discriminant clusters representing the three species.

PCA score plot for principal components (PCs), which are PC1 versus PC2, illustrated in Figure 3A accounts for 71% and 29% of the total variance, respectively. This perfectly results in the placement of the three *Jatropha* species in three different quadrants, which in turn highlights the evident dissimilarity between the three species. Both PC1 and PC2 effectively discriminate between *J. gossypifolia* and *J. roseae,* where the former lies in the left lower quadrant showing negative values for both PCs in contrast to the latter that is positioned in the upper right quadrant revealing positive values for both PCs. Regarding *J. intigrimma* that lies in the right lower quadrant in the PCA score plot, only PC1 could effectively discriminate between it and *J. gossypifolia,* as *J. intigrimma* showed positive values, while *J. gossypifolia* showed negative values for PC1. However, *J. intigrimma* and *J. roseae* could be discriminated only via PC2, where the former displayed negative values and the latter showed positive values. By careful analysis of the loading plot illustrated in Figure 3B, it was clearly obvious that heneicosane, phytol, nonacosane and silphiperfol-6-ene were the key markers for the discrimination of *J. roseae* from the other two species, while copaborneol constitutes the key marker for its differentiation from the other two species. Regarding *J. intigrimma*, hexatriacontane, octadecamethyl-cyclononasiloxane, 9,12,15-Octadecatrienoic acid, methyl ester and methyl linoleate represent the key markers. The results from chemometric analysis coupled with GC data allowed the clustering of samples, and this undoubtedly leads to better visualization of the differences among the essential oils obtained from different *Jatropha* species and in turn reflected the differences between their biological behaviors.

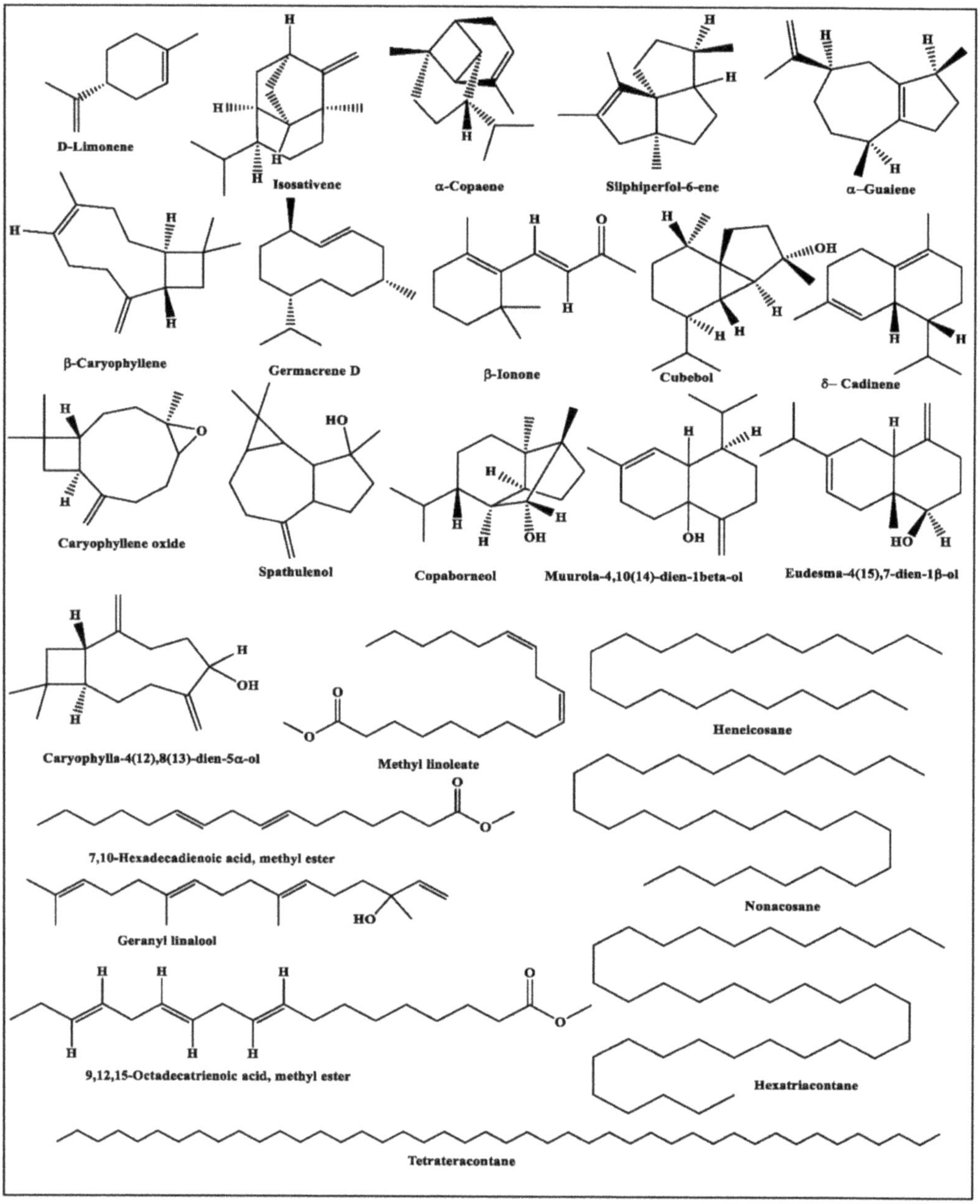

Figure 2. Major components identified in the essential oils obtained from *J. intigrimma, J. gossypifolia* and *J. roseae* leaves.

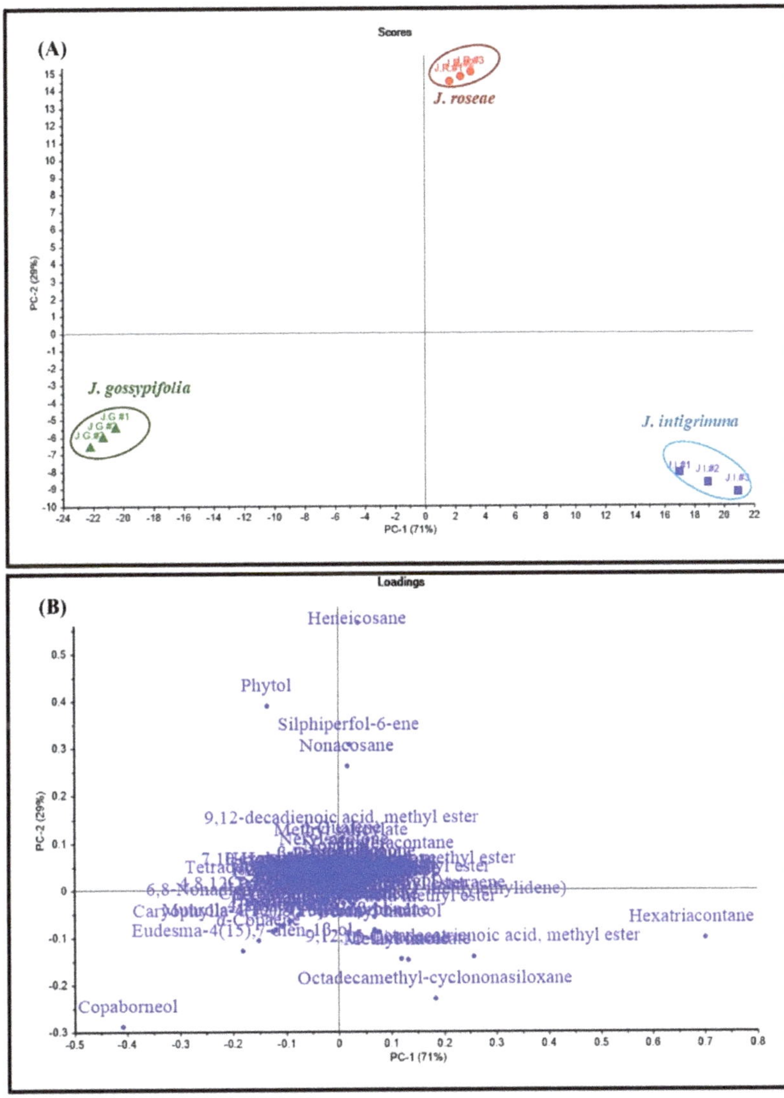

Figure 3. Score plot (**A**) and loading plot (**B**) of GC data collected from *J. intigrimma*, *J. gossypifolia* and *J. roseae* leaves essential oil analyses using unsupervised chemometric analysis (PCA).

2.3. Evaluation of Antibacterial and Anti-Biofilm Activity

2.3.1. Antibacterial Activities of *J. intigrimma*, *J. gossypifolia* and *J. roseae* Essential Oils

In vitro antibacterial activities of the essential oils obtained from the leaves of the three *Jatropha* species, *J. intigrimma*, *J. gossypifolia* and *J. roseae*, were investigated at doses of 10 mg/mL against the Gram-negative anaerobe *Escherichia coli*. The agar well diffusion method was adopted for calculating the mean diameter of inhibition zones produced by the three oil samples in comparison with the standard antimicrobial drug, Gentamicin. Furthermore, the minimum inhibition concentrations (MIC) values were estimated for the oil samples using the broth microdilution test. The essential oil obtained from *J. gossypifolia* leaves demonstrated the most potent antimicrobial activity against *E. coli*, demonstrating the largest inhibition zone (11.90 $\pm$ 0.46 mm) and the least minimum inhibitory concentration

(2.50 mg/mL), followed by the essential oil obtained from *J. intigrimma* leaves. The latter exhibited an inhibition zone of 9.57 ± 0.40 mm and an MIC of 5.00 mg/mL. The oil of *J. roseae* demonstrated the least inhibition zone (8.93 ± 0.60 mm) but had MIC values equal to those of *J. intigrimma* oil of 5.00 mg/mL. The results were compared to the standard gentamycin that exhibited a mean inhibition zone of 27.09 ± 0.01 mm at a dose of 4 µg/mL and demonstrated an MIC of 2 µg/mL. This experiment was repeated in triplicate and data were represented as mean ± S.D.

2.3.2. Antibiofilm Activities of *J. intigrimma*, *J. gossypifolia* and *J. roseae* Essential Oils

The essential oils of the different *Jatropha* species were further evaluated for their potential anti-adhesion properties against the Gram-negative *E. coli* biofilm. A modified method of biofilm inhibition spectrophotometric assay was adopted for the determination of the inhibitory activity of essential oils against the formation of *E. coli* biofilm and for the calculation of the minimum concentration required for complete inhibition of visible biofilm cell growth (MBIC). The essential oil of *J. intigrimma* demonstrated the most potent biofilm inhibitory activity, demonstrating the least minimum biofilm inhibitory concentration (MBIC) of 31.25 µg/mL. However, the essential oils of *J. roseae* and *J. gossypifolia* demonstrated less potent antibiofilm activities, demonstrating minimum biofilm inhibitory activities (MBIC) of 250 and above 1000 µg/mL, respectively, as displayed in Table 2.

Table 2. Mean biofilm inhibitory activity (µg/mL) of *J. intigrimma*, *J. gossypifolia* and *J. roseae* essential oils against *Escherichia coli* determined by modified method of biofilm inhibition spectrophotometric assay.

Sample Conc. (µg/mL)	Mean Biofilm Inhibitory Activity %		
	J. intigrimma	***J. gossypifolia***	***J. roseae***
7.81	52.14 ± 1.3	0	0
15.63	76.38 ± 2.5	0	16.31 ± 1.9
31.25	100 ± 0	0	38.82 ± 1.3
62.5	100 ± 0	0	62.25 ± 2.5
125	100 ± 0	5.08 ± 2.1	76.35 ± 0.72
250	100 ± 0	17.36 ± 1.5	100 ± 0
500	100 ± 0	28.14 ± 1.2	100 ± 0
1000	100 ± 0	39.25 ± 0.58	100 ± 0
MIC	31.25	>1000	250

Data are presented as means ± S.D. $n = 3$.

2.4. *Molecular Docking Studies of Adhesion Proteins with Major Constituents in Jatropha Essential Oils*

In silico molecular docking of the major compounds identified from *Jatropha* essential oils was performed within the active site of the adhesion proteins associated with *E. coli* that enable the bacterium to attach to the surfaces and consequently form its invasive biofilm as FimH (PDB ID 1TR7; 2.10 Å) downloaded from the protein data bank. The docking experiments were carried out using Discovery Studio 4.5 (Accelrys Inc., San Diego, CA, USA) using the C-Docker protocol. The results displayed in Table 3 revealed that heneicosane, followed by cubebol and methyl linoleate, displayed the best fitting score within the active center of *E. coli* adhesion protein FimH with free binding energies equal to −30.68, −8.92 and −4.55 Kcal/mole, respectively. Heneicosane forms two alkyl and π-alkyl bonds with Ile52 and Tyr48, in addition to the formation of Van der Waals interactions with many amino acid residues at the active center (Figure 4A). However, cubebol forms one conventional H-bond with Asp140, in addition to two alkyl and π-alkyl bonds with Ile13 and Phe142 together with Van der Waals bonds at the active site (Figure 4B). Regarding methyl linoleate, it forms two conventional H-bonds with Asn135 and Phe1, one alkyl bond with Ile52, in addition to three C-H bonds with Asp54 and Asn46, together with many Van der Waals interactions, as shown in Figure 4C.

Table 3. Free binding energies (kcal/mol) of major compounds in the active site of *E. coli* adhesion protein FimH using *in silico* studies.

Compound	Adhesion Protein FimH (1TR7)	Number of Formed Hydrogen Bonds	Number of Formed Alkyl and π-Alkyl Bonds
D-Limonene	21.10	-	3; Ile52, Ile13
Isosativene	52.27	-	3; Ile52, Ile13, Tyr48
α-Copaene	4.05	-	7; Ile52, Ile13, Tyr137, Tyr48
Silphiperfol-6-ene	52.27		9; Ile52, Ile13, Tyr137, Phe142
α-Guaiene	44.19	-	5; Ile52, Ile13, Phe142, Tyr48
β-Caryophyllene	17.04	-	4; Ile52, Ile13, Tyr48
Germacrene D	8.63	-	3; Phe142, Ile13, Tyr48
β-Ionone	10.36	1; Phe1	1; Ile13
Cubebol	−8.92	1; Asp140	2; Phe142, Ile13
δ-Cadinene	44.02	-	4; Ile52, Ile13, Tyr137
Caryophyllene oxide	0.82	1; Phe1	4; Phe142, Ile13, Tyr137
Spathulenol	47.82	1; Phe1	7; Ile52, Ile13, Tyr137, Tyr48, Phe142
Copaborneol	49.78	1; Asp140	4; Ile52, Ile13, Tyr137, Tyr48
Muurola-4,10(14)-dien-1β-ol	29.16	1; Asp140	6; Phe142, Ile13, Tyr137, Tyr48
Eudesma-4(15),7-dien-1β-ol	27.03	1; Phe1	6; Phe142, Ile13, Ile52, Tyr48
Caryophylla-4(12),8(13)-dien-5α-ol	12.29	2; Asp54, Phe1	3; Phe142, Ile13,
Methyl linoleate	−4.55	2; Asn135, Phe1	1; Ile52
7,10-Hexadecadienoic acid, methyl ester	−6.70	1; Phe1	1; Ile52
Geranyl linalool	42.39	1; Phe1	2; Tyr48, Tyr137
9,12,15-Octadecatrienoic acid, methyl ester	8.86	2; Asp47, Phe1	1; Ile52
Heneicosane	−30.68	-	2; Ile52, Tyr48
Nonacosane	FD	-	-
Tetrateracontane	FD	-	-
Hexatriacontane	FD	-	-

Positive values indicate unfavorable interaction. FD: fail to dock.

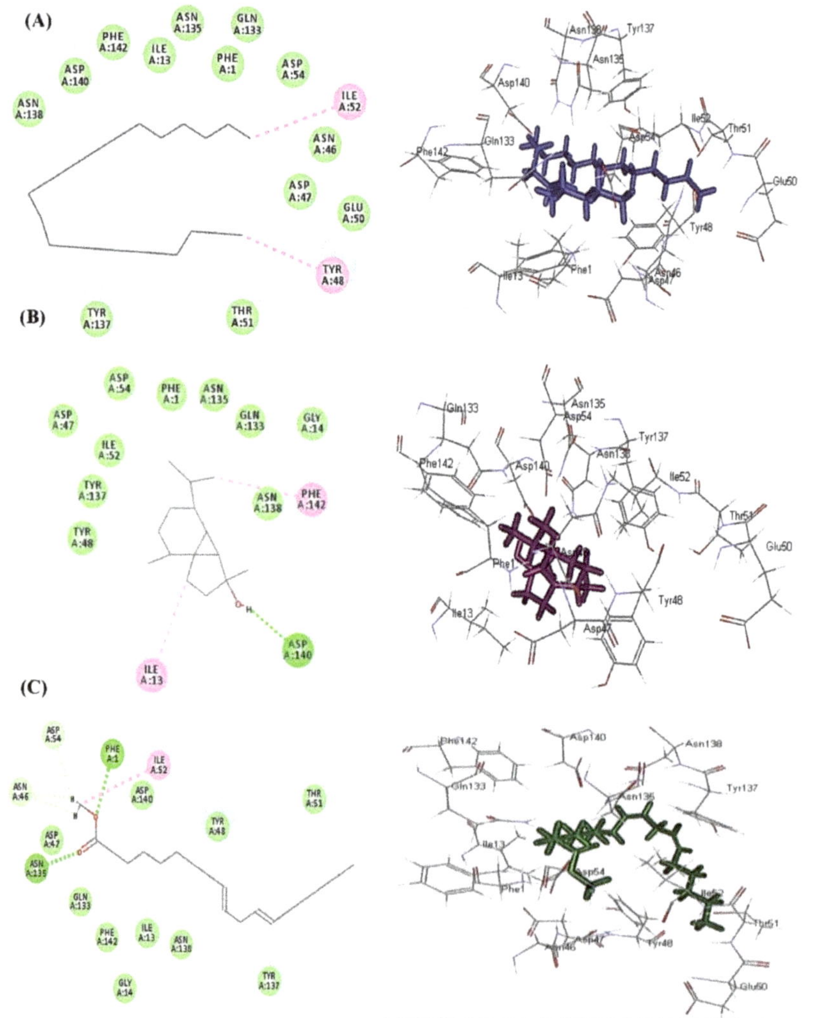

Figure 4. 2D and 3D binding modes of heneicosane (**A**), cubebol (**B**) and methyl linoleate (**C**) within the active center of *E. coli* adhesion protein FimH using *in silico* studies employing the C-docker protocol.

3. Discussion

This study represents the first report investigating the volatile constituents of *J. rosea* leaf oil. In addition, a comparative investigation of the volatile constituents of the three *Jatropha* species, *J. intigrimma*, *J. gossypifolia* and *J. roseae*, was conducted. A previous study on *J. intigrimma* leaves obtained from Nigeria reported β-ionone as one of the major volatile constituents of the oil, which was also identified in our study in a lower concentration. However, the other major constituents reported by Eshilokun et al., pentadecanal and 1,8-cineole, were absent in the present study [12]. Previous literature by Aboaba et al. on *J. gossypifolia* leaves grown in Nigeria reported the predominance of sesquiterpenes accounting for 74.3% of the oil content, which is almost relative to our study [14]. Meanwhile, fatty acids were reported by Ababa et al. in *J. gossypifolia* oil despite their scarcity in our *J. gossypifolia* oil. The major constituents reported by Aboaba et al., germacrene and hexahydrofarnesyl acetone, were identified in the present study but in negligible quantities. Another study in different regions of Nigeria [11] reported the predominance of phytol (33.4%) and linalool (9.81%) in *J. gossypifolia* oil, which were identified in the current study in different percentages.

The chemical composition variability among the essential oils of *Jatropha* species grown in Egypt and those in different regions of Nigeria or elsewhere are attributable to multiple exogenous and endogenous factors, such as seasonal variation, geographical region affecting soil, precipitation and light exposure, extraction method, and the age of the plant and its different chemotypes [15]. Additionally, chemometric analysis was adopted using an unsupervised pattern recognition technique represented by principal component analysis (PCA) based on the obtained GC data. PCA based upon the number, as well as the relative peak area, of volatile constituents obtained from GC spectra for different *Jatropha* species revealed the formation of three discriminant clusters due to the placement of the three *Jatropha* species in three different quadrants, which in turn highlights the evident dissimilarity between the three species evidenced by the score plot. By careful analysis of the loading plot, it was obvious that heneicosane, phytol, nonacosane and silphiperfol-6-ene, copaborneol, hexatriacontane, octadecamethyl-cyclononasiloxane, 9,12,15-Octadecatrienoic acid, methyl ester and methyl linoleate constituted the key markers for the differentiation of the three species.

Essential oils have long been known for their antimicrobial potential that made them crucial in different fields, including the food industry, preservation and medication [16]. To the best of our knowledge, few reports have addressed the antimicrobial activities of essential oils of different *Jatropha* species. Our current study presented the first report on the antimicrobial activities of *J. intigrimma* and *J. roseae* essential oils against *E. coli* Gram-negative bacterium and compared them with the activity of *J. gossypifolia* essential oil. The results of the *in vitro* antibacterial activities of the essential oils exhibited the superior potency of *J. gossypifolia* essential oil, followed by *J. intigrimma* essential oil, against the Gram-negative anaerobe *E. coli*. Results were in accordance with the previous literature reporting the bacteriostatic activity of *J. gossypifolia* leaf oil against the Gram-negative bacterium *E. coli* at a dose of 0.10 mg/mL [11]. It is worth mentioning that phytol, a major constituent of *J. gossypifolia* leaf oil constituting 10.33 % of the oil content, was previously reported for its potent antimicrobial activity against *Escherichia coli*, exhibiting growth inhibition at a minimum concentration (MIC) of 62.5 μg/mL [17,18]. Moreover, caryophylline oxide, identified in *J. gossypifolia* leaf oil, was reported to possess moderate inhibitory activity against Gram-negative *E. coli* with an estimated MIC of 60 ppm [19].

Bacterial biofilms are colonies of microorganisms lying in a matrix of polysaccharides attached to surfaces. They represent physical barriers that inhibit the penetration of antimicrobials to their target sites. Hence, bacterial biofilms represent rational biological risks in drinking water, food, and clinical and industrial environments [20]. Nowadays, increased interest has been directed toward investigating the different mechanisms of inhibiting bacterial biofilm formation and growth. Because attachment represents the initial step in almost all types of biofilm formation, the antiadhesive properties of natural products have recently become a prime interest of study in an aim for the early prevention of microbial biofilm and inhibiting the formation of micro colonies [21]. Furthermore, inhibiting the cell attachment of microbial biofilms has been found to be more readily achieved than preventing the growth of already established biofilms [22]. Hence, the three *Jatropha* oils were investigated for their inhibitory activities against *E. coli* adhesion. This study represents the first report of the antibiofilm activities of the essential oils of the three *Jatropha* species. The results demonstrated the superior antibiofilm activity of the essential oil of *J. intigrimma* leaves compared to the other two *Jatropha* oils. In addition, *in silico* molecular docking of the major compounds identified from *Jatropha* essential oils was performed within the active site of one of the adhesion proteins associated with *E. coli* that enables the bacterium to attach to the surfaces and consequently forms its invasive biofilm, FimH. The docking experiments revealed that heneicosane, followed by cubebol and methyl linoleate, displayed the best fitting score within the active center of *E. coli* adhesion protein FimH with free binding energies equal to −30.68, −8.92 and −4.55 Kcal/mole, respectively.

Previous literature has reported the anti-biofilm activities of different fatty acids and their methyl esters [23,24]. In addition, the monoterpene D-Limonene, an essential component of *J. intigrimma* essential oil, has been reported to possess strong anti-biofilm activity against the heterotrophic, Gram-negative, rod-shaped bacterium *Aeromonas hydrophila* isolated from fish [25]. The anti-adhesive properties and the capabilities of inhibiting the initial biofilm formation can be explained by the possibility of interference with the attraction forces that support the bacterial film with the affected surface or by interrupting the access of vital nutrients for bacterial growth and adhesion [26]. Thus, the antibiofilm of the essential oil could be attributed to the synergistic action of the existing compounds.

4. Materials and Methods

4.1. Plant Material

The fresh leaves of the three *Jatropha* species: *J. intigrimma* Jacq., *J. gossypifolia* L. and *J. roseae* Radcl.-Sm. (Euphorbiaceae) were collected from plants grown in Mazhar Botanical Garden, Giza, Egypt, on August 2020. The plants were kindly identified and authenticated by Eng. Terase Labib, Consultant of Plant Taxonomy at the Ministry of Agriculture and El-Orman Botanical Garden, Giza, Egypt. Voucher specimens of the authenticated plant with codes BMC-JI-MLA, BMC-JG-MLA and BMC-JR-MLA were kept at the Department of Pharmaceutical Sciences, Pharmacy Program, Batterjee Medical College.

4.2. Chemicals, Reagents and Strains

Dimethyl sulfoxide (DMSO), crystal violet and trypan blue dye were obtained from Sigma (St. Louis, MO, USA). Crystal violet stain was prepared as 1% using 0.5% (*w/v*) crystal violet and 50% methanol that were adjusted to volume using distilled water and subsequently filtered through a Whatman No.1 filter paper. The studied bacterial strain was *Escherichia coli*, ATCC 25922, obtained from the American type culture collection (ATCC).

4.3. Preparation of Essential Oils

The fresh leaves of the three *Jatropha* species were separately hydro-distillated for 6 h utilizing a Clevenger-type apparatus. The obtained essential oils were dried over anhydrous sodium sulfate and gathered in separate and sealed vials that were maintained at −30 °C until further analyses. The yield was calculated as % *v/w* after being determined in triplicate, where calculation was performed based on the initial plant weight.

4.4. Metabolic Profiling of the Essential Oils Obtained from J. intigrimma, J. gossypifolia and J. roseae Using GC/MS Analysis

Gas chromatography coupled with Mass Spectrometry (GC/MS) analyses were done on Shimadzu GCMS-QP 2010 (Shimadzu Corporation, Koyoto, Japan) accompanied by Rtx-5MS (30 m × 0.25 mm i.d. × 0.25 µm film thickness) capillary column (Restek, PA, USA) and attached to a Shimadzu mass spectrometer. An initial set of temperature of the column at 50 °C for 3 min was done that was gradually elevated from 50 °C to 300 °C at a rate of 5 °C/min, followed by isothermal maintenance for 10 min at 300 °C. The injector temperature was maintained at 280 °C, while the interface and the ion source temperature were kept at 220 and 280 °C, respectively. The flow rate of helium, which was used as a carrier gas, was 1.37 mL/min. One microliter was injected from the diluted sample with a concentration of 1% *v/v* through a split mode using a split ratio of 15:1. The mass spectrum was recorded using an EI mode of 70 eV in the range of *m/z* 35 to 500. Compound quantitation relied upon the normalization method, taking the reading of three chromatographic runs. Identification of compounds was achieved depending on the retention indices of the detected compounds with regard to a homologous series of *n*-alkanes (C8–C28) that were injected under the same conditions and via comparison mass spectra of the detected compounds with those recorded in the Wiley library database as well as the National Institute of Standards and Technology (NIST) and together with the literature [1,27–29].

4.5. Discrimination of the Three *Jatropha* Species Using GC Data Coupled with Chemometrics

Chemometric analysis using principal component analysis (PCA) as an unsupervised pattern recognition technique was done based on the obtained GC using CAMO's Unscrambler® X 10.4 software (Computer-Aided Modeling, As, Norway) as previously described [3,5]. This was done in an effort to allow the clustering of samples, which undoubtedly leads to better visualization of the differences among the essential oils obtained from different *Jatropha* species.

4.6. Evaluation of Antibacterial and Anti-Biofilm Activity

4.6.1. Susceptibility Test Using the Agar Well Diffusion Method

Susceptibility tests were performed according to NCCLS recommendations (National Committee for Clinical Laboratory Standards) [30]. Screening tests concerning the inhibition zone were performed employing the well diffusion assay previously conducted by Hindler et al. [31]. Preparation of the inoculum suspension was performed from cultures grown overnight on an agar plate that were concomitantly inoculated into Mueller–Hinton broth. A sterile swab was adopted for the inoculation of Mueller–Hinton agar plates (fungi using malt agar plates) after being immersed in the inoculum suspension. The examined samples at different concentrations (2.5, 5 and 10 mg/mL) were solubilized in dimethyl sulfoxide (DMSO) and the inhibition zones were determined around each well after 24 h at 37 °C where the control was prepared using DMSO.

4.6.2. Determination of the Minimum Inhibitory Concentration (MIC) Using the Broth Microdilution Method

The minimum inhibitory concentration (MIC) was determined as previously recommended by the Clinical and Laboratory Standards Institute (CLSI). Briefly, the dilution of a stock solution composed of 10% of the examined oil in the brain heart infusion broth (BHI) in two-fold serial dilutions was performed to obtain 0.02 to 25 mg/mL concentrations at a total volume of 100 mL per well in 96-well microtiter plates. One-hundred milliliters of each tested strain adopting a concentration of 1 × 106 CFU/mL were added to each well, followed by their incubation at 37 °C in appropriate conditions. The medium was used as the non-treated control, while 10% DMSO was employed as the negative control, whereas 0.1% (*w/v*) CHX was the positive control. MIC is the lowest concentration that completely prohibited growth when compared to the non-treated control. All experiments were repeated three times in duplicate.

4.6.3. Evaluation of Anti-Biofilm Activity

The volatile oil samples obtained from hydro-distillation of the three *Jatropha* species were evaluated for their inhibitory activity against biofilm formation of the Gram-negative anaerobe *Escherichia coli*. Biofilm inhibition assay was performed in 96-well plates adopting the modified method of biofilm inhibition spectrophotometric assay [32]. Briefly, 100 µL of an *Escherichia coli* cell suspension was added to a 96-well titer plate together with different concentrations of samples (1000, 500, 250, 125, 62.5, 31.25, 15.63 and 7.81 µg/mL); in addition, DMSO was added and incubated for 24 h at 37 °C. After incubation, the liquid suspension was removed, and 100 µL of 1% *w/v* aqueous solution of crystal violet was added. Removal of the excess crystal violet was achieved after 30 min of staining at room temperature followed by washing the wells thoroughly and the addition of 95% ethanol and incubation for 15 min. The reaction mixture was read spectrophotometrically at a wavelength of 570 nm using a microplate reader (TECAN, Inc.) after being shaken gently. The percent of inhibition of biofilm formation was determined according to the following equation:

$$\% \text{ inhibition} = \text{OD in control} - \text{OD in treatment} \times 100 \text{ OD in control}.$$

The relation between biofilm formation inhibitory % and drug concentration is plotted to obtain the inhibitory curve after treatment with the specified compound. MBIC was the concentration required to completely inhibit biofilm formation.

4.7. Molecular Docking Studies of Adhesion Proteins with Major Constituents in Jatropha Essential Oils

Molecular docking analysis was performed on the major constituents existing in *Jatropha* essential oils regarding adhesion proteins associated with *E. coli* that enable the bacterium to attach to the surfaces and consequently form its invasive biofilm as FimH (PDB ID 1TR7; 2.10 Å) [33]. This protein was downloaded from the protein data bank and docking experiments were carried out using Discovery Studio 4.5 (Accelrys Inc., San Diego, CA, USA) using the C-Docker protocol as previously reported [5,34–36], where binding energies (ΔG) were calculated from the following equation:

$\Delta G_{binding} = E_{complex} - (E_{protein} + E_{ligand})$ Where;

$\Delta G_{binding}$: The ligand–protein interaction binding energy,

$E_{complex}$: The potential energy for the complex of protein bound with the ligand,

$E_{protein}$: The potential energy of protein alone and

E_{ligand}: The potential energy for the ligand alone.

5. Conclusions

In conclusion, the essential oil obtained from *J. intigrimma, J. gossypifolia* and *J. roseae* leaves revealed considerable variation, as revealed by GC analyses. This variation becomes clearly obvious when coupled with chemometric analysis that results in the placement of the three *Jatropha* species in three different quadrants, which in turn highlights the evident dissimilarity between the three species. Moreover, the essential oils of *J. gossypifolia* and *J. intigrimma* leaves represent promising sources for a new generation of antibacterial drugs. Their distinctive antibacterial and antibiofilm activities are probably attributed to their major bioactive chemical constituents, as well as the possible synergistic effect among them. To further confirm the obtained results, *in silico* molecular docking of the major compounds identified from *Jatropha* essential oils was performed within the active center of *E. coli* adhesion protein FimH and results showed that heneicosane followed by cubebol and methyl linoleate displayed the best fitting score. Thus, additional in vivo studies and bioavailability studies are highly recommended to ascertain the obtained results.

Author Contributions: Conceptualization, methodology, software, validation, M.I.G.E.-D. and F.S.Y.; resources, A.E.A.; writing—original draft preparation, M.I.G.E.-D. and F.S.Y.; writing—review and editing, A.E.A. and M.L.A.; supervision, M.L.A.; funding acquisition, A.E.A. All authors have read and agreed to the published version of the manuscript.

Funding: This research was funded by the Deanship of Scientific Research (DSR) at King Abdulaziz University (KAU), Jeddah, Saudi Arabia, under grant number RG-25-166-43.

Institutional Review Board Statement: Not applicable.

Informed Consent Statement: Not applicable.

Data Availability Statement: Data are available in the manuscript.

Acknowledgments: The Deanship of Scientific Research (DSR) at King Abdulaziz University (KAU), Jeddah, Saudi Arabia, funded this project under grant no. RG-25-166-43. Therefore, all the authors acknowledge, with thanks, DSR for technical and financial support.

Conflicts of Interest: The authors declare no conflict of interest.

References

1. Youssef, F.S.; Hamoud, R.; Ashour, M.L.; Singab, A.N.; Wink, M. Volatile oils from the aerial parts of *Eremophila maculata* and their antimicrobial activity. *Chem. Biodivers.* **2014**, *11*, 831–841. [CrossRef] [PubMed]
2. Mokni, R.E.; Youssef, F.S.; Jmii, H.; Khmiri, A.; Bouazzi, S.; Jlassi, I.; Jaidane, H.; Dhaouadi, H.; Ashour, M.L.; Hammami, S. The Essential oil of Tunisian *Dysphania ambrosioides* and its antimicrobial and antiviral properties. *J. Ess.Oil Bear. Plants* **2019**, *22*, 282–294. [CrossRef]
3. Gamal El-Din, M.I.; Youssef, F.S.; Ashour, M.L.; Eldahshan, O.A.; Singab, A.N.B. Comparative analysis of volatile constituents of *Pachira aquatica* Aubl. and Pachira glabra Pasq., their anti-Mycobacterial and anti-Helicobacter pylori activities and their metabolic discrimination using chemometrics. *J. Ess.Oil Bear. Plants Bear. Plants* **2018**, *21*, 1550–1567. [CrossRef]
4. Bakkali, F.; Averbeck, S.; Averbeck, D.; Idaomar, M. Biological effects of essential oils—A review. *Food Chem. Toxicol.* **2008**, *46*, 446–475. [CrossRef] [PubMed]
5. Altyar, A.E.; Ashour, M.L.; Youssef, F.S. *Premna odorata*: Seasonal metabolic variation in the essential oil composition of its leaf and verification of its anti-ageing potential via *in vitro* assays and molecular Modelling. *Biomolecules* **2020**, *10*, 879. [CrossRef]
6. Jamal, M.; Ahmad, W.; Andleeb, S.; Jalil, F.; Imran, M.; Nawaz, M.A.; Hussain, T.; Ali, M.; Rafiq, M.; Kamil, M.A. Bacterial biofilm and associated infections. *J. Chin. Med.Assoc.* **2018**, *81*, 7–11. [CrossRef]
7. Hurlow, J.; Couch, K.; Laforet, K.; Bolton, L.; Metcalf, D.; Bowler, P. Clinical biofilms: A challenging frontier in wound care. *Adv. Wound Care.* **2015**, *4*, 295–301. [CrossRef]
8. Sutherland, I.W. The biofilm matrix–an immobilized but dynamic microbial environment. *Trends Microbiol.* **2001**, *9*, 222–227. [CrossRef]
9. Salmani, A.; Shakerimoghaddam, A.; Pirouzi, A.; Delkhosh, Y.; Eshraghi, M. Correlation between biofilm formation and antibiotic susceptibility pattern in Acinetobacter baumannii MDR isolates retrieved from burn patients. *Gene Rep.* **2020**, *21*, 100816. [CrossRef]
10. Devappa, R.K.; Makkar, H.P.; Becker, K. Jatropha diterpenes: A review. *J. Am. Oil Chem. Soci.* **2011**, *88*, 301–322. [CrossRef]
11. Okoh, S.O.; Iweriebor, B.C.; Okoh, O.O.; Nwodo, U.U.; Okoh, A.I. Antibacterial and antioxidant properties of the leaves and stem essential oils of *Jatropha gossypifolia* L. *BioMed Res. Int.* **2016**, *2016*, 9392716. [CrossRef] [PubMed]
12. Eshilokun, A.O.; Kasali, A.A.; Ogunwande, I.A.; Walker, T.M.; Setzer, W.N. Chemical composition and antimicrobial studies of the essential oils of *Jatropha integerrima* Jacq (leaf and seeds). *Nat. Prod. Commun.* **2007**, *2*, 1934578X0700200813. [CrossRef]
13. Youssef, F.S.; Mamatkhanova, M.A.; Mamadalieva, N.Z.; Zengin, G.; Aripova, S.F.; Alshammari, E.; Ashour, M.L. Chemical profiling and discrimination of essential oils from six Ferula species using GC analyses coupled with chemometrics and evaluation of their antioxidant and enzyme inhibitory potential. *Antibiotics* **2020**, *9*, 518. [CrossRef] [PubMed]
14. Aboaba, S.A.; Adebayo, M.A.; Ogunwande, I.A.; Olayiwola, T.O. Volatile constituents of *Jatropha gossypifolia* L. grown in Nigeria. *Am. J. Ess. Oils Nat. Prod.* **2015**, *2*, 8–11.
15. Barra, A. Factors affecting chemical variability of essential oils: A review of recent developments. *Nat. Prod. Comm.* **2009**, *4*, 1147–1154. [CrossRef]
16. Hammer, K.A.; Carson, C.F.; Riley, T.V. Antimicrobial activity of essential oils and other plant extracts. *J. Appl. Microbiol.* **1999**, *86*, 985–990. [CrossRef]
17. Ghaneian, M.T.; Ehrampoush, M.H.; Jebali, A.; Hekmatimoghaddam, S.; Mahmoudi, M. Antimicrobial activity, toxicity and stability of phytol as a novel surface disinfectant. *Environ. Health Eng. Manag. J.* **2015**, *2*, 13–16.
18. Islam, M.T.; Ali, E.S.; Uddin, S.J.; Shaw, S.; Islam, M.A.; Ahmed, M.I.; Shill, M.C.; Karmakar, U.K.; Yarla, N.S.; Khan, I.N. Phytol: A review of biomedical activities. *Food Chem. Toxicol.* **2018**, *121*, 82–94. [CrossRef]
19. Schmidt, E.; Bail, S.; Friedl, S.M.; Jirovetz, L.; Buchbauer, G.; Wanner, J.; Denkova, Z.; Slavchev, A.; Stoyanova, A.; Geissler, M. Antimicrobial activities of single aroma compounds. *Nat. Prod. Commun.* **2010**, *5*, 1365–1368. [CrossRef]
20. Donlan, R.M. Biofilms: Microbial life on surfaces. *Emerg. Infect. Dis.* **2002**, *8*, 881–890. [CrossRef]
21. Klemm, P.; Vejborg, R.M.; Hancock, V. Prevention of bacterial adhesion. *App. Microbiol. Biotechnol.* **2010**, *88*, 451–459. [CrossRef] [PubMed]
22. Cerca, N.; Martins, S.; Pier, G.B.; Oliveira, R.; Azeredo, J. The relationship between inhibition of bacterial adhesion to a solid surface by sub-MICs of antibiotics and subsequent development of a biofilm. *Res. Microbiol.* **2005**, *156*, 650–655. [CrossRef] [PubMed]
23. Lee, J.H.; Kim, Y.G.; Khadke, S.K.; Lee, J. Antibiofilm and antifungal activities of medium-chain fatty acids against *Candida albicans via* mimicking of the quorum-sensing molecule farnesol. *Microb. Biotechnol.* **2021**, *14*, 1353–1366. [CrossRef]
24. Vijay, K.; Kiran, G.S.; Divya, S.; Thangavel, K.; Thangavelu, S.; Dhandapani, R.; Selvin, J. Fatty acid methyl esters from the coral-associated bacterium *Pseudomonas aeruginosa* inhibit virulence and biofilm phenotypes in multidrug resistant *Staphylococcus aureus*: An *in vitro* approach. *Front. Microbiol.* **2021**, *12*, 394. [CrossRef] [PubMed]
25. Da Silva, E.G.; Bandeira Junior, G.; Cargnelutti, J.F.; Santos, R.C.V.; Gündel, A.; Baldisserotto, B. *In vitro* antimicrobial and antibiofilm activity of S-(-)-limonene and R-(+)-limonene against fish bacteria. *Fishes* **2021**, *6*, 32. [CrossRef]
26. Mishra, R.; Panda, A.K.; De Mandal, S.; Shakeel, M.; Bisht, S.S.; Khan, J. Natural anti-biofilm agents: Strategies to control biofilm-forming pathogens. *Front. Microbiol.* **2020**, *11*, 2640. [CrossRef]
27. Ayoub, I.M.; Youssef, F.S.; El-Shazly, M.; Ashour, M.L.; Singab, A.N.B.; Wink, M. Volatile constituents of *Dietes bicolor* (Iridaceae) and their antimicrobial activity. *Z. Naturforsch. C* **2015**, *70*, 217–225. [CrossRef] [PubMed]

28. Mamadalieva, N.Z.; Youssef, F.S.; Ashour, M.L.; Sasmakov, S.A.; Tiezzi, A.; Azimova, S.S. Chemical composition, antimicrobial and antioxidant activities of the essential oils of three Uzbek Lamiaceae species. *Nat. Prod. Res.* **2019**, *33*, 2394–2397. [CrossRef]
29. Youssef, F.S.; Ovidi, E.; Musayeib, N.M.A.; Ashour, M.L. Morphology, anatomy and secondary metabolites investigations of *Premna odorata* Blanco and evaluation of its anti-tuberculosis activity using *in vitro* and *in silico* studies. *Plants* **2021**, *10*, 1953. [CrossRef]
30. Kiehlbauch, J.A.; Hannett, G.E.; Salfinger, M.; Archinal, W.; Monserrat, C.; Carlyn, C. Use of the National Committee for Clinical Laboratory Standards guidelines for disk diffusion susceptibility testing in New York state laboratories. *J. Clin. Microbiol.* **2000**, *38*, 3341–3348. [CrossRef]
31. Hindler, J.; Howard, B.; Keiser, J. Antimicrobial agents and antimicrobial susceptibility testing. In *Clinical Pathogenic Microbiology*, 2nd ed.; Howard, B.J., Ed.; Mosby: St. Louis, MO, USA, 1994.
32. Regev-Shoshani, G.; Ko, M.; Miller, C.; Av-Gay, Y. Slow release of nitric oxide from charged catheters and its effect on biofilm formation by *Escherichia coli*. *Antimicrob. Agents Chemother.* **2010**, *54*, 273–279. [CrossRef] [PubMed]
33. Adnan, M.; Patel, M.; Deshpande, S.; Alreshidi, M.; Siddiqui, A.J.; Reddy, M.N.; Emira, N.; De Feo, V. Effect of Adiantum philippense extract on biofilm formation, adhesion with its antibacterial activities against foodborne pathogens, and characterization of bioactive metabolites: An *in vitro-in silico* approach. *Front. Microbiol.* **2020**, *11*, 823. [CrossRef] [PubMed]
34. Thabet, A.A.; Youssef, F.S.; El-Shazly, M.; El-Beshbishy, H.A.; Singab, A.N.B. Validation of the antihyperglycaemic and hepatoprotective activity of the flavonoid rich fraction of *Brachychiton rupestris* using in vivo experimental models and molecular modelling. *Food Chem. Toxicol.* **2018**, *114*, 302–310. [CrossRef] [PubMed]
35. Talaat, A.N.; Ebada, S.S.; Labib, R.M.; Esmat, A.; Youssef, F.S.; Singab, A.N.B. Verification of the anti-inflammatory activity of the polyphenolic-rich fraction of *Araucaria bidwillii* Hook. using phytohaemagglutinin-stimulated human peripheral blood mononuclear cells and virtual screening. *J. Ethnopharmacol.* **2018**, *226*, 44–47. [PubMed]
36. Labib, R.; Youssef, F.; Ashour, M.; Abdel-Daim, M.; Ross, S. Chemical composition of *Pinus roxburghii* Bark volatile oil and validation of its anti-inflammatory activity using molecular modelling and bleomycin-induced inflammation in albino Mice. *Molecules* **2017**, *22*, 1384. [CrossRef] [PubMed]

plants MDPI

Article

Deverra triradiata Hochst. ex Boiss. from the Northern Region of Saudi Arabia: Essential Oil Profiling, Plant Extracts and Biological Activities

Arbi Guetat [1,2,*], Abdelrahman T. Abdelwahab [1,3], Yassine Yahia [4], Wafa Rhimi [5,6], A. Khuzaim Alzahrani [7], Abdennacer Boulila [8], Claudia Cafarchia [6] and Mohamed Boussaid [2]

1 Department of Biological Sciences, College of Sciences, Northern Border University, Arar 92341, Saudi Arabia; abdelrhman.talha@nbu.edu.sa
2 Laboratory of Nanobiotechnology and Valorisation of Medicinal Phytoresources, National Institute of Applied Science and Technology, University of Carthage, Tunis 1080, Tunisia; mohamed.boussaid@insat.rnu.tn
3 Department of Botany and Microbiology, Faculty of Science, Al-Azhar University, Cairo 4293073, Egypt
4 Laboratoire d'Aridoculture et Cultures Oasiennes, Institut des Régions Arides de Médenine, Médenine 4119, Tunisia; yahia.yassine@gmail.com
5 Faculté des Sciences de Bizerte, Zarzouna, Université de Carthage, Carthage 7021, Tunisia; wafa_rhimi@hotmail.com
6 Dipartimento di Medicina Veterinaria, Università degli Studi di Bari, 70010 Valenzano, Italy; claudia.cafarchia@uniba.it
7 Faculty of Applied Medical Sciences, Northern Border University, Arar 92341, Saudi Arabia; ahmed.aldousi@nbu.edu.sa
8 Laboratory of Natural Substances LR10INRAP02, National Institute of Research and Physico-Chemical Analyses, Biotechnopole of Sidi Thabet, Ariana 2020, Tunisia; abdennacer.boulila@inrap.rnrt.tn
* Correspondence: arbi.guetat@fst.rnu.tn

Citation: Guetat, A.; Abdelwahab, A.T.; Yahia, Y.; Rhimi, W.; Alzahrani, A.K.; Boulila, A.; Cafarchia, C.; Boussaid, M. *Deverra triradiata* Hochst. ex Boiss. from the Northern Region of Saudi Arabia: Essential Oil Profiling, Plant Extracts and Biological Activities. *Plants* **2022**, *11*, 1543. https://doi.org/10.3390/plants11121543

Academic Editors: Hazem Salaheldin Elshafie, Ippolito Camele and Adriano Sofo

Received: 24 April 2022
Accepted: 2 June 2022
Published: 9 June 2022

Publisher's Note: MDPI stays neutral with regard to jurisdictional claims in published maps and institutional affiliations.

Abstract: *Devrra triradiata* Hochst. ex Boiss is an occasional plant species in the Northern region of Saudi Arabia. The shrub is favored on sandy desert wadis, gypsaceous substrate, and sandy gravel desert. In folk medicine, the plant is used for many purposes; to relieve stomach pains, against intestinal parasites, and for the regulation of menstruation. The present study describes the chemical composition of the essential oils (EOs) of different plant parts of *D. triradiata*. In vivo and in vitro biological activities of plant extracts and essential oils were also studied. Phenylpropanoids, elemicin (flowers: 100%), dillapiole (Stems: 82.33%; and seeds: 82.61%), and apiol (roots: 72.16%) were identified as the major compounds. The highest antioxidant activity was recorded for the EOs of roots and stems (IC_{50} = 0.282 µg/mL and 0.706 µg/mL, respectively). For plant extracts, ethyl acetate showed the highest antioxidant activities (IC_{50} = 2.47 and 3.18 µg/mL). EOs showed high antifungal activity against yeasts with low azole susceptibilities (i.e., Malassezia spp. and Candida krusei). The MIC values of EOs ranged between 3.4 mg/mL and 56.4 mg/mL. The obtained results also showed phytotoxic potential of plant extracts both on the germination features of *Triticum aestivum* seeds and the vegetative growth of seedlings.

Keywords: essential oils; plant extracts; *Deverra triradiata*; Saudi Arabia; biological activities

1. Introduction

The Apiaceae family (Umbelliferae), known under the name of the carrot or parsley family, is one of the most important families of flowering plants. This family encompasses aromatic plants of economic importance employed in foodstuffs, beverages, perfumery, pharmaceuticals, and cosmetics [1]. Species of this family include lianas, herbs, shrubs, or trees [2]. This family is nearly cosmopolitan, being diverse from tropical to temperate regions [3] and consisting of some 455 genera and more than 3700 species widely distributed across temperate regions, especially in Central Asia [4,5]. South-west Asia is an area of considerable diversity and endemism in this family [6].

In Saudi Arabia, the flora is one of the richest in biodiversity in the Arabian Peninsula and comprises very important genetic resources of crop and medicinal plants [7]. The flora of the country consists of more than 140 families, 835 genera, and 2250 species. It was estimated that the flora of Saudi Arabia is diverse and has a great number of medicinal species, which is expected to exceed 1200 (over 50%) [7] and about 471 species (20%) according to Aati et al. [8]. Among others, Asteraceae, Fabaceae, Lamiaceae, and Apiaceae are the most common families in the flora of Saudi Arabia [7,9,10]. Collenette [7] reported that Apiaceae is represented by about 24 genera and more than 45 species in the flora. However, Pimenov and Lenov [6] signaled that the family is present with 26 genera and 52 species.

In the Northern Region of the country (125,000 sq.km), two genera of the Apiaceae family are common: The genus *Deverra* with two species (*Deverra tortuosa* and *Deverra triradiata*) and the genus *Ducrosia* with two species (*Ducrosia anethifolia* and *Ducrosia flabellifolia*). These species are widespread in the Al Widyian region (Wadis), especially the Wadi Arar's tributaries. Taxa belonging to the genus *Deverra* in Saudi Arabia are poorly studied [10]. To the best of our knowledge, phytochemical investigation of *D. triradiata* (ex- *Pituranthos triradiatus* (Boiss.) Asch. & Schweinf) in Saudi Arabia is completely absent. The species is an occasional shrub in the Northern region of Saudi Arabia along stony plains and wadis (Al Widyian region). Once compared to *D. tortuosa*, *D. triradiata* is occasional and less frequent, and the two taxa are found in sympatric association. The species is favored on sandy desert wadis, gypsaceous substrate, and sandy gravel desert of Iraq, Egypt, Sinai, and Arabia [11]. The plant is in the bloom stage during the summer season from July until November (personal observation) [11].

The botanical description is reported by Chrtek et al. [12] and confirmed by Boulos [13]: *D. triradiata* Hochst. ex Borns is a perennial yellow-green plant, reaching 35–100 cm in height. The species is a glabrous shrub with erect stems, juncaceous, sparsely alternately branched, and stout branches. Cauline leaves are strongly reduced to short ovate-triangular sheaths, sometimes with remains of lamina. Umbels (2,3, or 6 flowers) are with glabrous unequal long rays. Bracts are early caducous; pedicels are unequal (2 mm long) and whitish shortly hairy; bracteoles are caducous. Fruits are narrowly ovoid or oblong (with 3 to 5 mm long) and densely hairy, hairs (±0.5 mm long) and the mericarp are three times longer than broad.

In Saudi Arabia, the species is known under the vernacular name of "Haza" or "Sousse". According to Halim et al. [14], the plant is used locally by the Bedouin population against stomach pain, intestinal parasites, haematuria, blood cough, and in the regulation of menstruation. Guetat (2022) [15], reported that the species is described to be a plant of interest in the restoration of sandy coastal areas [16]. According the same author [15], the ethnopharmacologial uses of the plant indicated that *D. triradiata* is used to relieve stomach pains, against intestinal parasites, and the regulation of menstruation [14,17–20]. A strong photodermatitis induced by the twigs of plants during the collection and furanocoumarins of the plant were reported to be the cause [15,17–20]. Bergapten is a linear furocoumarins, also known as 5-methoxypsoralen, with a wide range of pharmacological effects, including neuroprotection, organ protection, anticancer, anti-inflammatory, antimicrobial, and antidiabetes effects [21]. Furthermore, the compound was isolated many times (Ashkenazy and his collaborators) from *D. triradiata* [15]. In the last few decades, multidrug-resistant (MDR) organisms have increasingly become a serious issue in clinical practices and the emergence of drug resistance in pathogens is becoming progressively more common [22]. The term antimicrobial resistance (AMR) is used and referring to the development of resistance by parasites, protozoa fungi, viruses, and MDR bacteria. Many fungi are parasites for plants, animals, human, and other fungi. Plant pathogenic fungi are able to cause extensive damage and losses to agriculture and forestry and can cause serious diseases in humans, including aspergillosis, candidosis, coccidioidomycosis, etc. [23]. The emergence of antifungal resistance (AFR) has urged researchers to explore therapeutic alternatives, one of which includes the use of natural plant products such as EOs.

Studies focusing on essential oil profiling and bioactivities of phytochemicals from the species are sparse. This study bridges this gap as it seeks (i) to profile the essential oils of roots, seeds, stems, and flowers of Saudi Arabian *D. triradiata* and (ii) to study the antioxidant, antifungal, and allelopathic activities of EOs and plant extracts.

2. Material and Methods

2.1. Plant Material

In September 2018, *D. triradiata* Hochst. ex Boiss. was collected from the wild. The site of the collection is situated about 50 km to the East of Arar city (Northern region of Saudi Arabia) in Wadi Aboulkour (30° 44′ 9.816″; 41° 27′ 8.928″), a tributary of Wady Arar. The plant specimen was identified in the Department of Biology, College of Sciences, Northern Border University. A voucher specimen was deposited (Dtr891) in the herbarium of the College of Sciences.

2.2. Isolation of Essential Oil and Crude Collection

Using Clevenger-type apparatus, the essential oils (EOs) were extracted from plant materials (100 g) and dried at laboratory room temperature (25 °C). Flowers, stems, roots, and seeds were used for EOs extraction by hydrodistillation for 4 h. Anhydrous sodium sulfate was used for drying the obtained EOs. Oil yield was estimated based on the 100 g dry weight of plant material and stored in dark vials (4 °C) for further analysis.

For crude collection, powder of different plant parts (100 g) was macerated (7 days) in a glass container with organic solvents (petroleum ether, EtOAc, and MeOH) at room temperature (RT: 25 °C). Plant material was deposited in extraction thimbles and covered by an organic solvent (500 mL). One week later, plant extracts were filtered through Whatman paper and the obtained crudes were leaded. Petroleum ether lead to a crude extract of 800 mg and 4.43 g, respectively, for roots and aerial parts, ethyl acetate lead to a crude extract of 300 mg and 1.75 g, respectively, for roots and aerial parts, and finally, methanol lead to a crude extract of 1.93 g and 8.52 g, respectively, for roots and aerial parts.

2.3. Chemicals

All solvents and chemicals are of analytical grade. DPPH (2,2-diphenyl-2-picrylhydrazyl) and ABTS (2,2-azino-bis-(3-ethylbenzothiazoline-6-sulphonic acid)) were obtained from Sigma Aldrich, (Burlington, MA, USA).

2.4. Gas Chromatography

For GC analyses, a 0.2 µL sample of essential oil (diluted in dichloromethane: 1:100) was used. The analyses were carried out using Agilent gas chromatography (HP7890 GC). The apparatus was equipped with a flame ionization detector (FID) and an HP-5 fused silica column (30 m × 0.32 mm, 0.25 µm film thickness). As a carrier gas, nitrogen was used for the analysis of EOs samples. The injector and detector temperatures were 210 °C and 230 °C, respectively. The column oven temperature varied from 60 °C to 220 °C, with an increasing rate of 3 °C/min.

2.5. GC-MS Analyses

Analyses of EOs samples (0.2 µL) were performed on an Agilent mass spectrometer (Model HP 5975 C). The apparatus was coupled with an Agilent gas chromatograph HP-5MS capillary column (30 m × 250 µm coated with 5% phenyl methyl silicone, 95% dimethylpolysiloxane, and 0.25 µm film thickness). Helium was used as a carrier gas, and the flow rate was fixed to 0.8 mL/min. The oven temperature was initially programmed to vary from 60 to 220 °C with an increasing rate of 4 °C/min, and the transfer line temperature was 230 °C.

2.6. *Identification of Components*

The profiling of EOs components was performed according to the retention index (RI). The Mass Spectroscopy library (NIST/Wiley) and available literature [24] were referred to for recorded mass spectra. Two Databases were used for the identification of the EOs components synonyms (Database\NIST05a.L Minimum Quality: 90; Database\Wiley7Nist05.L). Common names are cited according to PubChem (https://pubchem.ncbi.nlm.nih.gov: last accessed date on 25 May 2022) and NISTBOOK (https://webbook.nist: last accessed date on 25 May 2022).

2.7. *Antioxidant Activities*

2.7.1. DPPH Radical Scavenging Assays

The used method was described by Guetat et al. [10], as described previously by Brand et al. [25], with slight modifications. Plant extracts and EOs samples extracts were prepared in methanol. Subsequently, 60 μL of oil/extracts (at different concentrations) were added to 2940 μL of the methanolic DPPH solution (100 μM). The mixture was conserved in the dark for 30 min and a spectrophotometer (V-630 UV-Vis Spectrophotometer from Jasco) was used to perform the analysis at 517 nm against the blank sample. As a negative control, methanol solution was used, and DPPH solution was referred to as a positive control. The radical-scavenging activity was calculated using the following equation:

$$DPPH_{ScA}\ (\%) = [(AB\text{-}AA)/AB] \times 100 \quad (1)$$

where $DPPH_{ScA}$ is the percentage of DPPH inhibition; AB and AA are, respectively, the optical density (OD) values of the positive control and the OD of the test sample. IC_{50} values were presented as results, where IC_{50} means the concentration of the antioxidant sufficient to scavenge 50% of DPPH present in the test solution. The experiment was replicated three times and IC_{50} values were reported as means ±SD.

2.7.2. ABTS Radical Cation Decolorization Assay

The method used was described by Guetat et al. [10] and Dorman and Hiltunen [26]; the total radical-scavenging capacity was evaluated by the ability of the sample to scavenge the ABTS radical (ABTS•+). The ABTS•+ solution was prepared by mixing 7 mM ABTS and 2.45 mM $K_2S_2O_8$, and the mixture was stored in darkness at RT for 12 h [10,27,28]. ABTS•+solution was diluted in order to obtain an absorbance of 0.7 ± 0.01 at 734 nm. 60 μL of diluted oil and the extract was added to 2940 μL ABTS•+ solution and the absorbance at 734 nm was then measured. The blank consisted of 60 μL of the solvent added to 2940 μL of the ABTS•+ solution. The absorbance was recorded at RT 10 min after the addition of the antioxidant. The experiment was performed in triplicate.

2.8. *Minimal Inhibitory Concentration (MIC) and Minimal Fungicidal (MFC) Concentration*

In the present study, Table 1 shows the used yeast strains (5 *Candida* spp. strains and 6 *Malassezia* spp. Strains). A non-selective isolation medium (Sabouraud Dextrose Agar: SAB) was used for *Candida* spp. isolation and cultivation. However, Dixon Agar (DXA) was used for *Malassezia* spp. strains. Formerly, fungal strains were suspended in 5 mL of sterile saline and vortexed for 15 s. Saline solution (0.85% NaCl) was added until obtaining 0.5 and 2.04 values in the McFarland scale, respectively, for *Candida* spp. and *Malassezia* spp.

Table 1. The used yeast strains in the present study.

Species	Collection Code	Origins
Candida krusei	ATCC 6258	American Type Culture Collection
Candida parapsilosis	ATCC 22019	American Type Culture Collection
Candida parapsilosis	ATCC 22020	American Type Culture Collection
Candida albicans6	CD1378	Cloaca of laying hens
Candida Guilliermondi	CD1379	Cloaca of laying hens
Malassezia pachydermatis	CD90	Dogs with otitis
Malassezia pachydermatis	CD 112	Dogs with dermatitis
Malassezia pachydermatis	CD 113	Dogs with dermatitis
Malassezia furfur	CBS1978	CBS-KNAW Collections
Malassezia furfur	CD1029	Human skin
Malassezia furfur	CD1006	Human blood

For *Candida* spp., EOs (3, 6 μL, 12, 24, and 48 μL/mL) were prepared in RPMI 1640 medium (pH 7). However, for *Malassezia* spp., natural products (EOs with the same concentrations used for Candida spp.) were prepared in SAB + 1% Tween 80. Aliquots of EOs samples (100 μL) and selected *Candida* and *Malassezia* species (100 μL) were bestowed into 96-well plates (Orange Scientific, Braine-l'Alleud). As a control, samples of EOs and yeast-free strains were also conferred. The 96-well plates were incubated at 37 °C for 48 h to 72 h. The MIC values were determined after the visualization of the resultant plate. MIC corresponds to the lower concentration of the antifungal agent in which no visible growth of the fungus was observed, compared to the control (strains grown without oil). The number of viable cells was assessed by the determination of the number of colonies forming units (CFUs) through several dilutions. The MFC values were defined as the antifungal agent concentration in which no CFUs were counted. The experiment was repeated in triplicate in three independent replications.

2.9. *D. triradiata and Phytotoxic Potential of the Plant Extracts*

The allelopathic activity of the roots and aerial parts extracts of *D. triradiata* was evaluated against wheat (*Triticum aestivum* L.) seeds germination and seedlings growth. Before starting the germination experiments, seeds were sterilized with 0.3% NaOCl for 5 min. Five replicates (100 seeds) were prepared for each concentration of the plant extracts. To examine the allelopathic effect of the extracts, various concentrations (0.2, 0.4, 0.6, 0.8, and 1 mg/mL) were prepared in organic solvents (petroleum ether, ethyl acetate, and methanol). Prior to the germination experiment, plant extract (5 mL) was added to Petri dishes (9 cm) lined with a filter paper Whatman No. 1 and air-dried at laboratory room temperature (25 °C) for 24 h. Then, 5 mL of distilled water was added to each Petri dish. The negative control consists of germinated seeds using distilled water and without plant extracts. The Petri dishes were wrapped with Parafilm tape and incubated in the darkness at 25 °C for 1 week [10]. Hypocotyl lengths, root lengths, fresh seedlings weight, and dry biomass weights per Petri dish were determined to evaluate the allelopathic activity of the plant extracts. The inhibitory or stimulatory effects were calculated using the following equation [28]:

$$\text{PytPot}_{D.\ triradiata}\ (\%) = [(\text{PEE-CE})/\text{CE}] \times 100 \quad (2)$$

where PEE: the effect of plant extract and CE: the null effect of the negative control.

2.10. *Statistical Analysis*

All results of biological activities (antioxidant, antifungal, and allelopathic) were expressed as mean values ±SD, and the experiments were replicated three times. ANOVA procedure was used to assess the quantitative differences. Duncan's multiple range test was referred to test the significance ($p < 0.05$) of the mean difference.

3. Results and Discussion

3.1. EOs Profiling

The obtained EOs from stems, seeds, and flowers were a light yellow to green color. EOs yields of these plant parts varied from 0.53, 0.42, and 0.36% (g EOs/100 g of dry plant material), respectively, for stems, seeds, and flowers. However, the oil of the roots was more viscous, colored dark golden and with the lowest yield (0.045% g of EOs/100 g of dry plant material). Nine phytochemicals representing 100% of the total oil were identified as constituents of the EOs by combined GC and GC/MS analyses (Table 2). The aromatic phenylpropanoid elemicin was identified as the unique compound of flowers' EOs (100%). However, this compound is present in the EOs of seeds with a low percentage (3.12%), and it is totally absent in the EOs of stems and roots (Table 2). Compared with the mass spectral data from literature [24,29], elemicin was found to be the unique pure compound isolated from the EOs of the flowers of *D. triradiada*. The compound has a molecular weight of 208.1 g/mol, and the mass spectrum (electron ionization) led to different fragments (Figure 1, Annex). The mass spectra of elemicin (Figure 1) showed a molecular ion peak with m/z 208 $[C_{11}H_{13}O_3]$ + (base peak), the molecular ion was fragmented by losing CH_3 radical to produce $[C_{11}H_{13}O_3]$ + ion at m/z 193 and this was followed by the most highest peak at m/z 177, 133, and 77, respectively. The fragmentation pattern was in agreement with that reported by Ekundayo et al. [29].

Table 2. Chemical composition of four parts of *D. triradiata* Hochst. ex Boiss essential oil from the Northern region of Saudi Arabia.

	Compounds	RI	Roots	Seeds	Stems	Flowers
1	Myristicin	1517	ND	2.03	3.14	ND
2	Seychellene	1458	ND	ND	1.01	ND
3	Dillapiol	1633	ND	82.61	82.33	ND
4	N-[3-(5-methyl-1,3-benzoxazol-2-yl)phenyl]formamide	1672	22.13	11.11	13.52	ND
5	Elemicin	1558	ND	3.12	ND	100
6	(E)-α-elemene	1469	0.84	ND	ND	ND
7	Apiol	1682	72.16	ND	ND	ND
8	7,8,9,10-tetrahydro-anthra[1,2-b]furanne	-	4.68			
9	Germacrene B	1562		1.13		
	Total		99.81%	100%	100%	100%

RI: Retention index relative to n alkanes (C_8–C_{24}) calculated on HP-5 capillary column. ND: not detected.

Alkenylbenzene, elemicin was described in a relatively low percentage in *Deverra tortuosa* (7.3%) [10]. Elemicin (3,4,5-trimethoxyallylbenzene), a fraction of nutmeg oil, is an active natural alkenylbenzene found in many medicinal plants, including several Apiaceae taxa such as *Petroselinum sativum, Ferula heuffelii, Petroselinum crispum*, etc. [30]. According to the same authors [30], elemicin shows extensive pharmacological effects, including antimicrobial, antioxidants, and anti-acetylcholinesterase [30].

For the EOs of seeds and stems, dillapiol was identified as a major component (82.61 and 82.33%), and apiol was the major compound for the roots' EOs with 72.16% (Table 2). The essential oils of *D. traradiata* have not been studied before, and to the best of our knowledge, this is the first report profiling the EOs of the species. Among the few taxa belonging to the genus *Deverra* DC (ex *Pituranthos* Viv.), *D. tortuosa* is a well-documented and studied species. Previous studies reported the EOs of *D. tortuosa* in Saudi Arabia [10], Tunisia [31,32], and Egypt [33–35]. The authors [10,31–33] reported a similar yield of EOs extracted from different parts of *D. tortuosa* not exceeding 0.63 [10]. The chemotype of EOs described herein is different from those described in the literature once compared to closely related species (*D. tortuosa*). Moreover, elemicin, apiol and dillapiol were found to be the major components of *D. triradiata's* EOs. However, *D. tortuosa's* EOs was described as a chemotype of apiol for all plant parts in Saudi Arabia [10] as a chemotype of sabinene, α-pinen, limonen, and terpinen-4-ol in Tunisia [29] and as a chemotype of terpinen-4-ol in Egypt [33] and a chemotype of dillapiole, elemicin, and myristicin [34].

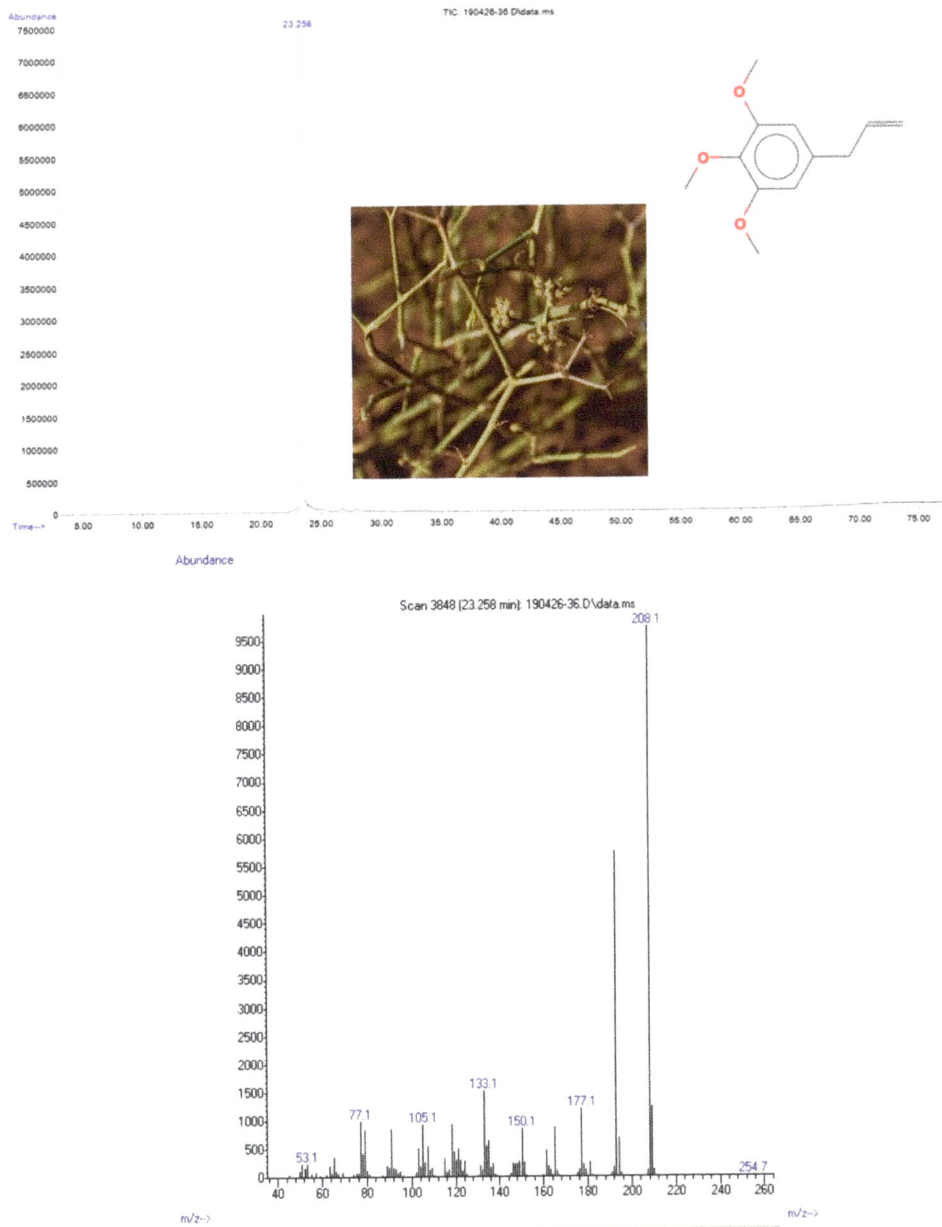

Figure 1. Mass spectra of elemicin as the unique pure compound isolated from flowers' EOs of *D. triradiata*.

3.2. *Antioxidant Activities as Assessed by EOs' ABTS System and Plant Extracts' DPPH System*

Table 3A,B show the ABTS values of EOs' IC_{50} and DPPH values of plant extracts (aerial parts and roots). The values of IC_{50} (ABTS assay system) for EOs are presented in Table 3A. The roots' EOs and those of stems showed the best antioxidant activity (IC_{50} = 0.282 µg/mL and 0.708 µg/mL, respectively). By comparing the obtained IC_{50} values of *D. triradiata* EOs with that of authentic ascorbic acid (0.413), it was found that the potential antioxidant activity of the EOs is high and not significantly different from the standard. However, the EOs of flowers presented showed a relatively low antioxidant activity (5.42 µg/mL) and were significantly different from ascorbic acid as a standard with the IC_{50} value of 0.413 µg/mL.

Table 3. Antioxidant activity evaluated by ABTS for the EOs (A) and DPPH test assays of plant extracts of *D. triradiata* samples (B).

		(A) ABTS IC_{50} (µg/mL)	
3A.		Samples	Concentration (µg/mL)
EOs		Flowers	5.42 ^{a} (±1.13)
		Stems	0.706 ^{b} (±0.14)
		Roots	0.282 ^{c} (±0.01)
		Ascorbic acid	0.413 ^{b,c} (±0.018)
		(B) DPPH IC_{50} (µg/µL)	
		Samples	Concentration (µg/mL)
Plant extracts	Arial parts	Petroleum ether extract	4.52 ^{a} (±0.95)
		Ethyl acetate extract	2.47 ^{c} (±0.2)
		Methanol extract	2.73 ^{b,c} (±0.32)
	Roots	Petroleum ether extract	3.70 ^{a,b} (±0.26)
		Ethyl acetate extract	3.18 ^{b,c} (±0.73)
		Methanol extract	4.36 ^{a} (±0.1)
		Ascorbic acid	5.41 ^{a} (±0.58)

Values are presented as mean ±SD (n = 3). Values followed by the same letter are not significantly different ($p > 0.05$).

For the aerial parts' extract samples, the results are presented in Table 3B. It is remarkable that in all plant extracts and for three extraction systems (petroleum ether, ethyl acetate, and methanol), IC_{50} values were less than 5 µg/mL. The highest antioxidant activities (DPPH assay system) were observed for the ethyl acetate extracts both for aerial parts and roots (2.47 and 3.18 µg/mL, respectively). The methanol extract of roots and petroleum ether aerial part extract showed the lower values of DPPH scavenging (4.36 µg/mL and 4.52 µg/mL, respectively). IC_{50} values of plant extracts were found to be with high antioxidant potential and are significantly different from ascorbic acid ($p > 0.05$). Moreover, the IC_{50} values of plant extract concentrations were found to be lower by 1.25 to 2 times compared to ascorbic acid concentration for the DPPH system assay (4.36/5.41 = 1.25 times and 2.47/5.41 = 2.19 times).

Our data on antioxidant activity showed values varying between 0.282 and 5.42 µg/mL for the EOs and concentrations ranging between 2.47 and 4.52 µg/mL for plant extracts. Compared to previous published work, antioxidant activity of hydroalcholic extract of *Ferula gummosa* [36] were reported to be moderate, at 579 µg/mL. Plant extracts from *Ferula lutea* showed that the strongest antioxidant activity was obtained for the ethyl acetate extract (IC_{50} = 12.8 µg/mL). However, a recent study focusing on *D. tortuosa* reported relatively comparable results with antioxidant activities for the flowers' EOS (IC_{50} = 0.175 µg/mL) [10].

According to the same authors, the samples of stems and roots exhibit lower antioxidant activity (IC_{50} = 0.201 µg/mL and 0.182 µg/mL, respectively). Methanol extracts of *D. tortuosa* exhibited the highest antioxidant activities (IC_{50} = 0.064 µg/mL) [10].

This is the first report describing the antioxidant activity of *D. triradiata*. However, the species was described to induce photosensitization [18] and photodermatitis [19]. As a personal observation, we report that the collected plant material of the studied species caused irritations, such as lines on the upper layer of the hand's skin during collection.

3.3. *D. triradiata's* EOs Activities against Selected Strains from *Malassezia spp.* and *Candida ssp*

The MIC values of 4 EOs (Roots, seeds, stems, and flowers) varied between 3 µL/mL and 48 µL/mL (Table 4 and Figure 2). *M. pachydermatis* (CBS 1978) seems to be the most sensitive fungal strain for the 4 EOs with an MIC value of 3 µL/mL. However, 3 *Candida* strains (*C. albicans* 6, *C. parapsilosis*: ACTT 22020 and *C. parapsilosis*: CD 1378) were found to be with the lowest inhibitory effect against the tested EOs (especially EOs of roots and seeds).

Table 4. Minimum inhibitory concentration (MIC) and minimum fungicidal concentration (MFC) (µg/mL) of *D. triradiata* EOS compared to Itraconazole (µg/mL) as a positive control.

	Roots		Seeds		Stems		Flower		Itraconazole	
Fungal Strains	**MIC**	**MFC**	**MIC**	**MFC**	**MIC**	**MFC**	**MIC**	**MFC**	**MIC**	**MFC**
C. guilliermondi (ACTT 6258)	13.6 mg/mL (12 µL/mL)	13.6 mg/mL (12 µL/mL)	14.1 mg/mL (12 µL/mL)	28.2 mg/mL (24 µL/mL)	6.95 mg/mL (6 µL/mL)	6.95 mg/mL (6 µL/mL)	7.15 mg/mL (6 µL/mL)	14.3 mg/mL (12 µL/mL)	32	64
C. albicans 6 (ACTT 22019)	54.4 mg/mL (48 µL/mL)	>54.4 mg/mL (48 µL/mL)	56.4 mg/mL (48 µL/mL)	56.4 mg/mL (48 µL/mL)	6.95 mg/mL (6 µL/mL)	13.8 mg/mL (12 µL/mL)	14.3 mg/mL (12 µL/mL)	14.3 mg/mL (12 µL/mL)	> 64	> 64
C. parapsilosis (ACTT 22020)	54.4 mg/mL (48 µL/mL)	> 54.4 mg/mL (48 µL/mL)	56.4 mg/mL (48µL/mL)	56.4 mg/mL (48 µL/mL)	6.95 mg/mL (6 µL/mL)	13.8 mg/mL (12 µL/mL)	14.3 mg/mL (12 µL/mL)	14.3 mg/mL (12 µL/mL)	0.32	0.64
*C. parapsilosis*17 (CD 1378)	54.4 mg/mL (48 µL/mL)	> 54.4 mg/mL (48 µL/mL)	56.4 mg/mL (48 µL/mL)	56.4 mg/mL (48 µL/mL)	6.95 mg/mL (6 µL/mL)	13.8 mg/mL (12 µL/mL)	14.3 mg/mL (12 µL/mL)	14.3 mg/mL (12 µL/mL)	0.32	0.64
C. krusei (CD 1379)	54.4 mg/mL (48 µL/mL)	> 54.4 mg/mL (48 µL/mL)	56.4 mg/mL (48 µL/mL)	56.4 mg/mL (48 µL/mL)	13.8 mg/mL (12 µL/mL)	13.8 mg/mL (12 µL/mL)	7.15 mg/mL (6 µL/mL)	7.15 mg/mL (6 µL/mL)	0.64	> 0.64
Candida ssp.	46.24 mg/mL (± 17.20)	> 46.24 mg/mL (17.20 ±)	47.94 mg/mL (±17.84)	48.76 mg/mL (± 11.54)	8.32 mg/mL (± 2.89)	12.43 mg/mL (± 2.89)	11.44 mg/mL (± 3.69)	12.87 mg/mL (± 3.01)	> 19.54 (± 26.78)	> 25.98 (± 32.72)
M. furfur (CD 90)	13.6 mg/mL (12 µL/mL)	27.2 mg/mL (24 µL/mL)	14.10 mg/mL (12 µL/mL)	14.10 mg/mL (12 µL/mL)	3.4 mg/mL (3 µL/mL)	6.95 mg/mL (6 µL/mL)	7.15 mg/mL (6 µL/mL)	14.3 mg/mL (12 µL/mL)	0.32	0.32
M. furfur (CD 112)	6.8 mg/mL (6 µL/mL)	13.6 mg/mL (12 µL)	14.10 mg/mL (12 µL/mL)	14.10 mg/mL (12 µL/mL)	6.95 mg/mL (6 µL/mL)	13.8 mg/mL (12 µL/mL)	7.15 mg/mL (6 µL/mL)	14.3 mg/mL (12 µL/mL)	0.32	0.32
M. furfur(CD 113)	6.8 mg/mL (6 µL/mL)	13.6 mg/mL (12 µL/mL)	14.10 mg/mL (12 µL)	14.10 mg/mL (12 µL/mL)	6.95 mg/mL (6 µL/mL)	13.8 mg/mL (12 µL/mL)	7.15 mg/mL (6 µL/mL)	14.3 mg/mL (12 µL/mL)	0.32	0.32
M. pachydermatis (CBS 1978)	3.4 mg/mL (3 µL/mL)	3.4 mg/mL (3 µL/mL)	3.52 mg/mL (3 µL/mL)	3.52 mg/mL (3 µL/mL)	3.4 mg/mL (3 µL/mL)	3.4 mg/mL (3 µL/mL)	3.57 mg/mL (3 µL)	3.57 mg/mL (3 µL/mL)	0.32	0.32
M. pachydermatis (CD 1029)	13.6 mg/mL (12 µL/mL)	27.2 mg/mL (24 µL/mL)	3.52 mg/mL (3 µL/mL)	7.05 mg/mL (6 µL/mL)	3.4 mg/mL (3 µL/mL)	3.4 mg/mL (3 µL/mL)	3.57 mg/mL (3 µL/mL)	3.57 mg/mL (3 µL/mL)	0.32	0.32
M. pachydermatis (CD 1006)	3.4 mg/mL (3 µL/mL)	3.4 mg/mL (3 µL/mL)	3.52 mg/mL (3 µL/mL)	7.05 mg/mL (6 µL/mL)	3.4 mg/mL (3 µL/mL)	6.95 mg/mL (6 µL/mL)	3.57 mg/mL (3 µL/mL)	3.57 mg/mL (3 µL/mL)	0.32	0.32
Malassezia ssp.	7.93 mg/mL (±4.43)	14.73 mg/mL (±10.18)	8.81 mg/mL (±5.53)	9.99 mg/mL (±4.47)	4.58 mg/mL (±1.75)	8.05 mg/mL (±4.51)	5.36 mg/mL (±1.87)	8.94 mg/mL (±5.6)	0.32 (±0.00)	0.32 (±0.00)

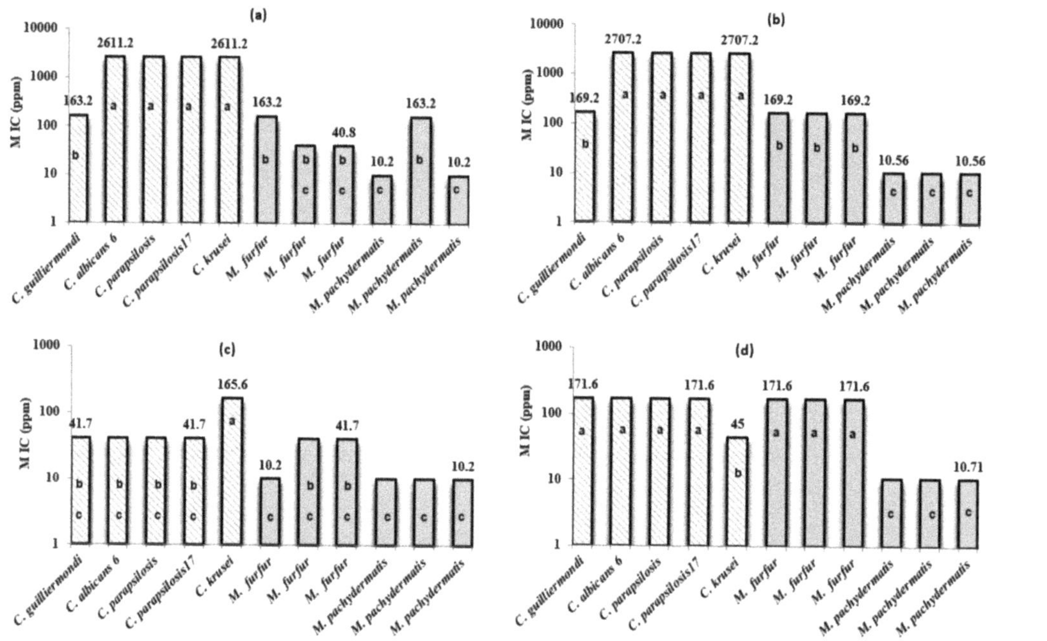

Values followed by the same letter are not significantly different ($p > 0.05$).

Figure 2. MIC values (ppm) of antifungal activity of the EOs of *D. triradiata*: roots (**a**), seeds (**b**), stems (**c**) and flowers (**d**) against selected strains from two fungus genera *Malassezia* (6 species) and *Candida* (5 species).

D. triradiata's EOs showed higher inhibitory activities against pathogenic *Malassezia* isolates compared to *Candida* ssp (Figure 2a–d). For *Malassezia* strains, the MIC values of EOs varied between 10.2 and 163.2 ppm for the Roots' EOs, 10.56 and 169.2 for the seeds' EOs, 10.2 and 41.7 for the stems' EOs, and the MIC values of the flowers' EOs varied from 10.71 to 171.6 ppm. Furthermore, the MIC values of *Candida* species ranged between 163.2 and 2611.2 ppm for Roots' EOs, 169.2 and 2707.2 for the seeds' EOs, 41.7 and 165.6 for the stems' EOs, and 45 to 171.6 for the flowers' EOs.

The mean MIC values of the two genera are reported in Figure 3. The genus *Malassezia* showed a mean MIC values less than 100 ppm for the 4 tested EOs (only 20.65 ppm for Stems' EOs). However, the mean MIC values, for candida genus, exceeded 2000 ppm respectively for Roots' EOs and Seeds' EOs; the mean MIC values were 66.48 and 146.28 ppm, respectively, for Stems' EOs and Flowers' EOs. The oil extracted from the stem was found to be the most potent EOs, with a mean MIC values less than 70 ppm for the two genera. However, pathogenic *Malassezia* and *Candida* isolates seemed to be less sensitive against the oils extracted from the roots and seeds. The chemical profiling of the 4 parts of the plant species (Table 2) showed that the EOs of seeds and stems were reported as a chemotype of dillapiole. This finding could support the effect of minor compounds (mainly seychellene) in the fraction of EOs extracted from the stem which could affect fungal pathogens in synergism with the major components. Moreover, seychellene was reported as a non-selective candidate for inhibitor cyclooxygenase on pre-osteoblast cells [37].

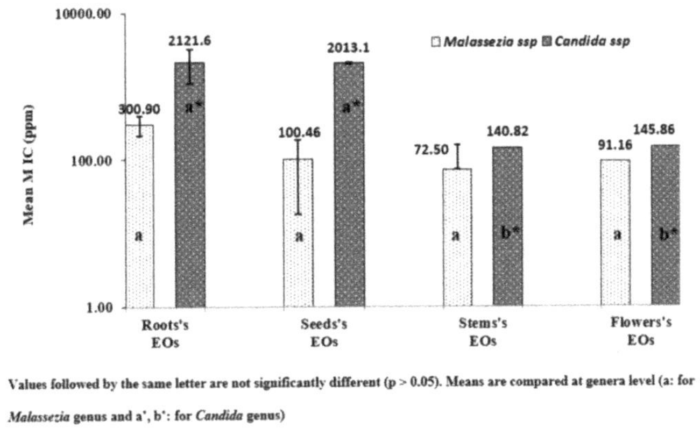

Figure 3. Mean MIC values of antifungal activity of the EOS from *D. triradiata* at genera level against selected strains from two fungus genera *Malassezia* and *Candida*.

Globally, *M. pachydermatis* was the most sensitive yeast to the tested EOs (MIC values ranged from 10.2 and 10.71 ppm). However, *C. parapsilosis*, *C. albicans* 6, and *C. krusei* strains seemed to be the less sensitive, and the highest MIC values were recognized for these three strains (7.15–56.4 mg/mL). The MFC values (Table 4 and Figure 4a–d); Figure 4a–d related to the 11 studied fungal strains corroborated entirely the MIC values for all strains belonging to the 2 genera.

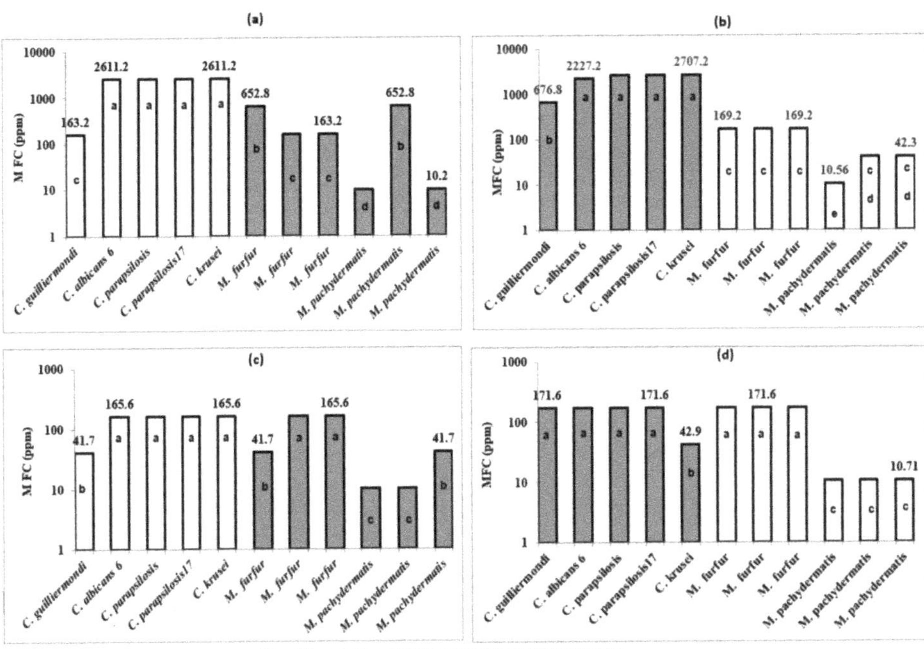

Figure 4. MFC values (ppm) of antifungal activity of the EOs of *D. triradiata*: roots (**a**), seeds (**b**), stems (**c**), and flowers (**d**) against selected strains from two fungus genera *Malassezia* (6 species) and *Candida* (5 species).

The review of the inhibitory potential of EOs on *Malassezia* and candida species by various plants has revealed the antifungal activity of this valuable natural products extracted from different plant taxa [38,39]. Our data corroborate this finding and the studied EOs exhibited inhibitory potential towards *Candida* and *Malassezia* spp. reference strains. However, the authors [38,39] reported the range of 1000 μg/mL and 2 μL/mL as thresholds to define an EOs as an antifungal inhibitor. Our results showed (in the case of excluding the concentration exceeding 1000 μg/mL) that the tested EOs of stems and flowers of the studied plant species displayed inhibitory potential towards fungal pathogens with concentrations varying between 10.2 to 171.6 ppm and 3 to 12 μL/mL.

Figure 3 showed that EOs possessed fungicidal activity against the tested pathogens belonging to *Candida* and *Malassezia*. The lowest MFC values were recorded against 3 strains of *M. pachydermatis*, and EOs of stems and flowers seem to be the most potent (Figure 4c,d). The finding of the MFC data corroborate those obtained for MIC and *Candida* ssp and were found to be less sensitive, especially towards EOs of roots and seeds. As shown by Figure 3a,b, the highest MFC values (2707.2, 2227.2, and 2611.2 ppm) were recorded against 4 *Candida* strains (*C. albicans* 6, 2 strains of *C. parapsilosis*, and *C. krusei*).

The MFC/MIC ratio was calculated to determine whether thymol has a fungistatic (MFC/MIC $\geq$ 4) or fungicidal activity (MFC/MIC < 4) [39–41]. The MFC/MIC ratio for four studied samples of EOs showed values ranging from to 1 to 2.08. This ratio was equal to 1 for 26 fungal strains out of 44 studied cases (11 fungal strains tested for each of the 4 analysed EOs). In general, the MFC values were found to be lower than the MIC values suggesting that the tested essential oils have fungicidal effects. Our results of MFC described for *D. triradiata* corroborate previous studies focusing on similar fungal strains [10,40–43]. Guetat et al. (2018) [10] reported that the MFC values of EOs of *D. totuosa* varied between 3 and 24 μL/mL, and *Candida* ssp were found to be less sensitive towards the two tested EOs (extracted from flowers and stems. The relationship between the effectiveness of the studied EOs and the antifungal activity was not discussed by the author [10] and the described chemotype, both for stems and flowers, was reported to be dominated by apiol as a major component. Dongmo et al. [44] reported that the EOs of four Cameroonian spices showed MFC values ranging between 256 and 4096 μL/mL.

3.4. Allelopathic Activity of D. triradiata Plant Extracts

The allelopathic effects of the root and aerial parts extracts are summarized in Table S1 (Supplementary Materials). The allelopathic influence on *T. aestivum* L. seeds germination and seedlings growth varied according to the plant extracts and concentration. The germination percentage varied between 65 and 88.23% (values not reported in Table S1). A significant promoting and inhibitory effect on the germination was found (Table S1). The general tendency of the experiment showed that root extracts and aerial part extracts with an inhibitory effect were found to promote germination. The *D. triradiata* ethyl acetate extract of aerial parts showed high phytotoxic effects against *T. aestivum* L. seeds, with an inhibition of seed germination of 17.26% at 0.2 mg/mL. However, the highest induction value of seed germination of the root extracts was exhibited by ethyl acetate extract (12.32% at 0.6 mg/mL).

The variance analysis performed on the allelopathic effects revealed significant differences among concentrations ($p < 0.05$). The effect of the aerial parts' extracts on the radicle growth varied from 22.73 as an induction factor (at 0.6 mg/mL; ethyl acetate) to −73.4% as inhibition (at 1 mg/mL, petroleum ether). For the root extracts, this character (radicle length) was induced by methanol extract (34.97% at 0.6 mg/mL). However, the radicle length was inhibited by ethyl acetate extract (−9.57% at 0.8 mg/mL).

The hypocotyl length was remarkably induced in the presence of different *D. triradiata* extracts at different concentrations (excepting petroleum ether extract of aerial parts at 1 mg/mL). The highest percentages of promoting were recorded in methanol extracts (from 44.23% at 0.6 mg/mL to 48.86% at 0.4 mg/mL). For the root extracts, the hypocotyl length was highly induced by methanol extracts (57.59% at 1 mg/mL). In this way petroleum ether root extracts also showed high values of hypocotyl length promoting (from 9.61 to 39.31%). The biomass production was considerably induced in the presence of different plant extracts at different concentrations, and the dry weight of the seedlings treated from different samples was highly increased. Globally the root extract increased more remarkably by the dry mass production. For petroleum ether at 1 mg/mL, the highest percentage of dry mass increasing was 66.6% for root extracts and 42.14% for aerial part extracts. The allelopathic effect is related to the plant extract composition and the target species *D. triradiata* extracts was efficient to promote and inhibit the seed germination and the seedling growth of *T. aestivum*. The phytotoxic effect can be mainly due to toxic compounds present in the plant extracts. Secondary plant metabolites such as terpenoids, steroids, phenols, coumarins, flavonoids, tannins, alkaloids, and cyanogenic glycosides, and their degradation products have been known to be involved in allelopathic phenomena [45].

Apiaceae plants are known to accumulate flavonoids mainly in the form of flavonols and flavones [5,46]. The allelochemicals present in *D. triradiata* extracts could, among others, refer to flavonoid metabolites. The action of allelochemicals can affect the respiration, photosynthesis, enzyme activity, water relations, stomatal opening, hormone levels, mineral availability, cell division and elongation, and structure and permeability of cell membranes and walls [46–48]. In addition, the allelopathic potential of *D. triradiata* species is not cited in the literature, however, our results corroborate those obtained by Guetat et al. [10] on *D. tortuosa* and by Znati et al. [35] on *Ferula lutea*. The authors [10] reported that the plant extracts of *D. tortuosa* from Saudi Arabia exhibited high phytotoxic effects against *T. aestivum* L. seeds and seedlings. Moreover, the highest germination inhibition (56%) was induced by petroleum ether extracts. The inhibition of the radicle growth was reported to reach 100% (petroleum ether extracts at 1 mg/mL). These findings about the allelopathic potentialities of *D. tortuosa* were supported by Fayed et al. [49]. The authors reported that the EOs showed a substantial allelopathic activity against the weed *Chenopodium murale*. Furthermore, the shoot growth, root growth, and germination was reduced (at higher dose: 100 µL/L) by 84.19, 74.45, and 53.57%, respectively. In addition, the authors ascribed that the EOs of *D. tortuosa* attained IC_{50} values of 94.96, 46.52, and 46.52 µL/L against germination, shoot, and root growth of *C. murale* [49].

4. Conclusions

In summary, this study on *D. triradiata* from Saudi Arabia showed that EOs and plant extracts have a variation in their chemical composition. Despite the low chemical diversity of extracted EOs from seeds, roots, stems, and flowers (only nine compounds), the chemicals in the EOs exhibited high biological activity potentials. The highest yield in the essential oils was recorded in stems. Among the tested plant extract samples, the highest antioxidant activity was observed in methanol extracts. Furthermore, the plant extracts also inhibited the shoot and root growth of *Triticum aestivum* seedlings. Moreover, EOs displayed a high inhibitory activity against selected fungal strains (*Malassezia* spp. and *Candida* species). Hence, natural products extracted from *D. triradiata* could be used as a natural herbicide as well as a good source for preventing yeast growth. Our results highlight the value-adding of this Saudi Arabian plant species, which can be a safe and renewable source of biological active compounds for multiple utilizations.

Supplementary Materials: The following supporting information can be downloaded at: https://www.mdpi.com/article/10.3390/plants11121543/s1, Table S1: Allelopathic effects of the plant extracts of *D. triradiata* samples at different concentrations on *Triticum aestivum* L. seedlings.

Author Contributions: All authors conceived and designed the experiment and performed all the in vitro experiments. W.R. and C.C. performed the in vitro experiments of antifungal activities. A.B. and Y.Y. performed antibacterial activity. A.T.A. performed data management and statistics and critically revised the manuscript. A.G., Y.Y., A.T.A. and M.B. contributed in writing and reviewing the manuscript. A.K.A. contributed in investigation and methodology. All authors have read and agreed to the published version of the manuscript.

Funding: This research was funded by the Deanship of Scientific Research in the Northern Border University, Saudi Arabia, grant project number "7344-SCI-2017-1-8-F7".

Institutional Review Board Statement: Not applicable.

Informed Consent Statement: Not applicable.

Data Availability Statement: Not applicable.

Acknowledgments: The authors extend their appreciation to the Deanship of Scientific Research in the Northern Border University, Saudi Arabia, for financial support of this research, grant project number "7344-SCI-2017-1-8-F7". We are very grateful to Khaled Jebahi, a certified teacher of English at King Abdul Aziz University (Saudi Arabia) for his fruitful discussion and for his kind help with English editing on a late version of the manuscript.

Conflicts of Interest: The authors declare that they have no conflict of interest.

References

1. Ngahang, K.; Ranjbarian, F.; Cianfaglione, K.; Sut, S.; Dall'Acqua, S.; Bruno, M.; Heshmati, F.; Afshar, R.; Iannarelli, R.; Benelli, G.; et al. Identification of highly effective antitrypanosomal compounds in essential oils from the Apiaceae family. *Ecotoxicol. Environ. Saf.* **2018**, *156*, 154–165. [CrossRef] [PubMed]
2. Christensen, L.P. Bioactivity of polyacetylenes in food plants. In *Bioactive Foods in Promoting Health*; Academic Press: Cambridge, MA, USA, 2010; pp. 285–306.
3. Calvino, A.; Federico, E.T.; Downie, S.R. The role of the Southern Hemisphere in the evolutionary history of Apiaceae, a mostly north temperate plant family. *J. Biogeogr.* **2016**, *43*, 398–409. [CrossRef]
4. Zhou, J.; Gong, X.; Downie, S.R.; Peng, H. Towards a more robust molecular phylogeny of Chinese Apiaceae subfamily Apioideae: Additional evidence from nrDNA ITS and cpDNA intron (rpl16 and rps16) sequences. *Mol. Phylogenet. Evol.* **2009**, *53*, 56–68. [CrossRef]
5. Sayed-Ahmad, B.; Thierry, T.; Saad, Z.; Hijazi, A.; Merah, O. The Apiaceae: Ethnomedicinal family as source for industrial uses. *Ind. Crops Prod.* **2017**, *109*, 661–671. [CrossRef]
6. Smart, N.; Fang, Z.Y.; Marwick, T.H. A practical guide to exercise training for heart failure patients. *J. Card. Fail.* **2003**, *9*, 49–58. [CrossRef]
7. Pimenov, M.G.; Lenov, M.V. The Asian Umbelliferae Biodiversity Database (ASIUM) with Particular Reference to South-West Asian Taxa. *Turk. J. Bot.* **2004**, *28*, 139–145.
8. Collenette, I.S. *Wildflowers of Saudi Arabia*; National Commission for Wildlife Conservation: Riyadh, Saudi Arabia, 1999; p. 799.
9. Aati, H.; El-Gamal, A.; Shaheen, H. Traditional use of ethnomedicinal native plants in the Kingdom of Saudi Arabia. *J. Ethnobiol. Ethnomed.* **2019**, *15*, 2. [CrossRef] [PubMed]
10. Guetat, A.; Boulila, A.; Boussaid, M. Phytochemical profile and biological activities of *Deverra tortuosa* (Desf.) DC.: A desert aromatic shrub widespread in Northern Region of Saudi Arabia. *Nat. Prod. Res.* **2018**, *16*, 2708–2713. [CrossRef]
11. *Deverra triradiate* . Available online: https://powo.science.kew.org/taxon/urn:lsid:ipni.org:names:841277-1 (accessed on 25 May 2022).
12. Chrtek, J.; Osbornova, J.; Sourkova, M. Notes on the genus *Deverra* (Umbelliferae). Poznamky k rodu *Deverra* (Umbelliferae). *Preslia* **1984**, *56*, 97–105.
13. Boulos, L. *Flora of Egypt*; AL-Hadara Publishing: Cairo, Egypt, 2000; Volume 2.
14. Halim, A.F.; Lahloub, M.F.; Saad, H.E.A.; Ahmed, A.F. Coumarins of roots of *Pituranthos triradiatus* growing in Egypt. *Mansoura J. Pharm. Sci.* **1991**, *7*, 402–413.
15. Guetat, A. The Genus *Deverra* DC. (Syn. *Pituranthos* Viv.): A natural valuable source of bioactive phytochemicals: A review of traditional uses, phytochemistry and pharmacological properties. *J. Ethnopharmacol.* **2002**, *284*, 114447. [CrossRef] [PubMed]
16. Bhatt, A.; Bhat, N.R.; Gallacher, D. Intra-plant Inflorescence and Seed Heterogeneity of *Deverra triradiata* (Apiaceae). *Natl. Acad. Sci. Lett.* **2020**, *43*, 463–465. [CrossRef]
17. Ashkenazy, D.; Friedman, J.; Kashman, Y. The furocoumarin composition of *Pituranthos triradiatus*. *Planta Med.* **1983**, *47*, 218–220. [CrossRef] [PubMed]
18. Ashkenazy, D.; Friedman, J.; Kashman, Y.; Egyed, M.N.; Shlosberg, A. Photosensitization in ducklings induced by *Pituranthus triradiatus*. *Vet. Hum. Toxicol.* **1984**, *26*, 118–120.
19. Ashkenazy, D.; Kashman, Y.; Nyska, A.; Friedman, J. Furocoumarins in shoots of *Pituranthos triradiatus* (Umbelliferae) as protectants against grazing by hyrax (Procaviidea: *Procavia capensis syriaca*). *J. Chem. Ecol.* **1985**, *11*, 231–239. [CrossRef] [PubMed]
20. Ashkenazy, D.; Eshel, A.; Kashman, Y.; Friedman, J. Morphological and chemical variations in natural populations of *Pituranthos triradiatus* in the Negev desert. *Biochem. Syst. Ecol.* **1987**, *15*, 453–458. [CrossRef]

21. Liang, Y.; Xie, L.; Liu, K.; Cao, Y.; Dai, X.; Wang, X.; Lu, J.; Yhang, X.; Li, X. Bergapten: A review of its pharmacology, pharmacokinetics, and toxicity. *Phytother. Res.* **2021**, *35*, 6131–6147. [CrossRef]
22. Spengler, G.; Gajdács, M.; Donadu, M.G.; Usai, M.; Marchetti, M.; Ferrari, M.; Mazzarello, V.; Zanetti, S.; Nagy, F.; Kovács, R. Evaluation of the antimicrobial and antivirulent potential of essential oils isolated from *Juniperus oxycedrus* L. ssp. *macrocarpa* aerial parts. *Microorganisms* **2022**, *10*, 758. [CrossRef] [PubMed]
23. Vandeputte, P.; Ferrari, S.; Coste, A.T. Antifungal resistance and new strategies to control fungal infections. *Int. J. Microbiol.* **2012**, *2012*, 713687. [CrossRef]
24. Adams, R.P. *Identification of Essential Oil Components by Gas Chromatography/Mass Spectrometry*; Allured Publishing Co.: Carol Stream, IL, USA, 2007.
25. Brand, W.; Culivier, M.E.; Berset, C. Use of a free radical method to evaluate antioxidant activity. *Lebensm. Wiss. Technol.* **1995**, *28*, 25–30. [CrossRef]
26. Dorman, H.J.D.; Hiltunen, R. Fe(III) reductive and free radical-scavenging properties of summer savory (*Satureja hortensis* L.) extract and subfractions. *Food Chem.* **2004**, *88*, 193–199. [CrossRef]
27. Re, R.; Pellegrini, N.; Proteggente, A.; Pannala, A.; Yang, M.; Rice-Evans, C. Antioxidant activity applying an improved ABTS radical cation decolorization assay. *Free Radic. Biol. Med.* **1999**, *26*, 1231–1237. [CrossRef]
28. El Hadj Ali, I.B.; Chaouachi, M.; Bahri, R.; Chaieb, I.; Boussaïd, M.; Harzallah-Skhiri, F. Chemical composition and antioxidant, antibacterial, allelopathic and insecticidal activities of essential oil of *Thymus algeriensis* Boiss. et Reut. *Ind. Crops Prod.* **2015**, *77*, 631–639. [CrossRef]
29. Ekundayo, O.; Laakso, I.; Adegbola, R.M.; Oguntimein, B.; Sofowora, A.; Hiltunen, R. Essential Oil Constituents of Ashanti Pepper (*Piper guineense*) Fruits (Berries). *J. Agric. Food Chem.* **1988**, *36*, 880–882. [CrossRef]
30. Wang, Y.K.; Yang, X.N.; Zhu, X.; Xiao, X.R.; Yang, X.W.; Qin, H.B.; Gonzales, F.J.; Li, F. Role of metabolic activation in elemicin-induced cellular toxicity. *J. Agric. Food Chem.* **2019**, *67*, 8243–8252. [CrossRef]
31. Abdelwahed, A.; Hayder, N.; Kilani, S.; Mahmoud, A.; Chibani, J.; Hammami, L.; Chekir-Ghedira, L.; Ghedira, K. Chemical composition and antimicrobial activity of essential oils from Tunisian *Pituranthos tortuosus* (Coss.) Maire. *Flavour Fragr. J.* **2006**, *21*, 129–133. [CrossRef]
32. Krifa, M.; El Mekdad, H.; Bentouati, N.; Pizzi, A.; Ghedira, K.; Hammami, M.; El Meshri, S.E.; Ghedira, C. Immunomodulatory and anticancer effects of *Pituranthos tortuosus* essential oil. *Tumor Biol.* **2015**, *36*, 5165–5170. [CrossRef]
33. Almadiy, A.A.; Nenaah, G.E.; Albogami, B.Z. Bioactivity of *Deverra tortuosa* essential oil, its nanoemulsion, and phenylpropanoids against the cowpea weevil, a stored grain pest with eco-toxicological evaluations. *Environ. Sci. Pollut. Res.* **2022**, 1–16. [CrossRef]
34. Azzazi, M.F.A.; Afifi, M.; Tammam, O.; Sheikh Alsouk, A.M. Chemical Composition and Antifungal Activity of the Essential Oil from *Deverra tortuosa* against Phytopathogenic Fungi. *Swift J. Agric. Res.* **2015**, *1*, 28–32.
35. Znati, M.; Ben Jannet, H.; Cazaux, S.; Bouajila, J. Chemical Composition, Biological and Cytotoxic Activities of Plant Extracts and Compounds Isolated from *Ferula lutea*. *Molecules* **2014**, *19*, 2733–2747. [CrossRef] [PubMed]
36. Ebrahimzadeh, M.A.; Nabavi, S.M.; Nabavi, S.F.; Dehpour, A.A. Antioxidant activity of hydroalcholic extract of *Ferula gummosa* Boiss roots. *Eur. Rev. Med. Pharmacol. Sci.* **2011**, *15*, 658–664.
37. Raharjo, S.J.; Mahdi, C.; Nurdiana, N.; Kikuchi, T.; Fatchiyah, F. In vitro and in silico: Selectivities of Seychellene compound as candidate cyclooxygenase isoenzyme inhibitor on pre-osteoblast cells. *Curr. Enzym. Inhib.* **2017**, *13*, 2–10. [CrossRef]
38. Siva Sai, C.; Mathur, N. Inhibitory Potential of Essential Oils on *Malassezia* strains by Various Plants. *Biol. Life Sci. Forum* **2020**, *4*, 46.
39. Potente, G.; Bonvicini, F.; Gentilomi, G.A.; Antognoni, F. Anti-*Candida* activity of essential oils from Lamiaceae plants from the Mediterranean area and the Middle East. *Antibiotics* **2020**, *9*, 395. [CrossRef] [PubMed]
40. Dias de Castro, R.; Pereira Andrade de Souza, T.M.; Dornelas Bezerra, L.M.; Ferreira, G.L.S.; Melo de Brito Costa, E.M.; Cavalcanti, A.L. Antifungal activity and mode of action of thymol and its synergism with nystatin against *Candida* species involved with infections in the oral cavity: An in vitro study. *BMC Complement. Altern. Med.* **2015**, *15*, 417–422. [CrossRef]
41. Siddiqui, Z.N.; Farooq, F.; Musthafa, T.M.; Ahmad, A.; Khan, A.U. Synthesis, characterization and antimicrobial evaluation of novel halopyrazole derivatives. *J. Saudi Chem. Soc.* **2013**, *17*, 237–243. [CrossRef]
42. Martins, M.R.; Tinoco, M.T.; Almeida, A.S.; Cruz Morais, J. Chemical composition, antioxidant and antimicrobial properties of three essential oils from Portuguese Flora. *J. Pharmac.* **2012**, *3*, 39–44.
43. Martins, N.; Ferreira, I.C.F.R.; Henriques, M.; Silva, S. In vitro anti-*Candida* activity of *Glycyrrhiza glabra* L. *Ind. Crops Prod.* **2016**, *83*, 81–85. [CrossRef]
44. Dongmo, S.C.M.; Njateng, G.S.S.; Tane, P.; Kuiate, J.R. Chemical composition and antimicrobial activity of essential oils from *Aframomum citratum*, *Aframomum daniellii*, *Piper capense* and *Monodora myristica*. *J. Med. Plants Res.* **2019**, *13*, 173–187.
45. Chung, I.M.; Kim, K.H.; Ahn, J.K.; Chun, S.C.; Kim, C.S.; Kim, J.T.; Kim, S.H. Screening of allelochemicals on barnyardgrass (*Echinochloa crus-galli*) and identification of potentially allelopathic compounds from rice (*Oryza sativa*) variety hull extracts. *Crop Prot.* **2002**, *21*, 913–920. [CrossRef]
46. Abdulmanea, K.; Elena, A.; Prokudina, P.L.; Vaníčková, L.; Koblovská, R.; Zelený, V.; Lapčík, O. Immunochemical and HPLC identification of isoflavonoids in the Apiaceae family. *Biochem. Syst. Ecol.* **2012**, *45*, 237–243. [CrossRef]

47. Reigosa, M.J.; Sánchez-Moreiras, A.; Gonzáles, L. Ecophysiological approach in allelopathy. *CRC Crit. Rev. Plant Sci.* **1999**, *18*, 577–608. [CrossRef]
48. Gatti, A.B.; Gui Ferreira, A.; Arduin, M.; Gualtieri de Andrade Perez, S.C. Allelopathic effects of aqueous extracts of *Artistolochia esperanzae* O. Kuntze on development of *Sesamum indicum* L. seedlings. *Acta Bot. Bras.* **2010**, *24*, 454–461. [CrossRef]
49. Fayed, E.M.; Abd-ElGawad, A.M.; Elshamy, A.I.; El-Halawany, E.S.F.; El-Amier, Y.A. Essential oil of *Deverra tortuosa* aerial parts: Detailed chemical profile, allelopathic, antimicrobial, and antioxidant activities. *Chem. Biodivers.* **2021**, *18*, e2000914. [CrossRef] [PubMed]

plants

MDPI

Article

Chemical and Biological Activity Profiling of *Hedyosmum strigosum* Todzia Essential Oil, an Aromatic Native Shrub from Southern Ecuador

Luis Cartuche *, James Calva, Eduardo Valarezo, Nayeli Chuchuca and Vladimir Morocho

Departamento de Química y Ciencias Exactas, Universidad Técnica Particular de Loja (UTPL), Calle M.Champagnat s/n, Loja 1101608, Ecuador
* Correspondence: lecartuche@utpl.edu.ec

Abstract: The present study aimed to determine the chemical composition, enantiomeric distribution and the biological profile of *Hedyosmum strigosum* essential oil (EO). The antioxidant properties and anticholinesterase effect were measured by spectroscopic methods and antimicrobial potency assessed against 8 bacteria and two fungi. *H. strigosum* is a native shrub, particularly found in Ecuador and Colombia at 2000 to 3500 m a.s.l. Chemical composition was determined by GC-MS and GC-FID. A total of 44 compounds were detected, representing more than 92% of the EO composition. The main compounds were thymol (24.35, 22.48%), α-phellandrene (12.15, 13.93%), thymol acetate (6.59, 9.39%) and linalool (8.73, 5.82%), accounting for more than 51% of the EO. The enantioselective analysis revealed the presence of 5 pure enantiomers and 3 more as a racemic mixture. The EO exerted a strong antioxidant capacity, determined by ABTS assay, with a SC_{50} of 25.53 μg/mL and a strong and specific antimicrobial effect against *Campylobacter jejuni* with a MIC value of 125 μg/mL. A moderate acetylcholinesterase inhibitory effect was also observed with an IC_{50} value of 137.6 μg/mL. To the best of our knowledge this is the first report of the chemical composition and biological profile of *H. strigosum* EO.

Keywords: enantiomeric distribution; chemical profiling; acetylcholinesterase; essential oil; thymol; α-thujene; α-pinene

Citation: Cartuche, L.; Calva, J.; Valarezo, E.; Chuchuca, N.; Morocho, V. Chemical and Biological Activity Profiling of *Hedyosmum strigosum* Todzia Essential Oil, an Aromatic Native Shrub from Southern Ecuador. *Plants* **2022**, *11*, 2832. https://doi.org/10.3390/plants11212832

Academic Editors: Hazem Salaheldin Elshafie, Ippolito Camele and Adriano Sofo

Received: 20 September 2022
Accepted: 17 October 2022
Published: 25 October 2022

1. Introduction

Chloranthaceae is the smallest family of vascular plants and it is the only family in the Chloranthales order. This family is composed mainly of lowering plants (angiosperms) and is considered one of the most primitive of its type [1]. The Chloranthaceae family consists of herbs, shrubs and trees, the species are characterized by the presence of secretory cells in the stems and leaves [2]. The family comprises 80 species belonging to the *Ascarina* (12 accepted species), *Chloranthus* (20 accepted species), *Hedyosmum* (44 accepted species) and *Sarcandra* genera (4 accepted species) [3]. Species of Chloranthaceae are naturally occurring in tropical and subtropical areas of Southeast Asia, Pacific Islands, New Zealand, Madagascar, Central and South America, and the West Indies [4]. In Ecuador, the Chloranthaceae family is represented only by the genus *Hedyosmum* [5].

Pharmacological investigations have revealed that the extracts of Chloranthaceae species are highly bioactive. Phytochemical studies showed the presence of terpenoids, flavonoids, coumarins, organic acids, amides, and sterols can be obtained from plants of this family [4]. Some species of the family Chloranthaceae have been used as folk medicine, especially in eastern Asia. A common use of these species includes: invigorating blood circulation and eliminating blood-stasis. The sarcandrae herb (*Sarcandra glabra*) is a folk medicine with a long history in China, widely used for its medicinal properties and it has been has been recorded in the Pharmacopoeia of the People's Republic of China [6].

The genus *Hedyosmum* consists of 44 species of shrubs or trees, with branches hinged at the height of the nodes or bulging. Its leaves are opposite, petiolate, simple, toothed and with pinnate veins [7]. Most of the species of this genus are aromatic, for this reason, the genus name finds its origin in the Greek words, hedy (pleasant) and osmum (smell) [8]. This genus is represented mainly in America, and only one species is registered in Southeast Asia [9]. *Hedyosmum* species are widespread in low and high mountain rain forests at altitudes of 500 to 3000 m a.s.l. [10]. The species of this genus are distributed mainly in the mountains from the state of Veracruz (Mexico) to Brazil (through southern Mexico, Central America and the Andes of South America) [11–13].

In Ecuador, this genus is represented by at least 15 species and H. anisodorum, H. cuatrecazanum, H. cumbalense, H. goudotianum, H. luteynii, H. purpurascens, H. racemosum, H. scabrum, H. spectabile, H. sprucei, H. strigosum, and H. translucidum are mainly found in Andean forests and subparamos areas. H. purpurascens is the only endemic species occurring in the high Andean forest in the south of the country. All the reported species for Ecuador grown mainly in cloud forests between 600 and 3000 m a.s.l. [14,15].

Previous studies have reported non-volatile and volatile secondary metabolites (essential oil) in species of the genus *Hedyosmum*. Among the identified non-volatile compounds are flavonoids, hydroxycinnamic acid derivatives, neolignans, sesquiterpenes and sesterterpenes. The essential oils (EO) have been extracted from leaves, flowers, and fruits from plants of this genus [12,13,16]. Traditional uses assigned to *Hedyosmum* genus include the use as sources of firewood, construction materials and food (fruits). The leaves of some species are used in infusion as a medicinal preparations or aromatic beverages [7]. The pharmacological effects associated with species of this genus include analgesic, anxiolytic, antibacterial, anticancer, antidepressant, antinociceptive, antiplasmodial neuroprotective and sedative properties [17]. Some members of this genus have been reported to possess antispasmodic and digestive effects, and also, they were reported for the treatment of kidney problems and anxious stomach. In addition, folk medicine report traditional uses such as, headaches, snake bites, rheumatic joint pain, fever, and cold symptoms [2,18].

Hedyosmum strigosum Todzia (class: Equisetopsida C. Agardh; subclass: Magnoliidae Novák ex Takht.; superorder: Austrobaileyanae Doweld ex M.W. Chase & Reveal; order: Chloranthales R. Br.; family: Chloranthaceae R. Br. ex Sims; genus: *Hedyosmum* Sw.) is a native shrub or tree of Ecuador, also found in Colombia. In Ecuador, this species is distributed in the Andean and Amazon region between 2000 to 3500 m a.s.l., especially in the northern andean provinces of Cotopaxi, Imbabura, Napo, Pichincha and Tungurahua, and the eastern province of Sucumbíos [5]. The plant is popularly known as "Quinillo" or "Olloco," and traditionally, the infusion of the leaves is drunk as an aromatic beverage, its fruits are food for birds, its branches and stems are used as fuel, and the stem is timber and is used for the construction of fences [18].

In south America, Ecuador is considered a megadiverse country because has many species per unit surface area. Currently, this country occupies the sixth position worldwide as a biodiversity hotspot [19]. However, there are few studies of its plant species, especially of the aromatic species of the *Hedyosmum* genus. The importance of the family and the genus, and the fact that only four species of this genus have been studied in Ecuador (*H. luteynii* [14], *H. racemosum* [2], *H. scabrum* [17] and *H. sprucei* [13]) encouraged us to determine a complete chemical profile of *H. strigosum* and validate its biological properties. To the best of our knowledge, this is the first report made for this specie, for that reason, the aim of this research was to determine the chemical composition, enantiomeric distribution, antimicrobial, antioxidant and anticholinesterase effect of the *H. strigosum* essential oil.

2. Results

The EO from *Hedyosmum strigosum* obtained by hydrodistillation exhibited a pale-yellow color and a yield of 0.66% (4.62 g from 700 g of plant material). The relative density was 0.9823 ± 0.004 g/mL, refractive index of 1.4893 ± 0.001 and a specific rotation of α_D^{20} of $+21.21 \pm 0.035$ (c 10.25, CH_2Cl_2).

2.1. Chemical Composition

According to the GC-MS analysis, a total of 44 compounds were identified, representing 97.74% for DB5-MS and 92.27% for HP-Innowax, of the total oil composition. The main chemical identified compounds were thymol (24.35, 22.48%), α-phellandrene (12.15, 13.93%), thymol acetate (6.59, 9.39%), linalool (8.73, 5.82%) accounting for more than 51% of the chemical composition. Oxygenated monoterpenes were the main chemical groups with 46.96% and 43.33%, followed by hidrocarbonated monterpenes (31.13 and 28.52%) respectively (Table 1).

Table 1. Chemical compounds present in the essential oil of the leaves of H. strigosum.

N°	Compounds	DB-5 ms			HP-INNOWax			CF
		LRI [a]	LRI [b]	%	LRI [a]	LRI [b]	%	
1	α-Thujene	923	924	0.17 ± 0.02	-	-	-	$C_{10}H_{16}$
2	α-Pinene	930	932	4.02 ± 0.02	1025	1028	3.92 ± 0.37	$C_{10}H_{16}$
3	Camphene	945	946	0.61 ± 0.01	1070	1075	0.51 ± 0.05	$C_{10}H_{16}$
4	Sabinene	968	969	0.95 ± 0.01	1113	1119	1.23 ± 0.14	$C_{10}H_{16}$
5	β-Pinene	973	974	1.43 ± 0.02	1125	1122	0.81 ± 0.09	$C_{10}H_{16}$
6	Myrcene	986	988	0.95 ± 0.04	-	-	-	$C_{10}H_{16}$
7	α-Phellandrene	1003	1002	12.15 ± 0.16	1168	1160	13.93 ± 1.09	$C_{10}H_{16}$
8	α-Terpinene	1013	1014	0.13 ± 0.01	1184	1186	0.21 ± 0.10	$C_{10}H_{16}$
9	*p*-Cymene <ρ->	1020	1020	3.88 ± 0.10	1277	1260	2.31 ± 0.35	$C_{10}H_{16}$
10	Limonene	1025	1024	3.70 ± 0.09	1204	1204	3.06 ± 0.39	$C_{10}H_{16}$
11	β-Phellandrene	1026	1025	2.03 ± 0.06	1212	1218	1.15 ± 0.32	$C_{10}H_{16}$
12	1,8-Cineole	1028	1026	4.7 ± 0.12	1215	1220	4.15 ± 0.45	$C_{10}H_{18}O$
13	(*Z*)-β-Ocimene	1032	1032	tr	1241	1233	0.58 ± 0.08	$C_{10}H_{16}$
14	(*E*)-β-Ocimene	1043	1044	0.32 ± 0.01	1258	1260	0.29 ± 0.04	$C_{10}H_{16}$
15	Terpinolene	1080	1086	0.11 ± 0.02	1286	1288	0.19 ± 0.02	$C_{10}H_{16}$
16	*p*-Cymenene	1086	1089	0.67 ± 0.02	1445	1438	0.33 ± 0.27	$C_{10}H_{14}$
17	Filifolone	-	-	-	1448	1445	0.55 ± 0.20	$C_{10}H_{14}O$
18	Linalool	1098	1095	8.73 ± 0.12	1571	1556	5.82 ± 0.19	$C_{10}H_{18}O$
19	Chrysanthenone	1117	1124	1.29 ± 0.13	1515	1489	5.02 ± 0.77	$C_{10}H_{14}O$
20	(*Z*)-*p*-Menth-2-en-1-ol	1121	1118	1.28 ± 0.02	1618	1638	0.41 ± 0.04	$C_{10}H_{18}O$
21	(*E*)-*p*-Menth-2-en-1-ol	1139	1136	0.47 ± 0.02	1587	1571	0.27 ± 0.05	$C_{10}H_{18}O$
22	Camphor	1142	1141	1.67 ± 0.04	1521	1522	1.85 ± 0.10	$C_{10}H_{16}O$
23	Pinocarvone	1157	1160	2.13 ± 0.05	1575	1580	1.19 ± 0.16	$C_{10}H_{14}O$
24	(*Z*)-Pinocamphone	1171	1172	0.40 ± 0.03	1550	1555	0.26 ± 0.09	$C_{10}H_{16}O$
25	(*Z*)-Piperitol	1191	1195	0.63 ± 0.03	1766	1758	0.34 ± 0.01	$C_{10}H_{18}O$
26	Thymol, methyl ether	1226	1232	1.50 ± 0.01	1603	1586	0.19 ± 0.03	$C_{11}H_{16}O$
27	Neral	1235	1235	0.69 ± 0.01	-	-	-	$C_{10}H_{16}O$
28	Geranial	1265	1264	0.62 ± 0.01	1749	1733	1.06 ± 0.08	$C_{10}H_{16}O$
29	Thymol	1289	1289	24.35 ± 0.27	2221	2189	22.48 ± 2.84	$C_{10}H_{14}O$
30	Thymol acetate	1341	1349	6.59 ± 0.23	1870	1840	9.39 ± 1.12	$C_{12}H_{16}O_2$
31	Methyl eugenol	1396	1403	4.15 ± 0.04	2039	2023	0.81 ± 0.14	$C_{11}H_{14}O_2$
32	α-Cedrene	1407	1410	0.23 ± 0.02	-	-	-	$C_{15}H_{24}$
33	β-Cedren	1410	1419	0.39 ± 0.01	-	-	-	$C_{15}H_{24}$
34	Germacrene D	1472	1480	3.17 ± 0.11	1700	1710	5.70 ± 0.44	$C_{15}H_{24}$
35	Bicyclogermacrene	1487	1500	0.84 ± 0.20	1725	1706	1.04 ± 0.11	$C_{15}H_{24}$
36	(*E*)-Methyl isoeugenol	1490	1491	0.04 ± 0.01	2196	2185	0.20 ± 0.02	$C_{11}H_{14}O_2$
37	δ-Amorphene	1510	1511	0.32 ± 0.02	-	-	-	$C_{15}H_{24}$
38	Elemol	1542	1548	0.36 ± 0.12	2099	2090	0.67 ± 0.07	$C_{15}H_{26}O$
39	(*E*)-Nerolidol	1557	1561	0.88 ± 0.02	2063	2050	1.04 ± 0.11	$C_{15}H_{26}O$
40	Spathulenol	1568	1577	0.79 ± 0.04	2139	2140	0.32 ± 0.04	$C_{15}H_{24}O$
41	(*E*)-Isoeugenol acetate	1601	1614	0.39 ± 0.15	-	-	-	$C_{12}H_{14}O_3$
42	α-Eudesmol	-	-	-	2235	2233	0.25 ± 0.08	$C_{15}H_{26}O$
43	β-Eudesmol	-	-		2243	2240	0.52 ± 0.15	$C_{15}H_{26}O$
44	α-Cadinol	-	-	-	2249	2225	0.22 ± 0.02	$C_{15}H_{26}O$

Table 1. *Cont.*

N°	Compounds	DB-5 ms			HP-INNOWax			
		LRI [a]	LRI [b]	%	LRI [a]	LRI [b]	%	CF
	ALM			31.13			28.52	
	OXM			46.96			43.40	
	HS			4.95			6.74	
	OS			2.03			3.02	
	Others			12.67			10.59	
	Total			97.74			92.27	

Percentage (%) is expressed as mean ± SD (standard deviation); LRI[a]: Calculated Linear Retention Index; LRI[b]: Linear Retention Index read in bibliography; tr: trace (<0.03). CF: Condensed formula. ALM. Aliphatic monoterpene hydrocarbons, OXM. Oxygenated monoterpenes, sesquiterpene hydrocarbons, Oxigenated sesquiterpenes.

2.2. Enantiomeric Analysis

The results obtained for the enantiomeric analysis from *H. strigosum* essential oil are depicted in Table 2.

Table 2. Enantiomeric analysis from *H. strigosum* essential oil.

LRI [a]	Component	Enantiomeric Distribution (%)	e.e. (%)
934	(1*R*,5*S*)-(−)-α-thujene	100	100
941	(1*S*,5*S*)-(−)-α-pinene	100	100
955	(1*R*,4*S*)-(−)-camphene	100	100
983	(1*R*,5*R*)-(+)-β-pinene	33.9	12.3
985	(1*S*,5*S*)-(−)-β-pinene	56.1	
989	(1*S*,5*S*)-(+)-sabinene	100	100
1006	(*R*)-(−)-α-phellandrene	33.7	32.7
1011	(*S*)-(+)-α-phellandrene	66.3	
1050	(*S*)-(−)-limonene	100	100
1194	(*R*)-(−)-linalool	44.1	11.9
1198	(*S*)-(+)-linalool	55.9	

LRI [a]: Linear Calculated Retention Index; e.e: enantiomeric excess.

2.3. Antimicrobial Activity

According to the results displayed in Table 3, the most relevant outcome was obtained for *Campylobacter jejuni*, and the two dermatophytes with a MIC value of 125 μg/mL. The remaining microorganisms exhibited weak or null inhibitory effects at concentrations higher than 1000 μg/mL.

Table 3. Minimum inhibitory concentration calculated for *H. strigosum* essential oil against ten human pathogenic microorganisms.

Microorganism	*H. strigosum* EO	Antimicrobial Agent
	MIC μg/mL	
	Gram negative bacteria	
Escherichia coli	2000.00	1.56
Pseudomonas aeruginosa	NA	<0.39
Salmonella enterica	4000.00	<0.39
Campylobacter jejuni	125.00	
	Gram positive bacteria	

Table 3. *Cont.*

Microorganism	*H. strigosum* EO	Antimicrobial Agent
	MIC µg/mL	
	Gram negative bacteria	
Listeria monocytogenes	NA	1.56
Enterococcus faecalis	4000.00	0.78
Enterococcus faecium	2000.00	<0.39
Staphylococcus aureus	1000.00	<0.39
	Fungi	
Trychophyton rubrum	125.00	<0.12
Trichophyton interdigitale	125.00	<0.12

NA. non active at the maximum dose tested of 4000 µg/mL

2.4. Antioxidant Capacity

The results obtained for the antioxidant capacity of the EO from H. strigosum are depicted in Table 4. Trolox was used as reference control with its corresponding IC_{50} value.

Table 4. Half scavenging capacity of *H. strigosum*, EO.

EO	ABTS	DPPH
	SC_{50} (µg/mL—µM *) ± SD	
Hedyosmum strigosum	25.53 ± 0.48	1313.73 ± 18.13
Trolox *	24.72 ±1.03	28.97 ± 1.24

* Half scavenging capacity of Trolox is expressed in micromolar units.

2.5. Anticholinesterase Activity

This is the first study of the inhibitory potency of H. strigosum, EO, against acetylcholinesterase, measured by the modified Ellman's method. Results showed a moderate inhibitory effect with an IC_{50} value of 137.6 ± 1.02 µg/mL (Figure 1). Donepezil hydrochloride was used as positive control with an IC_{50} of 13.6 ± 1.02 µM.

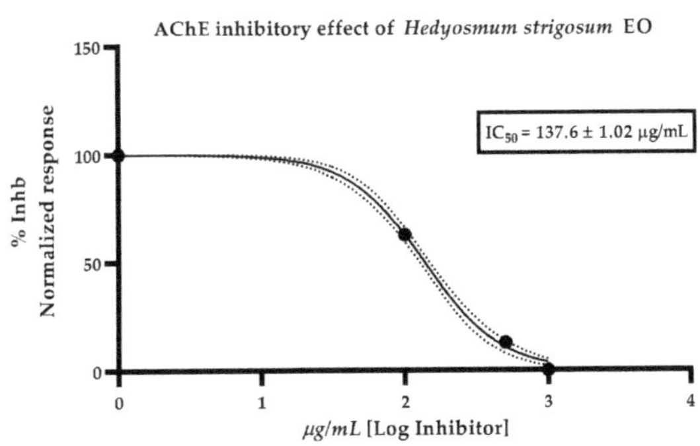

Figure 1. Inhibitory effect plot of *H. strigosum*, EO, against Acetylcholinesterase. Data represent the media of three replicas and three different experiments analyzed by non-linear regression method.

3. Discussion

There are no previous reports about the extraction of EO from *H. strigosum*, however, a study from a related species, *H. luteynii*, from Chimborazo-Ecuador, reported a very low yield of the EO of ca. 0.07% by using the same extraction method [14]. In contrast, by means of microwave radiation-assisted hydrodistillation from *H. translucidum*, the EO yield obtained was 1.2% [20], approximately twice the yield obtained in our study.

According to Radice et al. [8] the main chemical compounds occurring in essential oils from representative species of the genus *Hedyosmum* (≥10%) corresponds to estragole (55.8%, leaves, *H.* scabrum), (*E,E*)-α-farnesene (32%, leaves, *H. costaricense*), germacrene D (32%, aerial parts, *H.* costaricense), α-pinene (24%, aerial parts, *H. angustifolium*), β-pinene ((23.5%, aerial parts, *H. angustifolium*), 1,8-cineole (20.5%, leaves, male plant, *H.* scabrum), linalool (16.5%, leaves, female plant, *H.* scabrum), sabinene (15.8%, leaves, male plant, *H, bomplandianum*), β-caryophyllene (15.5%, fresh aerial part, *H. sprucei*), pinocarvone (14.2%, leaves, male plant, *H. scabrum*), D-germacrene-4-ol (12.6%, leaves, male plant, *H. scabrum*), δ-3-Carene (12.1%, aerial part, *H.scabrum*), α-phellandrene (11.4%, leaves, *H. arborscens*), α-eudesmol (11.4%, leaves, *H, translucidum*), (*E*)-β-ocimene (10.8%, leaves, *H, bomplandianum*), bicyclogermacrene (10.6%, leaves, *H. arborescens*), α-bisabolene (10.3%, leaves, *H. bomplandianum*) and α-terpineol (10.2%, leaves, *H. brasiliense*).

There are no reports about the antimicrobial activity of *H. strigosum*, however, *H. brasiliense,* a related species, showed a good profile of inhibition against several pathogenic bacteria such as *Bacillus subtillis*, *Staphylococcus aureus*, *Staphylococcus saprofiticus* and dermatophytic fungi such as *Microsporus canis*, *Trichophyton rubrum* and *Trichophyton mentagrophytes* with MIC values ranging from 0.31 to 0.12% *v/v* (3.12 to 1.25 mg/mL ca.) [9]. There are no extended criteria to value the potency of MIC values, however, Van Vuuren and Holl [21] suggested a detailed classification for extracts and essential oils and found that a MIC value between 101 to 500 μg/mL can be considered as a strong activity and according to our report, the activity for *C. jejuni* and the two dermatophytes lies between this range.

The *H. strigosum* EO as shown in Table 4, reveals a strong scavenging effect for ABTS radical, meanwhile a weak or null scavenging capacity for DPPH radical. This can be due to the mechanistic differences of both assays. ABTS cation radical is stabilized by electron transfer mechanism from antioxidants present in samples and for DPPH assays, this radical can be mainly stabilized by hydrogen atom transfer mechanism [22]. The occurrence of the aliphatic monoterpene α-phellandrene in concentrations higher to 28% in the species *H. racemosum*, another Ecuadorian related species, and its null antioxidant activity [2] could suggest at least in our research that, the antioxidant capacity could be related to the main compound, thymol, which demonstrated a strong antioxidant effect as described by Torres-Martínez et al. [23] against ABTS assay with a scavenging effect higher than 96% at a 0.1 mg/mL dose. However, according to Beena et al., thymol exerted a moderate antioxidant effect in the DPPH assay with a SC_{50} value of 167.57 μg/mL [24] which can agree with our results.

Enantiomeric analysis showed the presence of 5 pure enantiomers and 3 pairs found as a racemic mixture, other equally interesting studies showed enantiomeric compositions of EO from the *Hedyosmun* genus, such as; *H. scabrum* (Ruiz & Pavon) leaves collected in Ecuador and Bolivia, and *H. angustifolium* collected in Bolivia the presence of similar enantiomers such as β-pinene, sabinene, limonene and linalool [17,25].

Regarding to the anticholinesterase effect, the *H. strigosum* EO exerted a moderate inhibitory potency with an IC_{50} value of 137.6 μg/mL, much higher than the reported effect of *H. barisiliense* EO which exhibited an inhibition percentage of 69.82% at a dose of 1 mg/mL. Many species of *Hedyosmum* genus have been explored for their antibacterial or antioxidant potential and their chemical profile but little or nothing is said about their inhibitory potential against acetylcholinesterase, however, information related to chemical entities isolated from the EO of this genus and their inhibitory potential can be found in literature, such as thymol which is the characteristic occurring compound in *Thymus vulgaris* but, it was found as the majority compound in our EO. As reported by Jukic et al. [26], thymol presented an IC_{50} of 0.74 mg/mL and linalool was inactive at a dose higher than 1 mg/mL, surprisingly, carvacrol, the structural isomer of thymol, exerted a ten times higher potency with an IC_{50} of 0.063 mg/mL against acetylcholinesterase, suggesting that the hydroxyl functionality in -orto position, like occurring in carvacrol, plays an important role for the AChE inhibitory effect.

Despite its low concentration in the EO of *H. strigosum*, terpenes like terpinolene, terpinen-4-ol and *E*-caryophyllene (< to 1%) can be collaborating for the moderate AChE inhibitory effect observed. As reported by Bonesi et al., terpinolene presented an IC_{50} of

156.4 μg/mL [27], meanwhile, terpinen-4-ol and *E-caryophyllene* exhibited a moderate effect with inhibition percentages of 21.4% at 1.2 mM and 32% at 0.06 mM doses, respectively. There is no report about the AchE inhibitory effect of α-phellandrene, the second major compound occurring for *H. strigosum* EO, however, in the same study of Bonesi, the AChE inhibitory effect of β-phelandrene is reported with an IC_{50} value of 120.2 μg/mL. highlighting the importance of terpenes containing EO and their AChE inhibitory capacity.

4. Materials and Methods

4.1. Plant Material

The aerial parts of *Hedyosmum strigosum* were collecteced in Villonaco hill, Loja, province at 2720 m a.s.l. at 3°59′42.62″ S and 79°16′02.60″ W coordinates. The identified specimen was located at the herbarium of the Universidad Técnica Particular de Loja with voucher code HUTPL14299. The aerial parts were cleaned and dried at 34 °C for 6 days until the plant material was used for distillation.

4.2. Essential Oil Distillation

The aerial plant material was submitted to hydrodistillation in a Clevenger-type apparatus for a period of 3 h. Anhydrous sodium sulfate was added to the flask containing the EO with moisture to remove it and then, the EO properly labeled was stored at −4 °C until was used in the biological and chemical assays.

4.3. Physical Properties Determination

Three physical properties were determined for the EO. Relative density (RD_{20}), according to the AFNOR NFT75-111 standard, with an 1 mL pycnometer and an analytical balance (Mettler AC100 model), refractive index (RD_{20}), according to the ANFOR NF 75-112 25 standard in an ABBE refractometer (Boeco, Germany) and specific rotation $[\alpha^{20}_{514}]$, according to the ISO 592-1998 standard in a Hanon P 810 automatic polarimeter, were carried out. All measurements were performed by triplicate.

4.4. Chemical Profiling

4.4.1. GC-MS

Chemical analysis of *H. strigosum* EO was carried out using a gas chromatography-mass spectrometry (GC-MS) model Trace1310 gas chromatograph, coupled to a simple quadrupole mass spectrometry detector model ISQ 7000 (Thermo Fisher Scientific, Walthan, MA, USA). The mass spectrometer was operated in SCAN mode (scan range 35–350 m/z), with the electron ionization (EI)source 70 eV.

A non-polar column, based on 5%-phenyl-methylpolysiloxane, DB-5 ms (30 m long, 0.25 mm internal diameter, and 0.25 μm film thickness), was used for the qualitative and quantitative analysis of EO. The thermal program was as follow: initial temperature 60 °C for 5 min, followed by a gradient of 2 °C/min until 100 °C, then at 3 °C/min until 150 °C, and at 5 °C/min until 200 °C. Finally, a new gradient of 15 °C/min until 250 °C was applied. The final temperature was maintained for 5 min. GC purity grade helium 5.0 grade ultra-pure, was used as carrier gas, set at the constant flow of 1 mL/min [28].

4.4.2. GC-FID

Quantitative analysis was performed on an Agilent Gas Chromatograph (6890 series) with a flame ionization detector (CG-FID) the chromatographic conditions were the same as in the GC-MS analysis. The average percentage of each component EO was calculated from the area of the corresponding CG-FID peak with respect to the total area of peaks without applying any correction factor. The values and standard deviations were calculated from the results of the three injections [29].

4.4.3. Enantioselective Analysis

Enantioselective analysis was carried out through an enantioselective from Mega, MI, Italy capillary column, based on 2,3-diethyl-6-tert-butyldimethylsilyl-β-cyclodextrin (25 m × 250 μm internal diameter × 0.25 μm phase thickness). GC method was as follows: initial temperature was 60 °C for 2 min, followed at 2 °C/min until 220 °C, that was maintained for 2 min. The homologous series of n-alkanes (C9–C25) was also injected, in order to calculate the linear retention indices. The enantiomers were identified determined by injection of enantiomerically pure standards [30].

4.5. Antimicrobial Activity

The antimicrobial activity and Minimal inhibitory concentration values were determined by means of the broth microdilution technique with four gram positive bacteria (*Enterococcus faecalis* ATCC® 19433, *Enterococcus faecium* ATCC® 27270, *Staphylococcus aureus* ATCC® 25923, *Listeria monocytogenes* ATCC® 19115), four gram negative bateria (*Escherichia coli* (O157:H7) ATCC® 43888, *Pseudomonas aeruginosa* ATCC® 10145, *Salmonella enterica* serovar *Thypimurium* ATCC® 14028) and two fungi (*Trichophyton rubrum* ATCC® 28188, *Trichophyton interdigitale* ATCC® 9533) as microbial assay models. Two-fold serial dilution method was used to achieve concentrations ranging from 4000 to 31,25 μg/mL and a final inoculum concentration of 5 × 10^5 cfu/mL for bacteria, 2.5 × 10^5 cfu/mL for yeast and 5 × 10^4 spores/mL for sporulated fungi were used. Meller Hinton II (MH II) for bacteria and Sabouraud broth for fungi as assay media were used. The method followed was previously described for our research group. The culture of *Campylobacter jejuni*, was activated by adding horse serum at 5% in Tioglycolate medium and incubated at anaerobic conditions provided by a Campygen sachet (Thermo, scientific) at 37 °C for 48 h. The broth microdilution test was performed in MH II supplemented with 5% horse serum (Thermo) at the same conditions previously described as oil sample dilution and bacterial inoculum, and anaerobic conditions as described fully by our research group [31].

Antimicrobial commercial agents were used as positive control as follows: Ampicilin 1 mg/mL solution for *S.aureus*, *E. faecalis* and *E. faecium*. Ciprofloxacin 1 mg/mL solution for *P.*aeruginosa, *S. enterica*, *E. coli.* and *L. monocytogenes*, finally, Itraconazol 1 mg/mL for the two fungi.

4.6. Antioxidant Capacity

4.6.1. The 2,2-Diphenyl-1-Picril Hydrazyl Radical Scavenging Assay

The procedure was adapted from the methodology suggested by Thaipong et al. [32] and fully developed for our research group in previous reported jobs [31]. Briefly, a stable solution of DPPH in MeOH was prepared at a concentration of 625 μM. Working solution was prepared by adjusting an aliquot of the stable solution in methanol until an absorbance of 1.1 ± 0.02 was reached. An EPOCH microplate reader was used for absorbance reading at 515 nm and 96 microplates were used for the test. 270 μL of the working solution were allowed to react with 30 μL of the sample solution for 1 h at room temperature in darkness. The same procedure was followed for the blanks and positive control (Trolox). A solution representing 100% of the DPPH radical was prepared with 270 μL of working solution and 30 μL of MeOH. Sample solutions of essential oil were prepared by dissolving 10 mg of EO in MeOH. Three more 10 × dilutions were included to calculate the SC50 (half scavenging capacity).

4.6.2. The 2,2-Azino-Bis (3-Ethylbenzothiazoline-6-Sulfonic acid) Radical Scavenging Assay

The procedure was adapted from the methodology described by Thaipong et al., [32] and Arnao et al. [33] and fully developed for our research group in previous reported jobs [32] Briefly, a stable solution of ABTS* radical was prepared by reacting in darkness for 24 h, equal volumes of a 7.4 μM aqueous ABTS solution and 2.6 μM potassium peroxydisulfate. Working solution was prepared by adjusting an aliquot of the stable solution in methanol until an absorbance of 1.1 ± 0.02 was reached. An EPOCH microplate reader was used for absorbance reading at 734 nm and 96 microplates were used for the test. The same procedure as described for DPPH was followed for ABTS assay.

Half Scavenging capacity defined as the concentration of the EO required to scavenge 50% of the radical was calculated from the corresponding curve fitting of data (Radical scavenging percentage vs. EO concentrations). Radical scavenging percentages were calculated according to Equation (1):

$$\%\,RS = \left(1 - \left(\frac{\text{As}}{\text{Ac}}\right)\right) \times 100 \tag{1}$$

where,

Ac. Absorbance of the 100% radical adjusted solution.

As. Absorbance of the radical adjusted solution with the sample at different concentrations.

4.7. Anticholinesterase Assay

Acetylcholinesterase inhibitory effect of the essential oil was measured with a variation of the method described originally by Ellman et al. and reported in our previous jobs [28,33] and more recently in Andrade et al., [34]. Briefly, 20 μL of a 15 mM solution of acetylthiocoline in PBS pH 7.4 were placed along with 40 μL of Tris buffer pH 8.0 (containing 0.1 M NaCl and 0.02 M of $MgCl_2.6H_2O$), 100 μL of Ellman's reagent solution in Tris buffer (DNTB, 3 mM) and 20 μL of EO sample solution. The reaction was placed in an EPOCH 2 microplate reader and preincubated for 3 min at 25 °C. Finally, the addition of 20 μL of the enzyme solution (0.5 U/mL) started the reaction. The amount of product released was monitored at 405 nm for 60 min. EO sample solutions were made by dissolving 10 mg of EO in MeOH. Three more 10 × factor dilutions and two intermediate log based doses were included to obtain final concentrations of 1000, 500, 100, 50 and 10 μg/mL and the experiments were performed by triplicate.

IC_{50} value was calculated from the corresponding curve fitting of data obtained from the calculated rate of reactions with Graph pad Prism 8.0.1 (San Diego, CA. USA) Donepezil hydrochloride was used as positive control.

5. Conclusions

The chemical composition and enantiomeric distribution of the EO of *H. strigosum*, as well as, a biological activity profile, were determined for the first time. This research contributes to our knowledge about the native aromatic species from Ecuador, particularly of the *Hedyosum* genus which is the only genus of the Chlorantaceae family, represented in Ecuador. The importance of aromatic species is well supported with this study where, an important antibacterial and antifungal effect was observed. Also, the antioxidant and anticholinesterase effect observed for this species encourage us to continue exploring our native flora, searching for novel candidates for the pharmaceutical or cosmetic industry.

Author Contributions: Conceptualization, V.M., J.C., E.V. and L.C.; methodology, J.C., L.C. and N.C.; formal analysis, L.C., J.C. and N.C.; investigation, L.C., J.C., N.C.; data curation, J.C. V.M. and E.V.; writing—original draft preparation, L.C. and J.C.; writing—review and editing, V.M.; supervision, V.M. All authors have read and agreed to the published version of the manuscript.

Funding: This research received no external funding.

Acknowledgments: We are grateful to the Universidad Técnica Particular de Loja (UTPL) for supporting this investigation and open access publication.

Conflicts of Interest: The authors declare no conflict of interest.

References

1. Eklund, H.; Doyle, J.A.; Herendeen, P.S. Morphological Phylogenetic Analysis of Living and Fossil Chloranthaceae. *Int. J. Plant Sci.* **2004**, *165*, 107–151. [CrossRef]

2. Valarezo, E.; Morocho, V.; Cartuche, L.; Chamba-Granda, F.; Correa-Conza, M.; Jaramillo-Fierro, X.; Meneses, M.A. Variability of the chemical composition and bioactivity between the essential oils isolated from male and female specimens of *Hedyosmum racemosum* (Ruiz & Pav.) G. Don. *Molecules* **2021**, *26*, 4613. [CrossRef] [PubMed]
3. The Plant List. Chloranthaceae. Available online: http://www.theplantlist.org/ (accessed on 8 July 2022).
4. Cao, C.-M.; Peng, Y.; Shi, Q.-W.; Xiao, P.-G. Chemical Constituents and Bioactivities of Plants of Chloranthaceae. *Chem. Biodivers.* **2008**, *5*, 219–238. [CrossRef]
5. Jørgesen, P.M.; León-Yáñez, S. Catalogue of the Vascular Plants of Ecuador. Available online: http://legacy.tropicos.org/ProjectAdvSearch.aspx?projectid=2 (accessed on 11 July 2022).
6. State Pharmacopoeia Commission of the PRC. *Pharmacopoeia of The People's Republic of China: Volume I*; People's Medical Publishing House: Beijing, China, 2005; p. 946.
7. Todzia, C.A.; Keating, R.C. Leaf Architecture of the Chloranthaceae. In *Annals of the Missouri Botanical Garden*; Missouri Botanical Garden Press: St. Louis, MO, USA, 1991; Volume 78, pp. 476–496. [CrossRef]
8. Radice, M.; Tasambay, A.; Pérez, A.; Diéguez-Santana, K.; Sacchetti, G.; Buso, P.; Buzzi, R.; Vertuani, S.; Manfredini, S.; Baldisserotto, A. Ethnopharmacology, phytochemistry and pharmacology of the genus *Hedyosmum* (Chlorantaceae): A review. *J. Ethnopharmacol.* **2019**, *244*, 111932. [CrossRef]
9. Kirchner, K.; Wisniewski, A., Jr.; Cruz, A.B.; Biavatti, M.W.; Netz, D.J.A. Chemical composition and antimicrobial activity of *Hedyosmum brasiliense* Miq., Chloranthaceae, essential oil. *Rev. Bras. Farmacogn.* **2010**, *20*, 692–699. [CrossRef]
10. Bercion, S.; Couppe de K/Martin, M.-A.; Baltaze, J.-P.; Bourgeois, P. A new α-methylene γ-lactone sesquiterpene from *Hedyosmum arborescens*. *Fitoterapia* **2005**, *76*, 620–624. [CrossRef] [PubMed]
11. Sandoya, V.; Saura-Mas, S.; Granzow-de la Cerda, I.; Arellano, G.; Macía, M.J.; Tello, J.S.; Lloret, F. Contribution of species abundance and frequency to aboveground forest biomass along an Andean elevation gradient. *For. Ecol. Manag.* **2021**, *479*, 118549. [CrossRef]
12. Zhang, M.; Liu, D.; Fan, G.; Wang, R.; Lu, X.; Gu, Y.-C.; Shi, Q.-W. Constituents from Chloranthaceae plants and their biological activities. *Heterocycl. Commun.* **2016**, *22*, 175–220. [CrossRef]
13. Guerrini, A.; Sacchetti, G.; Grandini, A.; Spagnoletti, A.; Asanza, M.; Scalvenzi, L. Cytotoxic effect and TLC bioautography-guided approach to detect health properties of *amazonian Hedyosmum sprucei* essential oil. *Evid. Based Complement. Altern. Med.* **2016**, *2016*, 1638342. [CrossRef]
14. Torres-Rodríguez, S.H.; Tovar-Torres, M.C.; García, V.J.; Lucena, M.E.; Araujo Baptista, L. Composición química del aceite esencial de las hojas de *Hedyosmum luteynii* Todzia (Chloranthaceae). *Rev. Peru. De Biol.* **2018**, *25*, 173–178. [CrossRef]
15. León-Yánez, S.; Valencia, R.; Pitmam, N.; Endara, L.; Ulloa Ulloa, C.; Navarrete, H. Libro Rojo de Plantas Endémicas del Ecuador. Available online: https://bioweb.bio/floraweb/librorojo/home (accessed on 12 August 2022).
16. Murakami, C.; Cordeiro, I.; Scotti, M.T.; Moreno, P.R.H.; Young, M.C.M. Chemical Composition, Antifungal and Antioxidant Activities of *Hedyosmum brasiliense* Mart. ex Miq. (Chloranthaceae) Essential Oils. *Medicines* **2017**, *4*, 55. [CrossRef] [PubMed]
17. Herrera, C.; Morocho, V.; Vidari, G.; Bicchi, C.; Gilardoni, G. Phytochemical investigation of male and female *Hedyosmum scabrum* (Ruiz & Pav.) Solms leaves from Ecuador. *Chem. Biodivers.* **2018**, *15*, e1700423. [CrossRef]
18. De la Torre, L.; Navarrete, H.; Muriel, P.; Macía, M.J.; Balslev, H. Enciclopedia de las Plantas Útiles del Ecuador; Herbario QCA de la Escuela de Ciencias Biológicas de la Pontificia Universidad Católica del Ecuador/Herbario AAU del Departamento de Ciencias Biológicas de la Universidad de Aarhus. ISBN 978-9978-77-135-8: 2008.
19. Mestanza-Ramón, C.; Henkanaththegedara, S.M.; Vásconez-Duchicela, P.; Vargas-Tierras, Y.; Sánchez-Capa, M.; Constante Mejía, D.; Jimenez Gutierrez, M.; Charco Guamán, M.; Mestanza Ramón, P. In-situ and ex-situ biodiversity conservation in Ecuador: A review of policies, actions and challenges. *Diversity* **2020**, *12*, 315. [CrossRef]
20. Zamora-Burbano, A.M.; Arturo-Perdomo, D.E. Composición química del aceite esencial de hojas de *Hedyosmum translucidum* Cuatrec., Chloranthaceae (Granizo). *Boletín Latinoam. Y Del Caribe De Plantas Med. Y Aromáticas* **2016**, *15*, 192–198.
21. Van Vuuren, S.; Holl, D. Antimicrobial natural product research: A review from a South African perspective for the years 2009–2016. *J. Ethnopharmacol.* **2017**, *17*, 236–252. [CrossRef] [PubMed]
22. Danet, A.F. Recent Advances in Antioxidant Capacity Assays. In *Antioxidants-Benefits, Sources, Mechanisms of Action*; IntechOpen: London, UK, 2021. [CrossRef]
23. Torres-Martínez, R.; García-Rodríguez, Y.; Rios-Chavez, P.; Saavedra-Molina, A.; López-Meza, J.; Ochoa-Zarzosa, A.; Salgado-Garciglia, R. Antioxidant Activity of the Essential Oil and its Major Terpenes of *Satureja macrostema* (Moc. and Sessé ex Benth.) Briq. *Pharmacogn. Mag.* **2018**, *13*, S875–S880. [CrossRef] [PubMed]
24. Beena, M.; Kumar, D.; Rawat, D.S. Synthesis and antioxidant activity of thymol and carvacrol based Schiff bases. *Bioorg. Med. Chem. Lett.* **2013**, *23*, 641–645. [CrossRef]
25. Lorenzo, D.; Loayza, I.; Dellacassa, E. Composition of the essential oils from leaves of two *Hedyosmum* spp. from Bolivia. *Flavour. Fragran. J.* **2003**, *18*, 32–35. [CrossRef]
26. Jukic, M.; Politeo, O.; Maksimovic, M.; Milos, M.; Milos, M. In Vitro acetylcholinesterase inhibitory properties of thymol, carvacrol and their derivatives thymoquinone and thymohydroquinone. *Phytother. Res.* **2007**, *21*, 259–261. [CrossRef]
27. Bonesi, M.; Menichini, F.; Tundis, R.; Loizzo, M.R.; Conforti, F.; Passalacqua, N.G.; Statti, G.A.; Menichini, F. Acetylcholinesterase and butyrylcholinesterase inhibitory activity of *Pinus* species essential oils and their constituents. *J. Enzyme Inhib. Med. Chem.* **2010**, *25*, 622–628. [CrossRef]

28. Calva, J.; Cartuche, L.; González, S.; Montesinos, J.V.; Morocho, V. Chemical composition, enantiomeric analysis and anticholinesterase activity of *Lepechinia betonicifolia* essential oil from Ecuador. *Pharm. Biol.* **2022**, *60*, 206–211. [CrossRef]
29. Burneo, J.I.; Benítez, A.; Calva, J.; Velastegui, P.; Morocho, V. Soil and leaf nutrients drivers on the chemical composition of the essential oil of *Siparuna muricata* (Ruiz & Pav.) A. DC. from Ecuador. *Molecules* **2021**, *26*, 2949. [PubMed]
30. Jaramillo, D.; Calva, J.; Bec, N.; Larroque, C.; Vidari, G.; Armijos, C. Chemical characterization and biological activity of the essential oil from *Araucaria brasiliensis* collected in Ecuador. *Molecules* **2022**, *27*, 3793. [CrossRef] [PubMed]
31. Valarezo, E.; Ludeña, J.; Echeverria-Coronel, E.; Cartuche, L.; Meneses, M.A.; Calva, J.; Morocho, V. Enantiomeric composition, antioxidant capacity and anticholinesterase activity of essential oil from leaves of Chirimoya (*Annona cherimola* Mill.). *Plants* **2022**, *11*, 367. [CrossRef]
32. Thaipong, K.; Boonprakob, U.; Crosby, K.; Cisneros-Zevallos, L.; Hawkins-Byrne, D. Comparison of ABTS, DPPH, FRAP, and ORAC assays for estimating antioxidant activity from guava fruit extracts. *J. Food Compos. Anal.* **2006**, *19*, 669–675. [CrossRef]
33. Arnao, M.B.; Cano, A.; Acosta, M. The hydrophilic and lipophilic contribution to total antioxidant activity. *Food Chem.* **2001**, *73*, 239–244. [CrossRef]
34. Andrade, J.M.; Pachar, P.; Trujillo, L.; Cartuche, L. Suillin: A mixed-type acetylcholinesterase inhibitor from *Suillus luteus* which is used by Saraguros indigenous, southern Ecuador. *PLoS ONE* **2022**, *17*, e0268292. [CrossRef]

plants

Article

Antiviral Potentialities of Chemical Characterized Essential Oils of *Acacia nilotica* Bark and Fruits against Hepatitis A and Herpes Simplex Viruses: *In Vitro, In Silico,* and Molecular Dynamics Studies

Abd El-Nasser G. El Gendy [1], Ahmed F. Essa [2], Ahmed A. El-Rashedy [3], Abdelbaset M. Elgamal [4], Doaa D. Khalaf [5], Emad M. Hassan [1], Ahmed M. Abd-ElGawad [6,*], Abdallah M. Elgorban [7], Nouf S. Zaghloul [8], Salman F. Alamery [9] and Abdelsamed I. Elshamy [2,*]

1 Medicinal and Aromatic Plants Research Department, National Research Centre, Dokki, Giza 12622, Egypt
2 Department of Natural Compounds Chemistry, National Research Centre, Dokki, Giza 12622, Egypt
3 Natural and Microbial Products Department, National Research Centre, Dokki, Giza 12622, Egypt
4 Department of Chemistry of Microbial and Natural Products, National Research Centre, Dokki, Giza 12622, Egypt
5 Department of Microbiology and Immunology, National Research Centre, Dokki, Giza 12622, Egypt
6 Department of Botany, Faculty of Science, Mansoura University, Mansoura 35516, Egypt
7 Department of Botany and Microbiology, College of Science, King Saud University, P.O. Box 2455, Riyadh 11451, Saudi Arabia
8 Bristol Centre for Functional Nanomaterials, HH Wills Physics Laboratory, Tyndall Avenue, Bristol BS8 1FD, UK
9 Biochemistry Department, College of Science, King Saud University, P.O. Box 2455, Riyadh 11451, Saudi Arabia
* Correspondence: dgawad84@mans.edu.eg (A.M.A.-E.); elshamynrc@yahoo.com (A.I.E.)

Citation: El Gendy, A.E.-N.G.; Essa, A.F.; El-Rashedy, A.A.; Elgamal, A.M.; Khalaf, D.D.; Hassan, E.M.; Abd-ElGawad, A.M.; Elgorban, A.M.; Zaghloul, N.S.; Alamery, S.F.; et al. Antiviral Potentialities of Chemical Characterized Essential Oils of *Acacia nilotica* Bark and Fruits against Hepatitis A and Herpes Simplex Viruses: *In Vitro, In Silico,* and Molecular Dynamics Studies. *Plants* **2022**, *11*, 2889. https://doi.org/10.3390/plants11212889

Academic Editors: Hazem Salaheldin Elshafie, Ippolito Camele and Adriano Sofo

Received: 3 September 2022
Accepted: 23 October 2022
Published: 28 October 2022

Abstract: *Acacia nilotica* (synonym: *Vachellia nilotica* (L.) P.J.H.Hurter and Mabb.) is considered an important plant of the family Fabaceae that is used in traditional medicine in many countries all over the world. In this work, the antiviral potentialities of the chemically characterized essential oils (EOs) obtained from the bark and fruits of *A. nilotica* were assessed in vitro against HAV, HSV1, and HSV2. Additionally, the in silico evaluation of the main compounds in both EOs was carried out against the two proteins, 3C protease of HAV and thymidine kinase (TK) of HSV. The chemical profiling of the bark EOs revealed the identification of 32 compounds with an abundance of di- (54.60%) and sesquiterpenes (39.81%). Stachene (48.34%), caryophyllene oxide (19.11%), and spathulenol (4.74%) represented the main identified constituents of bark EO. However, 26 components from fruit EO were assigned, with the majority of mono- (63.32%) and sesquiterpenes (34.91%), where *trans*-caryophyllene (36.95%), Z-anethole (22.87%), and γ-terpinene (7.35%) represented the majors. The maximum non-toxic concentration (MNTC) of the bark and fruits EOs was found at 500 and 1000 μg/mL, respectively. Using the MTT assay, the bark EO exhibited moderate antiviral activity with effects of 47.26% and 35.98% and a selectivity index (SI) of 2.3 and 1.6 against HAV and HSV1, respectively. However, weak activity was observed via the fruits EO with respective SI values of 3.8, 5.7, and 1.6 against HAV, HSV1, and HSV2. The in silico results exhibited that caryophyllene oxide and spathulenol (the main bark EO constituents) showed the best affinities ($\Delta G = -5.62, -5.33, -6.90$, and -6.76 kcal/mol) for 3C protease and TK, respectively. While caryophyllene (the major fruit EO component) revealed promising binding capabilities against both proteins ($\Delta G = -5.31, -6.58$ kcal/mol, respectively). The molecular dynamics simulation results revealed that caryophyllene oxide has the most positive van der Waals energy interaction with 3C protease and TK with significant binding free energies. Although these findings supported the antiviral potentialities of the EOs, especially bark EO, the in vivo assessment should be tested in the intraoral examination for these EOs and/or their main constituents.

Keywords: gum arabic tree; volatile compounds; antiviral activity; in silico; stachene

1. Introduction

The herpes simplex virus (HSV, family: *Herpesviridae*) is a pathogenic virus that causes a wide range of human infections around the world [1,2]. The HSV types, including (HSV-1 and -2) cause several diseases such as oral cavity infections, eyes, pharynx, and mucous membrane, especially in immunodeficient patients [3,4]. The current usual anti-HSV drugs are described to have several side effects and a high prevalence of resistance [5]. This trouble strongly stimulated scientists to seek new therapeutic drugs and agents to overcome virus resistance with low or no side effects. Natural products, including different extracts, essential oils (EOs), and metabolites of herbal and medicinal plants, represented the main strategy for dissolving these issues [6,7].

The hepatitis A virus (HAV) is one of the causes of more than a million infections around the world every year. This virus is a liver-self-limiting and sometimes life-threatening disease [8]. HAV infections were primarily caused by foods and water, as well as direct contact with an infected person [8]. HAV is the most endemic disease in several developing countries [9].

Several plant extracts and metabolites have been documented to have the ability to protect and/or treat these types of viruses. Tannins such as chebulinic acid and chebulagic acid have been shown to have anti-HSV-2 activity nearly 20 times greater than acyclovir, with greater selectivity [10]. In addition, flavonoid aglycones including chrysin, fisetin, and galangin as well as prenylated flavonoids such as kuwanon T exhibited 4–3 more potent activity against HSV-1 than acyclovir [11–13]. Several EOs derived from ginger, thyme, hyssop, anise, sandalwood, and camomile demonstrated significant potencies against herpes simplex virus [14,15]. Additionally, geraniol, a metabolite derived from essential oils, showed a promising anti-HSV-2 activity similar to that of acyclovir [16].

The 3C protease enzyme is one of the main participants in the process of viral genome replication. This enzyme cleaves the viral polyprotein into capsid and nonstructural proteins and inhibits IRES-dependent translation, enabling subsequent viral replication [17]. This motivated researchers to target this protein to inhibit the viral activity. In addition, thymidine kinase, one of the virulence-related proteins in viruses, plays a vital role in the steps of the phosphorylation process associated with the nucleoside salvage pathway. Moreover, this protein can convert pro-antiviral nucleoside analogs such as acyclovir and its congeners to their active forms [18].

The EOs of plants are complicated mixtures of compounds of low molecular weights that are extracted from different medicinal and aromatic plants via different tools [19]. They included many metabolites belonging to different chemical classes, including terpenes and phenylpropanoids, representing the main components along with traces of other aliphatic and aromatic constituents [19,20]. Many biological properties of EOs have been discovered, including antimicrobial, cytotoxicity, anti-inflammatory, antipyretic, antiulcer, and antiviral properties [19].

Acacia plants (family: Leguminosae or Fabaceae) include around 1350 species or more that are widely distributed in warm and tropical areas. Several *Acacia* species have been shown to have pharmaceutical effects, including the treatment and inhibition of microbial infections, ulcerative colitis, skin and eye infections, wounds, mouthwash, and others [21–23]. Many compounds were documented from the plants belonging to the *Acacia* genus, especially di- and triterpenes, flavonoids, tannins, alkaloids, phenolic acids, and saponins [21,23]. The EOs were also reported from different *Acacia* plants, such as the Nigerian *Acacia nilotica* (L.) *P.J.H.Hurter and Mabb.*, *A. albida* (Delile) A.Chev. (synonym: *Faidherbia albida*) [24], *A. tortilis* (synonym: *Vachellia tortilis* (Forssk.) Galasso and Banfi) [25], *A. mearnsii* De Wild. (synonym: *A. mearnsii*) [26], and *A. cyanophylla* (synonym: *Acacia saligna* (Labill.) H.L.Wendl.) [27].

Acacia nilotica (common name: Gum Arabic tree) is an important traditional plant for the treatment of several diseases in several countries around the world [22]. Many components were isolated and identified from the extracts of the different organs of this plant comprising the terpenes, tannins, phenolic acids, and flavonoids [21,22,28,29]. Several reports described the interesting biological potentialities of the different extracts of this plant, such as the treatment of inflammation, free radicals, leishmanial, diabetes, cancers, plasmodial, and other infections, along with molluscicidal activity [30–32].

Little studies were carried out concerning the composition and pharmaceutical applications of *A. nilotica* EOs [24]. Thus, the current work aimed to (i) characterize the chemical profile of the EOs derived from the *A. nilotica* bark and fruits depending upon GC-MS techniques, (ii) evaluate the antiviral efficiency of these two EOs against the HSV1, HSV2, and HAV viruses, and (iii) study of the molecular docking of the main compounds in each oil on 3C protease of HAV (PDB ID: 1QA7) and thymidine kinase of HSV (PDB ID: 1KI3) proteins.

2. Results and Discussion

2.1. The Identification of the Chemical Constituents of Bark and Fruits of A. nilotica EOs

The EOs of the bark and fruits of *A. nilotica* were separately obtained via three hours of hydro-distillation over the Clevenger apparatus, where they yielded 0.072% and 0.056% (*v*/*w*) of the oil, respectively. The yield of the EO from bark was found to be more than the fruits, and both were more varied than those described for the Nigeria ecospecies [24]. The oils were analyzed via GC-MS (Figure 1).

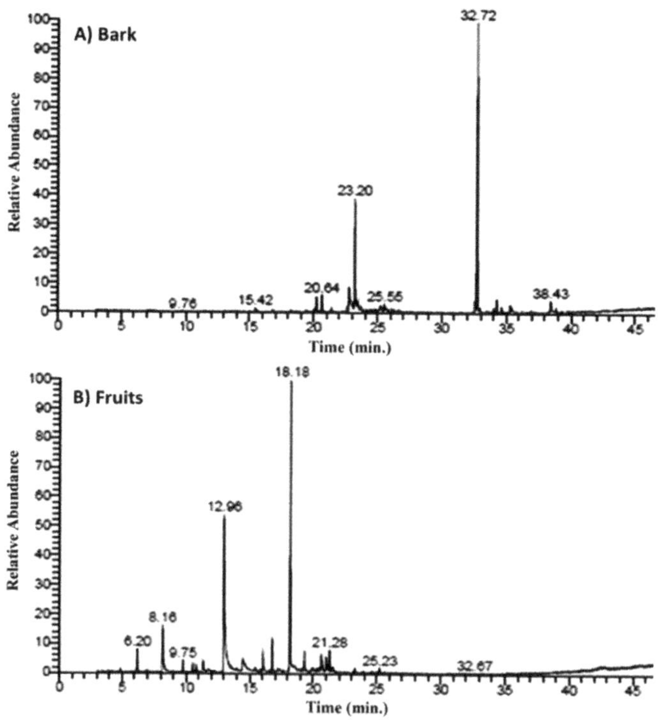

Figure 1. GC-MS ion chromatograms of bark (**A**) and fruits (**B**) of *Acacia nilotica* EOs.

The compounds' names, retention times, relative concentrations, and Kovats indexes are presented in Table 1. The analysis revealed that the terpenes represented the main constituents of the bark with a relative concentration of 95.25% of the hydrocarbons and oxygenated forms of mono-, sesquit-, and di-terpenes. Furthermore, the terpenes represented the major components of the fruits EO with a relative concentration of 98.23%, including the mono- and sesquiterpene hydrocarbons along with the oxygenated derivatives of monoterpenes. This preponderance of terpenes agreed with the published data from the Nigerian *A. nilotica* [24] and other *Acacia* species such as *A. albida* [24], *A. tortilis* [25], *A. mearnsii* [26], and *A. cyanophylla* [27].

Table 1. Essential oil compositions of EOs of bark and fruits of *Acacia nilotica*.

No	Rt [1]	Compound	Type	Relative Concentration (%)		KI	
				Bark	Fruit	Lit. [2]	Exp. [3]
1	3.89	α-Pinene	MH	-	0.11 ± 0.01	932	935
2	4.85	β-Pinene	MH	-	0.35 ± 0.02	974	979
3	6.20	1,8-Cineole	OM	-	3.52 ± 0.06	1026	1020
4	6.41	γ-Terpinene	MH	-	7.35 ± 0.08	1054	1059
5	6.63	α-Linalool	OM	-	1.82 ± 0.04	1095	1091
6	6.94	Camphor	OM	-	0.36 ± 0.02	1146	1144
7	9.75	Borneol	OM	-	2.31 ± 0.04	1169	1165
8	10.50	4-Terpineol	OM	-	1.15 ± 0.03	1177	1175
9	10.77	α-Terpineol	OM	-	1.16 ± 0.05	1186	1185
10	11.74	Cumin aldehyde	OM	-	0.49 ± 0.02	1238	1235
11	12.96	*Z*-Anethole	OM	-	22.87 ± 0.23	1249	1245
12	13.99	Bornyl acetate	OM	-	0.23 ± 0.02	1285	1283
13	14.47	2-Caren-10-al	OM	-	3.51 ± 0.05	1289	1287
14	15.45	α-Terpinyl acetate	OM	-	0.47 ± 0.02	1316	1314
15	15.82	Myrtenyl acetate	OM	0.84 ± 0.03	-	1324	1322
16	16.04	α-Elemene	SH	-	4.69 ± 0.06	1335	1339
17	16.13	α-Cubebene	SH	-	0.32 ± 0.01	1351	1346
18	16.54	α-Copaene	SH	0.35 ± 0.02	0.23 ± 0.01	1374	1370
19	16.67	β-Elemene	SH	-	3.72 ± 0.08	1389	1392
20	16.84	Methyl eugenol	OM	-	0.54 ± 0.02	1403	1401
21	18.18	*trans*-Caryophyllene	SH	0.24 ± 0.01	36.95 ± 0.18	1407	1405
22	18.35	Longifolene	SH	-	0.41 ± 0.01	1408	1413
23	18.58	Aromadendrene	SH	-	0.36 ± 0.02	1439	1433
24	19.40	α-Humulene	SH	0.27 ± 0.01	4.05 ± 0.06	1452	1455
25	19.85	γ-Muurolene	SH	-	0.26 ± 0.01	1478	1475
26	19.96	Germacrene-D	SH	0.14 ± 0.01	0.46 ± 0.02	1481	1484
27	20.02	α-Amorphene	SH	0.51 ± 0.02	-	1483	1486
28	20.12	α-Selinene	SH	-	0.54 ± 0.01	1498	1495
29	20.21	α-Muurolene	SH	2.42 ± 0.04	-	1500	1498
30	20.64	Bicyclogermacrene	SH	2.41 ± 0.06	-	1501	1504
31	21.35	δ-Cadinene	SH	0.75 ± 0.02	-	1522	1525
32	22.58	α-Calacorene	SH	1.18 ± 0.05	-	1545	1543
33	22.19	*E*-Nerolidol	OS	0.16 ± 0.01	-	1531	1535
34	22.76	Spathulenol	OS	4.74 ± 0.05	-	1577	1574
35	23.20	Caryophyllene oxide	OS	19.11 ± 0.09	-	1582	1585
36	23.45	Globulol	OS	1.01 ± 0.02	-	1590	1593
37	23.70	Veridiflorol	OS	0.83 ± 0.01	-	1596	1598
38	24.45	Neoclovenoxid	OS	0.55 ± 0.01	-	1608	1605
39	24.68	Isospathulenol	OS	0.14 ± 0.00	-	1630	1632
40	24.85	tau-Cadinol	OS	0.51 ± 0.02	-	1640	1642
42	25.22	Cubenol	OS	0.75 ± 0.03	-	1645	1644
43	25.28	Torreyol	OS	0.68 ± 0.03	-	1646	1648
44	25.55	α-Cadinol	OS	1.35 ± 0.06	-	1654	1657
45	25.68	Khusinol	OS	0.44 ± 0.01	-	1658	1659
46	32.58	Cryptomeridiol	OS	1.27 ± 0.07	-	1813	1816
47	32.72	Stachene	DH	48.34 ± 0.25	-	1931	1934

Table 1. *Cont.*

No	Rt [1]	Compound	Type	Relative Concentration (%)		KI	
				Bark	Fruit	Lit. [2]	Exp. [3]
48	34.26	Trachyloban	DH	2.25 ± 0.07	-	1965	1968
49	34.62	Isokaurene	DH	1.01 ± 0.04	-	1997	1999
50	35.31	Kaur-16-ene	DH	1.94 ± 0.06	-	2017	2015
51	34.02	Phytol	OD	0.76 ± 0.02	-	1942	1940
52	36.42	Sclareol	OD	0.14 ± 0.00	-	2223	2221
53	37.11	4,8,13-Duvatriene-1,3-diol	OD	0.16 ± 0.00	-	2400	2403
54	38.43	*n*-Nonacosane	Others	2.51 ± 0.06	-	2900	2900
56	45.21	*n*-Dotriacontane	Others	0.98 ± 0.03	-	3200	3200
		Monoterpene Hydrocarbons (MH)		0	7.81		
		Oxygenated Monoterpenes (OM)		0.84	55.51		
		Sesquiterpene Hydrocarbons (SH)		8.27	34.91		
		Oxygenated Sesquiterpenes (OS)		31.54	3.52		
		Diterpene Hydrocarbons (DH)		53.54	0		
		Oxygenated Diterpenes (OD)		1.06	0		
		Others		3.49	0		
		Total		98.74	98.23		

[1] Retention time (RT), [2] Published Kovats indexes (KI_p), [3] Calculated Kovats indexes (KIc).

The chemical characterization of the bark EO revealed that the diterpenes are the major constituents with a relative concentration of 54.60%, comprising of diterpene hydrocarbons (53.54%) and traces of oxygenated diterpene (1.06%). With a few exceptions, such as *Euphorbia mauritanica* L. [33], *Lactuca serriola* L. [34], and others [35], the phenomenon of diterpene abundance was rare in plants EOs. Four diterpene hydrocarbons were assigned as overall identified diterpenoids, including stachene (48.34%), trachyloban (2.25%), kaur-16-ene (1.94%), and isokaurene (1.01%). Stachene was reported as a major diterpene hydrocarbon in a number of plants, including *Chamaecyparis pisifera* (Siebold and Zucc.) Endl., *Chamaecyparis obtuse* (Siebold and Zucc.) Endl., and *Thuja orientalis* L. [36].

Trachylobane and kaur-16-ene were rarely found in *Acacia* EOs but were widely found in the EOs of other plants, such as *Croton zambesicus* Muell. Arg. [37], *Cryptomeria japonica* (L.f.) D.Don [38], *Alpinia galanga* (L.) Willd. [39], *Euphorbia heterophylla* L. [40], and *E. mauritanica* [33]. Three oxygenated diterpenes, phytol, sclareol, and 4,8,13-duvatriene-1,3-diol, were identified from the bark. The phytol was documented as the main component of EO derived from the leaves of *Acacia mearnsii* [26]. These three compounds are well-known and commonly reported compounds in the EOs of some plants, such as *Calotropis procera* (Aiton) W.T.Aiton [41], *Cyperus leavigatus* L. [42], and *Launaea mucronata* (Forssk.) Muschl. [43]. On the contrary, the GC-MS analysis of the EO derived from the fruits revealed the complete absence of the diterpenoid constituents. The phenomena of the diterpenes' disappearance from the EOs of *Acacia* species is commonly documented [19,24,25,27]; these data agree with the data of fruit oil but not with the bark oil.

Monoterpenes were found to be the major characterized components of *A. nilotica* fruits EO, with a relative concentration of 63.32% of mainly oxygenated compounds (55.51%) and hydrocarbons (7.81%). The abundance of the monoterpenoids in the fruits EO agreed with the previous data from Nigerian *A. nilotica* and other *Acacia* species [24,26,27,33]. Z-Anethole (22.87%), 2-caren-10-al (3.51%), and borneol (2.31%) were characterized as the main identified oxygenated monoterpenes, while γ-terpinene (7.35%) was the major monoterpene hydrocarbons while 1,8-cineol (3.52%) was assigned. All these compounds were reported in the EOs of *Acacia* species as traces and/or totally absent [19,24,26,33]. However, the bark EO was found to have traces of monoterpenes in oxygenated forms.

The bark and fruits EOs were found to include the sesquiterpenes with high relative concentrations (39.81% and 34.91%, respectively). The analysis of the bark EO revealed the presence of a low concentration of sesquiterpene hydrocarbons (8.27%) and major oxygenated sesquiterpenes (31.54%). The α-muurolene (2.42%) and bicyclogermacrene (2.41%) were identified as the major sesquiterpene hydrocarbons, while caryophyllene oxide (19.11%) and spathulenol (4.74%) are the main oxygenated sesquiterpenes of the bark EO. Furthermore, the sesquiterpene hydrocarbons were the only detected sesquiterpenes in the EO of the fruits with a relative concentration of 34.91% and a complete absence of oxygenated compounds. From all the assigned sesquiterpene hydrocarbons in fruits EO, *trans*-caryophyllene (36.95%), α-elemene (4.69%), α-humulene (4.05%), and β-elemene (3.72%) represented the main constituents. The majority of the sesquiterpenes were previously reported from EOs of the Nigerian *A. nilotica* [24], and other *Acacia* plants as *A. tortilis* [25]. Previous studies of EOs derived from *Acacia* species revealed the prevalence of muurolene, caryophyllene oxide, and spathulenol, where they are determined in considerable concentrations [24–26].

Ultimately, the other non-terpenoid compounds are represented as traces of the total mass of the bark EO (3.49%), while they are absent in fruit EO. In bark EO, *n*-nonacosane (2.51%) and *n*-dotriacontane (0.98%) were identified. Hydrocarbons are widely identified compounds in *Acacia* plants [25,26]. As observed in these findings, the net results revealed a significant variation in the quantity and quality as well as the chemical components of the two plant parts and also in Nigerian *A. nilotica* [24]. This phenomenon of variation might be attributed to the variation of the plant organ, genotypes, humidity, climate, weather, and environmental conditions [41,44].

2.2. In Vitro Antiviral Activity

The antiviral activities of EOs derived from the bark and fruits of *A. nilotica* were in vitro screened against HAV, HSV1, and HSV2, and the maximum non-toxic concentration (MNTC) used for the screening was determined. The results showed that MNTC of EOs of bark and fruits was 500 and 1000 μg/mL, respectively (Table 2). Using the MTT assay, the percentages of antiviral effects of bark and fruits EOs were determined by comparing the viability of cells treated by the virus only with the viability of cells treated by the virus and MNTC of the samples.

Table 2. Effect of EOs of bark and fruits of *Acacia nilotica* on hepatitis A (HAV) and herpes simplex (HSV1 and HSV2).

Acacia nilotica	MNTC (μg/mL) [a]	Antiviral Effect %			Selectivity Index (SI)		
		HAV	HSV1	HSV2	HAV	HSV1	HSV2
Bark EO	500 ± 6.2	47.26 ± 2.05	35.98 ± 1.31	9.07 ± 0.36	2.3	1.6	ND
Fruits EO	1000 ± 11.4	9.42 ± 0.62	14.26 ± 0.54	3.99 ± 0.15	3.8	5.7	1.6
Acyclovir						>387.63	12.24
Amantadine					51.62		

[a] MNTC: the maximum non-toxic concentration; SI = selectivity index (CC_{50}/IC_{50}).

The results revealed that bark EO has moderate antiviral effects against HAV with an effect of 47.26% alongside a selectivity index (SI) of 2.3, concerning amantadine as a reference drug with SI at 51.62. In addition, the bark EO demonstrated moderate antiviral abilities against HSV1 with an effect of 35.98% and SI of 1.6 compared with acyclovir as a positive control at SI > 387.63. However, this EO exhibited a very weak antiviral effect against HSV2 with a 9.07% comparison with acyclovir with SI at 12.24.

On the other side, the EO derived from the fruits showed weak potentialities against the three tested viruses. This oil showed weak anti-HAV with an effect % of 9.42% and SI of 3.8 compared with amantadine at SI of 51.62. Likewise, this oil also exhibited weak anti-HSV1 with effect% at 14.26% and SI of 5.7 compared with acyclovir as a standard drug. In addition, the results revealed that fruits EO has the lowest activity against HSV2 with an effect% of 3.99% as well as SI at 1.6.

The chemical components of the EOs and especially the main compounds played as the main mediators in their pharmaceutical and biological activities [8]. The present results revealed that the bark EO has moderate to weak antiviral potentialities, especially against the two viruses, HAV and HSV1. The GC-MS profiling of this oil revealed the abundance of terpenes, mainly di- and sesqui-terpenes, that were documented to have antiviral effects against several viruses [45,46]. Diterpenes, as the main components of the bark EO, were reported to display significant antiviral effects against HAV and HSV [45,47–49].

Plants' EOs enriched with diterpenes were stated to demonstrate significant antiviral agents such as the different extracts and EOs of *Croton lechleri* Müll.Arg. [2]. Mechanistically, the diterpenes could inhibit the viral replication process by blocking DNA polymerase activity, as reported with dolastane compounds [49]. Several diterpene skeletons, including kaurenes, have been shown to have significant anti-HSV, anti-HAV, and other antiviral properties [45,49]. In addition, the sesquiterpenes acted in a significant role as inhibitors of several viruses. Astani and his colleagues discovered that caryophyllene oxide has significant anti-HSV1 activity [47]. Many EOs derived from plants, such as *Eryngium alpinum* L., *Eryngium amethystinum* L. [50], *Melaleuca ericifolia* Sm., *Melaleuc leucadendron* (L.) L., *Melaleuc armillaris* (Sol. ex Gaertn.) Sm. and *Melaleuc stypheliοides* Sm. [51], have been shown to have antiviral activities due to the majority of caryophyllene oxide and spathulenol. The two sesquiterpenoids, α-muurolene and bicyclogermacrene, were also reported as major components of antiviral active EOs derived from *Glechon spathulata* Benth. and *Glechon marifolia* Benth. [52].

Previous studies hypothesized that the anti-HSV effectiveness greatly depends on the binding affinity of EO components to the surface of viruses, and this affinity might be affected by the polarity of compounds [15]. The anti-HSV effects of thymol-related monoterpenoids were reported to decrease with decreasing polarity [53].

Consequently, the present findings were in full agreement with the previous reports. The singular and/or synergetic effects of the major components, along with the minors of the bark EO, might be the main reasons for this anti-HAV and HSV1 efficiency. These results supported the ability of these compounds to cause abnormalities in the HSV membrane protein functions and structures that consequently decrease the penetration or binding of the virus into the cells [54].

On the other hand, the fruits' EO antiviral effects were in concordance with Astani et al. [47] who deduced the anti-HSV with β-caryophyllene and Z-anethole as singular compounds. The increase in the HSV inhibitory effects of the EOs was basically ascribed to the presence of the polyhydroxylated constituents [54]. With the exception of the mono-oxygenated monoterpene, *Z-anethole*, the other oxygenated monoterpenes were identified as traces that decrease the contribution and effect of each singular compound. Several plants, including *Mentha aquatic* L., *Mentha pulegium* L., *Mentha microphylla* C. Kock, Mentha x villosa (Hudson), *Micromeria thymifolia* (Scop.) Fritsch, and *Ziziphora clinopodioides* Lam., have demonstrated weak antiviral activity of EOs enriched with oxygenated monoterpenes [55].

2.3. Molecular Docking Studies

In our attempt to study the mechanistic action of each EO, the molecular operating environment (MOE) of the main three compounds in each EO toward the two proteins, HAV 3C protease and HSV TK, were evaluated. The results of in silico studies of the major compounds identified in both oils could rationalize the in vitro antiviral activities noticed by them. As shown in Table 3, caryophyllene oxide and spathulenol, the main constituents of oil of the bark, showed the best affinities (ΔG = −5.62, −5.33 and −6.90, −6.76 kcal/mol) for 3C protease and TK, respectively (Figures 2 and 3). Additionally, trans-caryophyllene, as a major constituent of fruit EO, revealed promising binding capabilities against both proteins (ΔG = −5.31, −6.58 kcal/mol, respectively) (Figures 2 and 3).

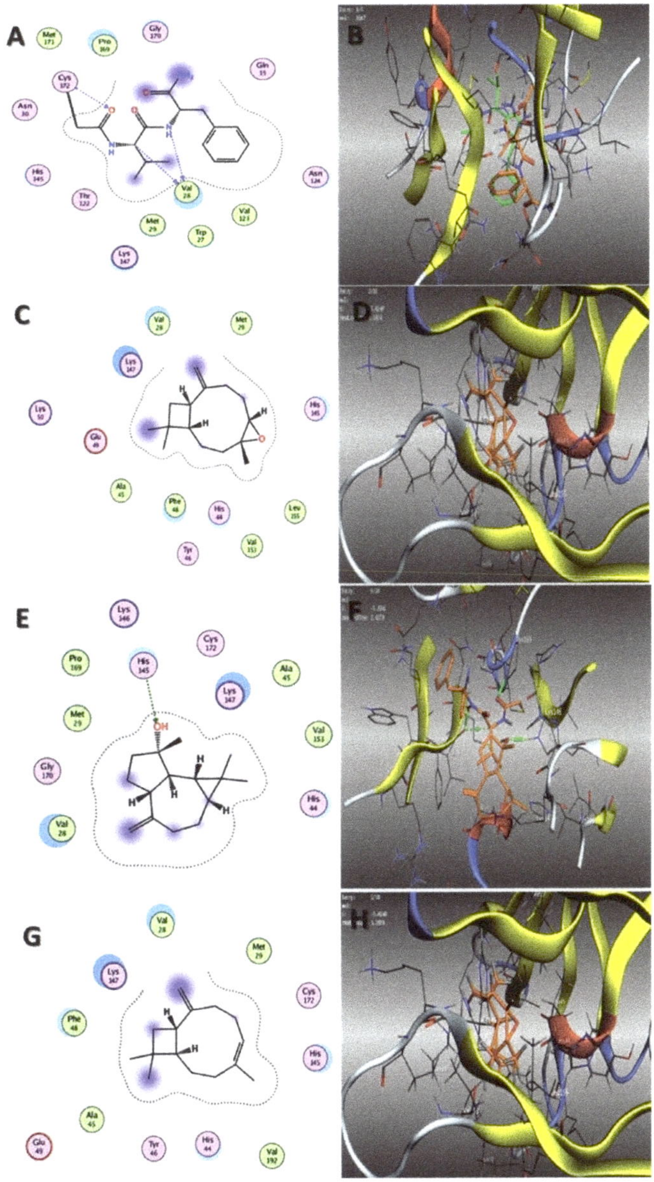

Figure 2. Two-dimensional binding mode of the co-crystallized ligand (**A**) and validation of docking study (**B**), 2D binding mode and 3D binding modes of caryophyllene oxide (**C**,**D**), spathulenol (**E**,**F**), and trans-caryophyllene (**G**,**H**) in the active site of 3C protease of HAV (PDB ID: 1QA7): overlay of the experimental (green) and docked conformation (orange).

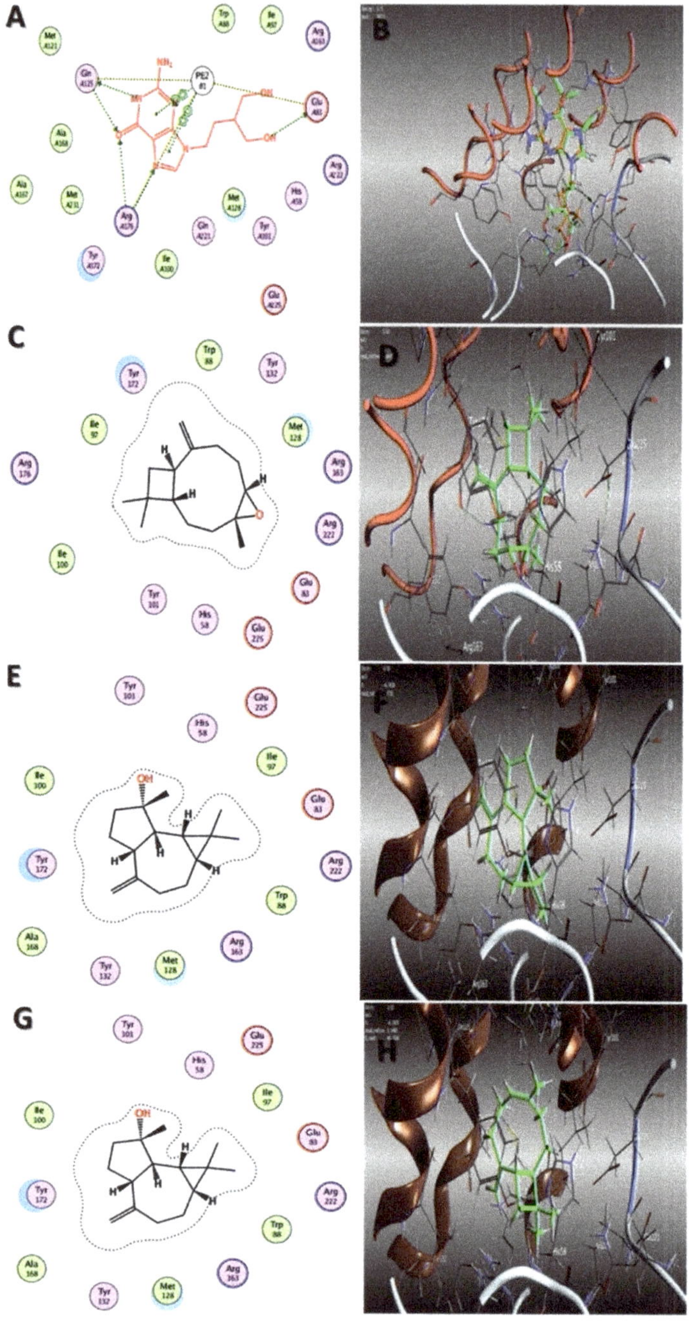

Figure 3. Two-dimensional binding mode of the co-crystallized ligand (**A**) and validation of docking study (**B**), 2D binding mode and 3D binding modes of caryophyllene oxide (**C**,**D**), spathulenol (**E**,**F**), and trans-caryophyllene (**G**,**H**) in the active site of thymidine kinase of HSV (PDB ID: 1KI3): overlay of the experimental (green) and docked conformation (orange).

Table 3. Results of docking simulations of the major compounds identified in EOs of both bark and fruits of *Acacia nilotica*.

Plant Part EO	Name of Phytoligands	ΔG * (kcal/mol)	
		HAV 3C Protease	HSV TK
Bark	Spathulenol	−5.23	−6.83
	Caryophyllene oxide	−5.43	−6.96
	Stachene	-	-
Fruits	γ-Terpinene	−4.85	−5.80
	Z-Anethole	−4.66	−5.20
	Trans-caryophyllene	−5.26	−6.79
Co-crystallized inhibitor of HAV 3C protease		−6.61	ND
Co-crystallized inhibitor of HSV TK 1		ND	−7.85

* binding free energy.

Previous studies have shown that the sesquiterpene caryophyllene and its oxides possess suitable antiviral activities with a high selectivity index [47]. The higher activity of the bark oil than the fruit oil might be attributed to the synergistic effect of caryophyllene oxide and spathulenol, as well as the other components. These data go in the same line with the in vitro results in which the bark EO is more active than the fruits EO.

2.4. Molecular Dynamic and System Stability

A molecular dynamic simulation was carried out to predict the inhibition performance and interaction of the caryophyllene oxide with the catalytic active site of both 3C protease of HAV and thymidine kinase of HSV targets as well as their stability throughout the simulation [56,57]. The validation of system stability is essential to trace disrupted motions and avoid artifacts that may develop during the simulation. The recorded average RMSD values for all frames of Apo-HAV and Complex-HAV were 1.53 Å and 1.37 Å, respectively. In addition, average RMSD values of 1.51 Å and 1.59 Å were observed for Apo-HSV and Complex-HSV, respectively (Figure 4A,C). In general, These results revealed that the caryophyllene oxide-bound protein complex system acquired a relatively more stable conformation than the other studied systems.

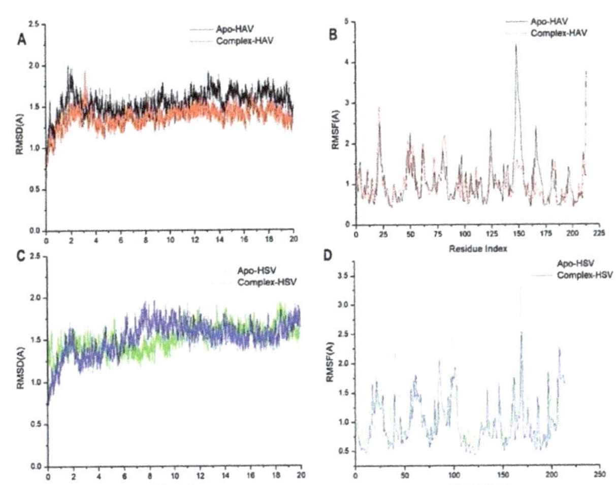

Figure 4. (**A**) RMSD of Cα atoms of the 3C protease of HAV protein backbone atoms. (**B**) RMSF of each residue of the 3C protease of HAV protein backbone Cα atoms (**C**) RMSD of Cα atoms of the thymidine kinase of HSV protein backbone atoms (**D**) RMSF of each residue of the thymidine kinase of HSV protein backbone Cα atoms.

During MD simulation, assessing protein structural flexibility upon ligand binding is critical for examining residue behavior and its connection with the ligand. The protein residue fluctuations were evaluated using the Root-Mean-Square Fluctuation (RMSF) algorithm to evaluate the effect of inhibitor binding toward the respective targets over 20 ns simulations. The computed average RMSF values were 1.21 and 1.00 for Apo-HAV, and caryophyllene oxide-bound to protein systems, respectively, while 1.17 Å and 1.02 Å were recorded for Apo-HSV and Complex-HSV, respectively. Figure 4B,D show the overall residue fluctuations of individual systems. These values reveal that the caryophyllene oxide-bound protein complex system has a lower residue fluctuation than the other systems.

Binding Interaction Mechanism Based on Binding Free Energy Calculation

A popular method for determining the free binding energies of small molecules to biological macromolecules is the molecular mechanics energy technique (MM/GBSA), which combines the generalized born and surface area continuum solvation, and it may be more trustworthy than docking scores. The MM-GBSA program in AMBER18 was used to calculate the binding free energies by extracting snapshots from the trajectories of the systems. As shown in Table 1, all the reported calculated energy components (except ΔG_{solv}) gave negative values, indicating favorable interactions. The results indicated that binding free energy (ΔG_{bind}) values −19.35 and −32.04 Kcal/mol were obtained for the interactions of caryophyllene oxide with 3C protease of HAV and thymidine kinase of HSV receptors, respectively.

A close look at the individual contributions of energy reveals that the more positive van der Waals energy components drive caryophyllene oxide interactions with 3C protease of HAV and thymidine kinase of HSV enzyme, resulting in the observed binding free energies (Table 4).

Table 4. The calculated energy binding for caryophyllene oxide against the 3C protease of HAV and thymidine kinase of HSV receptors.

Complex	Energy Components (kcal/mol)				
	ΔE_{vdW}	ΔE_{elec}	ΔG_{gas}	ΔG_{solv}	ΔG_{bind}
Caryophyllene oxide -HAV	−20.44 ± 0.27	−1.52 ± 0.12	−12.96 ± 0.33	3.60 ± 0.11	−19.35 ± 0.24
Caryophyllene oxide -HSV	−34.86 ± 0.081	−2.58 ± 0.07	−37.45 ± 0.11	5.40 ± 0.05	−32.04 ± 0.11

ΔEvdW = van der Waals energy; ΔEele = electrostatic energy; ΔGsolv = solvation free energy; ΔGbind = calculated total binding free energy.

3. Materials and Methods

3.1. Plant Materials and EOs Extraction

Acacia nilotica bark and fruits were collected in August 2020 from Naje Algamal, Sohag Governorate, Egypt (26°33′43.2″ N 31°40′39.7″ E). The specimen is authenticated and deposited in the herbarium of the National Research Center with specimen codes NRC-20xYD-AN-20-609.

The collected samples were dried at air temperature (25 °C ± 3) for 7 days and then crushed into a fine powder. The EOs of the air-dried powdered bark and fruits of *A. nilotica* (300 g, each) were subjected to hydro-distillation over a Clevenger apparatus for three hours. The oily layers were separated via *n*-hexane and then dried using Na_2SO_4 anhydrous (0.5 g). Three EO samples from each part were obtained by repeating the extraction process three times with the same sequence. All oil samples were saved in glass vials at 4 °C until further GC-MS and biological analysis.

3.2. Gas Chromatography-Mass Spectrometry (GC-MS) Analysis

The GC-MS technique was used for the analysis of the extracted EO samples, as reported in published studies [33,43]. The analysis was performed at the National Research Center using the GC-MS apparatus joined with the TRACE GC Ultra-Gas Chromatograph and quadrupole mass unit, model Thermo-Scientific™ EC, Waltham, MA, USA. The used GC-MS column dimension and film thicknesses were 0.32 mm × 30 m and 0.25 μm. The transporter gas (He) was used with 1 to 10 as the split ratio and 1.0 mL/min as the flow rate. As usual, the temperature was regulated as follows: 60 °C/1 min and elevated to 240 °C during 4 °C/min. The dilution of all oil samples was performed in 1 μL of n-hexane by the ratio of 1:10 (*v*/*v*) and injected where both of injector and detector were adjusted at 210 °C. The electron ionization (EI) at 70 eV and *m*/*z* 40–450 as spectral range were used for performing the mass spectral data of oil constituents. Finally, the authentication and identification of the chemical constituents occurred depending upon the Automated Mass spectral Deconvolution and Identification (AMDIS) software, (version: 1.0.0.13), NIST library database, Wiley spectral library collection, and retention indices relative to *n*-alkanes (C_8–C_{22}).

3.3. Antiviral Assays

3.3.1. Determination of the Maximum Non-Toxic Concentration (MNTC)

Solutions of 10 mg of EOs of bark and fruits of *A. nilotica* were prepared in 10% dimethylsulfoxide (DMSO) and then diluted with Dulbecco's Modified Eagle Medium (DMEM) (Lonza, Verviers, Belgium) to the working solutions. For the modified 3-(4, 5-dimethylthiazol-2-yl)-2,5-diphenyltetrazolium bromide (MTT) assay, the VERO-E6 cells (3×10^5 cells/mL) (Sigma-Aldrich, St. Louis, MO, USA) were seeded in 96-well plates (100 μL/well) and incubated in 5% CO_2 at 37 °C. After 24 h of incubation, cells were subjected to concentrations (31.25, 62.5, 125, 250, 500, and 1000 μg/mL) of EOs of bark and fruits of *A. nilotica* in triplicates for another 48 h. Cells were checked for any physical signs of toxicity, e.g., partial or complete loss of the monolayer, rounding, shrinkage, or cell granulation. Then, the cell monolayers were washed 3 times with 1 L sterile phosphate buffer saline (PBS) and treated with 20 μL MTT solution in PBS (5 mg/mL) for each well. After shaking, the plate was incubated for 4 h in 5% CO_2 at 37 °C. The formed formazan crystals in each well were dissolved with 200 μL of 0.04 M HCl in isopropanol. Optical density was determined at 560 nm after removing the background at 620 nm, where it should be directly correlated with cell quantity. The maximum non-toxic concentration (MNTC) of each extract was determined and was used for further biological studies. The MNTC was calculated from the plot of toxicity percent against sample concentration. The DMSO was tested as a control, and it did not show any activity.

3.3.2. Antiviral Effect Percent Determination

All viruses, HSV1, HSV2, and HAV, were obtained from the Microbiology Department, Faculty of Medicine for girls, Al-Azhar University. Vero-E6 cells were distributed in 96-well tissue culture plates (10^4/well) and incubated at a humidified 37 °C incubator under 5% CO_2 conditions. After 24 h, the cells were washed with 1 L of PBS. Equal volumes (1:1 *v*/*v*) of MNTC of tested samples and the virus (HSV1, HSV2, and HAV) suspension were incubated for one hour. In triplicate, 100 L of viral or sample suspension was propagated on cells and incubated for 24 h at 37 °C in 5% CO_2. A total of 20 μL MTT solutions in PBS (5 mg/mL) was added to each well and incubated for 1–5 h to allow MTT to be metabolized. As previously mentioned, the resultant was treated, and the optical density was determined. The antiviral effect percentage was calculated by dividing the viability of cells treated by the virus only/the viability of cells treated by the virus and the sample. Acyclovir was used as a reference drug for the viruses (HSV1 and HSV2), while amantadine was used as a positive control against the HAV virus. The calculation of the SI (selectivity index) values of the tested EO samples was performed from the equation: $SI = CC_{50}/IC_{50}$; where: CC_{50}: 50% cytotoxic concentration, and IC_{50}: 50% effective concentration [6,11].

3.4. Molecular Docking Studies

Molecular Operating Environment (MOE Vs. 2015) docking studies were performed on the catalytic domains of HAV 3C protease (PDB ID: 1QA7) [58] and HSV thymidine kinase

(PDB ID: 1KI3) [58] and thymidine kinase of HSV (PDB ID: 1KI3) [59]. The crystal structures of both proteins were retrieved from the protein data bank and processed as previously described [60], as well as the database file (mdb) of the major identified compounds in both EOs, stachene, caryophyllene oxide, spathulenol, *trans*-caryophyllene, *Z*-anethole, and γ-terpinene. The docking process was validated by re-docking the co-crystallized ligands in the binding site, which revealed their binding with crucial sub-pockets (Val 28 and Cys 172 and Gln A125, Arg A176, and Glu A83) in both proteins, respectively, at acceptable RMSD (Figures 2 and 3). After processing the study in triplicate (as shown in Table S1), the results of the docking process were presented as the ΔG (kcal/mol) with RMSD values ≤ 2 Å (Table 2). In addition, the interactions of the lowest energy pose with the binding pocket were two and three-dimensionally presented in (Figures 2 and 3).

3.5. Molecular Dynamics Stimulation Section

3.5.1. System Preparation

The crystal structures of the 3c proteinase from the hepatitis A virus receptor and the thymidine kinase from herpes simplex virus type I were retrieved from the protein data bank with codes 1QA7 [58] and 1KI3 [59], respectively. This structure was then prepared for molecular dynamics (MD) studies using UCSF Chimera [61]. Using PROPKA, pH was fixed and optimized to 7.5 (3). Caryophyllene oxide was drawn using ChemBioDraw Ultra 12.1. Altogether, all four prepared systems were subjected to 20 ns MD simulations as described in the simulation section.

3.5.2. Molecular Dynamic (MD) Simulations

The integration of molecular dynamic (MD) simulations into biological systems studies enables exploring the physical motion of atoms and molecules that cannot be easily accessed by any other means. The insight extracted from performing this simulation provides an intricate perspective into the biological systems' dynamical evolution, such as conformational changes and molecule association [61]. The MD simulations of all systems were performed using the GPU version of the PMEMD engine present in the AMBER 18 package [62].

The partial atomic charge of each compound was calculated with ANTECHAMBER's General Amber Force Field (GAFF) technique [63]. The Leap module of the AMBER 18 package implicitly solvated each system within an orthorhombic box of TIP3P water molecules within 10 Å of any box edge. The Leap module was used to neutralize each system by incorporating Na^+ and Cl^- counter ions. A 2000-step initial minimization of each system was carried out in the presence of a 500 kcal/mol applied restraint potential, followed by a 1000-step full minimization using the conjugate gradient algorithm without restraints.

During the MD simulation, each system was gradually heated from 0 K to 300 K over 500 ps, ensuring that all systems had the same amount of atoms and volume. The system's solutes were subjected to a 10 kcal/mol potential harmonic constraint and a 1 ps collision frequency. Following that, each system was heated and equilibrated for 500 ps at a constant temperature of 300 K. To simulate an isobaric-isothermal (NPT) ensemble, the number of atoms and pressure within each system for each production simulation were kept constant, with the system's pressure maintained at 1 bar using the Berendsen barostat [64].

For 20 ns, each system was MD simulated. The SHAKE method was used to constrain the hydrogen bond atoms in each simulation. Each simulation used a 2 fs step size and integrated an SPFP precision model. An isobaric-isothermal ensemble (NPT) with randomized seeding, constant pressure of 1 bar, a pressure-coupling constant of 2 ps, a temperature of 300 K, and a Langevin thermostat with a collision frequency of 1 ps was used in the simulations.

3.5.3. Post-MD Analysis

After saving the trajectories obtained by MD simulations every 1 ps, the trajectories were analyzed using the AMBER18 suite's CPPTRAJ [65] module. The Origin data analysis program and Chimera were used to create all graphs and visualizations.

3.5.4. Thermodynamic Calculation

The Poisson–Boltzmann or generalized born and surface area continuum solvation (MM/PBSA and MM/GBSA) approach has been found to be useful in the estimation of ligand-binding affinities [66,67]. The protein–ligand complex molecular simulations used by MM/GBSA and MM/PBSA compute rigorous statistical-mechanical binding free energy within a defined force field.

Binding free energy averaged over 500 snapshots extracted from the entire 50 ns trajectory. The estimation of the change in binding free energy (ΔG) for each molecular species (complex, ligand, and receptor) can be represented as follows [68]:

$$\Delta G_{bind} = G_{complex} - G_{receptor} - G_{ligand} \tag{1}$$

$$\Delta G_{bind} = E_{gas} + G_{sol} - TS \tag{2}$$

$$E_{gas} = E_{int} + E_{vdw} + E_{ele} \tag{3}$$

$$G_{sol} = G_{GB} + G_{SA} \tag{4}$$

$$G_{SA} = \gamma SASA \tag{5}$$

The terms Egas, Eint, Eele, and Evdw symbolize gas-phase energy, internal energy, Coulomb energy, and van der Waals energy. The Egas was directly assessed from the FF14SB force field terms. Solvation free energy (Gsol) was evaluated from the energy involvement of the polar states (GGB) and non-polar states (G). The non-polar solvation free energy (GSA) was determined from the solvent accessible surface area (SASA) [69,70] using a water probe radius of 1.4 Å. In contrast, solving the GB equation assessed the polar solvation (GGB) contribution. Items S and T symbolize the total entropy of the solute and temperature, respectively.

3.6. Statistical Analysis

The presented results were obtained from the mean values of three experimental processes independently performed. CC_{50} and IC_{50} values were determined from the curve of the dose response along with the regression analysis of the triplicates of the values. A one-way ANOVA followed by multiple Tukey's tests was applied for the comparison of the EOs with $p < 0.05$ significance in the antiviral assay.

4. Conclusions

The analysis of the EOs derived from the bark and fruits of *Acacia nilotica* revealed high relative concentrations of terpenoids in both oils. Stachene, caryophyllene oxide, spathulenol, trans-caryophyllene, Z-anethole, and γ-terpinene represented the main constituents. The EO of the bark exhibited moderate anti-HAV and anti-HSV1, while fruit EO showed weak effects against HAV, HSV1, and HSV2. Caryophyllene oxide and spathulenol exhibited the best affinities against the 3C protease and TK proteins. The molecular dynamics simulation proved the significant van der Waals energy of caryophyllene oxide with 3C protease of HAV and thymidine kinase of HSV enzyme. The present findings revealed the effects of the main constituents of *A. nilotica* EO. However, in vivo studies should be evaluated for these EOs and/or their major compounds, either in combination or singular, to determine the actual action mechanisms and safety.

Supplementary Materials: The following supporting information can be downloaded at: https://www.mdpi.com/article/10.3390/plants11212889/s1, Table S1: Three results of docking simulations of the major compounds identified in EOs of both bark and fruits of *Acacia nilotica*.

Author Contributions: Conceptualization, A.M.A.-E. and A.I.E.; Funding acquisition, S.F.A.; Investigation, A.E.-N.G.E.G., A.F.E., A.A.E.-R., A.M.E. (Abdelbaset M. Elgamal), D.D.K., A.M.A.-E., E.M.H. and A.I.E.; Methodology, A.E.-N.G.E.G., A.F.E., A.A.E.-R., A.M.E. (Abdelbaset M. Elgamal), D.D.K., E.M.H., A.M.A.-E. and A.I.E.; Visualization, A.F.E., A.M.A.-E. and A.I.E.; Writing—original draft, A.F.E., A.M.E. (Abdelbaset M. Elgamal), A.M.A.-E. and A.I.E.; Writing—review and editing, A.E.-N.G.E.G., A.F.E., A.A.E.-R., A.M.E. (Abdelbaset M. Elgamal), D.D.K., A.M.A.-E., E.M.H., A.M.E. (Abdallah M. Elgorban), N.S.Z., S.F.A. and A.I.E. All authors have read and agreed to the published version of the manuscript.

Funding: The authors extend their appreciation to The Researchers Supporting Project number (RSP-2021/241) King Saud University, Riyadh, Saudi Arabia.

Data Availability Statement: Not applicable.

Acknowledgments: The authors extend their appreciation to The Researchers Supporting Project number (RSP-2021/241) King Saud University, Riyadh, Saudi Arabia.

Conflicts of Interest: The authors declare no conflict of interest.

References

1. Ansari, M.; Sharififar, F.; Arabzadeh, A.M.; Mehni, F.; Mirtadzadini, M.; Iranmanesh, Z.; Nikpour, N. In vitro evaluation of anti-herpes simplex-1 activity of three standardized medicinal plants from Lamiaceae. *Anc. Sci. Life* **2014**, *34*, 33–38. [CrossRef] [PubMed]
2. Ubillas, R.; Jolad, S.; Bruening, R.; Kernan, M.; King, S.; Sesin, D.; Barrett, M.; Stoddart, C.; Flaster, T.; Kuo, J. SP-303, an antiviral oligomeric proanthocyanidin from the latex of *Croton lechleri* (Sangre de Drago). *Phytomedicine* **1994**, *1*, 77–106. [CrossRef]
3. Schnitzler, P.; Nolkemper, S.; Stintzing, F.; Reichling, J. Comparative in vitro study on the anti-herpetic effect of phytochemically characterized aqueous and ethanolic extracts of *Salvia officinalis* grown at two different locations. *Phytomedicine* **2008**, *15*, 62–70. [CrossRef] [PubMed]
4. Šudomová, M.; Berchová-Bímová, K.; Mazurakova, A.; Šamec, D.; Kubatka, P.; Hassan, S.T. Flavonoids target human herpesviruses that infect the nervous system: Mechanisms of action and therapeutic insights. *Viruses* **2022**, *14*, 592. [CrossRef]
5. Schuhmacher, A.; Reichling, J.; Schnitzler, P. Virucidal effect of peppermint oil on the enveloped viruses herpes simplex virus type 1 and type 2 in vitro. *Phytomedicine* **2003**, *10*, 504–510. [CrossRef] [PubMed]
6. De Logu, A.; Loy, G.; Pellerano, M.L.; Bonsignore, L.; Schivo, M.L. Inactivation of HSV-1 and HSV-2 and prevention of cell-to-cell virus spread by *Santolina insularis* essential oil. *Antivir. Res.* **2000**, *48*, 177–185. [CrossRef]
7. Reichling, J. Plant-microbe interactions and secondary metabolites with antibacterial, antifungal and antiviral properties. In *Functions Biotechnology of Plant Secondary Metabolites*; Wink, M., Ed.; Sheffield Academic Press: Sheffield, UK, 1999; Volume 39, pp. 214–347.
8. Jama-Kmiecik, A.; Sarowska, J.; Wojnicz, D.; Choroszy-Król, I.; Frej-Mądrzak, M. Natural products and their potential anti-HAV activity. *Pathogens* **2021**, *10*, 1095. [CrossRef] [PubMed]
9. Cuthbert, J.A. Hepatitis A: Old and New. *Clin. Microbiol. Rev.* **2001**, *14*, 38–58. [CrossRef] [PubMed]
10. Kesharwani, A.; Polachira, S.K.; Nair, R.; Agarwal, A.; Mishra, N.N.; Gupta, S.K. Anti-HSV-2 activity of *Terminalia chebula* Retz extract and its constituents, chebulagic and chebulinic acids. *BMC Complement. Altern. Med.* **2017**, *17*, 110. [CrossRef] [PubMed]
11. Van de Sand, L.; Bormann, M.; Schmitz, Y.; Heilingloh, C.S.; Witzke, O.; Krawczyk, A. Antiviral active compounds derived from natural sources against herpes simplex viruses. *Viruses* **2021**, *13*, 1386. [CrossRef]
12. Lyu, S.-Y.; Rhim, J.-Y.; Park, W.-B. Antiherpetic activities of flavonoids against herpes simplex virus type 1 (HSV-1) and type 2 (HSV-2) in vitro. *Arch. Pharmacal. Res.* **2005**, *28*, 1293–1301. [CrossRef] [PubMed]
13. Čulenová, M.; Sychrová, A.; Hassan, S.T.; Berchová-Bímová, K.; Svobodová, P.; Helclová, A.; Michnová, H.; Hošek, J.; Vasilev, H.; Suchý, P. Multiple In vitro biological effects of phenolic compounds from *Morus alba* root bark. *J. Ethnopharmacol.* **2020**, *248*, 112296. [CrossRef] [PubMed]
14. Koch, C.; Reichling, J.; Schneele, J.; Schnitzler, P. Inhibitory effect of essential oils against herpes simplex virus type 2. *Phytomedicine* **2008**, *15*, 71–78. [CrossRef] [PubMed]
15. Ma, L.; Yao, L. Antiviral effects of plant-derived essential oils and their components: An updated review. *Molecules* **2020**, *25*, 2627. [CrossRef]
16. Hassan, S.T.; Masarčíková, R.; Berchová, K. Bioactive natural products with anti-herpes simplex virus properties. *J. Pharm. Pharmacol.* **2015**, *67*, 1325–1336. [CrossRef] [PubMed]

17. Kanda, T.; Gauss-Müller, V.; Cordes, S.; Tamura, R.; Okitsu, K.; Shuang, W.; Nakamoto, S.; Fujiwara, K.; Imazeki, F.; Yokosuka, O. Hepatitis A virus (HAV) proteinase 3C inhibits HAV IRES-dependent translation and cleaves the polypyrimidine tract-binding protein. *J. Viral Hepat.* **2010**, *17*, 618–623. [CrossRef] [PubMed]
18. Xie, Y.; Wu, L.; Wang, M.; Cheng, A.; Yang, Q.; Wu, Y.; Jia, R.; Zhu, D.; Zhao, X.; Chen, S. Alpha-herpesvirus thymidine kinase genes mediate viral virulence and are potential therapeutic targets. *Front. Microbiol.* **2019**, *10*, 941. [CrossRef]
19. Raut, J.S.; Karuppayil, S.M. A status review on the medicinal properties of essential oils. *Ind. Crops Prod.* **2014**, *62*, 250–264. [CrossRef]
20. Carson, C.F.; Hammer, K.A. Chemistry and bioactivity of essential oils. *Lipids Essent Oils Antimicrob. Agents* **2011**, *25*, 203–238.
21. Abdallah, H.M.; Ammar, N.M.; Abdelhameed, M.F.; Gendy, A.E.-N.G.E.; Ragab, T.I.; Abd-ElGawad, A.M.; Farag, M.A.; Alwahibi, M.S.; Elshamy, A.I. Protective mechanism of *Acacia saligna* butanol extract and its nano-formulations against ulcerative colitis in rats as revealed via biochemical and metabolomic assays. *Biology* **2020**, *9*, 195. [CrossRef]
22. Rather, L.J.; Mohammad, F. *Acacia nilotica* (L.): A review of its traditional uses, phytochemistry, and pharmacology. *Sustain. Chem. Pharm.* **2015**, *2*, 12–30. [CrossRef]
23. Malviya, S.; Rawat, S.; Kharia, A.; Verma, M. Medicinal attributes of *Acacia nilotica* Linn.—A comprehensive review on ethnopharmacological claims. *Int. J. Pharm. Life Sci.* **2011**, *2*, 830–837.
24. Ogunbinu, A.; Okeniyi, S.; Flamini, G.; Cioni, P.; Ogunwande, I.; Babalola, I. Essential oil composition of *Acacia nilotica* Linn., and *Acacia albida* Delile (Leguminosae) from Nigeria. *J. Essent. Oil Res.* **2010**, *22*, 540–542. [CrossRef]
25. Ogunwande, I.A.; Matsui, T.; Matsumoto, K.; Shimoda, M.; Kubmarawa, D. Constituents of the essential oil from the leaves of *Acacia tortilis* (Forsk.) Hayne. *J. Essent. Oil Res.* **2008**, *20*, 116–119. [CrossRef]
26. Avoseh, O.N.; Oyedeji, O.-o.O.; Aremu, K.; Nkeh-Chungag, B.N.; Songca, S.P.; Oluwafemi, S.O.; Oyedeji, A.O. Chemical composition and anti-inflammatory activities of the essential oils from *Acacia mearnsii* De Wild. *Nat. Prod. Res.* **2015**, *29*, 1184–1188. [CrossRef]
27. El Ayeb-Zakhama, A.; Sakka-Rouis, L.; Bergaoui, A.; Flamini, G.; Ben Jannet, H.; Harzallah-Skhiri, F. Chemical composition and allelopathic potential of essential oils obtained from *Acacia cyanophylla* Lindl. cultivated in Tunisia. *Chem. Biodivers.* **2015**, *12*, 615–626. [CrossRef]
28. Singh, B.N.; Singh, B.R.; Sarma, B.; Singh, H. Potential chemoprevention of N-nitrosodiethylamine-induced hepatocarcinogenesis by polyphenolics from *Acacia nilotica* bark. *Chem. -Biol. Interact.* **2009**, *181*, 20–28. [CrossRef]
29. Singh, B.N.; Singh, B.; Singh, R.; Prakash, D.; Sarma, B.; Singh, H. Antioxidant and anti-quorum sensing activities of green pod of *Acacia nilotica* L. *Fooda Chem. Toxicol.* **2009**, *47*, 778–786. [CrossRef]
30. Ahmad, M.; Zaman, F.; Sharif, T.; Ch, M.Z. Antidiabetic and hypolipidemic effects of aqueous methanolic extract of *Acacia nilotica* pods in alloxan-induced diabetic rabbits. *Scand. J. Lab. Anim. Sci.* **2008**, *35*, 29–34.
31. Fatima, F.; Khalid, A.; Nazar, N.; Abdalla, M.; Mohomed, H.; Toum, A.M.; Magzoub, M.; Ali, M. In vitro assessment of anti-cutaneous leishmaniasis activity of some Sudanese plants. *Turk. Parazitol. Derg.* **2005**, *29*, 3–6.
32. El-Tahir, A.; Satti, G.M.; Khalid, S.A. Antiplasmodial activity of selected Sudanese medicinal plants with emphasis on *Acacia nilotica*. *Phytother. Res.* **1999**, *13*, 474–478. [CrossRef]
33. Essa, A.F.; El-Hawary, S.S.; Abd-El Gawad, A.M.; Kubacy, T.M.; AM El-Khrisy, E.E.D.; Elshamy, A.I.; Younis, I.Y. Prevalence of diterpenes in essential oil of *Euphorbia mauritanica* L.: Detailed chemical profile, antioxidant, cytotoxic and phytotoxic activities. *Chem. Biodivers.* **2021**, *18*, e2100238. [CrossRef] [PubMed]
34. Abd-ElGawad, A.M.; Elshamy, A.I.; El-Nasser El Gendy, A.; Al-Rowaily, S.L.; Assaeed, A.M. Preponderance of oxygenated sesquiterpenes and diterpenes in the volatile oil constituents of *Lactuca serriola* L. revealed antioxidant and allelopathic activity. *Chem. Biodivers.* **2019**, *16*, e1900278. [CrossRef]
35. Abd-ElGawad, A.M.; El Gendy, A.E.-N.G.; Assaeed, A.M.; Al-Rowaily, S.L.; Alharthi, A.S.; Mohamed, T.A.; Nassar, M.I.; Dewir, Y.H.; Elshamy, A.I. Phytotoxic effects of plant essential oils: A systematic review and structure-activity relationship based on chemometric analyses. *Plants* **2020**, *10*, 36. [CrossRef]
36. Jun, Y.; Lee, S.M.; Ju, H.K.; Lee, H.J.; Choi, H.-K.; Jo, G.S.; Kim, Y.-S. Comparison of the profile and composition of volatiles in coniferous needles according to extraction methods. *Molecules* **2016**, *21*, 363. [CrossRef] [PubMed]
37. Block, S.; Flamini, G.; Brkic, D.; Morelli, I.; Quetin-Leclercq, J. Analysis of the essential oil from leaves of *Croton zambesicus* Muell. Arg. growing in Benin. *Flavour Fragr. J.* **2006**, *21*, 222–224. [CrossRef]
38. Vernin, G.; Metzger, J.; Mondon, J.-P.; Pieribattesti, J.-C. GC/MS analysis of the leaf oil of *Cryptomeria japonica* D. Don from Reunion Island. *J. Essent. Oil Res.* **1991**, *3*, 197–207. [CrossRef]
39. Salasiah, M.; Alona, C.; Meekiong, K. Essential oil components in selected species of *Alpinieae* (zingiberaceae) from Sarawak and its taxonomic correlation. *J. Trop. For. Sci.* **2022**, *34*, 221–235. [CrossRef]
40. Elshamy, A.I.; Abd-ElGawad, A.M.; El Gendy, A.E.N.G.; Assaeed, A.M. Chemical characterization of *Euphorbia heterophylla* L. essential oils and their antioxidant activity and allelopathic potential on *Cenchrus echinatus* L. *Chem. Biodivers.* **2019**, *16*, e1900051. [CrossRef] [PubMed]
41. Al-Rowaily, S.L.; Abd-ElGawad, A.M.; Assaeed, A.M.; Elgamal, A.M.; Gendy, A.E. N.G.E.; Mohamed, T.A.; Dar, B.A.; Mohamed, T.K.; Elshamy, A.I. Essential oil of *Calotropis procera*: Comparative chemical profiles, antimicrobial activity, and allelopathic potential on weeds. *Molecules* **2020**, *25*, 5203. [CrossRef] [PubMed]

42. Nassar, M.I.; Yassine, Y.M.; Elshamy, A.I.; El-Beih, A.A.; El-Shazly, M.; Singab, A.N.B. Essential oil and antimicrobial activity of aerial parts of *Cyperus leavigatus* L.(Family: Cyperaceae). *J. Essent. Oil Bear. Plants* **2015**, *18*, 416–422. [CrossRef]
43. Elshamy, A.I.; Abd-ElGawad, A.M.; El-Amier, Y.A.; El Gendy, A.E.N.G.; Al-Rowaily, S.L. Interspecific variation, antioxidant and allelopathic activity of the essential oil from three *Launaea* species growing naturally in heterogeneous habitats in Egypt. *Flavour Fragr. J.* **2019**, *34*, 316–328. [CrossRef]
44. Abd-ElGawad, A.M.; El-Amier, Y.A.; Assaeed, A.M.; Al-Rowaily, S.L. Interspecific variations in the habitats of *Reichardia tingitana* (L.) Roth leading to changes in its bioactive constituents and allelopathic activity. *Saudi J. Biol. Sci.* **2020**, *27*, 489–499. [CrossRef]
45. Wardana, A.P.; Aminah, N.S.; Rosyda, M.; Abdjan, M.I.; Kristanti, A.N.; Tun, K.N.W.; Choudhary, M.I.; Takaya, Y. Potential of diterpene compounds as antivirals, a review. *Heliyon* **2021**, *7*, e07777. [CrossRef] [PubMed]
46. Zhang, G.-J.; Li, Y.-H.; Jiang, J.-D.; Yu, S.-S.; Wang, X.-J.; Zhuang, P.-Y.; Zhang, Y.; Qu, J.; Ma, S.-G.; Li, Y. Diterpenes and sesquiterpenes with anti-Coxsackie virus B3 activity from the stems of *Illicium jiadifengpi*. *Tetrahedron* **2014**, *70*, 4494–4499. [CrossRef]
47. Astani, A.; Reichling, J.; Schnitzler, P. Screening for Antiviral Activities of Isolated Compounds from Essential Oils. *Evid. -Based Complement. Altern. Med.* **2011**, *2011*, 253643. [CrossRef]
48. Ogawa, K.; Nakamura, S.; Hosokawa, K.; Ishimaru, H.; Saito, N.; Ryu, K.; Fujimuro, M.; Nakashima, S.; Matsuda, H. New diterpenes from *Nigella damascena* seeds and their antiviral activities against herpes simplex virus type-1. *J. Nat. Med.* **2018**, *72*, 439–447. [CrossRef]
49. Vallim, M.A.; Barbosa, J.E.; Cavalcanti, D.N.; De-Paula, J.C.; Silva, V.; Teixeira, V.L.; Paixão, I. In vitro antiviral activity of diterpenes isolated from the Brazilian brown alga *Canistrocarpus cervicornis*. *J. Med. Plants Res.* **2010**, *4*, 2379–2382.
50. Dunkić, V.; Vuko, E.; Bezić, N.; Kremer, D.; Ruščić, M. Composition and antiviral activity of the essential oils of *Eryngium alpinum* and *E. amethystinum*. *Chem. Biodivers.* **2013**, *10*, 1894–1902.
51. Farag, R.S.; Shalaby, A.S.; El-Baroty, G.A.; Ibrahim, N.A.; Ali, M.A.; Hassan, E.M. Chemical and biological evaluation of the essential oils of different *Melaleuca* species. *Phytother. Res.* **2004**, *18*, 30–35. [CrossRef]
52. Venturi, C.R.; Danielli, L.J.; Klein, F.; Apel, M.A.; Montanha, J.A.; Bordignon, S.A.; Roehe, P.M.; Fuentefria, A.M.; Henriques, A.T. Chemical analysis and in vitro antiviral and antifungal activities of essential oils from *Glechon spathulata* and *Glechon marifolia*. *Pharm. Biol.* **2015**, *53*, 682–688. [CrossRef]
53. Lai, W.-L.; Chuang, H.-S.; Lee, M.-H.; Wei, C.-L.; Lin, C.-F.; Tsai, Y.-C. Inhibition of herpes simplex virus type 1 by thymol-related monoterpenoids. *Planta Med.* **2012**, *78*, 1636–1638. [CrossRef]
54. Gavanji, S.; Sayedipour, S.S.; Larki, B.; Bakhtari, A. Antiviral activity of some plant oils against herpes simplex virus type 1 in Vero cell culture. *J. Acute Med.* **2015**, *5*, 62–68. [CrossRef]
55. Ćavar Zeljković, S.; Schadich, E.; Džubák, P.; Hajdúch, M.; Tarkowski, P. Antiviral activity of selected Lamiaceae essential oils and their monoterpenes against SARS-CoV-2. *Front. Pharmacol.* **2022**, *13*, 893634. [CrossRef] [PubMed]
56. Mirzaei, S.; Eisvand, F.; Hadizadeh, F.; Mosaffa, F.; Ghasemi, A.; Ghodsi, R. Design, synthesis and biological evaluation of novel 5, 6, 7-trimethoxy-N-aryl-2-styrylquinolin-4-amines as potential anticancer agents and tubulin polymerization inhibitors. *Bioorganic Chem.* **2020**, *98*, 103711. [CrossRef]
57. Hasanin, M.; Hashem, A.H.; El-Rashedy, A.A.; Kamel, S. Synthesis of novel heterocyclic compounds based on dialdehyde cellulose: Characterization, antimicrobial, antitumor activity, molecular dynamics simulation and target identification. *Cellulose* **2021**, *28*, 8355–8374. [CrossRef]
58. Bergmann, E.M.; Cherney, M.M.; Mckendrick, J.; Frormann, S.; Luo, C.; Malcolm, B.A.; Vederas, J.C.; James, M.N. Crystal structure of an inhibitor complex of the 3C proteinase from hepatitis A virus (HAV) and implications for the polyprotein processing in HAV. *Virology* **1999**, *265*, 153–163. [CrossRef]
59. Champness, J.N.; Bennett, M.S.; Wien, F.; Visse, R.; Summers, W.C.; Herdewijn, P.; De Clercq, E.; Ostrowski, T.; Jarvest, R.L.; Sanderson, M.R. Exploring the active site of herpes simplex virus type-1 thymidine kinase by X-ray crystallography of complexes with aciclovir and other ligands. *Proteins Struct. Funct. Bioinform.* **1998**, *32*, 350–361. [CrossRef]
60. Essa, A.F.; El-Hawary, S.S.; Emam, S.E.; Kubacy, T.M.; El-Khrisy, E.E.-D.A.; Younis, I.Y.; Elshamy, A.I. Characterization of undescribed melanoma inhibitors from *Euphorbia mauritanica* L. cultivated in Egypt targeting BRAFV600E and MEK 1 kinases via in-silico study and ADME prediction. *Phytochemistry* **2022**, *198*, 113154. [CrossRef]
61. Pettersen, E.F.; Goddard, T.D.; Huang, C.C.; Couch, G.S.; Greenblatt, D.M.; Meng, E.C.; Ferrin, T.E. UCSF Chimera—A visualization system for exploratory research and analysis. *J. Comput. Chem.* **2004**, *25*, 1605–1612. [CrossRef]
62. Hospital, A.; Goñi, J.R.; Orozco, M.; Gelpí, J.L. Molecular dynamics simulations: Advances and applications. *Adv. Appl. Bioinform. Chem.* **2015**, *8*, 37–47.
63. Wang, J.; Wang, W.; Kollman, P.A.; Case, D.A. Automatic atom type and bond type perception in molecular mechanical calculations. *J. Mol. Graph. Model.* **2006**, *25*, 247–260. [CrossRef]
64. Berendsen, H.J.; Postma, J.v.; van Gunsteren, W.F.; DiNola, A.; Haak, J.R. Molecular dynamics with coupling to an external bath. *J. Chem. Phys.* **1984**, *81*, 3684–3690. [CrossRef]
65. Roe, D.R.; Cheatham, T.E., III. PTRAJ and CPPTRAJ: Software for processing and analysis of molecular dynamics trajectory data. *J. Chem. Theory Comput.* **2013**, *9*, 3084–3095. [CrossRef] [PubMed]
66. Ylilauri, M.; Pentikäinen, O.T. MMGBSA as a tool to understand the binding affinities of filamin–peptide interactions. *J. Chem. Inf. Model.* **2013**, *53*, 2626–2633. [CrossRef]

67. Kollman, P.A.; Massova, I.; Reyes, C.; Kuhn, B.; Huo, S.; Chong, L.; Lee, M.; Lee, T.; Duan, Y.; Wang, W. Calculating structures and free energies of complex molecules: Combining molecular mechanics and continuum models. *Acc. Chem. Res.* **2000**, *33*, 889–897. [CrossRef]
68. Hou, T.; Wang, J.; Li, Y.; Wang, W. Assessing the performance of the MM/PBSA and MM/GBSA methods. 1. The accuracy of binding free energy calculations based on molecular dynamics simulations. *J. Chem. Inf. Model.* **2011**, *51*, 69–82. [CrossRef]
69. Sitkoff, D.; Sharp, K.A.; Honig, B. Accurate calculation of hydration free energies using macroscopic solvent models. *J. Phys. Chem.* **1994**, *98*, 1978–1988. [CrossRef]
70. Greenidge, P.A.; Kramer, C.; Mozziconacci, J.-C.; Wolf, R.M. MM/GBSA binding energy prediction on the PDBbind data set: Successes, failures, and directions for further improvement. *J. Chem. Inf. Model.* **2013**, *53*, 201–209. [CrossRef]

plants

MDPI

Article

Effect of *Lippia alba* (Mill.) N.E. Brown Essential Oil on the Human Umbilical Artery

Alex S. Borges [1,2], Carla M. S. Bastos [1,2], Debora M. Dantas [1,2], Cícera G. B. Milfont [2], Guilherme M. H. Brito [2], Luís Pereira-de-Morais [2], Gyllyandeson A. Delmondes [3], Renata E. R. da Silva [2], Emanuel Kennedy-Feitosa [4], Francisco P. A. Maia [5], Clara M. G. Lima [6], Talha Bin Emran [7,8], Henrique Douglas M. Coutinho [1,2,*], Irwin Rose A. Menezes [1,2], Marta R. Kerntopf [1,2], Gianluca Caruso [9,*] and Roseli Barbosa [1,2,*]

1 Biological Chemistry Department, Postgraduate Program in Biological Chemistry, Pimenta Campus, Regional University of Cariri, Crato 63105-010, Ceará, Brazil
2 Biological Sciences Department, Physiopharmacology of Excitable Cells Laboratory, Pimenta Campus, Regional University of Cariri, Crato 63105-010, Ceará, Brazil
3 Nursing Collegiate, Petrolina Campus, Federal University of The San Francisco Vale, Petrolina 56304-205, Pernambuco, Brazil
4 Health Science Department, Morphophysiopharmacology Laboratory, Federal Rural University of Semiarid, Mossoró 59625-900, Rio Grande do Norte, Brazil
5 CECAPE College, Juazeiro do Norte 63024-015, Ceará, Brazil
6 Department of Food Science, Federal University of Lavras, Lavras 37200-900, Minas Gerais, Brazil
7 Department of Pharmacy, BGC Trust University Bangladesh, Chittagong 4381, Bangladesh
8 Department of Pharmacy, Faculty of Allied Health Sciences, Daffodil International University, Dhaka 1207, Bangladesh
9 Department of Agricultural Sciences, University of Naples Federico II, 80055 Naples, Italy
* Correspondence: hdmcoutinho@gmail.com (H.D.M.C.); gcaruso@unina.it (G.C.); roselibarbo@gmail.com (R.B.)

Citation: Borges, A.S.; Bastos, C.M.S.; Dantas, D.M.; Milfont, C.G.B.; Brito, G.M.H.; Pereira-de-Morais, L.; Delmondes, G.A.; da Silva, R.E.R.; Kennedy-Feitosa, E.; Maia, F.P.A.; et al. Effect of *Lippia alba* (Mill.) N.E. Brown Essential Oil on the Human Umbilical Artery. *Plants* **2022**, *11*, 3002. https://doi.org/10.3390/plants11213002

Academic Editor: William N. Setzer

Received: 11 October 2022
Accepted: 4 November 2022
Published: 7 November 2022

Abstract: *Lippia alba* is popularly known as lemon balm, with its essential oil (EO) cited for displaying antimicrobial, sedative, and vasorelaxant effects. Yet, its action on isolated human vessels has not been described in the literature. Thus, we evaluated the vasorelaxant effect of essential oil of *L. alba* (EOLa) on human umbilical arteries (HUA) isolated in organ baths. HUA rings were isolated, subjected to contractions induced by potassium chloride (KCl), serotonin (5-HT), or histamine (HIST) to record the isometric tension, and then treated with EOLa (30–1000 μg/mL). The EOLa showed a more prominent inhibitory effect on the pharmacomechanical coupling contraction via HIST with an EC_{50} value of 277.1 ± 8.5 μg/mL and maximum relaxant effect at 600 μg/mL. The addition of tetraethylammonium (TEA) or 4-aminopyridine (4-AP) in HUA preparations did not inhibit EOLa total relaxant effect at 1000 μg/mL. In the presence of gliblenclamide (GLI), the oil relaxed the HUA rings by 90.8% at maximum concentration. The EOLa was also investigated for its effects on voltage-operated calcium channels (VOCCs), where the HUA preincubation with this oil at 1000 μg/mL inhibited $BaCl_2$ (0.1–30 mM)-induced contractions. This study demonstrates for the first time that EOla has a vasorelaxant effect on HUA and its particular blockade of VOCCs.

Keywords: essential oil; *Lippia alba*; human umbilical artery; vasorelaxant effect

1. Introduction

Essential oils (EOs) are aromatic complexes extracted from plants that contain a mixture of volatile organic compounds, which are present in the various organs of plants (roots, barks, leaves, flowers and fruits). These oils have a variety of pharmacological applications, as they have important biological properties for the plants themselves, for human beings and for other animals [1].

The species *Lippia alba* (Mill.) N.E. Brown belonging to the Verbenaceae family is found in Central America, South America, Africa, and Asia [2]. It is popularly known as lemon herb balm, Brazilian lemon balm, False-Melissa, Carmelite lemon balm, common lemon balm tea, field lemon balm and bush matgrass, among others [3,4]. The essential oil of this species is described as antibacterial [5–7], antifungal [8–11], anxiolytic, anticonvulsant [12,13], anesthetic [14–16], and smooth muscle relaxant [17–20]. Moreover, in folk medicine, this species is used to control blood pressure, as a calmant, and to treat respiratory diseases [21–23].

Searching for natural products that promote vasoactivity makes possible the discovery of therapeutic agents with a better efficacy and safety profile to treat disorders that cause hypertensive states. For example, pre-eclampsia is a hypertensive syndrome affecting the gestational period that increases the vascular umbilical resistance, resulting in a health risk to both the mother's and the baby's life [24,25].

The prevalence of pregnant women with hypertensive syndromes continues to increase, on the other hand, the most prescribed antihypertensive drugs are not recommended during pregnancy. There are few medications that are prescribed during pregnancy to control pre-eclampsia and eclampsia. However, the side effects of these are accounted for increasing the risk of fetal growth restriction and neonatal bradycardia, among other effects like depression for the mother [26,27].

Studies that prove the vasorelaxant potential of EOLa are only available on rodent tissue models [17–20]. Moreover, few are the studies that demonstrate the potential activity of natural products in human umbilical artery smooth musculature [28,29]. The use of human umbilical cord vessels as a model for screening substances with vasorelaxant properties is of great relevance for gestational hypertensive disorders, in addition to other cardiovascular diseases [30,31]. Herein, we show for the first time the vasorelaxant effect of *L. alba* essential oil on isolated human umbilical artery.

2. Results

2.1. *Effect of the Essential Oil of Lippia alba on Basal Vascular Tone in Human Umbilical Artery*

EOLa changed the final basal tone ($p < 0.017$), being significant at 1000 µg/mL compered to the control ($p < 0.001$) (Figure 1).

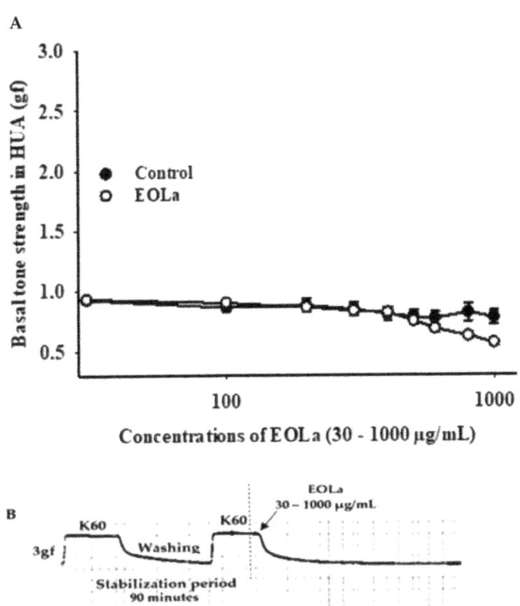

Figure 1. Effect of EOLa on basal vascular tone in HUA. (**A**) "Log-normal" plot for the effect of EOLa (30–1000 µg/mL) on basal HUA tone. The values are expressed as the mean ± SEM; n = 6 ($p < 0.001$,

one-way ANOVA). (**B**) Original register in LabChart Pro software for relaxant effect of EOLa on vascular basal tone in HUA.

2.2. Relaxant Effect of EOLa on Contractions Induced by KCl (60 mM), 5-HT (10 μM), and HIST (10 μM) in HUA

When evaluating the effect of EOLa on electromechanical coupling produced by KCl at 60 mM (K60), vasorelaxant activity was observed from 100 μg/mL ($p < 0.017$, one-way ANOVA) with EC_{50} at 377.0 ± 4.3 μg/mL (Figure 2).

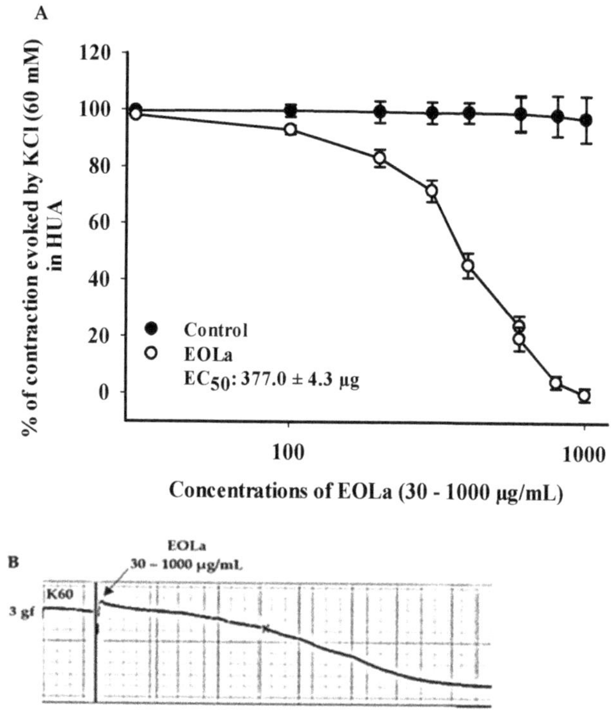

Figure 2. Relaxing effect of EOLa and evaluation of K60 electromechanical coupling in HUA. (**A**) "Log-normal" plot for EOLa (30–1000 μg/mL) on contractions induced by potassium chloride in HUA; values are expressed as the mean ± SEM; n = 6 ($p < 0.017$, one-way ANOVA). (**B**) Original registration made by the LabChart Pro software for the relaxing effect of EOLa on the electromechanical coupling induced by K60 in HUA.

When evaluating the pharmacomechanical pathway, it was observed that the EOLa promoted total blockage of 5-HT (10 μM)- and HIST (10 μM)-induced contractions in HUA rings. The blockade proved to be significant starting at a concentration of 200 μg/mL ($p < 0.002$, one-way ANOVA), with an EC_{50} value of 339.8 ± 4.5 μg/mL in experiments using the 5-HT agonist (Figure 3). When the HIST agonist was used, we observed a significant EOLa effect starting at a concentration of 200 μg/mL ($p < 0.001$ one-way ANOVA), with EC_{50} 277.1 ± 8.5 μg/mL, and total relaxation of the HUA rings at an EOLa concentration of 600 μg/mL. The control preparations—contracted using HIST—presented significant degradation; the effect of EOLa is also represented in the bar graph (Figures 4 and 5).

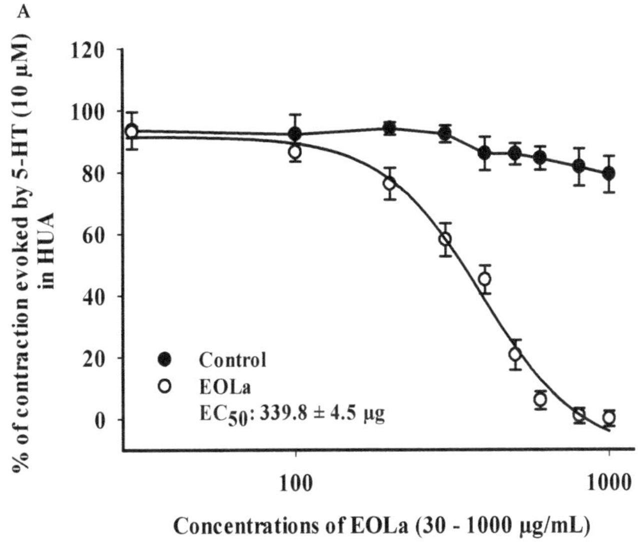

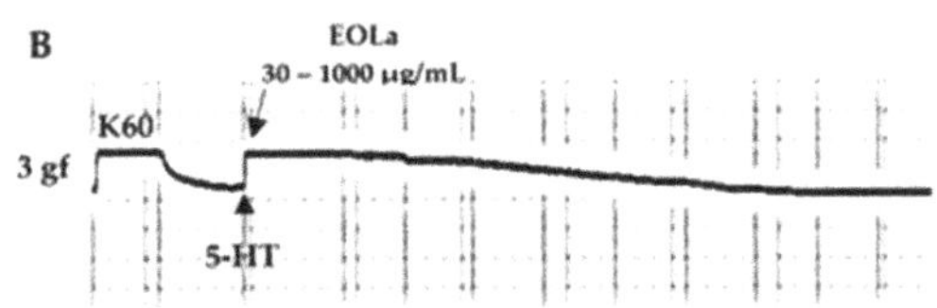

Figure 3. Relaxant effect of EOLa and evaluation of pharmacomechanical coupling by 5-HT (10 µM) in HUA. (**A**) "Log-normal" plot for the effect of EOLa (30–1000 µg/mL) on serotonin-induced contractions in HUA; values are expressed as the mean ± SEM; n = 6 ($p < 0.01$, one-way ANOVA). (**B**) Original recording in LabChart Pro software for relaxant effect of EOLa on the pharmacomechanical coupling induced by 5-HT in HUA.

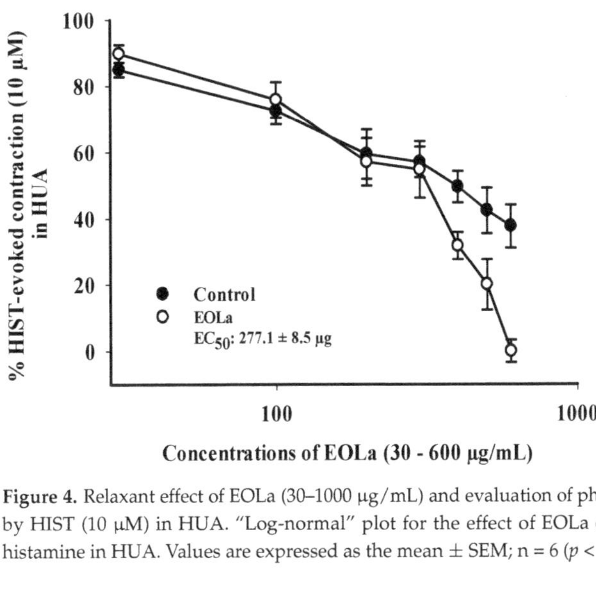

Figure 4. Relaxant effect of EOLa (30–1000 µg/mL) and evaluation of pharmacomechanical coupling by HIST (10 µM) in HUA. "Log-normal" plot for the effect of EOLa on contractions induced by histamine in HUA. Values are expressed as the mean ± SEM; n = 6 ($p < 0.01$, one-way ANOVA).

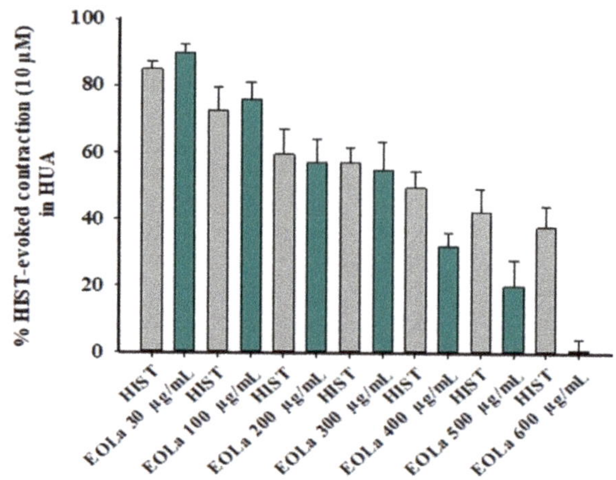

Figure 5. Relaxant effect of EOLa (30–600 μg/mL) and evaluation of pharmacomechanical coupling by HIST (10 μM) in HUA. Concentration–response bar graph for the effect of EOLa on contractions induced by histamine in HUA. Values are expressed as the mean ± SEM; n = 6 ($p < 0.01$, one-way ANOVA).

2.3. Relaxant Effect of EOLa on High Conductance K^+ Channels Activated by Ca^{2+} (BKCa), Voltage-Operated K^+ Channels (Kv) and ATP-Sensitive K^+ Channels (K_{ATP})

EOLa (30–1000 μg/mL) relaxed 100% of the preincubated HUA preparations with TEA. The EOla initial significant concentration was 200 μg/mL for both TEA concentrations used: TEA (1 mM)—$p < 0.023$, one-way ANOVA; EC_{50} of 471.2 ± 5.8 μg/mL; TEA (10 mM)—$p < 0.017$, one-way ANOVA; EC_{50} of 526.4 ± 4.0 μg/mL (Figure 6). The EC_{50} values for EOla obtained in preincubated HUA rings with 4-AP (1 mM) or GLI (10 μM) preparations were 494.0 ± 8.8 μg/mL and 530.8 ± 6.5 μg/mL, respectively. The concentrations statistically significant started at 200 μg/mL and 400 μg/mL ($p < 0.001$, one-way ANOVA), respectively. In the presence of 4-AP, EOLa achieved 100% of relaxation with a higher EC_{50} value. However, in the presence of GLI, EOLa performed 90.8% of relaxation, demonstrating some participation of K_{ATP} channels in its vasodilatory response (Figure 7).

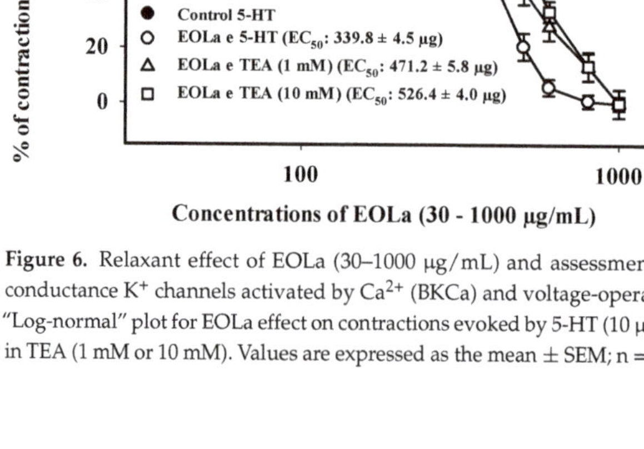

Figure 6. Relaxant effect of EOLa (30–1000 μg/mL) and assessment of the participation of high conductance K^+ channels activated by Ca^{2+} (BKCa) and voltage-operated K^+ channels (*Kv*) in HUA. "Log-normal" plot for EOLa effect on contractions evoked by 5-HT (10 μM) in HUA rings preincubated in TEA (1 mM or 10 mM). Values are expressed as the mean ± SEM; n = 6 ($p < 0.05$, one-way ANOVA).

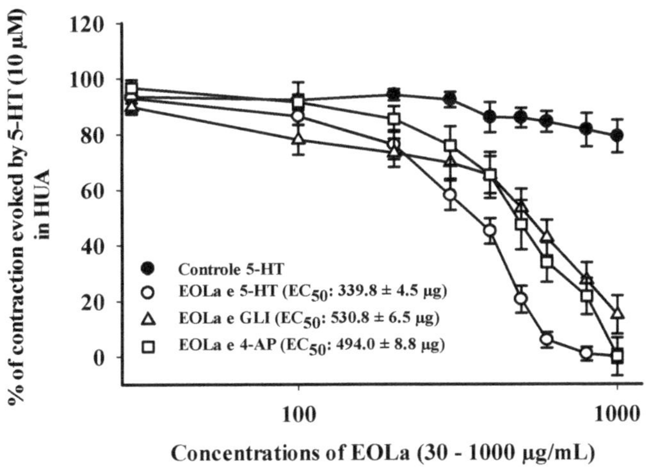

Figure 7. Relaxant effect of EOLa (30–1000 μg/mL) and evaluation of the participation of voltage-operated K^+ channels (*Kv*) and ATP-sensitive K^+ channels (K_{ATP}) in HUA. "Log-normal" plot for the effect of EOLa on stimulated contractions using 5-HT (10 μM) in HUA rings preincubated with 4-AP (1 mM) or GLI (10 μM). Values are expressed as the mean ± SEM; n = 6 ($p < 0.05$, one-way ANOVA).

2.4. *Effect of EOLa on Voltage-Operated Ca^{2+} Channels (VOCCs)*

Preincubation with EOLa (1000 μg/mL) reduced the contraction response of HUA rings to $BaCl_2$ (0.1–30 mM). This reduction was similar to that observed in rings preincubated with nifedipine (10 μM), a selective L-type VOCCs (Figure 8). Table 1 presents the main findings of this study, with EC_{50} values of the EOLa effect in the presence of each contractile agonist, as well as K^+ channel blockers.

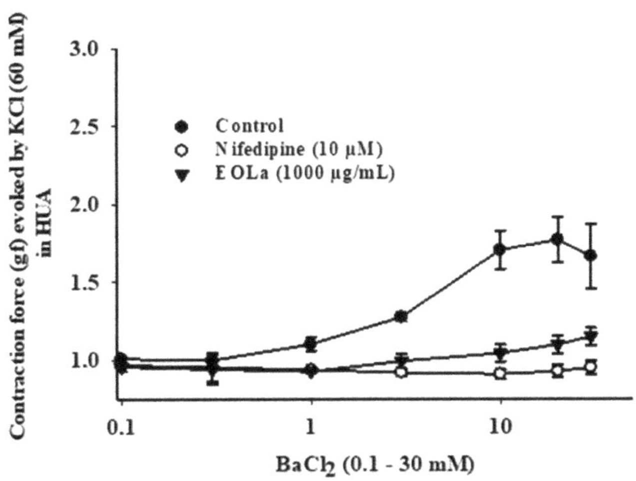

Figure 8. Relaxant effect of EOLa (1000 μg/mL) and assessment of participation of voltage-operated calcium channels (VOCCs) in HUA. Control: "Log-normal" plot for the contractile effect by cumulative addition of $BaCl_2$ (0.1–30 mM) in HUA preparations. Nifedipine: Graph for inhibitory effect of nifedipine (10 μM) on induced contraction by cumulative addition of $BaCl_2$ (0.1–30 mM); positive control. EOLa: Graph for the contraction-blocking effect of EOLa (1000 μg/mL) on accumulative addition of $BaCl_2$ (0.1–30 mM). Values are expressed as the mean ± SEM; n = 6 ($p < 0.05$, one-way ANOVA).

Table 1. Contraction agonists and blocking agents (substances); concentration values capable of triggering 50% of relaxing response (EC_{50}); values of statistically significant initial concentrations (SSIC) and total relaxation (%) promoted by the *Lippia alba* essential oil in the presence of differing contractile agents and channel blockers.

Substances	EC_{50} of EOLa	SSIC, Total Relaxation (%)
Potassium chloride (60 mM)	377.0 ± 4.3 μg/mL	100 μg/mL ($p < 0.017$), 100%
Serotonin (10 μM)	339.8 ± 4.5 μg/mL	200 μg/mL ($p < 0.001$), 100%
Histamine (10 μM)	277.1 ± 8.5 μg/mL	200 μg/mL ($p < 0.001$), 100%
Tetraethylammonium (1 mM)	471.2 ± 5.8 μg/mL	200 μg/mL ($p < 0.023$), 100%
Tetraethylammonium (10 mM)	526.4 ± 4.0 μg/mL	200 μg/mL ($p < 0.017$), 100%
Glibenclamide (10 μM)	530.8 ± 6.5 μg/mL	200 μg/mL ($p < 0.008$), 90,8%
4-Aminopyridine (1 mM)	494.0 ± 8.8 μg/mL	400 μg/mL ($p < 0.002$), 100%

3. Discussion

Lippia alba is well known for its essential oils, and various concentrations of compounds have been isolated and identified deriving from different parts of the plant. Terpenes and its derivatives are the major constituents, generally known for their applications as flavors and fragrances. The structures of these compounds are variable, and their classification is based on the number of carbon chains and chemical function presented. The main types of terpenes found in *L. alba* are the monoterpenes and sesquiterpenes [2].

This is the first study investigating the effect of EOLa on the contractility of human umbilical arteries. EOLa under study has as major constituents the monoterpenoid citral (a mixture of two aldehyde geometric isomers, E-geranial (41.81%) and Z-neral (34.11%)), the monoterperne 1-limonene (a cyclic unsaturated hydrocarbonate), and the monoterpenoid carvone (a cyclic unsaturated oxygenated ketone).

Published data with different rat isolated smooth muscles show that EOLa has a relaxant effect with possible main mechanism of action by blocking L-type voltage-operated Ca^{2+} channels (VOCCs) [17,19,20,32], thus supporting our results. Regarding their various electrophysiological and pharmacological properties, VOCCs were initially classified as types L, N, P/Q, R, and T [33]. L-type VOCCs are designated as CaV1.1–1.4, N-type VOCCs are designated as CaV2.2, the P/Q type is designated as CaV 2.1, and type R is designated as CaV2.3, which are insensitive to dihydropyridine and activated by high voltage; lastly, T-type channels CaV3.1–3.3 are activated by low voltage [34,35]. In smooth muscle cells, L-type VOCCs are the most studied, since they have great importance in contraction processes [36,37]. In this study, EOLa presented a response similar to the positive control nifedipine (10 μM), a selective L-type VOCC blocker. Hence, we believe that these channels are prominently involved in the relaxant effect of EOLa on the electromechanical coupling contraction.

Carvalho et al. [17], used EOLa, citral, or 1-limonene on rat trachea demonstrating that the oil and citral alone had a similar relaxant effect over tracheal smooth muscle contracted by K60 electromechanical coupling, including with blockade of VOCCs similar to the positive control nifedipine (1 μM), thus inhibiting contractions by $BaCl_2$, as well as by acetylcholine on pharmacomechanical coupling. On the other hand, 1-limonene did not display great relaxant effect, while also not being unable to inhibit contractions evoked by $BaCl_2$.

Pereira-de-Morais et al. [19], used EOLa, citral, or 1-limonene on rat uterus, observing that the oil or each of its main terpene constituents produced a relaxant effect on contractions induced by various contractile agents. For both electromechanical and pharmacomechanical coupling via oxytocin, serotonin, and acetylcholine, all had the same efficacy on total contraction inhibition at the same concentration. However, analyzing the inhibitory concentrations (IC_{50}) used, there was a difference amongst them, with EOLa having lower IC_{50} values than citral and 1-limonene, suggesting that the other compounds that make up a smaller proportion of the composite essential oil may be active, possibly contributing to the relaxant effect.

Seres-Bokor et al. [38] suggested an alternative mechanism for the relaxing effect of citral on rat uterus. A 14 day treatment in vivo with this monoterpenoid induced an increase of aquaporin channel (AQP5) expression in uterine smooth muscle, reducing hypertonic stress activation on the transient receptor potential cation channel receptor type 4 (TRPV4).

Silva et al. [20] used OELa and citral on rat aorta artery, corroborating the findings of our study, since the effect observed for EOLa was quite significant in the electromechanical pathway, presenting even greater effect in the thoracic artery (EC_{50}: 83.30 μg/mL) than in the HUA (EC_{50}: 377.0 ± 4.3 μg/mL). However, these differences can be explained by the fact that the blood vessels of experimental animals and humans, specifically the HUA, are distinct from each other in terms of physiology, receptor types, and activity of specific signaling pathways [39].

Maynard et al. [32] used EOLa on rat mesenteric artery rings, showing an endothelium-independent vasorelaxant effect on both electromechanical by KCl (80 mM) and pharmacomechanical coupling by phenylephrine (1 μM), possibly due to an inhibition on the Ca^{2+} influx through the L-type VOCCs. The EO applied in this study also has the constituent citral in majority.

Few studies have reported the vasorelaxant activities of plant natural products in human umbilical vessels. Among them, one that shows *Cymbopogon citratus* extract, rich in polyphenols (chlorogenic acid, iso-orientin, and swertiajaponin), is able to inhibit vasoconstriction induced by the thromboxane A2 receptor agonist U46619 [40]. Another with octylmethoxycinnamate that activates soluble guanylate cyclase and inhibits VOCCs, subsequently causing vasorelaxation [41]. This mechanism was also observed for genistein on HUA, a natural phytoestrogen belonging to the isoflavone group [29].

It is known that HUA vascular responsiveness occurs more effectively to contracting agents than to relaxing agents [42]. Since umbilical blood vessels are not innervated, control of umbilical blood flow depends on vasoactive substances released locally or already existing in circulation such serotonin, histamine, oxygen, nitric oxide (NO), thromboxane, and ions (e.g., calcium and potassium) [43–46]. Furthermore, 5-HT and HIST are the most potent vasoconstrictors in umbilical human arteries, and increased release of these mediators, as well as increased sensitivity in HUA to these agents, has already been related to certain pathological processes disturbing the umbilical cord circulation (e.g., preeclampsia) [47,48]. Thus, mechanisms that regulate HUA smooth muscle contractility are of paramount importance for placental exchange of both gases and nutrients with the fetus.

The EOLa was able to relax the smooth musculature in HUA with greater potency on 5-HT-evoked contractions (EC50 = 339.8 μg/mL ± 4.5 μg/mL) than on K60-evoked contractions (EC50 = 377.0 μg/mL ± 4.3 μg/mL). An even more promising relaxing effect of EOLa was demonstrated by the histaminergic coupling protocol, with an EC_{50} of 277.1 ± 8.5 μg/mL. This difference suggests that it also acts on distal targets of intracellular cascades activated on the pharmacomechanical pathway. Silva et al. [20] demonstrated that EOLa and citral equally inhibited contraction evoked by phorbol-12,13-dibutyrate—an activator of intracellular calcium-binding proteins, such as protein-kinase C (PKC), which are active in contraction of smooth muscle—in rat aorta arteries isolated in zero-calcium medium.

Potassium channels play a key role in membrane potential regulation and cell excitability, with smooth muscle contraction dependent on the equilibrium between increasing K^+ ion conductance, leading to hyperpolarization, and decreasing K^+ conductance (leading to depolarization) [49,50]. In smooth muscle, the basal tone can be regulated by several subtypes of K^+ channels. Using TEA (10 mM) as a nonselective K^+ channel blocker, including blockade of large conductance K^+ channels activated by Ca^{2+} (BKCa), and TEA (1 mM), as a more sensible blockade on voltage-operated K^+ channels (K*v*), an increase in the EC_{50} values was observed; however, the oil promoted total relaxation at the maximum concentration used. A similar effect was seen for the monoterpenoid carveol on human umbilical arteries [28].

Comparing the relaxing potential of EOla in the presence of 4-AP (a K*v* blocker) or GLI (a blocker for ATP-sensitive K^+ channels (K_{ATP})) we observed an expressed increase for EC_{50} value when using GLI. In addition, there was a 9.2% decrease in the relaxant potential of the EOLa at maximum concentration. This indicates the influence of these pathways, especially K_{ATP}, on the relaxant effect of EOLa in human umbilical arteries.

Potassium channels activated by intracellular metabolites such as ATP can hyperpolarize the membrane, inhibiting Ca^{2+} influx by blocking VOCCs and, consequently, favoring relaxation of smooth muscle cells. K_{ATP} is designed to have the ability to decrease its activity when the intracellular levels of ATP increase. This activity can also be modulated by ATP-independent signaling pathways [51]. In a study carried out with labdane-302, a diterpene isolated from *Xylopia langsdorfian* A. St.-Hil. & Tul., the involvement of K_{ATP} in promoting its relaxant effect on guinea pig ileum was also verified [52].

The actions of K_{ATP} are ambiguous, and its blocking or activation can be beneficial or harmful depending on the pathology and the channel subunit expressed in the target tissue. The absence and/or dysfunction of this channel and its subunit SUR2 are implicated in increased cardiovascular risks and poor prognosis in cases of infarction and stroke, while the increase in its expression and its adequate function together bring cardio and vasoprotective effects. In neurodegenerative diseases, such as Alzheimer's and Parkinson's, for example, inhibition of this channel seems to be beneficial, with improvement of the underlying neuroinflammatory markers of these pathologies [53,54]. There is no consensus on the role of this channel in HUA contraction/relaxation process, and the clinical importance is still unknown relating to pregnancy-specific hypertensive diseases [51].

Many EOs are generally recognized as safe [55]. However, the use of herbs and EOs is still a highly controversial matter, and their use in clinical practice is still restricted due to their physicochemical properties (e.g., limited bioavailability) and/or their toxicity. On the other hand, they can often be excellent leads for the development of new drugs. Modifying and/or isolating the structure of these products is a strategic way to increase pharmacological action, as well as improve absorption, distribution, metabolism, and excretion properties, thus decreasing toxicity and side-effects [56]. In this context, basic research initially evaluates the pharmacological potential of such compounds, favoring decision making for clinic research aiming for their therapeutic applicability [57].

Investigating bioactive and umbilical cord artery smooth muscle opens perspectives for potential agents capable of modulating the contractile activity. The vascular smooth muscle is important for homeostasis, and several pathologies may be related to changes in vascular tone. These include hypertensive disorders, such as those afflicting pregnant women, in addition to atherosclerosis, heart failure, and ischemia [20].

We conclude that *Lippia alba* (Mill.) N.E. Brown essential oil has a vasorelaxant effect on the contractility of human umbilical artery smooth muscle. EOLa was able to inhibit contractions induced by physiological contracting agents of HUA, 5-HT and HIST, with greater potency on the pharmacomechanical coupling via the histaminergic pathway. We highlight its inhibitory action on L-type VOCCs in electromechanical coupling by K60, similar to the standard drug nifedipine. Our study also showed that K_{ATP} channels play a discreet mediation in the relaxing effect of EOLa on HUA.

It is worth to mention that, in tissue viability investigations at the end of each experiment, full contractions were again induced by the same agonists, thus indicating that EOLa has no acute toxic effects on HUA contractility and survival. This same procedure was conducted by Carvalho et al. [17], Pereira-de-Morais et al. [19], and Silva et al. [20] using EOLa, citral, and 1-limonenedone on rodent smooth muscle, again corroborating our results of EOLa on HUA.

In conclusion, our results demonstrate the relaxing effect of EOLa on human umbilical cord arteries, encouraging further studies to be conducted on HUA using the major constituents isolated from EOLa, citral, and 1-limonene. Therefore, the hypothesis is that citral is primarily responsible for the relaxing effect of *L. alba* essential oil on HUA.

4. Materials and Methods

4.1. Solutions and Drugs

The drugs and reagents used presented analytical purity, being obtained from Sigma Chemical Corporation (St. Louis, MI, USA) or Merck (Darmstadt, Germany), and they were maintained under conditions indicated by the manufacturer. To prepare the solutions in this study, the salts used were potassium chloride (KCl), sodium chloride (NaCl), magnesium sulfate ($MgSO_4$), calcium chloride ($CaCl_2$), glucose ($C_6H_{12}O_6$), potassium phosphate (KH_2PO_4), sodium carbonate ($NaHCO_3$), barium chloride ($BaCl_2$), ethylenediaminetetraacetic acid (EDTA), and 2-[4-(2-hydroxyethyl)piperazin-1-yl]-ethanesulfonic acid (HEPES). The concentrations were expressed in millimole/liter (mM/L).

The substances serotonin and tetraethylammonium were dissolved in distilled water, and nifedipine was diluted in ethanol; the solutions obtained were kept at 0–4 °C and only withdrawn at the moment of the experiment.

4.2. *Lippia alba (Mill.) N.E. Brown Essential Oil*

The oil, 99% pure, was purchased from Dr. Sergio Horta at the experimental farm of the Federal University of Ceará (UFC). A dried-up sample of *L. alba* species was deposited in the Prisco Bezerra Herbarium of UFC under identification code #EAC-08474.

Leaves (1 kg) were collected on the same day and hour, from flowering plants, and were steam distilled for 1 h for EO acquisition. The chemical constitution was evaluated in the Natural Products Laboratory by Dr. Afrânio Aragão Craveiro at the Technological Development Park (PADETEC) of UFC using gas chromatography coupled to mass spectrometry (GC–MS, Hewlett-Packard 6971, Harris County, TX, USA). Analysis conditions were as follows: dimethylpolysiloxane DB-1 fused silica capillary column (30 m × 0.25 mm; 0.1 μm); helium (1 mL/min) as carrier gas; injector temperature, 250 °C; detector temperature, 280 °C; column temperature, 35–180 °C at 4 °C/min and 180–250 °C at 10 °C/min; mass spectrometry electronic impact, 70 eV. The compounds were identified using mass spectral library MS searches and Kovats indices as a preselection aid [58], and visual mass spectra were compered to data from the literature for confirmation [59–61].

The following major constituents were encountered: citral (75.92%) (a mixture of E-geranial (41.81%) and Z-neral (34.11%)), 1-limonene (9.85%), and carvone (8.92%) (Table 2).

Table 2. *Lippia alba* essential oil constituents.

Constituent	Content (%)
E-Geranial	41.81
Z-Neral	34.11
Limonene	9.85
Carvone	8.92
Gamma-terpinene	2.05
Benzene-1-methyl-3-(1-methylethyl)	1.00
6-methyl-5-hepten-2-one	0.72
Alpha-humulene	0.58
Linalool	0.50
Beta-pinene	0.47

To analyze its activity on HUA, the EOLa was prepared as a solution, diluted directly in Krebs Henseleit with 3% Tween 80® [28]. Administration was given hypertonically, analyzing the volume in each cuvette to obtain the final concentrations used inside of the organ bath chambers. A response curve was performed for EOLa concentrations of 30, 100, 200, 300, 400, 600, 800, and 1000 μg/mL.

4.3. Tissue Preparation and Isolation

Collection and processing of samples was approved by the Ethics Committee in Human Research of the Regional University of Cariri—URCA (n° 3.832.881), and by the Hospital and Maternity Camilo Ethics Committee. Fragments of approximately 10 cm of human umbilical cord (portions that would be destined for biological disposal), were obtained with the consent of healthy, normotensive mothers/donors and without any cord abnormalities, after childbirth (normal or cesarean). Samples were collected and stored in Krebs modified solution (composition in mM: NaCl, 125; KCl, 4.8; $CaCl_2$, 1; $MgSO_4$, 1.2; $NaHCO_3$, 25; KH_2PO4, 1.2; $C_6H_{12}O_6$, 11; HEPES, 25; EDTA, 0.3) refrigerated and transported to the Excitable Cells Physiopharmacology Laboratory of URCA. The strand segments were stored in a refrigerator and kept at 4–8 °C, being usable for 48 h after collection [30]. The human umbilical arteries were isolated from its covering tissue, i.e., Wharton's jelly, and cut into 3–4 mm rings.

4.4. Determination of Tension Exerted on the HUA Rings

After removal from Wharton's jelly and isolation, the HUAs were sectioned into rings. In thermostated organ bath equipment, measurements were performed using a stem connected to the force transducer (MLT0420, ADInstruments Bridge Amps, ADInstruments, Sydney, Australia) which allowed capturing measurements of the isometric tension produced by the HUA rings as converted into electrical signals. The transducer was connected to an amplifier (ADInstruments Bridge Amps), and this to the input of an analog-to-digital converter board (BCN/Pod port) installed in a computer. The collected data were converted into strokes and stored in LabChart Pro software (ADInstruments) files for later analysis.

The rings were individually suspended on stainless-steel hooks inserted into their lumens and mounted using an isometric tension of 3 g. This assembly was performed in glass chambers with 10 mL of Krebs Henseleit solution, kept at 37 °C, and bubbled with a carbogenic mixture (95% O_2; 5% CO_2). After the artery rings were assembled, the solution was renewed every 15 min.

After a stabilization period of 90 min, all protocols began with two subsequent contractions, produced by addition of 60 mM potassium chloride (K60) in hypertonic mode to the studied HUA rings, and, after reaching stable values, the maximum response obtained was considered the maximum contraction of each ring. Then the contraction inducers KCl (60 mM) or 5-HT (10 µM) or HIST (10 µM) were added to the preparations, followed by the addition of increasing and cumulative concentrations of EOLa (30, 100, 200, 300, 400, 600, 800, and 1000 µg/mL). For each new EOLa concentration, sufficient time was allowed for the response to reach a steady state, normally 5 to 15 min. Only experiments with reproducible contractions were considered viable for the experimental series.

Experimental Series

Series 1: EOLa Effect on the Basal Tone of HUA

To evaluate the effect of EOLa on the basal HUA tone, after verification of tissue viability, a concentration–effect curve was performed by addition of increasing and cumulative concentrations of EOLa (30–1000 µg/mL) to the HUA preparations to obtain a concentration–response curve.

Series 2: Effect of EOLa on contractions induced by KCl (60 mM), 5-HT (10 µM), and HIST (10 µM) in HUA

To investigate the effect of EOLa on the electromechanical coupling of the HUA rings, after verifying tissue viability, contractions were induced using KCl (60 mM), and increasing and cumulative concentrations of EOLa (30–1000 µg/mL) were added to the HUA preparations to obtain a single concentration–response curve.

To evaluate the oil on pharmacomechanical coupling, we used two classical contracting agents—5-HT (10 µM) and HIST (10 µM)—one for each experiment [39]. After verification of tissue viability, contraction was induced by one of the agents, and then increasing and cumulative concentrations of EOLa (30–1000 μg/mL) were added to the HUA preparations to obtain a curve of the concentration-response.

At the end of each experiment, washings were performed for 30 consecutive minutes, and then contractions were again induced by the same agonists to demonstrate tissue viability throughout the experiment.

Series 3: Effect of EOLa on K^+ channels in HUA

To evaluate the participation of K^+ channels in this series, three blockers were used—TEA (1 mM or 10 mM), 4-AP (1 mM), or GLI (10 μM). An experiment was carried out for each blocking agent and its respective standard action concentration(s) for the subtype of K^+ channel targeted.

TEA at a concentration of 10 mM is a nonselective potassium channel blocker, such as the high-conductance K^+ channels activated by Ca^{2+} (BKCa). Yet, at a concentration of 1 mM, TEA selectively blocks voltage-operated K^+ channels (Kv). In both experiments, at their respective concentrations, TEA was added and incubated for 30 min. At the end of the incubation, the HUA rings were subjected to contractions induced by 5-HT (10 μM), and EOLa in concentrations of 30–1000 μg/mL was added to obtain a concentration-response curve.

The experiments with 4-AP and GLI, i.e., selective voltage-operated (Kv) and ATP-sensitive K^+ channel (K_{ATP}) blockers, respectively, followed the protocol of the TEA series as to incubation time, contraction agent, and EOLa concentrations used.

Series 4: Effect of EOLa on voltage-operated Ca^{2+} channels (VOCCs)

To investigate the involvement of L-type voltage-operated calcium channels, after verifying the tissue viability, HUA preparations were kept in Krebs Henseleit zero-calcium solution incubated with EOLa (1000 μg/mL) for 30 min before receiving $BaCl_2$ (0.1–30 mM). Barium ion (Ba^{2+}) is a selective voltage-operated Ca^{2+} channel agonist and induces contraction depending on concentration. In this series, the control preparations, without the presence of EOLa, reached maximum contraction at 30 mM of $BaCl_2$.

Nifedipine (10 μM), a dihydropyridine channel blocker, was also included as positive control and added to HUA preparations, without the presence of EOLa; after 10 min, a cumulative concentration–response curve was performed by the addition of $BaCl_2$ (0.1–30 mM).

4.5. Statistical Analysis

Data were expressed as the mean ± SEM. Sigma Plot 11.0 software (Systat Software Inc.; San Jose, CA, USA) was used for statistical analysis and graph production. The results that were considered statistically significant presented a null hypothesis probability of less than 5% ($p < 0.05$). Student's t-test and analysis of variance (one-way ANOVA) were followed (when appropriate) by Holm–Sidak and Bonferroni t-tests. The EC_{50} values were determined as the substance concentration able to produce 50% of the maximum effect obtained from the higher concentration used in each protocol. Normal logarithmic interpolation calculations were performed for each experiment when fitted.

The scale presented in the Y-axis graph sometimes had a broader range than the exact value used as the standard 100% contraction plateau in the stabilization period. This should be taken into account when analyzing Figure 1, where the Y-axis displays a top value of 3.0, which can be interpreted as 100% contraction made by the 3 g force upon the HUA rings. This is simply an adjustment made by the Sigma Plot 11.0 software to produce a better image for the graph, considering the value of 1 as 100% contraction obtained from the 3 g force applied to the HUA rings.

5. Conclusions

Lippia alba essential oil presents the vasorelaxant effect on human umbilical arteries, modulating both electromechanical and pharmacomechanical coupling. EOLa presents higher action via the pharmacomechanical pathway, as shown by the minor EC_{50} values recorded for both protocols using contraction agonists 5-HT and HIST, whereas its effect seems to be related mostly to the electromechanical pathway, as seen in its ability to block voltage-operated calcium channels.

Author Contributions: Conceptualization, A.S.B. and R.B.; methodology, C.M.S.B., D.M.D., and C.G.B.M.; software, L.P.-d.-M. and G.A.D.; validation, G.M.H.B. and R.E.R.d.S.; investigation, F.P.A.M. and C.M.G.L.; resources, E.K.-F. and T.B.E.; writing—original draft preparation, I.R.A.M. and M.R.K.; supervision, R.B.; project administration, H.D.M.C. and G.C. All authors have read and agreed to the published version of the manuscript.

Funding: This research received no external funding.

Institutional Review Board Statement: Approved by the Ethics Committee in Human Research of the Regional University of Cariri—URCA (n° 3.832.881).

Informed Consent Statement: The research obtained the consent of all participants and was approved by the Ethics Committee in Human Research of the Regional University of Cariri—URCA (n° 3.832.881).

Data Availability Statement: Not applicable.

Acknowledgments: The authors acknowledge the Cearense Foundation to Support Scientific and Technological Development (FUNCAP) Funding code BPI: BP3. 00139-00072.0200/18, Coordination for the Improvement of Higher Education Personnel (CAPES), National Council for Scientific and Technological Development (CNPq), and Regional University of Cariri (URCA).

Conflicts of Interest: The authors declare no conflict of interest.

References

1. Baser, K.H.C.; Buchbauer, G. *Handbook of Essential Oils: Science, Technology and Applications*, 1st ed.; CRC Press: London, UK; Taylor and Francis: London, UK, 2010; pp. 1–949.
2. Malik, S.; Odeyemi, S.; Pereira, G.C.; Junior, L.M.F.; Abdul-Hamid, H.; Atabaki, N.; Makhzoum, A.; Bezerra de Almeida, E., Jr.; Dewar, J.; Abiri, R. New insights into the biotechnology and therapeutic potential of *Lippia alba* (Mill.) N.E.Br. ex P. Wilson. *J. Essent. Oil Res.* **2021**, *33*, 523–535. [CrossRef]
3. Matos, F.J.A. *Medicinal Plants: Selection Chart and Employment Plants Used in Phytotherapy in Northeast Brazil*, 2nd ed.; University Press: Fortaleza, Bazil, 2000; pp. 1–346.
4. Pascual, M.E.; Slowing, K.; Carretero, E.; Mata, D.S.; Villar, A. *Lippia*: Traditional uses, chemistry and pharmacology: A review. *J. Ethnopharmacol.* **2001**, *76*, 201–214. [CrossRef]
5. Porfírio, E.M.; Melo, H.M.; Pereira, A.M.G.; Cavalcante, T.T.A.; Gomes, G.A.; De Carvalho, M.G.; Costa, R.A.; Júnior, F.E.A.C. In Vitro Antibacterial and Antibiofilm Activity of *Lippia alba* Essential Oil, Citral, and Carvone against Staphylococcus aureus. *Sci. World J.* **2017**, *2017*, 2496–2707. [CrossRef] [PubMed]
6. Teixeira de Oliveira, G.; Ferreira, J.M.S.; Lima, W.G.; Alves, L.F.; Duarte-Almeida, J.M.; Lima, L.A.R.D.S. Phytochemical characterisation and bioprospection for antibacterial and antioxidant activities of *Lippia alba* Brown ex Britton & Wilson (Verbenaceae). *Nat. Prod. Res.* **2018**, *32*, 723–731. [PubMed]
7. Mamun-Or-Rashid, A.N.M.; Islam, M.R.; Dash, B.K. In vitro Antibacterial Effect of Bushy Matgrass (*Lippia alba* Mill.) Extracts. *Res. J. Med. Plant* **2012**, *6*, 334–340.
8. Singh, G.; Rao, G.P.; Kapoor, P.S. Chemical constituents and antifungal activity of *Lippia alba* Mill. Leaf essential oil. *JMAPS* **2000**, *22*, 701–703.
9. Tomazoni, E.Z.; Pansera, M.R.; Pauletti, G.F.; Moura, S.; Ribeiro, R.T.; Schwambach, J. In vitro antifungal activity of four chemotypes of *Lippia alba* (Verbenaceae) essential oils against *Alternaria solani* (Pleosporeaceae) isolates. *An. Acad. Bras. Ciências* **2016**, *88*, 999–1010. [CrossRef]
10. Oliveira, G.T.; Ferreira, J.M.; Rosa, L.H.; Siqueira, E.P.; Johann, S.; Lima, L.A. In vitro antifungal activities of leaf extracts of *Lippia alba* (Verbenaceae) against clinically important yeast species. *Rev. Soc. Bras. Med. Trop.* **2014**, *47*, 247–250. [CrossRef]
11. Sales, G.; Medeiros, S.; Soares, I.; Sampaio, T.; Bandeira, M.; Nogueira, N.; Queiroz, M. Antifungal and Modulatory Activity of Lemon Balm (*Lippia alba* (MILL.) N. E. BROWN) Essential Oil. *Sci. Pharm.* **2022**, *90*, 31. [CrossRef]
12. Hatano, V.Y.; Torricelli, A.S.; Giassi, A.C. Anxiolytic effects of repeated treatment with an essential oil from *Lippia alba* end (R)-(-)-carvone in the elevated T-maze. *Braz. J. Med. Biol. Res.* **2012**, *3*, 238–243. [CrossRef]
13. Viana, G.S.; Vale, T.G.; Silva, C.M.; Matos, F.J. Anticonvulsant activity of essential oils and active principles from chemotypes of *Lippia alba* (Mill.) N.E. Brown. *Biol. Pharm. Bull.* **2000**, *23*, 1314–1317.
14. Kampke, E.H.; Barroso, M.E.S.; Marques, F.M.; Fronza, M.; Scherer, R.; Lemos, M.F.; Campagnaro, B.P.; Gomes, L.C. Genotoxic effect of *Lippia alba* (Mill.) N. E. Brown essential oil on fish (*Oreochromis niloticus*) and mammal (*Mus musculus*). *Environ. Toxicol. Pharmacol.* **2018**, *59*, 163–171. [CrossRef] [PubMed]

15. Souza, C.D.F.; Baldissera, M.D.; Bianchini, A.E.; Silva, E.G.; Mourão, R.H.V.; Silva, L.V.F.; Schmidt, D.; Heinzmann, B.M.; Baldisserotto, B. Citral and linalool chemotypes of *Lippia alba* essential oil as anesthetics for fish: A detailed physiological analysis of side effects during anesthetic recovery in silver catfish (*Rhamdia quelen*). *Fish Physiol. Biochem.* **2018**, *44*, 21–34. [CrossRef] [PubMed]
16. Sousa, D.G.; Sousa, S.D.; Silva, R.E.; Silva-Alves, K.S.; Ferreira-da-Silva, F.W.; Kerntopf, M.R.; Menezes, I.R.; Leal-Cardoso, J.H.; Barbosa, R. Essential oil of *Lippia alba* and its main constituent citral block the excitability of rat sciatic nerves. *Braz. J. Med. Biol. Res.* **2015**, *48*, 697–702. [CrossRef] [PubMed]
17. Carvalho, P.M.M.; Macêdo, C.A.F.; Ribeiro, T.F.; Silva, A.A.; Da Silva, R.E.R.; de Morais, L.P.; Kerntopf, M.R.; Menezes, I.R.A.; Barbosa, R. Effect of the *Lippia alba* (Mill.) N.E. Brown essential oil and its main constituents, citral and limonene, on the tracheal smooth muscle of rats. *Biotechnol. Rep.* **2018**, *17*, 31–34. [CrossRef]
18. Blanco, M.A.; Colareda, G.A.; Baren, C.V.; Bandoni, A.L.; Ringuelet, A.E.; Consolini, A.E. Antispasmodic effects and composition of the essential oils from two South American chemotypes of *Lippia alba*. *J. Ethnopharmacol.* **2013**, *149*, 803–809. [CrossRef]
19. Pereira-de-Morais, L.; Silva, A.A.; Da Silva, R.E.R.; Costa, R.H.S.; Monteiro, A.B.; Santos, C.R.; Amorim, T.S.; Menezes, I.R.A.; Kerntopf, M.R.; Barbosa, R. Tocolytic activity of the *Lippia alba* essential oil and its major constituents, citral and limonene, on the isolated uterus of rats. *Chem.-Biol. Interact.* **2019**, *297*, 155–159. [CrossRef]
20. Silva, R.E.R.; De Morais, L.P.; Silva, A.A.; Bastos, C.M.S.; Gonçalves, A.P.; Kerntopf, M.R.; Menezes, I.R.A.; Leal-Cardoso, J.H.; Barbosa, R. Vasorelaxant effect of the *Lippia alba* essential oil and its major constituent, citral, on the contractility of isolated rat aorta. *Biomed Pharm.* **2018**, *108*, 792–798. [CrossRef]
21. Hennebelle, T.; Sahpaz, S.; Joseph, H.; Bailluel, F. Ethnopharmacology of *Lippia alba*. *J. Ethnopharmacol.* **2008**, *116*, 211–222. [CrossRef]
22. Hennebelle, T.; Sahpaz, S.; Gressier, B. Antioxidant and neurosedative properties of polyphenols and iridoids from *Lippia alba*. *Phytother. Res.* **2008**, *22*, 256–258. [CrossRef]
23. Borges, M.S.; Dars, M.F.; Silva, P.C.; Citadini-Zanette, V.; Dal, S.B.; Amaral, P.A. Ethnobotanical study of selected medicinal plants used for the treatment of respiratory diseases in Southern Brazil. *J. Med. Plants Res.* **2021**, *15*, 22–34.
24. Burton, G.J.; Jauniaux, E. Pathophysiology of placental-derived fetal growth restriction. *Am. J. Obstet. Gynecol.* **2018**, *218*, 745–761. [CrossRef] [PubMed]
25. Poon, L.C.; Shennan, A.; Hyett, J.A.; Kapur, A.; Hadar, E.; Divakar, H.; McAuliffe, F.; Costa, F.D.S.; Von Dadelszen, P.; McIntyre, H.D.; et al. The International Federation of Gynecology and Obstetrics (FIGO) initiative on pre-eclampsia: A pragmatic guide for first-trimester screening and prevention. *Int. J. Gynaecol. Obstet.* **2019**, *145*, 1–33. [CrossRef] [PubMed]
26. Raia-Barjat, T.; Osasere, E.; Ainle, F.N. Preeclampsia and Venous Thromboembolism: Pathophysiology and Potential Therapy. *Front. Cardiovasc. Med.* **2022**, *9*, 856–923. [CrossRef]
27. Wilkerson, R.G.; Ogunbodede, A.C. Hypertensive Disorders of Pregnancy. *Emerg. Med. Clin. N. Am.* **2019**, *37*, 301–316. [CrossRef]
28. Silva, R.E.R.; Silva, A.D.A.; Pereira-De-Morais, L.; Almeida, N.D.S.; Iriti, M.; Kerntopf, M.R.; Menezes, I.R.A.; Coutinho, H.D.M.; Barbosa, R. Relaxant Effect of Monoterpene (-)-Carveol on Isolated Human Umbilical Cord Arteries and the Involvement of Ion Channels. *Molecules* **2020**, *25*, 2681.
29. Speroni, F.; Rebolledo, A.; Salemme, S.; Roldán-Palomo, R.; Rimorini, L.; Añón, M.C.; Spinillo, A.; Tanzi, F.; Milesi, V. Genistein effects on Ca^{2+} handling in human umbilical artery: Inhibition of sarcoplasmic reticulum Ca^{2+} release and of voltage-operated Ca^{2+} channels. *J. Physiol. Biochem.* **2009**, *65*, 113–124. [CrossRef] [PubMed]
30. Fei, J.Q.; Zhou, H.B.; Shen, Y.L.; Chen, X.Z.; Wang, L.L. A comparison study on the responses of umbilical arteries and thoracic aortics to the adrenergic receptor agonists. *Cell Biol. Int.* **2008**, *32*, 55. [CrossRef]
31. Medina-Leyte, D.J.; Domínguez-Pérez, M.; Mercado, I.; Villarreal-Molina, M.T.; Jacobo-Albavera, L. Use of human umbilical vein endothelial cells (HUVEC) as a model to study cardiovascular disease: A review. *Appl. Sci.* **2020**, *10*, 938. [CrossRef]
32. Maynard, L.G.; Santos, K.C.; Cunha, P.S.; Barreto, A.S.; Peixoto, M.G.; Arrigoni-Blank, F.; Blank, A.F.; Alves, P.B.; Bonjardin, L.R.; Santos, M.R. Chemical composition and vasorelaxant effect induced by the essential oil of *Lippia alba* (Mill.) N.E. Brown. (Verbenaceae) in rat mesenteric artery. *Indian J. Pharmacol.* **2011**, *43*, 694–698.
33. Catterall, W.A.; Perez-Reyes, E.; Snutch, T.P.; Striessnig, J. International Union of Pharmacology. XLVIII. Nomenclature and structure-function relationships of voltage-gated calcium channels. *Pharmacol. Rev.* **2005**, *57*, 411–425. [CrossRef] [PubMed]
34. Wang, D.; Ragnarsson, L.; Lewis, R.J. T-type Calcium Channels in Health and Disease. *Curr. Med. Chem.* **2020**, *27*, 3098–3122. [CrossRef] [PubMed]
35. Alexander, S.P.; Mathie, A.; Peters, J.A. Guide to Receptors and Channels (GRAC). *Br. J. Pharmacol.* **2007**, *150*, 1–168. [CrossRef] [PubMed]
36. Nelson, E.J.; Li, C.C.; Bangalore, R.; Benson, T.; Kass, R.S.; Hinkle, P.M. Inhibition of L-type calcium-channel activity by thapsigargin and 2,5-t-butylhydroquinone, but not by cyclopiazonic acid. *Biochem. J.* **1994**, *302*, 147–154. [CrossRef]
37. Gollasch, M.; Nelson, M.T. Voltage-dependent Ca2+ channels in arterial smooth muscle cells. *Kidney Blood Press. Res.* **1997**, *20*, 355–371. [CrossRef]
38. Seres-Bokor, A.; Kemény, K.K.; Taherigorji, H.; Schaffer, A.; Kothencz, A.; Gáspár, R.; Ducza, E. The Effect of Citral on Aquaporin 5 and Trpv4 Expressions and Uterine Contraction in Rat—An Alternative Mechanism. *Life* **2021**, *11*, 897. [CrossRef]
39. Araújo, I.G.A. Hypotensive and Vasorelaxant Effects of *Lippia microphylla* Cham Essential Oil and Its Major Constituent Thymol: Involvement of Calcium Channels Blockage. Ph.D. Thesis, Federal University of Paraíba, João Pessoa, Brazil, 2011.

40. Campos, J.; Schmeda-Hirschmann, G.; Leiva, E.; Guzmán, L.; Orrego, R.; Fernández, P.; González, M.; Radojkovic, C.; Zuñiga, F.A.; Lamperti, L. Lemon grass (*Cymbopogon citratus* (D.C) Stapf) polyphenols protect human umbilical vein endothelial cell (HUVECs) from oxidative damage induced by high glucose, hydrogen peroxide and oxidised low-density lipoprotein. *Food Chem.* **2014**, *151*, 175–181. [CrossRef]
41. Lorigo, M.; Quintaneiro, C.; Lemos, M.C.; Martinez-De-Oliveira, J.; Breitenfeld, L.; Cairrão, E. UV-B filter octylmethoxycinnamate induces vasorelaxation by Ca^{2+} channel inhibition and guanylyl cyclase activation in human umbilical arteries. *Int. J. Mol. Sci.* **2019**, *20*, 1376. [CrossRef]
42. Santos-Silva, A.J.; Cairrão, E.; Verde, I. Study of the mechanisms regulating human umbilical artery contractility. *Health* **2010**, *2*, 321–331. [CrossRef]
43. Mildenberger, E.; Biesel, B.; Siegel, G.; Versmold, H.T. Nitric oxide and endothelin in oxygen-dependent regulation of vascular tone of human umbilical vein. *Am. J. Physiol. Circ. Physiol.* **2003**, *285*, 1730–1737. [CrossRef]
44. Tufan, H.; Ayan-Polat, B.; Tecder-Unal, M.; Polat, G.; Kayhan, Z.; Oğüş, E. Contractile responses of the human umbilical artery to KCl and serotonin in Ca-free medium and the effects of levcromakalim. *Life Sci.* **2003**, *72*, 1321–1329. [CrossRef]
45. Milesi, V.; Raingo, J.; Rebolledo, A.; Grassi De Gende, A.O. Potassium channels in human umbilical artery cells. *J. Soc. Gynecol. Investig.* **2003**, *10*, 339–346. [CrossRef]
46. Leung, S.W.S.; Quan, A.; Lao, T.T.; Man, R.Y.K. Efficacy of different vasodilators on human umbilical arterial smooth muscle under normal and reduced oxygen conditions. *Early Hum. Dev.* **2006**, *82*, 457–462. [CrossRef] [PubMed]
47. Bolte, A.C.; Van Geijn, H.P.; Dekker, G.A. Management and monitoring of severe preeclampsia. *Eur. J. Obstet. Gynecol. Reprod. Biol.* **2001**, *96*, 8–20. [CrossRef]
48. Gupta, S.; Hanff, L.M.; Visser, W.; Steegers, E.A.; Saxena, P.R.; Vulto, A.G.; Maassen Van Den Brink, A. Functional reactivity of 5-HT receptors in human umbilical cord and maternal subcutaneous fat arteries after normotensive or pre-eclamptic pregnancy. *J. Hypertens.* **2006**, *24*, 1345–1353. [CrossRef]
49. Knot, H.J.; Zimmermann, P.A.; Nelson, M.T. Extracellular $K^{(+)}$-induced hyperpolarizations and dilatations of rat coronary and cerebral arteries involve inward rectifier $K^{(+)}$ channels. *J. Physiol.* **1996**, *492*, 419–430. [CrossRef]
50. Lin, A.L.; Shangari, N.; Chan, T.S.; Remirez, D.; O'Brien, P.J. Herbal monoterpene alcohols inhibit propofol metabolism and prolong anesthesia time. *Life Sci.* **2006**, *79*, 21–29. [CrossRef]
51. Lorigo, M.; Oliveira, N.; Cairrão, E. Clinical Importance of the Human Umbilical Artery Potassium Channels. *Cells* **2020**, *9*, 4–8. [CrossRef]
52. Macêdo, C.L. Involvement of Potassium Channels in the Spasmolytic Action of 8,12E, 14-Labdatrien-18 Oic Acid (Labdane-302), Obtained from *Xylopia langsdorfiana* A. St.-Hil. & Tul. on Guinea-Pig Ileum. Master's Thesis, Federal University of Paraíba, João Pessoa, Brazil, 2008.
53. Maqoud, F.; Scala, R.; Hoxha, M.; Zappacosta, B.; Tricarico, D. ATP-sensitive Potassium Channel Subunits in Neuroinflammation: Novel Drug Targets in Neurodegenerative Disorders. *CNS Neurol. Disord.-Drug Targets* **2022**, *21*, 130–149. [CrossRef]
54. Scala, R.; Maqoud, F.; Zizzo, N.; Mele, A.; Camerino, G.M.; Zito, F.A.; Ranieri, G.; Mcclenaghan, C.; Harter, T.M.; Nichols, C.G.; et al. Pathophysiological Consequences of K_{ATP} Channel Overactivity and Pharmacological Response to Glibenclamide in Skeletal Muscle of a Murine Model of Cantù Syndrome. *Front. Pharmacol.* **2020**, *11*, 604–885. [CrossRef]
55. Dosoky, N.S.; Setzer, W.N. Maternal Reproductive Toxicity of Some Essential Oils and Their Constituents. *Int. J. Mol. Sci.* **2021**, *22*, 2380. [CrossRef] [PubMed]
56. Seca, A.M.I.; Pinto, D.C.G.A. Plant Secondary Metabolites as Anticancer Agents: Successes in Clinical Trials and Therapeutic Application. *Int. J. Mol. Sci.* **2018**, *19*, 263. [CrossRef]
57. Schor, N.F. Why our patients (and we) need basic science research. *Neurology* **2013**, *80*, 2070–2075. [CrossRef] [PubMed]
58. Alencar, J.W.; Craveiro, A.A.; Matos, F.J.A.; Machado, M.I.L. Kovats Indices Simulation in Essential Oil Analysis. *Química Nova* **1990**, *13*, 282–283.
59. Adams, R.P. *Identification of Essential Oils by Ion Trap Mass Spectroscopy*; Academic Press, Inc.: Waco, TX, USA, 1989.
60. Stenhagen, E.; Abrahmsson, S.; McLafferty, F.C. *Registry of Mass Spectral Data*; John Wiley Sons: New York, NY, USA, 1974.
61. Zechmeister, L. *Progress in the Chemistry of Organic Natural Products*; Springer: New York, NY, USA, 1979.

plants

MDPI

Article

Characterization and In Vivo Anti-Inflammatory Efficacy of Copal (*Dacryodes peruviana* (Loes.) H.J. Lam) Essential Oil

Lupe Carolina Espinoza [1,2], Eduardo Valarezo [1], María José Fábrega [3], María José Rodríguez-Lagunas [4,5], Lilian Sosa [6,*], Ana Cristina Calpena [2,7] and Mireia Mallandrich [2,7]

1 Departamento de Química, Universidad Técnica Particular de Loja, Loja 1101608, Ecuador
2 Institute of Nanoscience and Nanotechnology (IN2UB), University of Barcelona, 08028 Barcelona, Spain
3 Department of Experimental and Health Sciences, Parc of Biomedic Research of Barcelona, Pompeu Fabra University, 08003 Barcelona, Spain
4 Department of Biochemistry and Physiology, Faculty of Pharmacy and Food Sciences, University of Barcelona, 08028 Barcelona, Spain
5 Nutrition and Food Safety Research Institute (INSA-UB), 08921 Santa Coloma de Gramenet, Spain
6 Pharmaceutical Technology Research Group, Faculty of Chemical Sciences and Pharmacy, National Autonomous University of Honduras (UNAH), Tegucigalpa 11101, Honduras
7 Department of Pharmacy, Pharmaceutical Technology and Physical Chemistry, Faculty of Pharmacy and Food Sciences, University of Barcelona, 08028 Barcelona, Spain
* Correspondence: lilian.sosa@unah.edu.hn

Abstract: Essential oils are natural aromatic substances that contain complex mixtures of many volatile compounds frequently used in pharmaceutical and cosmetic industries. *Dacryodes peruviana* (Loes.) H.J. Lam is a native species from Ecuador whose anti-inflammatory activity has not been previously reported, thus the aim of this study was to evaluate the anti-inflammatory activity of *D. peruviana* essential oil. To that end, essential oil from *D. peruviana* fruits was isolated by hydrodistillation and characterized physically and chemically. The tolerance of the essential oil was analyzed by cytotoxicity studies using human keratinocytes. The anti-inflammatory activity was evaluated by an arachidonic acid-induced edema model in mouse ear. The predominant compounds in *D. peruviana* essential oil were α-phellandrene, limonene, and α-pinene, with the three compounds reaching approximately 83% of the total composition. Tolerance studies showed high biocompatibility of this essential oil with human keratinocytes. In vivo studies demonstrated a moisturizing effect and an alleviation of several events occurred during the inflammatory process after topical treatment with *D. peruviana* essential oil such as decline in skin edema; reduction in leukocytic infiltrate; and decrease in inflammatory cytokines TNFα, IL-8, IL-17A, and IL-23. Therefore, this essential oil could be an attractive treatment for skin inflammation.

Keywords: inflammation; essential oil; *Dacryodes peruviana*; α-phellandrene; limonene

Citation: Espinoza, L.C.; Valarezo, E.; Fábrega, M.J.; Rodríguez-Lagunas, M.J.; Sosa, L.; Calpena, A.C.; Mallandrich, M. Characterization and In Vivo Anti-Inflammatory Efficacy of Copal (*Dacryodes peruviana* (Loes.) H.J. Lam) Essential Oil. *Plants* **2022**, *11*, 3104. https://doi.org/10.3390/plants11223104

Academic Editors: Hazem Salaheldin Elshafie, Ippolito Camele and Adriano Sofo

Received: 20 October 2022
Accepted: 7 November 2022
Published: 15 November 2022

1. Introduction

Acute inflammation occurs as a host immune response to external changes or cellular injury. Common signs of the inflammatory process are redness, heat, pain, and swelling [1]. The cutaneous immune system contains a diverse profile of inflammatory response mediators including antimicrobial peptides, complement proteins, and phagocytes such as neutrophils and macrophages [2]. This inflammatory process constitutes an essential defense mechanism to protect the integrity of the body against invaders including foreign substances, microorganisms, or cancer cells [3]. However, an inappropriate or excessive inflammatory response can mistakenly lead to damage of normal tissues, owing to a high production of reactive oxygen species, nitric oxide, and pro-inflammatory cytokines, which can trigger chronic inflammation implicated in the pathogenesis of skin disorders including psoriasis, dermatitis, and rosacea [3,4].

Volatile oils, essential oils (EOs) or simply essences, are the natural aromatic substances responsible for the fragrances of flowers, leaves, and other plant organs [5]. EOs are also known as volatile secondary metabolites, volatile compounds, or volatile fractions. EOs are especially abundant in the families Apiaceae, Asteraceae, Burseraceae, Lamiaceae, Myrtaceae, and Rutaceae [6]. Various functions are attributed to them in plants, such as protection against insects and herbivores and adaptation to water stress, and they are of great importance in pollination, as they constitute elements of chemical communication owing to their volatility and marked odor [5].

EOs can be extracted from flowers, fruits, leaves, roots, trunk, and other parts of the plant, and constitute values < 0.01% up to values > 3% of the dry weight of the plant [7]. In terms of their chemical composition, EOs are generally complex mixtures (from a few to more than 100 compounds) of highly variable constituents that belong almost exclusively to the group of terpenes and, to a lesser extent, to the group of aromatic compounds derived from phenylpropane (cinnamic aldehyde, eugenol, anethole, anisic aldehyde, and safrole, among others) [8]. The heterogeneity of essential oil (EO) compounds provides them with different biological properties, such as analgesic, anti-inflammatory, antioxidant, bactericidal, fungicidal, insecticidal, larvicidal, and repellent action, among others [9].

Dacryodes Vahl is a genus of small or large trees, with indehiscent fruits, with an oily mesocarp and a cartilaginous and smooth pyrene. The species of this genus are distributed in the tropical forests of Africa, America, and Asia. The generic name is from the Greek dakruon meaning "tear, drop", referring to how resin droplets form on the bark surface. *Dacryodes peruviana* (Loes.) H.J. Lam is a native species from Ecuador known as "copal", "copal comestible", and "anime" (Spanish language); "wigonkawe", "wiñimonkawe", "winkayamogeinka", and "witakeño" (Wao tededo, dialect of the Amazon region); "kunchay" and "wichilla kupall" (Kichwa language); "ccovi shasha" and "shasha" (A'ingae language, spoken by the Cofán people); and "kunchai" and "shiríkip" (Shuar chicham language) [10,11]. This species is widely distributed in the Amazonian and Andean Ecuadorian regions between 0 and 2500 m a.s.l.; in this country, it can be found especially in the Amazonian provinces of Morona-Santiago, Napo, Pastaza, and Zamora-Chinchipe [12].

The *Dacryodes peruviana* plant is a 20–25 m tall tree that produces fruits annually. Copal fruits are smooth, ovoid-shaped, greenish-red capsules, averaging 3.3 cm long (2 to 4 cm) and 1.25 cm in diameter, with a thin pericarp about 0.4 cm thick. The fruits open in three or four valves and contain one to four seeds approximately 1.5 cm long. The fruit is used as food for birds and monkeys. In the Ecuadorian Amazon, the seeds and the mesocarp of the fruit are edible, raw, and passed in hot water or on the coals. The resin of this species is used as mosquito repellent and as incense, as well as to smoke around the houses to eliminate evil spirits, and is smoked to treat the "mal aire" (mythical disease) [11]. The EO of copal fruits collected in the Ecuadorian Amazon exerted a moderate activity against Staphylococcus aureus and a repellent activity class 4 against mosquitoes (Diptera: Culicidae) at concentrations of 3%, 2%, and 1%, and class 3 for a concentration of 0.5%, and the same EO showed a weak antioxidant activity through the DPPH and ABTS methods [7].

Currently, Ecuador occupies the sixth position worldwide in the number of plant species per unit surface area, in 0.17% of the planet's land surface hosts' some 20,000 species, which makes this country a biodiversity hotspot [13]. However, the fact that there are few studies of its aromatic plant species, especially of the aromatic species of the Burseraceae family, and that anti-inflammatory activity of the *Dacryodes peruviana* EO has not been previously reported in the literature, have stimulated our interest in investigating the EO extracted from this species. Considering these remarkable findings, the aim of this research was to evaluate the anti-inflammatory activity of the *D. peruviana* fruit EO (*D. peruviana* EO).

2. Results

2.1. Essential Oil Isolation

Through hydrodistillation in a Clevenger-type apparatus, 240 mL of EO was obtained from 5 kg of *D. peruviana*, which represents a yield of 4.5 ± 0.3% (*v/w*) or 45 mL/Kg, considered as high yields, which makes the *D. peruviana* EO suitable for industrial applications.

2.2. Physical Properties of Essential Oil

The fruits of *D. peruviana* provided an EO with a density of ρ^{20} = 0.8456 ± 0.0023 g/cm^3, refractive index of n^{20} = 1.4751 ± 0.0002, and specific rotation of $[\alpha]^{20}$ = +12.2 ± 0.7, as previously published [7].

2.3. Essential Oil Compounds' Identification

The identification of volatile compounds present in *Dacryodes peruviana* EO was carried out by means of GC-FID and GC-MS using capillary nonpolar column DB-5MS. The results obtained are summarized in Table 1. In *D. peruviana* EO, twenty-four chemical constituents were identified, representing 99.78% of the total composition.

Table 1. Chemical composition of *Dacryodes peruviana* essential oil.

CN	Rt	Compound	RI	RIf	%	SD	Type	CF	MM (Da)
1	5.83	α-Thujene	926	924	1.18	0.05	MH	$C_{10}H_{16}$	136.13
2	6.08	α-Pinene	932	932	8.45	0.63	MH	$C_{10}H_{16}$	136.13
3	6.7	Camphene	947	946	0.16	0.01	MH	$C_{10}H_{16}$	136.13
4	7.6	Sabinene	969	969	0.80	0.07	MH	$C_{10}H_{16}$	136.13
5	7.76	β-Pinene	973	974	3.39	0.05	MH	$C_{10}H_{16}$	136.13
6	8.29	Myrcene	986	988	0.82	0.02	MH	$C_{10}H_{16}$	136.13
7	9.07	α-Phellandrene	1005	1002	52.35	3.14	MH	$C_{10}H_{16}$	136.13
8	9.28	δ-3-Carene	1010	1008	0.08	0.01	MH	$C_{10}H_{16}$	136.13
9	9.4	α-Terpinene	1013	1014	0.31	0.03	MH	$C_{10}H_{16}$	136.13
10	9.73	p-Cymene	1021	1020	5.24	0.56	MH	$C_{10}H_{14}$	134.1
11	9.89	Limonene	1025	1024	22.51	1.68	MH	$C_{10}H_{16}$	136.13
12	11.12	γ-Terpinene	1055	1054	0.12	0.01	MH	$C_{10}H_{16}$	136.13
13	12.23	Terpinolene	1082	1086	2.08	0.17	MH	$C_{10}H_{16}$	136.13
14	14.61	Camphor	1140	1141	0.35	0.03	OM	$C_{10}H_{16}O$	152.12
15	15.84	Terpinen-4-ol	1170	1174	0.17	0.01	OM	$C_{10}H_{18}O$	154.14
16	17.15	γ-Terpineol	1202	1199	0.98	0.08	OM	$C_{10}H_{18}O$	154.14
17	18.54	Ascaridole	1236	1234	0.13	0.01	OM	$C_{10}H_{16}O_2$	168.12
18	22.73	δ-Elemene	1338	1335	0.06	0.00	SH	$C_{15}H_{24}$	204.19
19	23.87	α-Copaene	1366	1374	0.08	0.01	SH	$C_{15}H_{24}$	204.19
20	25.88	trans-Caryophyllene	1415	1417	0.21	0.02	SH	$C_{15}H_{24}$	204.19
21	27.36	α-Humulene	1451	1452	tr		SH	$C_{15}H_{24}$	204.19
22	28.38	Germacrene D	1476	1480	0.13	0.01	SH	$C_{15}H_{24}$	204.19
23	29.74	δ-Amorphene	1509	1511	0.05	0.00	SH	$C_{15}H_{24}$	204.19
24	29.94	β-Curcumene	1514	1514	0.14	0.01	SH	$C_{15}H_{24}$	204.19
		Monoterpene hydrocarbons (MH)			97.48				
		Oxygenated monoterpenes (OM)			1.63				
		Sesquiterpene hydrocarbons (SH)			0.67				
		Total identified			99.78				

CN: compound number; Rt: retention time; RI: calculated retention indices; RIf: references retention indices; SD: standard deviation; CF: chemical formula; MM: monoisotopic mass; tr: traces (<0.5%); MH: monoterpene hydrocarbons; OM: oxygenated monoterpenes; SH: sesquiterpene hydrocarbons.

It was determined that the vast majority (97.48%) of *D. peruviana* EO compounds are monoterpene hydrocarbons (MHs) in nature; furthermore, a low number of oxygenated monoterpenes (OMs, 1.63%) and sesquiterpene hydrocarbons (SHs, 0.67%) were identified. The presence of oxygenated sesquiterpene (OS) and diterpenes was not determined. The principal constituents (>5%) are found to be MH α-phellandrene (CN: 7, CF: C10H16, MM: 136.13 Da) with 52.35 ± 3.14%, limonene (CN: 11, CF: C10H16, MM: 136.13 Da) with 22.51 ± 1.68%, α-pinene with 8.45 ± 0.63%, and ρ-cymene with 5.24 ± 0.56%.

2.4. Tolerance Studies: Cytotoxicity Assay

Figure 1 shows the results of the MTT cytotoxicity assay using human keratinocytes HaCaT cell line. After 24 h of incubation, cell viability greater than 80% was observed in the assayed dilutions from 1/200 to 1/2000 (*v/v* in phosphate buffered solution—PBS). Dilutions from 1/400 showed cell viability close to 100% in relation to the control. Therefore, these results suggest that *D. peruviana* EO does not cause toxicity in human keratinocytes.

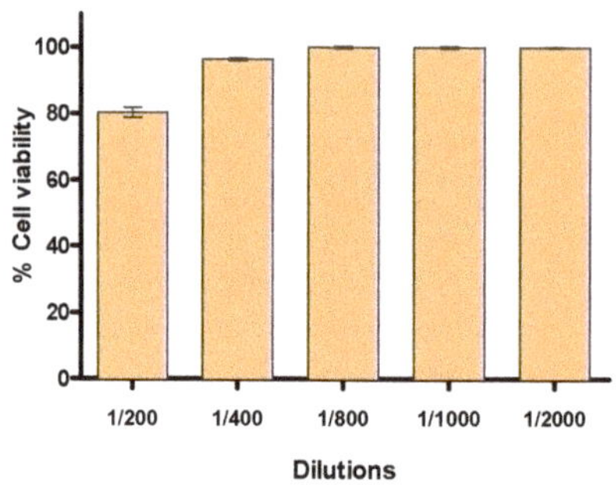

Figure 1. Percentage of cell viability of the HaCaT keratinocyte cell line exposed to different concentrations of *D. peruviana* EO.

2.5. In Vivo Anti-Inflammatory Efficacy Studies: Arachidonic Acid (AA)-Induced Inflammation

2.5.1. Stratum Corneum Hydration (SCH) and Thickness Evaluation

Anti-inflammatory activity of *D. peruviana* EO was evaluated using a mouse model of inflammation induced by topical application of araquidonic acid (AA). After inducing inflammation, a slight decrease, but without significant statistical differences with respect to the basal state, was observed in the SCH levels. However, topical treatment with *D. peruviana* EO and ibuprofen gel on mice ears induced a significant increase in the skin hydration, ever higher than the basal state. The thickness of mice ears was markedly greater after inducing inflammation, whereas the treatment with *D. peruviana* EO and ibuprofen gel significantly reduced this parameter to the basal state and consequently alleviated the skin edema. Figure 2 shows the results for the stratum corneum hydration (SCH) and the variations in the ear thickness, representing the inflammatory process and the inflammatory reduction.

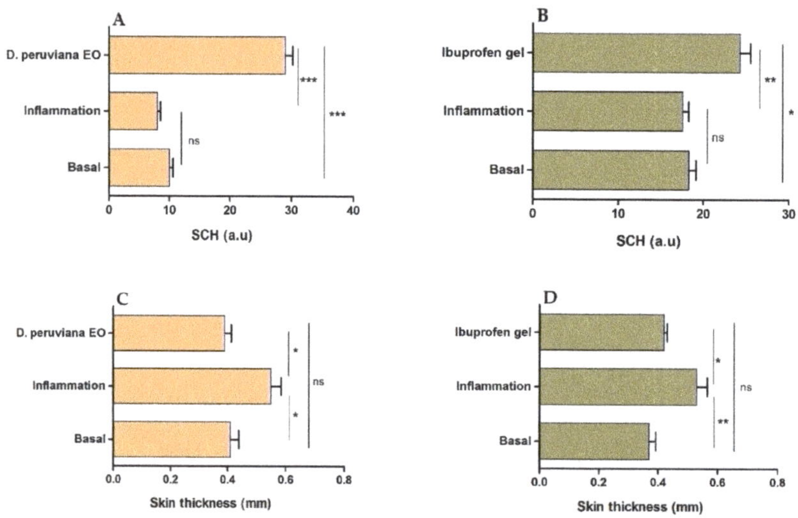

Figure 2. Stratum corneum hydration (SCH) and skin thickness of the mouse ears. (**A**) SCH of skin

treated with *D. peruviana* EO, (**B**) SCH of skin treated with reference drug (Ibuprofen gel), (**C**) thickness of skin treated with *D. peruviana* EO, and (**D**) thickness of skin treated with reference drug (Ibuprofen gel). Mean ± SD of three replicates. Statistically significant difference: * = $p < 0.05$, ** = $p < 0.01$, *** = $p < 0.001$, ns: no significant.

2.5.2. Histological Analysis

The histomorphological analysis of the mice ears was performed to assess the anti-inflammatory activity of the *D. peruviana* EO. Representative images of assayed treatments are shown in Figure 3.

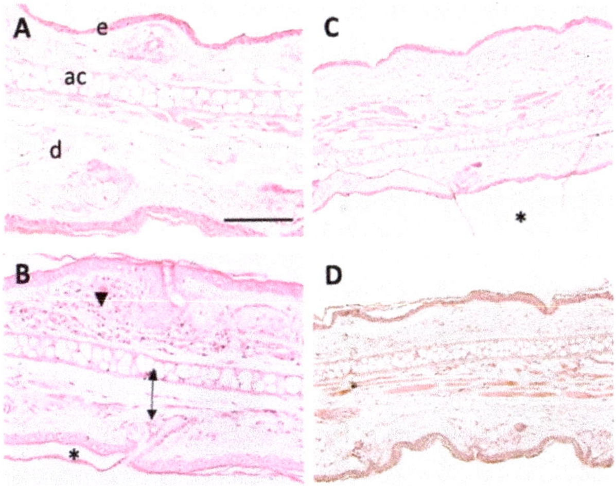

Figure 3. treated with *D. peruviana* EO, (**B**) SCH of skin treated with reference drug (Ibuprofen gel), Micrographs of mouse ear samples (×100 magnification). (**A**) Negative control, (**B**) positive control, (**C**) reference drug (ibuprofen gel), and (**D**) treatment with *D. peruviana* EO. Scale bar = 200 μM. e: epidermis, d: dermis, ac: auricular cartilage. Arrowhead indicates neutrophilic infiltrates, arrows show edema, and asterisks indicate stratum corneum loss.

A mild inflammation was observed in ears treated with AA (Figure 3B), characterized by a slight edema, increased leucocytic infiltrate, and loss of stratum corneum. The treatment with *D. peruviana* EO (Figure 3D) prevented the appearance of those inflammatory indicators and showed similar morphology as in negative control (Figure 3A) and treatment with ibuprofen gel (Figure 3C).

2.5.3. Pro-Inflammatory Cytokines' Determination

A statistically significant increase in the expression of cytokines TNF-α, IL-8, IL-17A, and IL-23 was observed in the positive control as result of the inflammation produced by AA when compared with the negative group. Topical treatment with *D. peruviana* EO on the inflamed area significantly reduced the expression of these pro-inflammatory cytokines to comparable levels to the negative control and with the ibuprofen gel treatment. These results suggest that *D. peruviana* EO could be acting in the regulation of inflammatory processes (Figure 4).

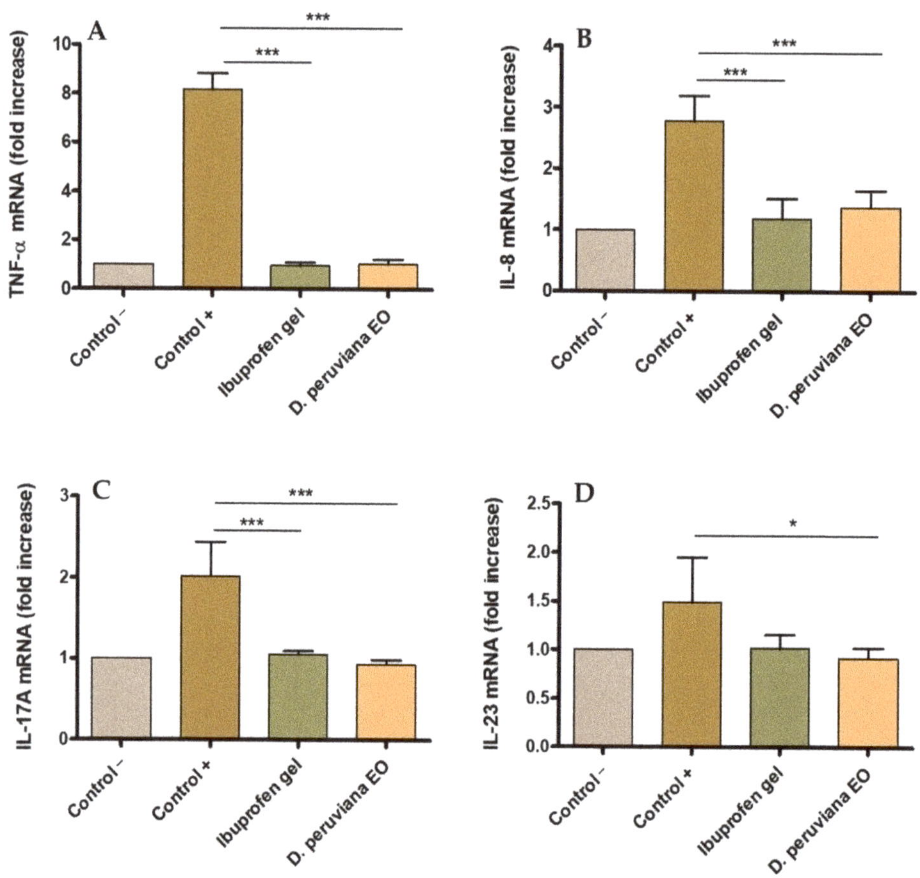

Figure 4. Expression of pro-inflammatory interleukins obtained by quantitative reverse transcription polymerase chain reaction (RT-qPCR): (**A**) tumor necrosis factor-alpha (TNF-α); (**B**) interleukin-8 (IL-8); (**C**) interleukin-17A (IL-17A); and (**D**) interleukin-23 (IL-23). Negative control (control −), positive (control +), treatment with reference drug (Ibuprofen gel), treatment with *D. peruviana* EO. Results are shown as mean ± SD ($n = 3$). Statistically significant difference: * = $p < 0.05$, *** = $p < 0.001$.

3. Discussion

The EO yield of *D. peruviana* fruit, 45 mL/Kg, is considered as high yield (>10 mL/Kg), according to the categorization proposed by the CYTED (Science and Technology for Development) [14]. The excellent yield of this species, added to the fact that the fruits are used instead of the wood, make the *D. peruviana* fruit a suitable option for use at industrial level. The predominant compounds in the chemical composition of *D. peruviana* EO were α-phellandrene, limonene (mixture of D- and L-), and α-pinene, with the three compounds reaching approximately 83% of the total composition. More than 50% (~52%) of the EO of *D. peruviana* fruits is constituted by α-phellandrene; this compound with CAS 99-83-2 is a cyclic monoterpene, which has been found as the main, major, or minor compound in various EOs [15–17]. α-phellandrene has low toxicity in rats and minimal risk of irritation sensitization [18]. Some studies have shown that α-phellandrene, limonene (CAS 138-86-3), and α-pinene (CAS 80-56-8) have some biological properties and activities such as anticancer, antimicrobial, anti-inflammatory, antidepressive, anti-leishmania, antioxidant, antihyperalgesic, antinociceptive, insecticidal, gastroprotective, and neuroprotective [18–20]. EOs containing α-phellandrene as one of the main compounds present antiacetylcholinesterase, antidepressive, antioxidant, antimicrobial, antihyperalgesic, and repellent activity [15,17,21,22].

The tolerance of *D. peruviana* EO was analyzed by in vitro studies, which are useful to screen the tolerability of pharmaceutical products prior to the pre-clinical and clinical assays. Cell lines are widely used for this type of study owing to their easiness to cultivate and sensitivity to toxic irritation [23]. The results obtained in this study showed that the analyzed dilutions of *D. peruviana* EO did not provoke cytotoxic effects on human keratinocytes after 24 h of incubation, suggesting high biocompatibility of this EO for skin application.

The anti-inflammatory potential of *D. peruviana* was studied using an AA-induced edema model in mouse ear. Inflammatory processes are a complex and biologically natural pathophysiological response initiated by vascular tissues; it helps the body to defend itself against pathogens, possible cell damage, and irritant damage. Cutaneous inflammation increases skin thickness, causes its dryness, promotes infiltration of inflammatory cells, and stimulates the release of various inflammatory mediators [24]. Increased skin thickness is one of the first signs manifested during inflammation, which is indicative of several events including edema, increased vascular permeability, and proliferation of keratinocytes [25]. In this study, topical application of AA caused an increase of 34.15% in the skin thickness as a result of the inflammatory process, whereas topical treatment with *D. peruviana* EO reduced this parameter to the basal state, showing a similar result to the reference anti-inflammatory drug—ibuprofen gel (Figure 2C,D). Additionally, SCH was determined in order to analyze the skin barrier function. The results of this evaluation showed only a slight decrease in SCH after inducing inflammation; however, a moisturizing action was noticeable in mice treated with *D. peruviana* EO (Figure 2A). These results were corroborated by histological analysis that showed the presence of a slight edema, leukocytic infiltrate, and loss of stratum corneum in the positive control. Conversely, topical treatment with *D. peruviana* EO aided in the improvement of these signs (Figure 3).

Inflammation begins with the activation of Phospholipase A2, which degrades lipids in cell membranes, resulting in the release of arachidonic acid and other inflammatory mediators such as eicosanoids, serotonin, histamine, and interleukins [26]. Interleukins consist of a large group of proteins that can cause many reactions in cells and tissues by binding to high-affinity receptors on cell surfaces. The main function of interleukins is thus to modulate growth, differentiation, and activation during inflammatory and immune responses [27]. In the present study, TNF-α, IL-8, IL-17A, and IL-23 were evaluated. TNF-α is a primary inflammatory factor released in the skin after trauma, injury, or infection and triggers the expression of other pro-inflammatory cytokines including IL-6, IL-8, and IL-1β [28]. IL-8 is produced by several types of cells, including monocytes, fibroblasts, endothelial cells, keratinocytes, and chondrocytes. Previous studies have shown that IL-8 levels increased with stimulus such as irradiation in keratinocyte cultures [29]. IL-17A acts on epithelial and endothelial cells. The main effects of IL-17A are the release of proinflammatory cytokines specifically in allergic processes [30]. Macrophages and dendritic cells mainly synthesize IL-23 acting on T cells, causing the maintenance of IL-17-producing T cells, and it increases in the induction of chronic intestinal inflammation, in autoimmune diseases, cancer, and psoriasis [31]. In this research, the increase in the expression of cytokines TNF-α, IL-8, IL-17A, and IL-23 in mice ears inflamed by topical application was counteracted by *D. peruviana* EO treatment, which restored the mRNA values and, consequently, this EO reduced the production of these pro-inflammatory cytokines to values similar to those of the negative control and ibuprofen gel treatment (Figure 4A–D). The restorative benefits of *D. peruviana* in the clinical, histological, and immunological levels of skin inflammation may be due to the mixture of substances present in this EO, specifically by α-Phellandrene (52.35%) and D-limonene (22.51%). Specifically, α-Phellandrene exerts an anti-inflammatory action through different mechanisms, such as the inhibition of neutrophil migration towards the site of inflammation and the decrease in the production of proinflammatory cytokines induced by TNF-α, preventing the release of inflammatory mediators in the area. It is reported that α-Phellandrene has the potential to be used for the treatment of inflammatory diseases, such as rheumatoid arthritis, osteoarthritis, and allergic diseases [32]. Our results are in line with other investigations carried out in EOs, in which a reduction in the pro-inflammatory cytokins was observed when they tested EOs from Pinus spp in murine macrophages. α-pinene and α-phellandrene were the major compounds found in Pinus EOs among 45 compounds identified [33]. On the other hand, previous studies have reported the role of D-limonene administered subcutaneously in the reduction in proinflammatory cytokines that produce dermatitis as well as inhaled D-limonene in the decrease in lung inflammation caused by allergies [34,35]. D-limonene reduces the expression of TNF-α, IL-1β, and IL-6, and has been suggested as a stimulator of the production of IL-10, which is a powerful anti-inflammatory cytokine [36,37]. D-limonene demonstrated anti-inflammatory effects in colitis by decreasing cytokines (NF-κB, TNF-α, IL-1β, and IL-6) when administered orally in rats [38]. Therefore, these results suggest that *D. peruviana* EO acts in different events occurring during inflammatory processes and support its possible use in the treatment of inflammatory skin diseases.

4. Materials and Methods

4.1. Materials

Alliphatic hydrocarbons standard (code M-TPH6 × 4-1ML and name Diesel Range Organics Mixture #2-GRO/DR) was provided from CHEM SERVICE (West Chester, PA, USA). Arachidonic acid sodium salt, dichloromethane (DCM), and sodium sulfate anhydrous were purchased from Sigma-Aldrich (San Luis, MO, USA). Components for histological analysis were purchased from Thermo Fisher Scientific (Waltham, MA, USA). HaCaT cell line was purchased from Cell Line Services (Eppelheim, Germany). Helium (gas carrier) was obtained from INDURA (Quito, Ecuador). Reagents for the MTT ((3-[4,5-dimethylthiazol-2-yl]-2,5 diphenyl tetrazolium bromide) assay were obtained from Invitrogen Alfagene (Carcavelos, Lisbon, Portugal). Reagents used for cell cultures were purchased from Gibco (Carcavelos, Lisbon, Portugal). Transcutol-P (diethylene glycol monoethyl ether) was purchased by Gattefossé (Saint-Priest, France). All chemicals were of analytical grade.

4.2. Plant Material

The *Dacryodes peruviana* fruits in state of maturation were collected in La Paz parish. This parish, belonging to the Yacuambi canton, is located north of the Zamora Chinchipe province, in the Ecuadorian Amazon. The collection coordinates were a latitude of 3°36′40″ S and longitude of 78°39′38″ W. Copal fruits were harvested between the months of February and April. The collection conditions were a temperature of 25 °C and a pressure of 0.87 atm and the fruits were transported in closed plastic containers at an average temperature of 20 °C. The botanical specimens were identified by Nixon Cumbicus at the herbarium of the Universidad Técnica Particular de Loja (HUTPL). A voucher specimen is preserved in the HUTPL.

4.3. Essential Oil Isolation

The fruits were processed fresh, immediately after arriving at the laboratory, between 3 and 4 h after being collected. The postharvest treatment consists of the separation of foreign material and degraded fruits. The EO of copal fruits was extracted according to the procedures described by Valarezo, et al. (2020) [7] with some modifications, for which two consecutive procedures are used, the release and the isolation. The release was carried out in a patented device called "Device for the release of EO from a plant matrix by crushing by immersion centrifugal force" Title No. PI-2022-012, for which whole fruits were placed in the device, with a 1:1 ratio (fruit: water), and crushed for 45 s, until EO was released. This first step was carried out because the epicarp does not allow the exit of the EO found in the mesocarp, which is why the epicarp must be broken in adequate conditions to avoid the loss or degradation of the EO. In addition, this procedure called release allows to increase the contact surface between the raw material and the steam (in the distillation phase), which decreases the extraction times. After release isolation was performed, the isolation of EO was carried out by hydrodistillation, for which a semi-pilot distiller Clevenger type of 80 L of capacity was used. The distillation was carried out for 2 h and the EO collected was dried with anhydrous sodium sulphate and stored in amber sealed vials at 4 °C until being used in the subsequent analysis.

4.4. Determination of Physical Properties of Essential Oil

The physical properties of the EO determined were density, refraction index, and optical rotation, according to the standard method AFNOR NF T 75-111, AFNOR NF T 75-112, and ISO 592:1998, respectively. An analytical balance (Mettler AC 100, Mettler Toledo, Columbus, OH, USA) and a pycnometer were used to determine the density, a refractometer (ABBE, Hamburg, Germany) for the refraction index, and an automatic polarimeter (AP-81, MRC, Holon, Israel) for optical rotation. All measurements were performed at 20 °C.

4.5. Determination of the Chemical Composition of Essential Oil

The compounds present in the EO were determined qualitatively and quantitatively using a gas chromatograph (GC) (model 6890N series, Agilent Technologies, Santa Clara, CA, USA) according to the procedures described by Valarezo, et al. (2021) [39], with minimal modifications to some parameters. In the case of qualitative analyses, the GC was equipped with a mass spectrometer (type quadrupole) detector (MS) (model Agilent series 5973 inert, Agilent Technologies, Santa Clara, CA, USA) and, for quantitative analyses, GC was coupled to a flame ionization detector (FID). In both cases, an automatic injector (Agilent 7683, Agilent Technologies, Santa Clara, CA, USA) in split mode and a nonpolar column DB-5 ms were used. The samples are prepared with a ratio of 1/100 (*v/v*) of EO/DCM and then injected with a split ratio of 1:50. The temperature ramp was 50 °C for 3 min, then 2.5 °C/min until 210 °C, and 3 min at this temperature. The injector temperature was 210 °C and 250 °C for both detectors. The retention index (IR) was determined based on the comparison of retention times of the EO compounds and of the aliphatic hydrocarbons of standard injection under the same conditions. The compounds were identified based on a comparison of mass spectrum data and IRs with those published in the literature [40,41].

4.6. Tolerance Studies: Cytotoxicity Assay

Human immortalized keratinocytes (HaCaT) obtained from ATCC were used for in vitro experiments. They were maintained in Dubelco's modified Eagle's medium (DMEM) supplemented with 10% of fetal bovine serum (FBS), 1% non-essential amino acids, 100 U/mL of penicillin, and 100 g/mL of streptomycin. To perform the assays, cells were incubated at 37 °C and 5% of CO2 until 80–90% of growth. Methylthiazolydiphenyltetrazolium bromide (MTT) assay was used to evaluate the cytotoxicity effect of *D. peruviana* EO on HaCaT cells. For this purpose, 2×10^5 cells/mL were cultured in 96-well plates (Corning) for 2 days. Then, different dilutions of *D. peruviana* EO (1/100, 1/400, 1/800, 1/1000 and 1/2000) in PBS (*v/v*) were incubated for 24 h. Afterwards, keratinocytes were washed with PBS 1× (Thermofisher) and incubated with MTT reagent at a concentration of 5 mg/mL (Sigma Aldrich) during 2 h maintaining 37 °C and 5% of CO_2. The media was aspirated and MTT purple crystals were dissolved with Dimethyl sulfoxide (DMSO). Cytotoxicity levels were measured by absorbance at 570 nm using an Infinity Tecan 200 Pro Microplate Reader. As control, non-treated cells were used. The results were determined as the percentage of cell viability relative to the control (100% viability).

4.7. Anti-Inflammatory Efficacy Studies: Arachidonic Acid (AA)-Induced Edema Model in Mouse Ear

4.7.1. Study Protocol

In vivo studies were carried out in accordance with the Bioethics Committee of the University of Barcelona (CEEA/UB ref. 4/16 and Generalitat ref. 8756. Date: 28 January

2016) using male BALB/c mice (n = 12, 4–5 months old). The inflammatory process was induced topically by applying 60 µL AA (at a concentration of 5 mg/mL diluted in PBS) on the right ears. A group of three inflamed animals were used as positive control. A second group (n = 3) was treated with 60 µL of *D. peruviana* EO 5% dissolved previously in Transcutol-P/water (1:1, *v/v*) after 20 min of AA application (*D. peruviana* EO group), whereas a third group (n = 3) was treated with a known anti-inflammatory drug (60 mg of ibuprofen gel 50 mg/g; reference: 886192.7). Transcutol-P was used to dissolve *D. peruviana* EO owing to its high skin biocompatibility. Finally, a group of untreated healthy mice (n = 3) was used to compare the results (negative group). The animals were sacrificed by cervical dislocation after 20 min and the right ears were cut in order to perform histological analysis and to evaluate the expression of pro-inflammatory cytokines.

4.7.2. Stratum Corneum Hydration (SCH) and Thickness Evaluation

SCH and skin thickness of the mice ears were determined before the inflammation (basal state), after inflammation with AA, and after treatment with *D. peruviana* EO or ibuprofen gel. Corneometer CM-825 (Courage & Khazaka Electronics GmbH, Köln, Germany) and a digital thickness gauge of 0–10 mm (Mitutoyo Corp, Kawasaki, Japan) were used to measure these parameters, respectively.

4.7.3. Histological Analysis

Histological evaluation was carried out by hematoxylin and eosin staining. To that end, the samples of mice ears were rinsed with PBS (pH 7.4) and stored overnight in 4% buffered formaldehyde. The tissues were dehydrated, embedded in paraffin wax, cut into transversal sections of 5 µm, and stained with hematoxylin and eosin. Finally, these samples were mounted on DPX (Sigma Aldrich, St. Louis, MO, USA) and observed under light microscope (BX41 microscope and XC50 camera, Olympus Hamburg, Hamburg, Germany).

4.7.4. Pro-Inflammatory Cytokines Determination

TRI-Reagent® (Sigma Aldrich) was used to extract total RNA from small fragments of mouse ear tissue. Purity and RNA concentration were measured on the NanoDrop One spectrophotometer (Thermo Scientific, Waltham, MA, USA) by the absorbance ratio at 260 and 280 nm. RNA was reverse-transcribed using the cDNA Reverse Transcription Kit (Thermo Fisher Scientific, Waltham, MA, USA) and the ProFlex PCR System (Applied Biosystems, Waltham, MA, USA). RT-qPCR reactions were carried out on the QuantStudio 7 Flex Real-Time PCR System (Applied Biosystems) using SYBR Green PCR Master Mix (Applied Biosystems) and cell-type specific primers for IL-17A, IL-8, IL-23, and TNFα (Table 2). The formula $2^{-\Delta\Delta Ct}$ was used to normalize the expression results. The values of the housekeeping GAPDH gene were used to standardize the values obtained for the studied genes.

Table 2. Primer sequences used for the RT-qPCR assay.

Gene	Primer Sequence (5′ to 3′)	Gene Accession Number
GAPDH	FW: AGCTTGTCATCAACGGGAAG RV: TTTGATGTTAGTGGGGTCTCG	BC023196.2
IL-8	FW: GCTGTGACCCTCTCTGTGAAG RV: CAAACTCCATCTTGTTGTGTC	X53798.1
IL-17A	FW: TTTTCAGCAAGGAATGTGGA RV: TTCATTGTGGAGGGCAGAC	NM_010552.3

Table 2. Cont.

Gene	Primer Sequence (5′ to 3′)	Gene Accession Number
TNFα	FW: AACTAGTGGT-GCCAGCCGAT RV: CTTCACAGAGCAAT-GACTCC	NM_013693.3

GAPDH = glyceraldehyde-3-phosphate dehydrogenase; IL-8 = interleukin-8; IL-17A = interleukin-17A; TNFα = tumor necrosis factor alpha; FW = forward primer; RV = reverse primer.

4.7.5. Statistical Analyses

These experiments were carried out in triplicate. The results are presented as the mean ± SD. Statistical analyses were performed by one-way ANOVA followed by Tukey's test using the GraphPad Prism, v5.0 software (GraphPad Software Inc., San Diego, CA, USA). p-values less than 0.05 were considered statistically significant.

5. Conclusions

In conclusion, the present research provides evidence of the therapeutic benefits of that *D. peruviana* EO thanks to its content rich in terpenes, mainly α-phellandrene and limonene, in different aspects of skin inflammation including clinical, histological, and immunological. Its role in the inflammatory process was strongly demonstrated throughout the studies, showing a significant decline in edema, improvement in biophysical parameters (SCH), and reduction in leukocytic infiltrate as well as in the expression of proinflammatory cytokines including TNF-α, IL-8, IL-17A, and IL-23. Therefore, this EO could be used as a promising therapeutic treatment of local inflammation via topical application on affected area.

Author Contributions: Conceptualization, L.C.E. and A.C.C.; methodology, L.C.E., E.V., M.J.F., and M.J.R.-L.; formal analysis, M.M.; investigation, L.C.E. and L.S.; data curation, L.C.E. and M.J.F.; writing—original draft preparation, L.C.E., E.V., and L.S.; writing—review and editing, L.C.E. and M.M.; supervision, A.C.C.; project administration, L.C.E. All authors have read and agreed to the published version of the manuscript.

Funding: This research received no external funding.

Data Availability Statement: Not applicable.

Acknowledgments: The authors acknowledge the support of the Universidad Técnica Particular de Loja and the Secretaría de Educación Superior, Ciencia, Tecnología e Innovación (SENESCYT—Ecuador).

Conflicts of Interest: The authors declare no conflict of interest.

References

1. Zeng, W.J.; Tan, Z.; Lai, X.F.; Xu, Y.N.; Mai, C.L.; Zhang, J.; Lin, Z.J.; Liu, X.G.; Sun, S.L.; Zhou, L.J. Topical delivery of l-theanine ameliorates TPA-induced acute skin inflammation via downregulating endothelial PECAM-1 and neutrophil infiltration and activation. *Chem. Biol. Interact.* **2018**, *284*, 69–79. [CrossRef] [PubMed]
2. Kendall, A.C.; Nicolaou, A. Bioactive lipid mediators in skin inflammation and immunity. *Prog. Lipid Res.* **2013**, *52*, 141–164. [CrossRef] [PubMed]
3. Pireddu, R.; Caddeo, C.; Valenti, D.; Marongiu, F.; Scano, A.; Ennas, G.; Lai, F.; Fadda, A.M.; Sinico, C. Diclofenac acid nanocrystals as an effective strategy to reduce in vivo skin inflammation by improving dermal drug bioavailability. *Colloids Surf. B Biointerfaces* **2016**, *143*, 64–70. [CrossRef] [PubMed]
4. Dainichi, T.; Hanakawa, S.; Kabashima, K. Classification of inflammatory skin diseases: A proposal based on the disorders of the three-layered defense systems, barrier, innate immunity and acquired immunity. *J. Derm. Sci.* **2014**, *76*, 81–89. [CrossRef] [PubMed]
5. Ali, B.; Al-Wabel, N.A.; Shams, S.; Ahamad, A.; Khan, S.A.; Anwar, F. Essential oils used in aromatherapy: A systemic review. *Asian Pac. J. Trop. Biomed.* **2015**, *5*, 601–611. [CrossRef]
6. Ibrahim, T.A.; El-Hela, A.A.; El-Hefnawy, H.M.; Al-Taweel, A.M.; Perveen, S. Chemical Composition and Antimicrobial Activities of Essential Oils of Some Coniferous Plants Cultivated in Egypt. *Iran. J. Pharm. Res. IJPR* **2017**, *16*, 328–337.

7. Valarezo, E.; Ojeda-Riascos, S.; Cartuche, L.; Andrade-González, N.; González-Sánchez, I.; Meneses, M.A. Extraction and Study of the Essential Oil of Copal (*Dacryodes peruviana*), an Amazonian Fruit with the Highest Yield Worldwide. *Plants* **2020**, *9*, 1658. [CrossRef]
8. Perveen, S. Introductory Chapter. In *Terpenes and Terpenoids*; IntechOpen: London, UK, 2018. [CrossRef]
9. Frangedakis, E.; Shimamura, M.; Villarreal, J.C.; Li, F.-W.; Tomaselli, M.; Waller, M.; Sakakibara, K.; Renzaglia, K.S.; Szövényi, P. The hornworts: Morphology, evolution and development. *New Phytol.* **2021**, *229*, 735–754. [CrossRef]
10. Pérez, A.J.; Hernández, C.; Romero-Saltos, H.; Valencia, R. Árboles Emblemáticos de Yasuní, Ecuador. Available online: https://bioweb.bio/floraweb/arbolesyasuni/FichaEspecie/Dacryodes%20peruviana (accessed on 29 September 2020).
11. Torre, L.d.l.; Navarrete, H.; Muriel, M.P.; Macía Barco, M.J.; Balslev, H. *Enciclopedia de las Plantas Útiles del Ecuador*; Herbario QCA de la Escuela de Ciencias Biológicas de la Pontificia Universidad Católica del Ecuador and Herbario AAU del Departamento de Ciencias Biológicas de la Universidad de Aarhus: Quito, Ecuador; Aarhus, Denmark, 2008.
12. Jørgesen, P.M.; León-Yáñez, S. Catalogue of the Vascular Plants of Ecuador. Available online: http://legacy.tropicos.org/ProjectAdvSearch.aspx?projectid=2 (accessed on 11 July 2020).
13. Mestanza-Ramón, C.; Henkanaththegedara, S.M.; Vásconez Duchicela, P.; Vargas Tierras, Y.; Sánchez Capa, M.; Constante Mejía, D.; Jimenez Gutierrez, M.; Charco Guamán, M.; Mestanza Ramón, P. In-Situ and Ex-Situ Biodiversity Conservation in Ecuador: A Review of Policies, Actions and Challenges. *Diversity* **2020**, *12*, 315. [CrossRef]
14. Molares, S.; González, S.B.; Ladio, A.; Agueda Castro, M. Etnobotánica, anatomía y caracterización físico-química del aceite esencial de Baccharis obovata Hook. et Arn. (Asteraceae: Astereae). *Acta Bot. Bras.* **2009**, *23*, 578–589. [CrossRef]
15. Badalamenti, N.; Bruno, M.; Gagliano Candela, R.; Maggi, F. Chemical composition of the essential oil of *Elaeoselinum* asclepium (L.) Bertol subsp. meoides (Desf.) Fiori (Umbelliferae) collected wild in Central Sicily and its antimicrobial activity. *Nat. Prod. Res.* **2020**, *36*, 789–797. [CrossRef] [PubMed]
16. Park, I.-K.; Lee, S.-G.; Choi, D.-H.; Park, J.-D.; Ahn, Y.-J. Insecticidal activities of constituents identified in the essential oil from leaves of Chamaecyparis obtusa against *Callosobruchus chinensis* (L.) and *Sitophilus oryzae* (L.). *J. Stored Prod. Res.* **2003**, *39*, 375–384. [CrossRef]
17. Piccinelli, A.C.; Santos, J.A.; Konkiewitz, E.C.; Oesterreich, S.A.; Formagio, A.S.N.; Croda, J.; Ziff, E.B.; Kassuya, C.A.L. Antihyperalgesic and antidepressive actions of (R)-(+)-limonene, α-phellandrene, and essential oil from Schinus terebinthifolius fruits in a neuropathic pain model. *Nutr. Neurosci.* **2015**, *18*, 217–224. [CrossRef]
18. Tisserand, R.; Young, R. *Essential Oil Safety: A Guide for Health Care Professionals*, 2nd ed.; Churchill Livingstone; Elsevier: London, UK, 2014.
19. Erasto, P.; Viljoen, A.M. Limonene—A Review: Biosynthetic, Ecological and Pharmacological Relevance. *Nat. Prod. Commun.* **2008**, *3*, 1934578X0800300728. [CrossRef]
20. Allenspach, M.; Steuer, C. α-Pinene: A never-ending story. *Phytochemistry* **2021**, *190*, 112857. [CrossRef] [PubMed]
21. Lopez, S.; Lima, B.; Agüero, M.B.; Lopez, M.L.; Hadad, M.; Zygadlo, J.; Caballero, D.; Stariolo, R.; Suero, E.; Feresin, G.E.; et al. Chemical composition, antibacterial and repellent activities of Azorella trifurcata, Senecio pogonias, and Senecio oreophyton essential oils. *Arab. J. Chem.* **2018**, *11*, 181–187. [CrossRef]
22. Souza, A.D.; Lopes, E.M.C.; Silva, M.C.d.; Cordeiro, I.; Young, M.C.M.; Sobral, M.E.G.; Moreno, P.R.H. Chemical composition and acetylcholinesterase inhibitory activity of essential oils of Myrceugenia myrcioides(Cambess.) O. Berg and Eugenia riedelianaO. Berg, Myrtaceae. *Rev. Bras. Farmacogn.* **2010**, *20*, 175–179. [CrossRef]
23. Espinoza, L.C.; Sosa, L.; Granda, P.C.; Bozal, N.; Diaz-Garrido, N.; Chulca-Torres, B.; Calpena, A.C. Development of a Topical Amphotericin B and Bursera graveolens Essential Oil-Loaded Gel for the Treatment of Dermal Candidiasis. *Pharmaceuticals* **2021**, *14*, 1033. [CrossRef] [PubMed]
24. Sarango-Granda, P.; Silva-Abreu, M.; Calpena, A.C.; Halbaut, L.; Fabrega, M.J.; Rodriguez-Lagunas, M.J.; Diaz-Garrido, N.; Badia, J.; Espinoza, L.C. Apremilast Microemulsion as Topical Therapy for Local Inflammation: Design, Characterization and Efficacy Evaluation. *Pharmaceutics* **2020**, *13*, 484. [CrossRef]
25. Espinoza, L.C.; Vera-Garcia, R.; Silva-Abreu, M.; Domenech, O.; Badia, J.; Rodriguez-Lagunas, M.J.; Clares, B.; Calpena, A.C. Topical Pioglitazone Nanoformulation for the Treatment of Atopic Dermatitis: Design, Characterization and Efficacy in Hairless Mouse Model. *Pharmaceutics* **2020**, *12*, 255. [CrossRef]
26. Jang, Y.; Kim, M.; Hwang, S.W. Molecular mechanisms underlying the actions of arachidonic acid-derived prostaglandins on peripheral nociception. *J. Neuroinflamm.* **2020**, *17*, 30. [CrossRef] [PubMed]
27. Akdis, M.; Burgler, S.; Crameri, R.; Eiwegger, T.; Fujita, H.; Gomez, E.; Klunker, S.; Meyer, N.; O'Mahony, L.; Palomares, O.; et al. Interleukins, from 1 to 37, and interferon-gamma: Receptors, functions, and roles in diseases. *J. Allergy Clin. Immunol.* **2011**, *127*, 701–721.e70. [CrossRef]
28. Lin, Z.M.; Ma, M.; Li, H.; Qi, Q.; Liu, Y.T.; Yan, Y.X.; Shen, Y.F.; Yang, X.Q.; Zhu, F.H.; He, S.J.; et al. Topical administration of reversible SAHH inhibitor ameliorates imiquimod-induced psoriasis-like skin lesions in mice via suppression of TNF-alpha/IFN-gamma-induced inflammatory response in keratinocytes and T cell-derived IL-17. *Pharm. Res.* **2018**, *129*, 443–452. [CrossRef] [PubMed]
29. Kondo, S.; Kono, T.; Sauder, D.N.; McKenzie, R.C. IL-8 gene expression and production in human keratinocytes and their modulation by UVB. *J. Investig. Dermatol.* **1993**, *101*, 690–694. [CrossRef]

30. Dhaouadi, T.; Chahbi, M.; Haouami, Y.; Sfar, I.; Abdelmoula, L.; Ben Abdallah, T.; Gorgi, Y. IL-17A, IL-17RC polymorphisms and IL17 plasma levels in Tunisian patients with rheumatoid arthritis. *PLoS ONE* **2018**, *13*, e0194883. [CrossRef] [PubMed]
31. Guerra, E.S.; Lee, C.K.; Specht, C.A.; Yadav, B.; Huang, H.; Akalin, A.; Huh, J.R.; Mueller, C.; Levitz, S.M. Central Role of IL-23 and IL-17 Producing Eosinophils as Immunomodulatory Effector Cells in Acute Pulmonary Aspergillosis and Allergic Asthma. *PLoS Pathog.* **2017**, *13*, e1006175. [CrossRef] [PubMed]
32. Valsalan Soba, S.; Babu, M.; Panonnummal, R. Ethosomal Gel Formulation of Alpha Phellandrene for the Transdermal Delivery in Gout. *Adv. Pharm. Bull.* **2021**, *11*, 137–149. [CrossRef]
33. Basholli-Salihu, M.; Schuster, R.; Hajdari, A.; Mulla, D.; Viernstein, H.; Mustafa, B.; Mueller, M. Phytochemical composition, anti-inflammatory activity and cytotoxic effects of essential oils from three *Pinus* spp. *Pharm. Biol.* **2017**, *55*, 1553–1560. [CrossRef]
34. d'Alessio, P.A.; Mirshahi, M.; Bisson, J.-F.; Bene, M.C. Skin repair properties of dLimonene and perillyl alcohol in murine models. Anti-Inflammatory Anti-Allergy Agents. *Med. Chem.* **2014**, *13*, 29–35.
35. Hansen, J.S.; Norgaard, A.W.; Koponen, I.K.; Sorli, J.B.; Paidi, M.D.; Hansen, S.W.; Clausen, P.A.; Nielsen, G.D.; Wolkoff, P.; Larsen, S.T. Limonene and its ozone-initiated reaction products attenuate allergic lung inflammation in mice. *J. Immunotoxicol.* **2016**, *13*, 793–803. [CrossRef]
36. Ku, C.M.; Lin, J.Y. Anti-inflammatory effects of 27 selected terpenoid compounds tested through modulating Th1/Th2 cytokine secretion profiles using murine primary splenocytes. *Food Chem.* **2013**, *141*, 1104–1113. [CrossRef] [PubMed]
37. Lappas, C.M.; Lappas, N.T. D-Limonene modulates T lymphocyte activity and viability. *Cell Immunol.* **2012**, *279*, 30–41. [CrossRef] [PubMed]
38. Yu, L.; Yan, J.; Sun, Z. D-limonene exhibits anti-inflammatory and antioxidant properties in an ulcerative colitis rat model via regulation of iNOS, COX-2, PGE2 and ERK signaling pathways. *Mol. Med. Rep.* **2017**, *15*, 2339–2346. [CrossRef] [PubMed]
39. Valarezo, E.; Morocho, V.; Cartuche, L.; Chamba-Granda, F.; Correa-Conza, M.; Jaramillo-Fierro, X.; Meneses, M.A. Variability of the Chemical Composition and Bioactivity between the Essential Oils Isolated from Male and Female Specimens of Hedyosmum racemosum (Ruiz & Pav.) G. Don. *Molecules* **2021**, *26*, 4613. [CrossRef] [PubMed]
40. Adams, R.P. *Identification of Essential Oil Components by Gas Chromatography/Mass Spectrometry*, 4th ed.; Allured Publishing Corporation: Carol Stream, IL, USA, 2007.
41. NIST 05. *Mass Spectral Library (NIST/EPA/NIH)*; National Institute of Standards and Technology: Gaithersburg, MD, USA, 2005.

plants

MDPI

Article

Green Synthesized of *Thymus vulgaris* Chitosan Nanoparticles Induce Relative WRKY-Genes Expression in *Solanum lycopersicum* against *Fusarium solani*, the Causal Agent of Root Rot Disease

Sawsan Abd-Ellatif [1,†], Amira A. Ibrahim [2,†], Fatmah A. Safhi [3], Elsayed S. Abdel Razik [4], Sanaa S. A. Kabeil [5], Salman Aloufi [6], Amal A. Alyamani [6], Mostafa M. Basuoni [7], Salha Mesfer ALshamrani [8] and Hazem S. Elshafie [9,*]

1 Bioprocess Development Department, Genetic Engineering and Biotechnology Research Institute, City of Scientific Research and Technology Applications, Alexandria 21934, Egypt
2 Botany and Microbiology Department, Faculty of Science, Arish University, Al-Arish 45511, Egypt
3 Department of Biology, College of Science, Princess Nourah bint Abdulrahman University, Riyadh 11671, Saudi Arabia
4 Plant Protection and Biomolecular Diagnosis Department, Arid Lands Cultivation Research Institute, City of Scientific Research and Technology Applications, Alexandria 21934, Egypt
5 Protein Research Department, Genetic Engineering and Biotechnology Research Institute, City of Scientific Research and Technology Applications, Alexandria 21934, Egypt
6 Department of Biotechnology, Faculty of Sciences, Taif University, Taif 21944, Saudi Arabia
7 Botany and Microbiology Department, Faculty of Science (Boys), Al-Azhar University, Cairo 11884, Egypt
8 Department of Biology, College of Science, University of Jeddah, Jeddah 21493, Saudi Arabia
9 School of Agricultural, Forestry, Food and Environmental Sciences (SAFE), University of Basilicata, 85100 Potenza, Italy
* Correspondence: hazem.elshafie@unibas.it; Tel.: +39-0971-205522; Fax: +39-0971-205503
† These authors contributed equally to this work.

Citation: Abd-Ellatif, S.; Ibrahim, A.A.; Safhi, F.A.; Abdel Razik, E.S.; Kabeil, S.S.A.; Aloufi, S.; Alyamani, A.A.; Basuoni, M.M.; ALshamrani, S.M.; Elshafie, H.S. Green Synthesized of *Thymus vulgaris* Chitosan Nanoparticles Induce Relative WRKY-Genes Expression in *Solanum lycopersicum* against *Fusarium solani*, the Causal Agent of Root Rot Disease. *Plants* **2022**, *11*, 3129. https://doi.org/10.3390/plants11223129

Academic Editor: Kwang-Hyun Baek

Received: 4 October 2022
Accepted: 9 November 2022
Published: 16 November 2022

Abstract: *Fusarium solani* is a plant pathogenic fungus that causes tomato root rot disease and yield losses in tomato production. The current study's main goal is testing the antibacterial efficacy of chitosan nanoparticles loaded with *Thyme vulgaris* essential oil (ThE-CsNPs) against *F. solani* in vitro and in vivo. GC-MS analysis was used to determine the chemical constituents of thyme EO. ThE-CsNPs were investigated using transmission electron microscopy before being physicochemically characterized using FT-IR. ThE-CsNPs were tested for antifungal activity against *F. solani* mycelial growth in vitro. A pot trial was conducted to determine the most effective dose of ThE-CsNPs on the morph/physiological characteristics of *Solanum lycopersicum*, as well as the severity of fusarium root rot. The relative gene expression of WRKY transcript factors and defense-associated genes were quantified in root tissues under all treatment conditions. In vitro results revealed that ThE-CsNPs (1%) had potent antifungal efficacy against *F. solani* radial mycelium growth. The expression of three WRKY transcription factors and three tomato defense-related genes was upregulated. Total phenolic, flavonoid content, and antioxidant enzyme activity were all increased. The outfindings of this study strongly suggested the use of ThE-CsNPs in controlling fusarium root rot on tomatoes; however, other experiments remain necessary before they are recommended.

Keywords: WRKY transcription factor; defense-related genes; biotic stress; fusarium root rot; antioxidant enzymes; antifungal activity

1. Introduction

Tomato (*Solanum lycopersicum* L.) is one of the most widely cultivated vegetable crops worldwide. For those participating in its value chains, the tomato industry provides a significant source of revenue since it is represent a superior source of vitamins, potassium, other minerals, antioxidants and fibers [1–3]. A number of biotic and abiotic stress conditions severely limit the growth and economic productivity of tomatoes [4,5]. Numerous soil-borne pathogens of various horticultural and food crops, including *Fusarium* pathogens, cause lethal vascular wilts, rots, and damping-off diseases [6]. Some *Fusarium* species can produce mycotoxins in food and agricultural products, in addition to the losses caused by pre/postharvest losses [7]. *Fusarium oxysporum* f. sp. *lycopersici* (Fol) and *F. solani* (Fol) are two of the most important causes contributing to economic loss in tomato production [8].

Fusarium root rot disease has been found in numerous regions and is rather frequent (25–55%) in tomatoes [9,10]. Under favorable fungal infection weather conditions, the potential economic losses in tomatoes could be enhanced by up to 80% [11]. According to the literature, fungal mycotoxins have a substantial phytotoxic effect, cause pathogenicity, and are required for root rot disease [12]. According to El-Saadony et al. [13,14], the exposure to mycotoxin metabolites may have a moderately damaging effect on animal cells and opportunistic infections in humans.

Nanotechnology is one of the most active fields of research being an effective antimicrobial activity tool [15–19]. Additionally, chitosan treatment controls the number of defense genes in plants, particularly the activation of signaling pathways, which results in the acquisition of plant antitoxin and pathogenesis-related (PR) proteins [20]. Chitosan has been studied to control several pre- and postharvest diseases [21]. *Thymus* sp., belonging to the *Lamiaceae* family, contains over 300 evergreen species that are naturally grown in Southern Europe and Asia. Much research has supported the therapeutic potential of thyme EO in the treatment of cancer, immune deficiency syndrome, infection, and retinal neovascularization [22–24].

Gene expression changes cause and occupy these alterations, according to Saijo et al. [25]. It has been proposed that plant transcription factors, which belong to several families, take part in stress reduction or other coping mechanisms when exposed to stress by altering the gene expression patterns. According to Vives-Peris et al. [26], the WRKY transcription factors (Tfs) are a group of plant-specific zinc finger-type regulatory proteins playing a critical role in plant development and the defense response to various abiotic and biotic stresses. The WRKY proteins modulate downstream target genes, other genes (encoding Tfs), or their own expression in order to control the expression of the genes either directly or indirectly [27]. WRKY TFs are essential for controlling a wide range of biological functions, but they are particularly important for controlling how plants react to biotic and abiotic stresses in tomatoes [28–30]. Dynamic changes in the quantities of accumulated WRKY transcripts inside the cell can be used to establish how the plant defense regulation involving WRKY proteins is regulated [31]. The main objective of the current research was to determine the most efficient doses of ThE-CsNPs able to inhibit the growth of *F. solani*, causing root rot disease for *S. lycopersicum*. There may be alterations in total phenols, certain antioxidant enzyme activity, and seedling germination investigating the function and mechanism of WRKY family members and their responses to fusarium root rot disease resistance. It will be estimated by examining the expression levels of defense-related genes in tomato roots, such as glucanase A, defensin, and chitinase, as well as transcript factors WRKY4, WRKY31, and WRKY37.

2. Results

2.1. GC-MS Composition of *T. vulgaris* EO

The GC-MS analysis results of *T. vulgaris* EO showed bioactive compounds, as listed below in Table 1. In particular, the main constituents are: thymol, α-Pinene, camphene, carvacrol, caryophyllen, carvacrol, myrcene, α-terpineol (Figure S1). The highest percentages of constituents presented in *T. vulgaris* EO were camphene (35.97%), cyclohexane (10.1%), myrcene (7.6%), and α-pinene (6.5%).

Table 1. The list of main constituents present in *T. vulgaris* EO using GC-MS analysis.

Quantitative ID	Component Identified	Retention Time (min)	Retention Index (RI)	Area (%)	Identification
1	Cyclohexane	4.11	1215	10.12	RI, MS *
2	Myrcene	5.13	106	7.55	RI, MS
3	Caryophyllene	9.14	974	3.25	RI, MS
4	α-Pinene	12.39	993	6.49	RI, MS
5	Camphene	16.81	1429	35.97	RI, MS
6	β-Myrcene	17.74	974	2.77	RI, MS
7	Carene	18.56	1042	2.86	RI, MS
8	p-Cymene	20.35	938	2.19	RI, MS
9	γ-Terpinene	21.16	1062	2.68	RI, MS
10	α-Terpineol	21.53	1138	2.92	RI, MS
11	Linalool	26.13	1126	2.84	RI, MS
12	Thymol	27.85	1074	2.31	RI, MS
13	α-Thymol	29.14	1261	5.17	RI, MS
14	Carvacrol	29.38	1062	2.21	RI, MS
15	Caryophyllen	32.75	1211	4.91	RI, MS
16	Total	-	-	94.24	-

(*) RI: retention index, MS: mass spectroscopy.

2.2. Characterization of ThE-CsNPs

2.2.1. Transmission Electron Microscope

A vital characterization method for obtaining quantitative estimates of particle size, size distribution, and morphology of nanomaterials is transmission electron microscopy (TEM). The generated ThE-CsNPs by using TEM was illustrated in Figure 1, where the majority of the particles were spherical and irregular. ThE-CsNPs had an average size ranging between 20 and 80 (±0.84) nm.

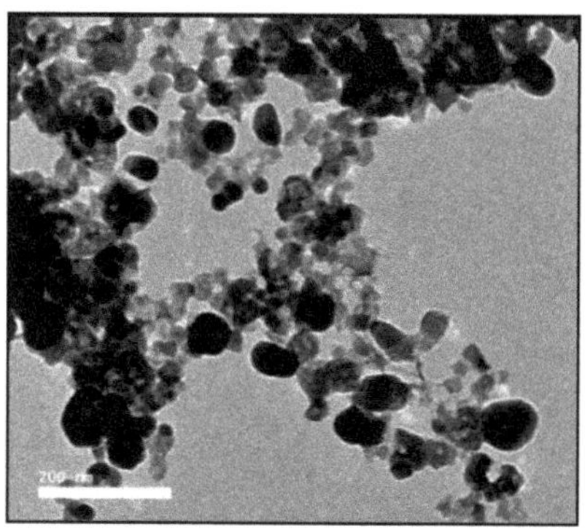

Figure 1. TEM microphotography of ThE-CsNPs.

2.2.2. Fourier Transform Infrared Spectroscopy

The chemical structure of the components involved in ThE-CsNPs was characterized using the Fourier transform infrared (FT-IR) technology, as illustrated in Figure S2, which demonstrates the presence of numerous functional groups in its structure.

2.3. *In Vitro Antifungal Activity*

In vitro antifungal activities of different concentrations (0.2, 0.4, 0.6, 0.8, and 1.0%) of ThE-CsNPs were assessed against the *F. solani* pathogen. The average reduction in *F. solani* radial growth in response to treatment is illustrated in Figure 2. All tested concentrations exhibited varied inhibitory activity as compared with the positive control fungicide nystatin (0.05%) and the untreated experimental control. Inhibition of mycelium growth increased with time. After 7 days of incubation at 28 ± 2 °C, the highest growth reduction percentage

(89.6 ± 0.0) was observed at ThE-CsNPs (1%) solution treatment, followed by (69.0 ± 0.0 and 31.0 ± 0.0) at 0.8 and 0.6% ThE-CsNPs, respectively, as compared with (62.1 ± 0.1) reduction by using nystatin (0.05%). The minimum growth inhibition (20.8 ± 0.0) was obtained at 0.2% ThE-CsNPs treatment.

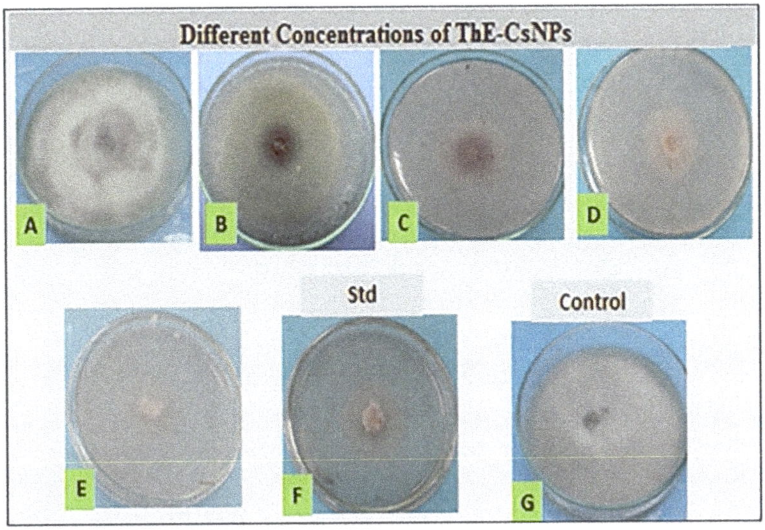

Figure 2. The antifungal activity of ThE-CsNPs at different concentrations against *F. solani*. Where (**A**) 0.2, (**B**) 0.4, (**C**) 0.6, (**D**) 0.8, (**E**) ThE-CsNPs (1%), (**F**) nystatin (0.05%), and (**G**) negative control.

2.4. *In Vivo Trial*

2.4.1. Pathogenicity Assay

The first signs of the root rot disease of fusarium appeared 18 days after infection (DAF) (Figure 3). In particular, the highest disease severity (DS) was recorded as 92.4% in case of positive disease control. In contrast, the lowest DS (4.3%) was observed in case of ThE-CsNPs infected seedlings compared to ThE-CsNPs primed seedlings (2.3%) (Figure 4).

Figure 3. Effect of ThE-CsNPs (1%) on *S. lycopersicum* seedling growth under fusarium root rot disease (20 DAF).

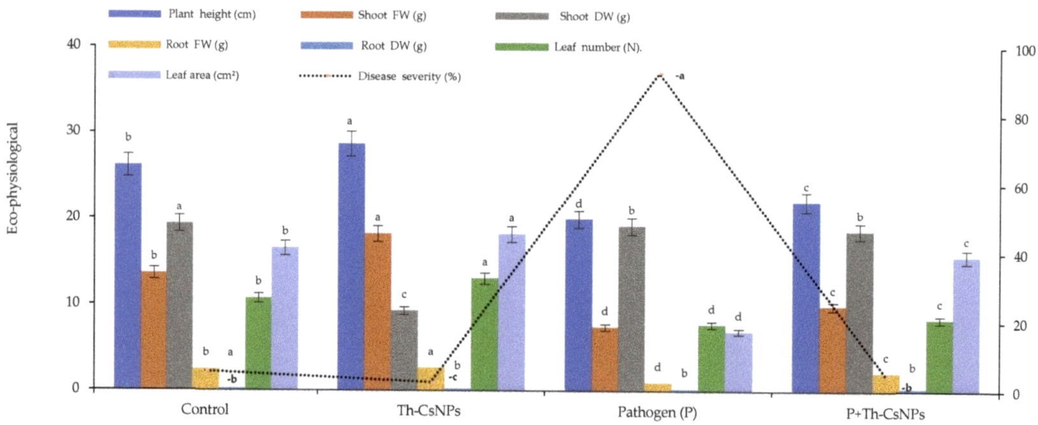

Figure 4. Effects of ThE-CsNPs on disease severity and eco-physiological parameters of *S. lycopersicum* seedlings. Bars with different letters indicate significant differences between treatments at $p \leq 0.05$. Data are expressed as mean values of three replicates (±SDs).

2.4.2. Eco-Physiological Parameters

It was found that the priming of tomato seedlings with ThE-CsNPs (1%) resulted in the profuse growth of the tomato plants. Morphological characters differed by using ThE-CsNPs treatment after 20 days; plant height was increased after ThE-CsNPs with a value of 28.6 cm. Shoot dry weights were also increased after ThE-CsNPs treatment to 18.18 g. Root fresh and root dry weights were 2.6 and 0.3 g, respectively after treatment as presented in Figure 4.

2.4.3. Physiological and Biochemical Characteristics

Total chlorophyll recorded the highest value at 55.2 and 57.2 μg/g FW in case of using ThE-CsNPs after 48 and 96 h, respectively. Anthocyanin content was 1.3 and 1.4 Ug/g FW after 48 and 96 h, respectively, in case of using ThE-CsNPs. The highest malondialdehyde (MDA) contents (92.4, 106.4) were observed after 48 h and 96 h, respectively, of *F. solani* inoculation under P+Th-CsNPs primed. The total phenolic contents (TPC) were recorded for control, and ThE-CsNPs primed tomato plant under *F. solani* (F.s) infestation condition, and the results showed a gradual increase in TPC level at 48 h and significant elevations at 96 h post inoculation with values 42.6 and 52.9 μmol/g FW, respectively, under P+Th-CsNPs primed. The total flavonoid contents (TFC) were determined in the control and ThE-CsNPs-primed tomato plant under *F. solani* infestation condition, and the results showed significant elevations at 96 h of infestation as compared with control and gradually increased at 48 h post infestation. The highest TFC content was 136.1 μmol/g FW observed after 96 h of *F. solani* inoculation under P+Th-CsNPs primed and at 48 h of *F. solani* inoculation under P+Th-CsNPs primed was 121.1 μmol/g FW. Protein content (PC) in tomato plants was increased under P+Th-CsNPs at 48 and 96 h with values of 19.7 and 21.7 nmol/g FW, respectively, as presented in Figure 5.

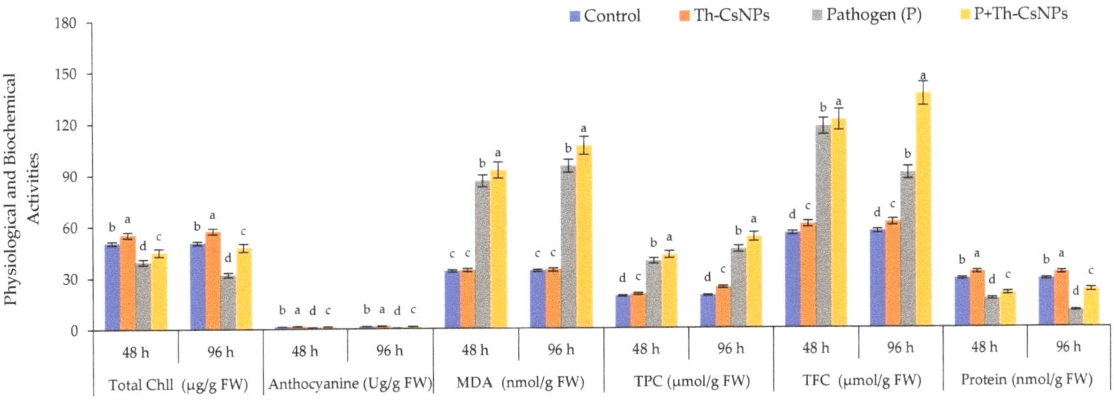

Figure 5. Effects of ThE-CsNPs on total chlorophyll, anthocyanin, MDA, TPC, TFC, and PC level in *S. lycopersicum* seedling under fusarium root rot disease infection. Bars with different letters indicate significant differences between treatments at $p \leq 0.05$. Data are expressed as mean values of three replicates (±SDs).

2.4.4. Antioxidant Enzymes Activities

Data presented in Figure 6 represent the effects of ThE-CsNPs priming on tomato superoxide dismutase (SOD), catalase (CAT), and ascorbate peroxidase (APX) antioxidant enzymes under *F. solani* inoculation at different times intervals (48 and 96 h post fungal inoculation). Data obtained indicated that infection of tomato plants with *F. solani* to an induction in the activities of all studied enzymes when compared with ThE-CsNPs primed and untreated control plants at 48 and 96 h. The highest SOD enzyme activity was 67.1 and 70.2 U/mg protein observed after 48 and 96 h, respectively, under P+ThE-CsNPs primed tomato seedlings. The maximum CAT enzyme activity values 88.3 and 92.5 U/mg were observed after 48 and 96 h of F.s inoculation under P+ThE-CsNPs primed and tomato seedling. The maximum APX enzyme activity values were 28.4 and 36.1 U/mg protein after 48 and 96 h, respectively, of F.s inoculation under P+ThE-CsNPs primed tomato seedlings as compared with the control.

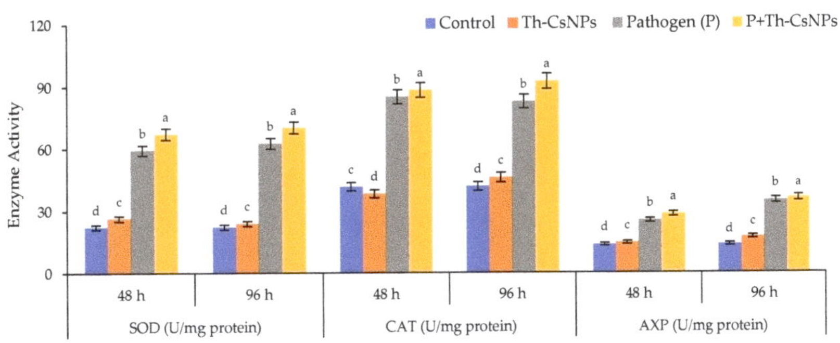

Figure 6. Effects of ThE-CsNPs on SOD, CAT, and AXP enzyme activity in *S. lycopersicum* seedling under fusarium root rot disease infection. Bars with different letters indicate significant differences between treatments at $p \leq 0.05$. Data are expressed as mean values of three replicates (±SDs).

2.5. Molecular Analysis

With the help of mRNA, qRT-PCR was used to measure the expression levels of the several WRKY transcription factors, which are crucial for both biotic and abiotic tolerance and include WRKY4, WRKY31, WRKY31, and WRKY37. Additionally, chitinase (PR3) and defensin (PR12) genes were represented by qRT-PCR as the relative expression levels of defense-related proteins in tomato plant roots two weeks after *Fusarium* inoculation and after the application of ThE-CsNPs. Hierarchical clustering heat map of transcriptional expression of the investigated WRKYs transcripts factors (WRKY4, WRKY31, and WRKY37) and defense-related (β-1,3-glucanase, defensin, chitinase) genes in tomato roots (Figure 7). As represented from the heat map, all tested treatments are grouped into two main clusters, the first one represents the untreated control plants, and ThE-CsNPs primed plants at 48 and 96 h post *F. solani* inoculation, while the other represents the Fs infected plants that primed with ThE-CsNPs solution or not at 48 and 96 h.

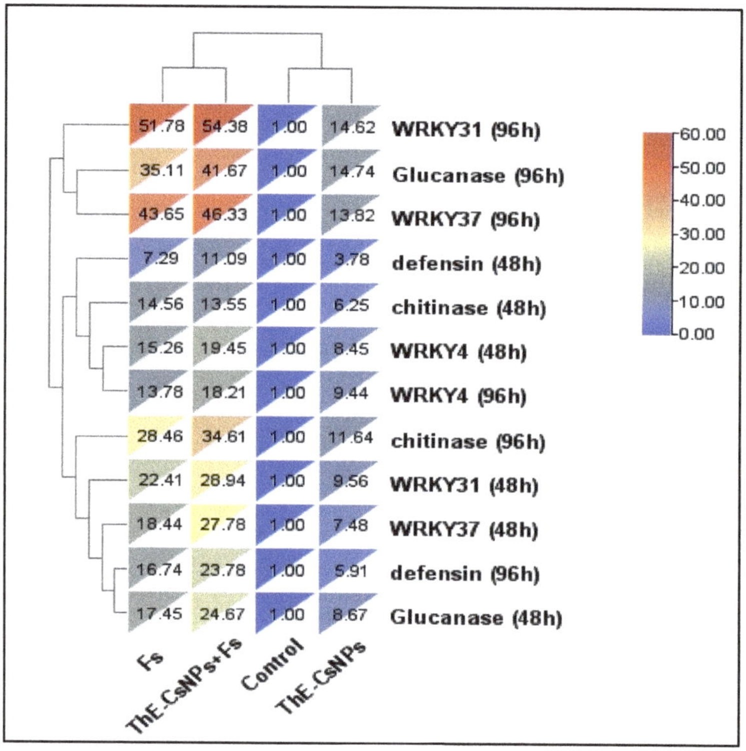

Figure 7. Hierarchical clustering heat map of transcriptional expression of three WRKY transcription factors and some defense-related genes in tomato plant infected with *F. solani* and/or primed with ThE-CsNPs (1%) after 2 and 4 days post *F. solani* inoculation. Where, C: untreated control, Fs: infected with *F. solani*, ThE-CsNPs: thyme extract loaded chitosan nanoparticles solution at 1.0%, and ThE-CsNPs+ Fs: infected tomato plants with *F. solani* and primed with ThE-CsNPs (1%).

In our study, the highest expression mRNA level (28.9 and 54.4) at 48 and 96 h post inoculation with ThE-CsNPs priming, respectively, was recorded in the WRKY31 gene in roots, followed by (27.8 and 46.3) at 48 and 96 h post inoculation with ThE-CsNPs priming, respectively recorded in WRKY37 gene in roots. Overall, the WRKY TFs genes (WRKY1, WRKY4, WRKY31, and WRKY53) in roots showed positive expression levels (upregulation) under infection.

3. Discussion

Producing crops, especially vegetables, can be severely hampered by soil-borne illnesses brought on by a variety of fungi, including *F. solani*. On the remains of the host plant, in the organic matter of the soil, or as free-living organisms, soil-borne infections frequently persist for protracted durations. Even with standard methods, they are frequently challenging to control. In addition, these soil-borne fungi developed resistance to chemical fungicides. Crop productivity can be significantly hampered by soil-borne illnesses brought on by a variety of fungi, including *F. solani*, especially for vegetable crops. On the remains of the host plant, in the soil's organic matter, or as free-living organisms, soil-borne infections frequently last for a very long time. Even when using traditional techniques, they are frequently challenging to be managed. Therefore, finding innovative, safe, and efficient fungicide solutions to manage plant infections such as tomato root rot is a serious challenge. Since the focus is now on creating non-toxic, safe nanoparticles, the idea of using nanoparticles as an antibacterial agent is relatively new [32].

The thought of using nanoparticles as antifungal agents is relatively new, as the emphasis has since changed toward creating secure, non-toxic nanoparticles [33]. Root rot disease in tomato crops can be controlled in many ways by utilizing chemical and biological approaches [34]. The latest management trend for reducing the negative effects of chemicals (fungicides) is a biological control for pathogenic fungi [35,36]. Microorganisms, semi-chemical products, naturally derived goods derived from plants, and living microorganisms are the four different forms of biocontrol management [37–40].

Additionally, plants essential oils are efficient biocontrol agents against a range of pathogenic bacteria and fungi. Plant EOs are much potential because of their diverse origins and abilities to prevent mycelial growth and spore germination and regulate the vitality of conidia. In comparison to traditional pesticides, nano-emulsions created from these plant EOs may be more effective in controlling pathogens [41–44]. Researchers assessing the potential phytotoxicity of plant extracts have demonstrated that systemic administration has an impact on Fol [45].

The mode of action of any control agents is crucial and aids in enhancing their efficacy when we encounter efficacy problems. The current study demonstrated in vitro and in vivo effects ThE-CsNPs against root rot disease of tomato infected with *F. solani*. Results of in vitro assay showed that ThE-CsNPs (1%) has greatly inhibited the linear growth of the tested pathogenic fungi. Chitosan changes the rhizosphere's balance, favoring helpful microbes, including *Bacillus*, *Pseudomonas fluorescens*, *Actinomycetes*, *Mycorrhiza*, and *Rhizobacteria*, while disadvantages microbiological pathogens [46,47]. It has been demonstrated that treating soil with chitin and/or chitosan reduced the rate of nematode infection of plant roots [46,47] and increased the suppressiveness against soil-borne diseases.

In fact, chitosan/microbe interaction causes a variety of changes in cell permeability as follows: (1) bringing electronegative charges on the outer surface of fungal membrane; (2) addition of polycationic character on the chitosan amino group, interfering with homeostasis ions (K+, Ca^{2+}) and, hence, enhancing the outflow of small molecules that interfere with fungal respiration [48]; (3) microbial starvation due to the chelation process between chitosan-metals-vital nutrients; (4) inhibition of the synthesis of mRNA and proteins, which is related to their capacity to pass through the microbe cell membrane and subsequently bind to DNA [48]. The efficiency of chitosan is most frequently attributed to changes in the composition and/or the activity of soil microbiota [49].

Generally, the antimicrobial activity of plant EOs has been verified by several researchers all over the world, where they explained that EOs are able to disrupt the cellular enzyme system and causing the stiffness loss of the hyphal cell wall [50]. Furthermore, the presence of terpenes, alcohols, and phenolic compounds in the composition of EOs are important to improve the physiological and biochemical parameters of tomato seedlings [51,52].

The obtained results confirmed that thyme EO has antifungal effect against fusarium infection at all tested concentrations. In particular, the highest tested concentration (1%) was the lethal dose increased the mycelium growth reduction up to 89.7%. On the other hand, thyme EO has significantly reduced *Botrytis cinerea* colonization on pretreatment detached leaves and significantly decreased the severity of fusarium wilt up to 30.8% particularly 7 days after treatment [53].

According to the chemical analysis, the obtained results showed that the studied thyme EO is a thymol chemotype in agreement with Al-Asmari et al. [54]. In particular, the current study also explicated that the main constituents of thyme EO are camphene (36.1%) and cyclohexane (10.12%). Consequently, Micucci et al. [55] reported that thyme EO is of the thymol type, where carvacrol and p-cymene are the main components. Terpenoids such as thymol and carvacrol are frequently present in the EO and play a significant role in its biological activity [56,57]. The Food and Drug Administration (FDA) has approved them for use in food [57]. Thymol and carvacrol are extremely efficient against foodborne bacteria such as *Salmonella* spp. and *Staphylococcus aureus* [58,59]. The same authors reported that the mechanism of antimicrobial action of this EO could be due to the synergic effect between thymol, citral, and carvacrol, which can transport across microbe membranes [60,61]. In a different study, Isham et al. [62] showed that linalool had an antifungal effect on *Candida albicans* inhibiting the development of germ tubes and biofilm.

Diverse regulatory mechanisms are present in members of the WRKY family. Through the cis-acting mechanism, their protein can be effectively coupled with W-box elements and bind to acting regions to activate or inhibit the transcription of downstream target genes [63]. Thus, WRKY as a transcription factor contributes significantly to plant defense against various pathogen attacks. This reaction either directly or indirectly activates the expression of resistance genes. According to literature review, WRKY DNA binding proteins bind to the promoter region of the plant defense system-activating Arabidopsis natriuretic peptide receptor 1 (NPR1) [64].

The current findings also showed that overexpression of chitinase, defensin, and WRKY transcripts contributing the plant resistance against *F. solani* and reduce its severity. Additionally, pathogen infection substantially elevated WRKY transcripts and PR3 and PR12 genes, which are already regarded as markers for the plant-microbe interaction. Our findings are in agreement with our previous research, which concluded that WRKY3 and WRKY4 encode two structurally related WRKY proteins and that both of these proteins' expressions were sensitive to stressful environments. Pathogen infection further increased the expression of WRKY4, which was induced by stress but not WRKY3. These findings imply that WRKY4 regulates crosstalk between SA and JA/ET-mediated signaling pathways and, hence, its role in resistance to *F. solani* root rots disease. It is interesting to underline that some WRKY genes, including WRKY4, WRKY33, redundant WRKY18, WRKY40, and WRKY60, contribute to plant resistance to necrotrophic diseases [65,66].

4. Materials and Methods

4.1. Sample Collection, Identification, and Preparation

The Department of Plant Protection and Biomolecular Diagnosis, Arid Lands Cultivation Research Institute, Egypt, provided the single spore culture (monosporic culture) of *F. solani* (KJ831188) isolate. The *F. solani* isolate was cultivated on potato dextrose agar (PDA) media in plate and slant at 28 °C for 10 days, followed by microscopic analysis with lactophenol cotton blue stain. Up to the next bioassay, the fungus was kept at 4 °C. Tomato seeds (Dania commercial hybrid) variety with suitable appearance and uniform size were purchased from the Egyptian Ministry of Agriculture.

4.2. Medicinal Plant Material

On the basis of their ethno-medicinal importance and literature, the *T. vulgaris* plant was collected from an aromatic ornamental garden at Al Nubaria, Egypt (longitudes 30°10′ E and latitudes 30°52′ N) in May 2022. The collected plants were identified and authenticated by the Department of Botany, Faculty of Science—Mansoura University, Egypt, using standard references.

4.2.1. Extractions of Essential Oils

T. vulgaris fresh leaves had been previously harvested and sterilized for 30 min with sodium hypochlorite (2%) [67], then were rinsed with sterile distilled water. Fresh leaves had been air-dried, ground, and kept in sealed vials under darkness until use. Hydro-distillation of leaf powders has been conducted for 3 h using a Clevenger-style device.Ltd., India's New Delhi [68].

4.2.2. GC-MS of Essential Oils

The volatile components of thyme EO have been screened using GC-MS-QP2010 Ultra analysis equipment. The oven temperature schedule started at 50 °C, held for 3 min, then rose by 8 °C/min to 250 °C, holding for 10 min. In electron impact mode, the spectrophotometer was used. The injector, interface, and ion source were maintained at 250, 250, and 220 °C, respectively. Helium served as the carrier gas for the split injection, which used a split ratio of 1:20 and a column flow rate of 1.5 mL/min to inject a 1 µL diluted sample in *n*.hexane (1:1, *v*/*v*). The main single components were identified using WILEY and National Institute of Standards and Technology (NIST08) libraries based on their relative indices and mass spectra.

4.2.3. Preparation of ThE-CsNPs

ThE-CsNPs have been prepared using the ionic gelation process and sodium tripolyphosphate (TPP) as a crosslinking agent as follows. Low molecular weight (LMW) chitosan (0.2 g) at 5% was dissolved in 40 mL of acetic acid (1% *v*/*v*) by stirring at 1000 rpm overnight at room temperature. The pH was then raised to 4.6 by adding 1M NaOH. A 0.45 µm syringe filter was then used to filter the chitosan solution. About 200 µL of *T. vulgaris* EO was added to the CS solution. After that, 8 mL of TPP (2 mg/mL) was added dropwise (1 mL/min) and adjusted to pH = 4.0, with stirring, at 900 rpm for 1 h, then centrifuged at 10,000× *g* rpm for 10 min. The precipitate was collected and oven-dried at 40 °C and kept at room temperature.

4.2.4. Fourier Transform Infrared Analysis

The functional groups of ThE-CsNPs were characterized using FT-IR analysis in order to confirm their synthesis. ThE-CsNPs powder was mixed with potassium bromide (KBr) in a 1:100 ratio, and a Shimadzu FT-IR (Shimadzu 8400S, Kyoto, Japan) apparatus was used to record the spectra.

4.2.5. Transmission Electron Microscopy

The form and size of ThE-CsNPs nanoparticles were investigated using transmission electron microscopy (JEOL JEM-2100 equipment, JEOL Ltd., Tokyo, Japan). The sample was prepared by placing a drop on a copper grid, coating it with carbon, and drying it under a lamp.

4.3. *In Vitro Antifungal Activity of ThE-CsNPs*

The in vitro antifungal efficacy of ThE-CsNPs was carried out in the Bacteriology-Phytopathology Laboratory in the School of Agricultural, Forestry, Food and Environmental Sciences (SAFE), University of Basilicata, Potenza, Italy. The radial mycelia growth of *F. solani* was assessed using the incorporation method [69]. Different concentrations of ThE-CsNPs were used as 0.2, 0.4, 0.6, 0.8 and 1.0% were prepared to determine the most efficient dose with the highest antifungal effect. Nystatin (0.05%) was used as a synthetic (8 µL/150 mL PDA Petri). Sterile distilled water was used as a negative control. About 0.5 cm Ø of *F. solani* fresh culture was inoculated in the center of Petri dishes. The Petri dishes were incubated at 24 °C, then the diameter of mycelium was measured in cm, and the growth reduction percentage (GR %) was calculated using Formula (1). The whole experiment was carried out twice with triplicates.

$$\mathrm{GR}\ (\%) = \frac{\text{Diameter of mycelium in treated plates (cm)}}{\text{Diameter of myclium in contraol plates (cm)}} \times 100 \quad (1)$$

4.4. *In Vivo Greenhouse Trial*

4.4.1. Seeds Treatments

Tomato seeds were surface sterilized in sodium hypochlorite for 30 min, followed by five sterile water washes. In the first experiment, tomato seeds were treated by soaking on 1.0% ThE-CsNPs as a priming solution for 5 h prior to germination in peat moss, while the control treatment was steeped in dH_2O. Hoagland solution at 1/4 strength was frequently applied to the peat moss.

4.4.2. Pathogenicity Assay

Fungal suspension of *F. solani* (10^5–10^6 spores/mL), from 7 days fresh culture, was used for the pathogenicity test for root rot disease. Fresh tomato seedlings (20 d age) were uprooted either from control and ThE-CsNPs-treated ones, washed, surface sterilized with 0.1% mercuric chloride, and immersed in the prepared *F. solani* spore suspension for 60 min. Plants were grown in greenhouses (12 h light/dark and temperature from 18 to 30 °C) and were irrigated daily. The symptoms of root rot disease were first noticed 2–3 weeks after inoculation.

4.4.3. Pot Trials

Four pot groups were arranged in a randomized complete block design with five replications and regularly irrigated with 1/4 strength Hoagland solution as necessary and kept under natural daylight and humidity at 65% until the end of the experiment. The four groups were: (1) tomato seedlings were grown under controlled conditions (control, C); (2) inoculated tomato seedlings only with *F. solani* (pathogen, P); (3) the control tomato seedlings foliar-treated with ThE-CsNPs (1%) contains one drop of tween-40 in the surrounding areas of plant stems (C+ ThE-CsNPs); (4) the infected seedlings (15 days after infection) were foliar-treated with ThE-CsNPs (1%) solution (P+ThE-CsNPs). All seedlings were irrigated regularly for 20 days.

4.4.4. Disease Assessments

Disease severity (DS) of fusarium root rot was assessed, 15 days after infection, using the scale described by Filion et al. [70] (Formula (2)).

$$\text{Disease severity } (\%) = \frac{\Sigma \text{ab}}{AK} \times 100 \quad (2)$$

where: (a) number of diseased plants with the same infection degree; (b) infection degree; (*A*) total number of the assessed plants; and (*K*) the greatest infection degree.

4.4.5. Eco-Physiological Parameters

Sampling was performed 10 to 20 days of treatment with ThE-CsNPs (1%). Morphological traits for both treated and untreated plants were measured. Three plants of each experiment were harvested and transferred to the laboratory and carefully uprooted and were measured for plant height, leaf number, and leaf area. After that, the plants were measured for shoot and root fresh weight and shoot and root dry weight after oven drying at 40 °C for 48 h.

4.4.6. Biochemical Assessment

Malondialdehyde content (MDA) in fresh tomato leaves from treated seedlings with ThE-CsNPs (1%) and control ones were measured following the method described by Heath and Packer [71]. Briefly, 0.5 g fresh tomato leaf was homogenized with 10 mL ethanol and centrifuged at 12,000× *g* for 15 min. After that, 1 mL of supernatant was added to 2 mL of thiobarbituric acid (0.65%) and trichloroacetic acid (20%) mixture. This mixture was boiled for 30 min, cooled rapidly, and centrifuged at 12,000× *g* for 10 min. The MDA content was determined in the supernatant at an absorbance of 532 and 600 nm using a UV-VIS spectrometer (Jenway, Essex, UK). Total phenolic content (TPC) for all treatments was determined by dissolving 5 mg of air-dried powder of leaf in 10 mL methanol using the Folin–Ciocalteu reagent protocol [72]. The total flavonoid content (TFC) for all treatments was measured using the aluminum chloride colorimetry method [73]. The content of soluble protein was estimated for all treatments following Bradford [74].

For total chlorophyll analysis, extracts were produced in triplicate by adding 2 mL of 90% MeOH containing 10% water (*v*/*v*) to 20 mg of each sample, sonicated for 1 h at 4 °C. The crude extract was centrifuged for 20 min at 4000× *g* rpm/4 °C to separate plant debris and supernatant. After that, the supernatant was filtered using a 0.45 m filter, and then 1.5 mL of filtrate was collected for the measurement of functional components. About 1.5 mL of a 10 diluted solution and 150 µL of the 1.5 mL filtered supernatant were combined with 90% MeOH containing 10% H_2O for analysis. Using a spectrophotometer (Cary 60 UV-Vis, Agilent Technologies, Santa Clara, CA, USA), the absorbance of the diluted sample solution was measured at 665.2, 652.4, and 470.0 nm wavelengths, as described previously [75]. The spectrophotometer's specifications include a Xenon Flash Lamp (80 Hz) as a light source, measuring wavelengths ranging from 190 to 1100 nm with a resolution of 1.5 nm. Total chlorophyll a (Chla) and total chlorophyll b (Chlb) were calculated using absorbance (A) at each wavelength (Formulas (3) and (4)).

$$\text{Chla}\left(\mu\text{g mL}^{-1}\right) = 16.82\ \text{A}665.2 - 9.28\ \text{A}652.4 \tag{3}$$

$$\text{Chlb}\left(\mu\text{g mL}^{-1}\right) = 36.92\ \text{A}652.4 - 16.54\ \text{A}665.2 \tag{4}$$

The total anthocyanin analysis has been carried out following the method described by Yang et al. [76] with some minor modifications. The sample extracts were made in triplicate. Briefly, 20 mg of the sample was sonicated for 1 h at 60 °C in 2 mL of acidic MeOH containing 1% HCl (*v*/*v*). The crude extract was centrifuged and filtered in the same manner as samples were pre-processed for total chlorophylls. Then, 300 µL of the 1.5 mL filtered supernatant was diluted with MeOH containing 1% HCl, yielding a 5 diluted solution of 1.5 mL volume. The solution's absorbance was measured at 530 and 600 nm wavelengths, and the results were used to compute the total anthocyanin concentrations using Formula (5) [76].

$$\text{y}\ (\mu\text{g g}^{-1}) = (\text{A}530 - \text{A}600)\frac{V\ x\ n\ x\ Mw}{\varepsilon x\ m} \tag{5}$$

where A530 is the absorbance at 530 nm; A600 is the absorbance at 600 nm; *V* is the total volume of the extracted solution; *n* is the dilution ratio; *Mw* is the molecular weight of cyanidin-3-glucoside (i.e., 449.4); ε is the anthocyanin molar extinction coefficient (29.60 $M^{-1}{\cdot}cm^{-1}$); and *m* is the sample mass.

4.4.7. Assay of Antioxidant Enzymes

Antioxidant enzymes were extracted by homogenizing 1 g fresh tomato leaf tissue in chilled 50 mM phosphate buffer (pH 7.0) supplemented with 1% polyvinyl pyrolidine and 1 mM EDTA using a prechilled pestle and mortar. Centrifuging at 18,000× *g* for 30 min at 4 °C, the supernatant was used for enzyme assay.

The activity of superoxide dismutase (SOD, EC 1.15.1.1) and NBT photochemical reductions were recorded at 560 nm in a 1.5 mL assay mixture containing sodium phosphate buffer (50 mM, pH 7.5), 100 µL EDTA, L-methionine, 75 µM NBT, riboflavin, and 100 µL enzyme extract. After 15 min of incubation, the light was switched off, and the activity was expressed as EU mg^{-1} protein.

The catalase activity assay (CAT, EC1.11.1.6) was carried out according to Luck [77], where the change in absorbance was monitored at 240 nm for 2 min, and the extinction coefficient of 39.4 $mM^{-1}{\cdot}cm^{-1}$ was used for the calculation.

Ascorbate peroxidase activity assay (APX, EC 1.11.1.11) was carried out by monitoring absorption change at 290 nm for 3 min in 1 mL reaction mixture containing potassium phosphate buffer (pH 7.0), 0.5 mM ascorbic acid, hydrogen peroxide, and enzyme extract. The calculation of the extinction coefficient of 2.8 mM/cm was used.

4.5. Molecular Analysis

4.5.1. RNA Extraction and cDNA Synthesis

The relative expression of distinctly upregulated WRKY transcripts factors and defense-related genes under each treated condition was carried out in the central lab of the Faculty of Science, Arish University (August 2022) and measured quantitatively in root tissues at different time intervals (24 and 48 h) after Fs infestation. Total RNA was isolated from 0.5 g from fresh leaves of tomato plants for each treatment after 10 days from infestation by using the TRIZOL reagent (Invitrogen, Waltham, MA, USA) according to the manufacturer's protocol. The purified RNA was analyzed on 1% agarose gel. For each sample, to obtain cDNA, 10 µL total RNA was treated with DNAse RNAse-free (Fermentas, Waltham, MA, USA), 5 µL of which was reverse transcribed in a reaction mixture consisting of oligo dT primer (10 pml/µL), 2.5 µL 5× buffer, 2.5 µL $MgCl_2$, 2.5 µL 2.5 mM dNTPs, 4 µL from oligo (dT), 0.2 µL (5 unit/µL) reverse transcriptase (Promega, Germany) and 2.5 µL RNA. RT-PCR amplification was performed in a thermal cycler PCR at 42 °C for 1 h and 80 °C for 15 min.

4.5.2. Real-Time Quantitative PCR Analysis

The quantitative study was performed to evaluate the temporal expression of accumulated WRKY transcripts factors and other defense-related genes in all four treatment conditions. Quantitative real-time PCR was carried out on 1 µL diluted cDNA by triplicate using the real-time analysis using (Rotor-Gene 6000, Germany) system, and the primer sequences used in qRT-PCR were given in Table S1. Primers of three defense-related genes (chitinase, defensin, and β-1,3-Glucanase) and three WRKY TFs genes (WRKY4, WRKY31, WRKY37) and housekeeping gene (reference gene) as listed in Table S1 were used for gene expression. Analysis used an SYBR® Green-based method; a total reaction volume of 20 µL was used. The reactions mixture consists of 2 µL of template, 10 µL of SYBR Green Master Mix, 2 µL of reverse primer, 2 µL of forward primer, and sterile distilled water for a total volume of 20 µL. PCR assays were performed using the following conditions: 95 °C for 15 min followed by 40 cycles of 95 °C for 30 s and 58 °C for 30 s. The CT of each sample was used to calculate ΔCT values (target gene CT subtracted from β-Actin gene CT). The relative gene expression was determined using the $2^{-\Delta\Delta CT}$ method [78].

4.6. Statistical Analysis

The obtained data were expressed as the mean ± standard deviation. Statistical analysis was performed using the statistical package SPSS for social sciences (SPSS) 16. All outfindings were analyzed using a one-way ANOVA analysis of variance performed for the significant difference at the $p < 0.05$ level. In contrast, the heatmap was created to compare and contrast the examined treatment responses against different used genes using the TB tools software [79].

5. Conclusions

This study concluded that ThE-CsNPs explicated a potential antifungal activity against *F. solani*, the causal agent of root rot disease infecting tomato plants. Additionally, the use of EO showed a promising growth promotion effect for tomato seedlings by enhancing their physiological, biochemical, and antioxidant activities. The three transcription factors: WRKY4, WRKY31, and WRKY37, as well as the other three defense-related genes glucanase A, defensin, and chitinase, have been upregulated due to the effect of ThE-CsNPs (1%). As an overall conclusion, the use of biological fungicides based on *T. vulgaris* EOs with chitosan nanoparticles might be used as a natural alternative to promote tomato growth and control serious phytopathogens.

Supplementary Materials: The following supporting information can be downloaded at: https://www.mdpi.com/article/10.3390/plants11223129/s1, Figure S1: GC-MS profile of *T. vulgaris* EO; Figure S2: FT-IR spectrum of ThE-CsNPs; Table S1: Oligonucleaotide primers used in qRT-PCR analysis.

Author Contributions: Conceptualization, A.A.I. and E.S.A.R.; data curation, A.A.I., E.S.A.R., S.S.A.K., and S.M.A.; formal analysis, S.A.-E. and A.A.I.; investigation, A.A.I., F.A.S., S.A., A.A.A., M.M.B., S.M.A., and H.S.E.; methodology, S.A.-E., E.S.A.R., M.M.B., and H.S.E.; resources, E.S.A.R., S.A. and A.A.A.; supervision, H.S.E.; validation, S.S.A.K.; visualization, S.A.-E. and F.A.S.; writing—original draft, A.A.I., E.S.A.R., S.S.A.K., S.M.A., A.A.A., M.M.B., S.A., and H.S.E.; writing—review and editing, F.A.S. All authors have read and agreed to the published version of the manuscript.

Funding: This research received no external funding.

Institutional Review Board Statement: Not applicable.

Informed Consent Statement: Not applicable.

Data Availability Statement: Relevant data applicable to this research are within the paper.

Acknowledgments: The authors extend their appreciation to City of Scientific Research and Technology Applications (Alexandria, Egypt), and Arish University (Al-Arish, Egypt) for providing all facilities of the practical work. Authors also acknowledge Princess Nourah bint Abdulrahman University Researchers Supporting, Riyadh, Saudi Arabia. The authors are also grateful to the School of Agricultural, Forestry, Food and Environmental Sciences, University of Basilicata (Potenza, Italy) for the collaboration and scientific support.

Conflicts of Interest: The authors declare no conflict of interest.

References

1. Bihon, W.; Ognakossan, K.E.; Tignegre, J.-B.; Hanson, P.; Ndiaye, K.; Srinivasan, R. Evaluation of Different Tomato (*Solanum lycopersicum* L.) Entries and Varieties for Performance and Adaptation in Mali, West Africa. *Horticulturae* **2022**, *8*, 579. [CrossRef]
2. EL-Mansy, A.B.; Abd El-Moneim, D.; ALshamrani, S.M.; Safhi, F.A.; Abdein, M.A.; Ibrahim, A.A. Genetic Diversity Analysis of Tomato (*Solanum lycopersicum* L.) with Morphological, Cytological, and Molecular Markers under Heat Stress. *Horticulturae* **2021**, *7*, 65. [CrossRef]
3. Fufa, F.; Hanson, P.; Dagnoko, S.; Dhaliwal, M. AVRDC—The world vegetable center tomato breeding in Sub-Saharan Africa:Lessons from the past, present work, and future prospects. *Acta Hortic.* **2011**, *911*, 87–98. [CrossRef]
4. Ayenan, M.A.T.; Danquah, A.; Hanson, P.; Asante, I.K.; Danquah, E.Y. Tomato (*Solanum lycopersicum* L.) Genotypes Respond Differently to Long-Term Dry and Humid Heat Stress. *Horticulturae* **2022**, *8*, 118. [CrossRef]
5. Soliman, S.A.; Hafez, E.E.; Al-Kolaibe, A.M.G.; Abdel Razik, E.-S.S.; Abd-Ellatif, S.; Ibrahim, A.A.; Kabeil, S.S.A.; Elshafie, H.S. Biochemical Characterization, Antifungal Activity, and Relative Gene Expression of Two *Mentha* Essential Oils Controlling *Fusarium oxysporum*, the Causal Agent of *Lycopersicon esculentum* Root Rot. *Plants* **2022**, *11*, 189. [CrossRef]
6. Bodah, E.T. Root rot diseases in plants: A review of common causal agents and management strategies. *Agri. Res. Tech.* **2017**, *5*, 555661.
7. Srinivas, C.; Devi, D.N.; Murthy, K.N.; Mohan, C.D.; Lakshmeesha, T.R.; Singh, B.; Srivastava, R.K.; Niranjana, S.R.; Hashem, A. *Fusarium oxysporum* f. sp. *lycopersici* causal agent of vascular wilt disease of tomato: Biology to diversity—A review. *Saudi J. Biol. Sci.* **2019**, *26*, 1315–1324.
8. Kabaş, A.; Fidan, H.; Batuhan, D.M. Identification of new sources of resistance to resistance-breaking isolates of tomato spotted wilt virus. *Saudi J. Biol. Sci.* **2021**, *28*, 3094–3099. [CrossRef]
9. Arora, H.; Sharma, A.; Poczai, P.; Sharma, S.; Haron, F.F.; Gafur, A.; Sayyed, R.Z. Plant-Derived Protectants in Combating Soil-Borne Fungal Infections in Tomato and Chilli. *J. Fungi.* **2022**, *8*, 213. [CrossRef]
10. Haq, I.U.; Sarwar, M.K.; Faraz, A.; Latif, M.Z. Synthetic chemicals: Major component of plant disease management. In *Plant Disease Management Strategies for Sustainable Agriculture through Traditional and Modern Approaches*; Haq, I.U., Ijaz, S., Eds.; Springer: Berlin/Heidelberg, Germany, 2020; pp. 53–81.
11. Aamir, M.; Singh, V.K.; Dubey, M.K.; Kashyap, S.P.; Zehra, A.; Upadhyay, R.S.; Singh, S. Structural and functional dissection of differentially expressed tomato WRKY transcripts in host defense response against the vascular wilt pathogen (*Fusarium oxysporum* f. sp. *lycopersici*. *PLoS ONE* **2018**, *13*, e0193922. [CrossRef]
12. Singh, P.; Shekhar, S.; Rustagi, A.; Sharma, V.; Kumar, D. Insights into the Role of WRKY Superfamily of Protein Transcription Factor in Defense Response. In *Molecular Aspects of Plant-Pathogen Interaction*; Singh, A., Singh, I., Eds.; Springer: Singapore, 2018; pp. 185–202. [CrossRef]
13. El-Saadony, M.T.; Alkhatib, F.M.; Alzahrani, S.O.; Shafi, M.E.; Abdel-Hamid, S.E.; Taha, T.F.; Aboelenin, S.M.; Soliman, M.M.; Ahmed, N.H. Impact of mycogenic zinc nanoparticles on performance, behavior, immune response, and microbial load in *Oreochromis niloticus*. *Saudi J. Biol. Sci.* **2021**, *28*, 4592–4604. [CrossRef] [PubMed]

14. El-Saadony, M.T.; Desoky, E.S.M.; Saad, A.M.; Eid, R.S.; Selem, E.; Elrys, A.S. Biological silicon nanoparticles improve *Phaseolus vulgaris* L. yield and minimize its contaminant contents on a heavy metals-contaminated saline soil. *J. Environ. Sci.* **2021**, *106*, 1–14. [CrossRef] [PubMed]
15. El-Saadony, M.T.; Saad, A.M.; Najjar, A.A.; Alzahrani, S.O.; Alkhatib, F.M.; Shafi, M.E.; Selem, E.; Desoky, E.S.; Fouda, S.E.E.; El-Tahan, A.M.; et al. The use of biological selenium nanoparticles to suppress *Triticum aestivum* L. crown and root rot diseases induced by *Fusarium* species and improve yield under drought and heat stress. *Saudi J. Biol. Sci.* **2021**, *28*, 4461–4471. [CrossRef]
16. El-Saadony, M.T.; Saad, A.M.; Taha, T.F.; Najjar, A.A.; Zabermawi, N.M.; Nader, M.M. Selenium nanoparticles from *Lactobacillus paracasei* HM1 capable of antagonizing animal pathogenic fungi as a new source from human breast milk. *Saudi J. Biol. Sci.* **2021**, *28*, 6782–6794. [CrossRef]
17. Elizabath, A.; Babychan, M.; Mathew, A.M.; Syriac, G.M. Application of nanotechnology in agriculture. *Int. J. Pure Appl. Biosci.* **2019**, *7*, 131–139. [CrossRef]
18. Elieh-Ali-Komi, D.; Hamblin, M.R. Chitin and chitosan: Production and application of versatile biomedical nanomaterials. *Int. J. Adv. Res.* **2016**, *4*, 411–427.
19. Abd El-Hack, M.E.; El-Saadony, M.T.; Shafi, M.E.; Zabermawi, N.M.; Arif, M.; Batiha, G.E.; Khafaga, A.F.; Abd El-Hakim, Y.M.; Al-Sagheer, A.A. Antimicrobial and antioxidant properties of chitosan and its derivatives and their applications: A review. *Int. J. Biol. Macromol.* **2020**, *164*, 2726–2744. [CrossRef]
20. Pichyangkura, R.; Chadchawan, S. Biostimulant activity of chitosan in horticulture. *Sci. Hortic.* **2015**, *196*, 49–65. [CrossRef]
21. Hassan, O.; Chang, T. Chitosan for eco-friendly control of plant disease. *Asian J. Plant Pathol.* **2017**, *11*, 53–70. [CrossRef]
22. Ocaña, A.; Reglero, G. Effects of Thyme Extract Oils (from *Thymus vulgaris*, *Thymus zygis*, and *Thymus hyemalis*) on Cytokine Production and Gene Expression of oxLDL-Stimulated THP-1-Macrophages. *J. Obes.* **2012**, *2012*, 104706. [CrossRef]
23. Ramachandran, L.; Nair, C.K. Therapeutic potentials of silver nanoparticle complex of α-lipoic acid. *Nanomater. Nanotechnol.* **2011**, *1*, 17–24. [CrossRef]
24. Tohidi, B.; Rahimmalek, M.; Arzani, A. Essential oil composition, total phenolic, flavonoid contents, and antioxidant activity of *Thymus* species collected from different regions of Iran. *Food Chem.* **2017**, *220*, 153–161. [CrossRef] [PubMed]
25. Saijo, Y.; Loo, E.P.; Yasuda, S. Pattern recognition receptors and signaling in plant-microbe interactions. *Plant J.* **2018**, *93*, 592–613. [CrossRef] [PubMed]
26. Vives-Peris, V.; Marmaneu, D.; Gómez-Cadenas, A.; Pérez-Clemente, R.M. Characterization of Citrus WRKY transcription factors and their responses to phytohormones and abiotic stresses. *Biol. Plant* **2018**, *62*, 33–44. [CrossRef]
27. Shankar, A.; Pandey, A.; Pandey, G.K. WRKY transcription factor: Role in abiotic and biotic stress. *Plant Stress* **2013**, *7*, 26–34.
28. Bai, Y.; Sunarti, S.; Kissoudis, C.; Visser, R.G.F.; van der Linden, C.G. The role of tomato WRKY genes in plant responses to combined abiotic and biotic stresses. *Front. Plant Sci.* **2018**, *9*, 801. [CrossRef]
29. Huang, S.; Gao, Y.; Liu, J.; Peng, X.; Niu, X.; Fei, Z. Genome-wide analysis of WRKY transcription factors in *Solanum lycopersicum*. *Mol. Genet. Genomics* **2012**, *287*, 495–513. [CrossRef]
30. Karkute, S.G.; Gujjar, R.S.; Rai, A.; Akhtar, M.; Singh, M. Genome wide expression analysis of WRKY genes in tomato (*Solanum lycopersicum*) under drought stress. *Plant Gene* **2018**, *13*, 8–17. [CrossRef]
31. Chi, Y.; Yang, Y.; Zhou, Y.; Zhou, J.; Fan, B.; Yu, J.Q. Protein—Protein interactions in the regulation of WRKY transcription factors. *Mol. Plant* **2013**, *6*, 287–300. [CrossRef]
32. Shenashen, M.; Derbalah, A.; Hamza, A.; Mohamed, A.; El Safty, S. Recent trend in controlling root rot disease of tomato caused by *Fusarium Solani* using aluminasilica nanoparticles. *Int. J. Adv. Res. Biol. Sci.* **2017**, *4*, 105–119.
33. Wani, A.H.; Shah, M.A. A unique and profound effect of MgO and ZnO nanoparticles on some plant pathogenic fungi. *J. Appl. Pharm. Sci.* **2012**, *2*, 40–44.
34. Ozbay, N.; Newman, S.E. Fusarium crown and root rot of tomato and control methods. *Plant Pathol. J.* **2004**, *3*, 9–18. [CrossRef]
35. Cosentino, C.; Labella, C.; Elshafie, H.S.; Camele, I.; Musto, M.; Paolino, R.; Freschi, P. Effects of different heat treatments on lysozyme quantity and antimicrobial activity of jenny milk. *J. Dairy Sci.* **2016**, *99*, 5173–5179. [CrossRef] [PubMed]
36. Elshafie, H.S.; Viggiani, L.; Mostafa, M.S.; El-Hashash, M.A.; Bufo, S.A.; Camele, I. Biological activity and chemical identification of ornithine lipid produced by *Burkholderia gladioli* pv. *agaricicola* ICMP 11096 using LC-MS and NMR analyses. *J. Biol. Res.* **2017**, *90*, 96–103. [CrossRef]
37. Elshafie, H.S.; Devescovi, G.; Venturi, V.; Camele, I.; Bufo, S.A. Study of the regulatory role of N-acyl homoserine lactones mediated quorum sensing in the biological activity of *Burkholderia gladioli* pv. *agaricicola* causing soft rot of *Agaricus* spp. *Front. Microbiol.* **2019**, *10*, 2695. [CrossRef]
38. Sofo, A.; Elshafie, H.S.; Scopa, A.; Mang, S.M.; Camele, I. Impact of airborne zinc pollution on the antimicrobial activity of olive oil and the microbial metabolic profiles of Zn-contaminated soils in an Italian olive orchard. *J. Trace Elem. Med. Biol.* **2018**, *49*, 276–284. [CrossRef]
39. Elshafie, H.S.; Racioppi, R.; Bufo, S.A.; Camele, I. In vitro study of biological activity of four strains of *Burkholderia gladioli* pv. *agaricicola* and identification of their bioactive metabolites using GC–MS. *Saudi J. Biol. Sci.* **2017**, *24*, 295–301. [CrossRef]
40. Al-Harbi, N.A.; Al Attar, N.M.; Hikal, D.M.; Mohamed, S.E.; Abdel Latef, A.A.H.; Ibrahim, A.A.; Abdein, M.A. Evaluation of Insecticidal Effects of Plants Essential Oils Extracted from Basil, Black Seeds and Lavender against *Sitophilus oryzae*. *Plants* **2021**, *10*, 829. [CrossRef]

41. Raveau, R.; Fontaine, J.; Lounès-Hadj Sahraoui, A. Essential oils as potential alternative biocontrol products against plant pathogens and weeds: A Review. *Foods* **2020**, *9*, 365. [CrossRef]
42. Elshafie, H.S.; Caputo, L.; De Martino, L.; Grul'ová, D.; Zheljazkov, V.D.; De Feo, V.; Camele, I. Biological investigations of essential oils extracted from three *Juniperus* species and evaluation of their antimicrobial, antioxidant and cytotoxic activities. *J. Appl. Microbiol.* **2020**, *129*, 1261–1271. [CrossRef]
43. Camele, I.; Grul'ová, D.; Elshafie, H.S. Chemical composition and antimicrobial properties of *Mentha piperita* cv. 'Kristinka' essential oil. *Plants* **2021**, *10*, 1567. [CrossRef] [PubMed]
44. Ali, H.M.; Elgat, W.; El-Hefny, M.; Salem, M.Z.M.; Taha, A.S.; Al Farraj, D.A.; Elshikh, M.S.; Hatamleh, A.A.; Abdel-Salam, E.M. New Approach for Using of *Mentha longifolia* L. and *Citrus reticulata* L. Essential Oils as Wood-Biofungicides: GC-MS, SEM, and MNDO Quantum Chemical Studies. *Materials* **2021**, *14*, 1361. [CrossRef] [PubMed]
45. Werrie, P.Y.; Durenne, B.; Delaplace, P.; Fauconnier, M.L. Phytotoxicity of Essential Oils: Opportunities and Constraints for the Development of Biopesticides. A Review. *Foods* **2020**, *9*, 1291. [CrossRef]
46. Hjort, K.; Bergstrom, M.; Adesina, M.F.; Jansson, J.K.; Smalla, K.; Sjoling, S. Chitinase genes revealed and compared in bacterial isolates, DNA extracts and a metagenomic library from a phytopathogen-suppressive soil. *FEMS Microbiol. Ecol.* **2010**, *71*, 197–207. [CrossRef] [PubMed]
47. Bell, A.A.; Hubbard, J.C.; Liu, L.; Davis, R.M.; Subbarao, K.V. Effects of chitin and chitosan on the incidence and severity of *Fusarium* yellows in celery. *Plant Dis.* **1998**, *82*, 322–328. [CrossRef]
48. Hosseinnejad, M.; Jafari, S.M. Evaluation of different factors affecting antimicrobial properties of chitosan. *Int. J. Biol. Macromol.* **2016**, *85*, 467–475. [CrossRef]
49. Cretoiu, M.S.; Korthals, G.W.; Visser, J.H.M.; van Elsas, J.D. Chitin amendment increases soil suppressiveness toward plant pathogens and modulates the actinobacterial and oxalobacteraceal communities in an experimental agricultural field. *Appl. Environ. Microbiol.* **2013**, *79*, 5291–5301. [CrossRef]
50. Sharma, N.; Tripathi, A. Effects of Citrus sinensis (L.) Osbeck epicarp essential oil on growth and morphogenesis of *Aspergillus niger* Van Tieghem. *Microbiol. Res.* **2008**, *163*, 337–344. [CrossRef]
51. Elshafie, H.S.; Ghanney, N.; Mang, S.M.; Ferchichi, A.; Camele, I. An in vitro attempt for controlling severe phytopathogens and human pathogens using essential oils from Mediterranean plants of genus *Schinus*. *J. Med. Food* **2016**, *19*, 266–273. [CrossRef]
52. Elshafie, H.S.; Sakr, S.; Mang, S.M.; De Feo, V.; Camele, I. Antimicrobial activity and chemical composition of three essential oils extracted from Mediterranean aromatic plants. *J. Med. Food* **2016**, *19*, 1096–1103. [CrossRef]
53. Ben-Jabeur, M.; Ghabri, E.; Myriam, M.; Hamada, W. Thyme essential oil as a defense inducer of tomato against gray mold and Fusarium wilt. *Plant Physiol. Biochem.* **2015**, *94*, 35–40. [CrossRef] [PubMed]
54. Al-Asmari, A.K.; Athar, M.T.; Al-Faraidy, A.A.; Almuhaiza, M.S. Chemical Composition of Essential Oil of *Thymus vulgaris* Collected from Saudi Arabian Market. *Asian Pac. J. Trop. Biomed.* **2017**, *7*, 147–150. [CrossRef]
55. Micucci, M.; Protti, M.; Aldini, R.; Frosini, M.; Corazza, I.; Marzetti, C.; Mattioli, L.B.; Tocci, G.; Chiarini, A.; Mercolini, L.; et al. *Thymus vulgaris* L. Essential Oil Solid Formulation: Chemical Profile and Spasmolytic and Antimicrobial Effects. *Biomolecules* **2020**, *10*, 860. [CrossRef] [PubMed]
56. Hyldgaard, M.; Mygind, T.; Meyer, R.L. Essential Oils in Food Preservation: Mode of Action, Synergies, and Interactions with Food Matrix Components. *Front. Microbiol.* **2012**, *3*, 12. [CrossRef] [PubMed]
57. Igoe, R.S.; Hui, Y.H. Substances for use in foods: Listing under Title 21 of the Code of Federal Regulations. In *Dictionary of Food Ingredients*; Igoe, R.S., Hui, Y.H., Eds.; Springer: Boston, MA, USA, 1996; pp. 159–185. ISBN 978-1-4615-6838-4.
58. Casarin, L.S.; Casarin, F.D.O.; Brandelli, A.; Novello, J.; Ferreira, S.O.; Tondo, E.C. Influence of Free Energy on the Attachment of *Salmonella Enteritidis* and *Listeria Monocytogenes* on Stainless Steels AISI 304 and AISI 316. *LWT-Food Sci. Technol.* **2016**, *69*, 131–138. [CrossRef]
59. Kostaki, M.; Chorianopoulos, N.; Braxou, E.; Nychas, G.-J.; Giaouris, E. Differential Biofilm Formation and Chemical Disinfection Resistance of Sessile Cells of *Listeria Monocytogenes* Strains under Monospecies and Dual-Species (with *Salmonella enterica*) Conditions. *Appl. Environ. Microbiol.* **2012**, *78*, 2586–2595. [CrossRef]
60. Nguefack, J.; Tamgue, O.; Dongmo, J.L.; Dakole, C.; Leth, V.; Vismer, H.; Zollo, P.A.; Nkengfack, A. Synergistic action between fractions of essential oils from *Cymbopogon citratus*, *Ocimum gratissimum* and *Thymus vulgaris* against Penicillium expansum. *Food Control.* **2012**, *23*, 377–383. [CrossRef]
61. Arnal-Schnebelen, B.; Hadji-Minaglou, F.; Peroteau, J.F.; Ribeyre, F.; de Billerbeck, V.G. Essential oils in infectious gynaecological disease: A statistical study of 658 cases. *Int. J. Aromatherapy.* **2004**, *14*, 192–197. [CrossRef]
62. Isham, C.R.; Bossou, A.R.; Negron, V.; Fisher, K.E.; Kumar, R.; Marlow, L.; Bible, K.C. Pazopanib enhances paclitaxel-induced mitotic catastrophe in anaplastic thyroid cancer. *Sci. Transl. Med.* **2013**, *5*, 166ra3. [CrossRef]
63. Phukan, U.J.; Jeena, G.S.; Shukla, R.K. WRKY transcription factors: Molecular regulation and stress responses in plants. *Front. Plant Sci.* **2016**, *7*, 760. [CrossRef]
64. Després, C.; Chubak, C.; Rochon, A.; Clark, R.; Bethune, T.; Desveaux, D.; Fobert, P. The Arabidopsis NPR1 disease resistance protein is a novel cofactor that confers redox regulation of DNA binding activity to the basic domain/leucine zipper transcription factor TGA1. *Plant Cell* **2003**, *15*, 2181–2191. [CrossRef] [PubMed]
65. Park, C.Y.; Lee, J.H.; Yoo, J.H.; Moon, B.C.; Choi, M.S.; Kang, Y.H.; Lee, S.M.; Kim, H.S.; Kang, K.Y.; Chung, W.S. WRKY group IId transcription factors interact with calmodulin. *FEBS Lett.* **2005**, *579*, 1545–1550. [CrossRef] [PubMed]

66. Xu, X.; Chen, C.; Fan, B.; Chen, Z. Physical and Functional Interactions between Pathogen-Induced *Arabidopsis* WRKY18, WRKY40, and WRKY60 Transcription Factors. *Plant Cell* **2006**, *18*, 1310–1326. [CrossRef] [PubMed]
67. Parnes, C.A. Efficacy of sodium hypochlorite bleach and "alternative" products in preventing transfer of bacteria to and from inanimate surfaces. *Environ. Health* **1997**, *59*, 14–20.
68. Périno, S.; Chemat-Djenni, Z.; Petitcolas, E.; Giniès, C.; Chemat, F. Downscaling of industrial turbo-distillation to laboratory turbo-clevenger for extraction of essential oils. Application of concepts of green analytical chemistry. *Molecules* **2019**, *24*, 2734. [CrossRef]
69. Elshafie, H.S.; Sakr, S.H.; Sadeek, S.A.; Camele, I. Biological investigations and spectroscopic studies of new Moxifloxacin/Glycine-Metal complexes. *Chem. Biodiver.* **2019**, *16*, e1800633. [CrossRef]
70. Filion, M.; St-Arnaud, M.; Jabaji-Hare, S.H. Quantification of *F. solani* f. sp. phaseoli in mycorrhizal bean plants and surrounding mycorrhizosphere soil using realtime polymerase chain reaction and direct isolations on selective media. *Phytopathology* **2003**, *93*, 229–235. [CrossRef]
71. Heath, R.L.; Packer, L. Photoperoxidation in isolated chloroplasts. *Arch. Biochem. Biophys.* **1968**, *125*, 189–198. [CrossRef]
72. Slinkard, K.; Singleton, V.L. Total phenol analysis: Automation and comparison with manual methods. *Am. J. Enol. Vitic.* **1977**, *28*, 49–55.
73. Chavan, J.J.; Gaikwad, N.B.; Kshirsagar, P. Total phenolics, flavonoids and antioxidant properties of three Ceropegia species from Western Ghats of India. *S. Afr. J. Bot.* **2013**, *88*, 273–277. [CrossRef]
74. Bradford, M.M. A rapid and sensitive method for the quantitation of microgram quantities of protein utilizing the principle of protein-dye binding. *Anal. Biochem.* **1976**, *72*, 248–254. [CrossRef]
75. Lichtenthaler, H.K.; Buschmann, C. Chlorophylls and carotenoids: Measurement and characterization by UV-VIS spectroscopy. *Curr. Protoc. Food Anal. Chem.* **2001**, *1*, F4.3.1–F4.3.8. [CrossRef]
76. Yang, Y.C.; Sun, D.W.; Pu, H.; Wang, N.N.; Zhu, Z. Rapid detection of anthocyanin content in lychee pericarp during storage using hyperspectral imaging coupled with model fusion. *Postharvest. Biol. Technol.* **2015**, *103*, 55–65. [CrossRef]
77. Luck, H. *Catalase in Methods of Enzymatic Analysis*; Bergmeyer, J., Grabi, M., Eds.; Academic Press: New York, NY, USA, 1974; Volume II.
78. Livak, K.J.; Schmittgen, T.D. Analysis of relative gene expression data using real-time quantitative PCR and the 2-DDCT Method. *Methods* **2001**, *25*, 402–408. [CrossRef]
79. Chen, C.; Chen, H.; Zhang, Y.; Thomas, H.R.; Frank, M.H.; He, Y.; Xia, R. TBtools: An integrative toolkit developed for interactive analyses of big biological data. *Mol. Plant* **2020**, *13*, 1194–1202. [CrossRef]

plants MDPI

Article

Citrus reticulata Leaves Essential Oil as an Antiaging Agent: A Comparative Study between Different Cultivars and Correlation with Their Chemical Compositions

Nouran M. Fahmy [1], Sameh S. Elhady [2], Douha F. Bannan [3], Rania T. Malatani [3] and Haidy A. Gad [1,*]

[1] Department of Pharmacognosy, Faculty of Pharmacy, Ain Shams University, Cairo 11566, Egypt
[2] Department of Natural Products, Faculty of Pharmacy, King Abdulaziz University, Jeddah 21589, Saudi Arabia
[3] Department of Pharmacy Practice, Faculty of Pharmacy, King Abdulaziz University, Jeddah 21589, Saudi Arabia
* Correspondence: haidygad@pharma.asu.edu.eg

Abstract: The mass-based metabolomic approach was implemented using GC-MS coupled with chemometric analysis to discriminate between the essential oil compositions of six cultivars of *Citrus reticulata*. The antiaging capability of the essential oils were investigated through measurement of their ability to inhibit the major enzymes hyaluronidase, collagenase, and amylase involved in aging. GC-MS analysis resulted in the identification of thirty-nine compounds including β-pinene, D-limonene, γ-terpinene, linalool, and dimethyl anthranilate as the main components. Multivariate analysis using principal component analysis (PCA) and hierarchal cluster analysis (HCA) successfully discriminated the cultivars into five main groups. In vitro antiaging activity showed that Kishu mandarin (Km) (2.19 ± 0.10, 465.9 ± 23.7, 0.31 ± 0.01 µg/mL), Cara mandarin (Cm) (3.22 ± 0.14, 592.1 ± 30.1, 0.66 ± 0.03 µg/mL), and Wm (8.43 ± 0.38, 695.2 ± 35.4, 0.79 ± 0.04%) had the highest inhibitory activity against hyaluronidase, collagenase, and amylase, respectively. Molecular docking studies on the major compounds validated the activities of the essential oils and suggested their possible mechanisms of action. Based on our result, certain cultivars of *Citrus reticulata* can be proposed as a promising candidate in antiaging skin care products.

Keywords: *Citrus reticulata*; GC-MS; antiaging; molecular docking; chemometric analysis; industries development; drug discovery

Citation: Fahmy, N.M.; Elhady, S.S.; Bannan, D.F.; Malatani, R.T.; Gad, H.A. *Citrus reticulata* Leaves Essential Oil as an Antiaging Agent: A Comparative Study between Different Cultivars and Correlation with Their Chemical Compositions. *Plants* **2022**, *11*, 3335. https://doi.org/10.3390/plants11233335

Academic Editors: Adriano Sofo, Ippolito Camele and Hazem Salaheldin Elshafie

Received: 7 November 2022
Accepted: 28 November 2022
Published: 1 December 2022

1. Introduction

Genus *Citrus* belongs to family Rutaceae (Rue family) and comprises about 17 species and distributed all over the tropical and temperate regions with numerous health benefits [1]. Citrus essential oil (EO) is present in different plant parts such as peels, leaves, and flowers. Terpenes, sesquiterpenes, aldehydes, alcohols, esters, and sterols constitute the major classes of active compounds present in citrus essential oils. They may also be described as mixtures of hydrocarbons, oxygenated compounds and nonvolatile residues. It is worth noting that the essential oil composition of different citrus species is affected by various factors such as the harvest year, due to climate change throughout the years [2], season [3], cultivar [4], the use of different rootstocks [5], as well as the extraction technique [6,7]. Mostly, they are consumed as aroma flavor in the food industry, including alcoholic and nonalcoholic beverages, marmalades, gelatins, sweets, soft drinks, ice creams, dairy products, jams, candies, and cakes [8–10]. Moreover, citrus essential oils exhibited significant importance due to their wide range of biological activities such as antimicrobial [11–13], antifungal [14], antioxidant [15], antidiabetic [16], antihyperlipidemic [17], anti-inflammatory, anti-allergic [18], anticancer [19], anxiolytic [20], and insecticidal activity [21].

Citrus reticulata Blanco, one of the most commercially important species, is commonly known as mandarin fruit [22]. It is native to China, East Asia, and Southeast Asia [23]. The name 'Mandarin' was provided to the *C. reticulata* by the Portuguese from its Chinese name Guānhuà, to reflect its country of origin [24]. Different cultivars of *C. reticulata* were traditionally used in folk medicine for the treatment of various ailments such as fever, snakebite, stomachache, edema, cardiac diseases, bronchitis, and asthma [25]. Mandarin oil is well-known for its broad spectrum antibacterial and antifungal actions [26–30]. Antiproliferative [31], antioxidant [26,32], antidiabetic [30], and schistosomicidal effects [33] were also reported.

On the other hand, nature remains a substantial source for drug discovery for treatment of major ailments as well as in the development of cosmetic products. Currently, the use of natural ingredients in cosmetics is widely spread compared with the synthetic alternative, due to their wide safety margin, potential antioxidant, anti-inflammatory, and skin soothing effects. Interestingly, many of the commercial and medical skin aging creams contain essential oils, herbal extracts, and other natural ingredients which meet the increased demand in the market for herbal-based cosmetics [34]. Moreover, the promising antioxidant activity of various citrus essential oils encouraged us to study the anti-collagenase, anti-hyaluronidase, and anti-elastase capabilities to prove their antiaging potential in an attempt incorporate it safely as a natural ingredient in management of age-related skin problem [35–37]. Additionally, the essential oil composition of different *C. reticulata* cultivars showed a significant difference in their chemical constituents [38,39]; thus, a comparative metabolic profiling of poorly studied cultivars is needed for the selection of a good quality cultivar that can be used in improving and upgrading of essential oil composition. The present study targeted a comparative metabolic profiling of the essential oil composition of six cultivars of *Citrus reticulata* leaves cultivated in Egypt using GC-MS and the study of their potential antiaging activity aiming to discover a natural ingredient that can be incorporated safely in antiaging skin care products.

2. Results and Discussion

2.1. Extraction and Distillation of the Essential Oils

Hydrodistillation of *C. reticulata* fresh leaves cultivars yielded a pale-yellow oil. The yield was expressed as the weight of the oil per 500 g fresh leaves and varied from 0.078 to 0.232% w/w (Table 1).

Table 1. *Citrus reticulata* cultivars, codes, and yield % (w/w).

Citrus reticulata Cultivar	Cultivar Code	Yield % (w/w)
Avana Apriena mandarin	Am	0.232
Balady mandarin	Bm	0.220
Cara mandarin	Cm	0.124
Willow leaf mandarin	Wm	0.078
Sunburst mandarin	Sm	0.130
Kishu mandarin	Km	0.151

2.2. Metabolic Profile of the Essential Oils

The chemical composition of the essential oils obtained from six cultivars of *C. reticulata* leaves were analyzed using GC-MS. The chemical compounds identified, Kovats indices, and percentages (average of three replicates for each sample) for each cultivar are displayed in Table 2. Monoterpenes hydrocarbons were predominant in all cultivars (36.38–86.45%), oxygenated monoterpenes and sesquiterpenes were present at a lower percentage, other classes of compounds were reported in a high percentage represented chiefly by dimethyl anthranilate in Am (56.51 ± 1.06%), Bm (58.77 ± 2.04%), and Wm (49.06 ± 1.94%) cultivars.

Table 2. Metabolic profiles of *Citrus reticulata* leaves cultivar EO by GC-MS.

No.	R_t	Compound Name	KI_{exp} a	KI_{rep} b	Content%						Molecular Formula
					Am	Bm	Cm	Wm	Sm	Km	
1.	7.19	α-Thujene	911	911	0.97 ± 0.05	0.67 ± 0.17	1.37 ± 0.03	1.55 ± 0.63	0.73 ± 0.12	1.08 ± 0.07	$C_{10}H_{16}$
2.	7.38	α-Pinene	918	918	2.54 ± 0.12	1.85 ± 0.49	3.76 ± 0.07	3.72 ± 0.98	2.93 ± 0.65	3.77 ± 0.17	$C_{10}H_{16}$
3.	7.82	Camphene	933	933	0.03 ± 0.01	0.01 ± 0.02	0.06 ± 0.01	0.05 ± 0.01	0.09 ± 0.01	0.18 ± 0.01	$C_{10}H_{16}$
4.	8.61	β-Thujene	962	962	-	-	-	-	-	3.34 ± 0.10	$C_{10}H_{16}$
5.	8.68	β-Pinene	965	965	2.48 ± 0.11	1.80 ± 0.38	6.34 ± 0.06	3.53 ± 0.68	34.24 ± 3.50	19.62 ± 0.37	$C_{10}H_{16}$
6.	9.13	β-Myrcene	981	981	0.44 ± 0.02	0.16 ± 0.15	1.13 ± 0.07	1.10 ± 0.21	5.16 ± 0.21	0.89 ± 0.11	$C_{10}H_{16}$
7.	9.52	α-Phellandrene	995	995	0.06 ± 0.00	0.02 ± 0.03	0.13 ± 0.04	0.04 ± 0.00	0.20 ± 0.00	0.13 ± 0.09	$C_{10}H_{16}$
8.	9.72	3-Carene	1002	1002	0.06 ± 0.00	0.04 ± 0.01	0.03 ± 0.03	0.01 ± 0.01	0.02 ± 0.01	-	$C_{10}H_{16}$
9.	9.90	2-Carene	1008	1008	0.28 ± 0.01	0.20 ± 0.04	1.03 ± 0.04	0.45 ± 0.03	3.54 ± 0.16	0.69 ± 0.04	$C_{10}H_{16}$
10.	10.08	*p*-Cymene	1014	1014	2.45 ± 0.10	1.74 ± 0.31	2.29 ± 0.08	3.11 ± 0.02	-	2.29 ± 0.09	$C_{10}H_{14}$
11.	10.17	*m*-Mentha-6,8-diene	1017	1019	-	-	-	-	4.12 ± 0.18	-	$C_{10}H_{16}$
12.	10.29	D-Limonene	1020	1021	10.22 ± 0.18	7.85 ± 1.50	10.03 ± 0.27	12.49 ± 0.21	-	6.34 ± 0.04	$C_{10}H_{16}$
13.	10.59	trans-β-Ocimene	1030	1031	0.06 ± 0.01	0.10 ± 0.06	0.12 ± 0.01	0.41 ± 0.02	0.31 ± 0.01	0.06 ± 0.01	$C_{10}H_{16}$
14.	10.92	cis-β-Ocimene	1041	1041	0.36 ± 0.02	0.27 ± 0.02	8.29 ± 0.25	1.99 ± 0.01	11.31 ± 0.62	2.81 ± 0.10	$C_{10}H_{16}$
15.	11.23	γ-Terpinene	1051	1052	19.37 ± 0.3	21.04 ± 0.36	48.56 ± 1.01	19.42 ± 0.50	6.19 ± 0.31	17.92 ± 0.26	$C_{10}H_{16}$
16.	11.53	trans-Sabinenehydrate	1060	1062	-	-	-	-	0.22 ± 0.02	0.03 ± 0.01	$C_{10}H_{18}O$
17.	12.15	α-Terpinolene	1080	1082	0.61 ± 0.05	0.63 ± 0.03	3.31 ± 0.14	1.20 ± 0.14	1.12 ± 0.09	1.78 ± 0.07	$C_{10}H_{16}$
18.	12.28	*p*-Cymenene	1085	1085	-	-	-	-	-	0.41 ± 0.14	$C_{10}H_{12}$
19.	12.54	Linalool	1093	1093	-	-	5.18 ± 0.34	0.53 ± 0.06	16.8 ± 1.49	22.20 ± 0.43	$C_{10}H_{18}O$
20.	13.22	Cis-Sabinenehydrate	1115	1116	-	-	-	-	0.40 ± 0.05	0.02 ± 0.00	$C_{10}H_{18}O$
21.	14.85	Isocamphopinone	1167	1168	-	-	-	-	-	0.03 ± 0,00	$C_{10}H_{16}O$
22.	14.95	Terpinen-4-ol	1170	1171	0.24 ± 0.02	-	0.08 ± 0.13	0.19 ± 0.03	10.91 ± 1.35	0.81 ± 0.01	$C_{10}H_{18}O$
23.	15.38	α-Terpineol	1184	1184	-	-	-	0.17 ± 0.04	0.50 ± 0.05	0.79 ± 0.01	$C_{10}H_{18}O$
24.	16.64	Anisole	1227	1227	-	-	-	0.20 ± 0.02	-	1.11 ± 0.01	C_7H_8O
25.	18.38	Thymol	1288	1288	-	-	-	-	-	10.98 ± 0.27	$C_{10}H_{14}O$
26.	18.50	Carvacrol	1292	1292	-	-	-	0.36 ± 0.04	-	-	$C_{10}H_{14}O$
27.	19.59	σ-Elemene	1330	1330	-	-	-	0.01 ± 0.01	-	0.07 ± 0.01	$C_{15}H_{24}$
28.	21.12	β-Elemene	1384	1385	0.28 ± 0.01	0.24 ± 0.02	5.54 ± 0.10	0.11 ± 0.01	0.64 ± 0.06	0.06 ± 0.01	$C_{15}H_{24}$
29.	21.62	Dimethyl anthranilate	1401	1402	56.51 ± 1.06	58.77 ± 2.04	0.02 ± 0.02	49.06 ± 1.94	-	-	$C_9H_{11}NO_2$
30.	21.90	β-Caryophyllene	1412	1413	2.90 ± 0.09	4.38 ± 0.51	1.36 ± 0.01	-	0.30 ± 0.03	0.82 ± 0.01	$C_{15}H_{24}$
31.	22.82	α-Caryophyllene	1448	1448	0.09 ± 0.01	0.19 ± 0.09	0.60 ± 0.01	0.19 ± 0.01	0.07 ± 0.01	0.09 ± 0.01	$C_{15}H_{24}$
32.	23.37	β-Chamigren	1470	1473	-	-	0.03 ± 0.00	-	-	-	$C_{15}H_{24}$
33.	23.53	Germacrene D	1476	1476	-	-	-	-	-	0.02 ± 0.00	$C_{15}H_{24}$
34.	23.69	β-Selinene	1482	1483	-	-	0.07 ± 0.01	-	-	-	$C_{15}H_{24}$
35.	23.94	δ-Guaiene	1492	1493	0.03 ± 0.00	0.04 ± 0.01	0.30 ± 0.01	0.08 ± 0.01	-	1.06 ± 0.02	$C_{15}H_{24}$

Table 2. *Cont.*

No.	R_t	Compound Name	KI_{exp} [a]	KI_{rep} [b]	Content%						Molecular Formula
					Am	Bm	Cm	Wm	Sm	Km	
36.	24.18	α-Selinene	1501	1501	-	-	0.23 ± 0.01	-	-	-	$C_{15}H_{24}$
37.	24.26	α-Farnesene	1505	1505	-	-	-	-	-	0.22 ± 0.02	$C_{15}H_{24}$
38.	24.59	σ-Cadinene	1517	1517	-	-	0.04 ± 0.00	-	-	0.12 ± 0.00	$C_{15}H_{24}$
39.	26.18	Caryophyllene oxide	1579	1579	-	-	-	0.02 ± 0.00	-	0.02 ± 0.02	$C_{15}H_{24}O$
	Monoterpene hydrocarbons (%)				39.93	36.38	86.45	49.07	69.96	61.31	
	Oxygenated monoterpenes (%)				0.24	-	5.26	1.25	28.83	34.86	
	Sesquiterpene hydrocarbons (%)				3.30	4.85	8.16	0.39	1.01	2.46	
	Oxygenated sesquiterpenes (%)				-	-	-	0.02	-	0.02	
	Others (%)				56.51	58.77	0.02	49.26	-	1.11	
	Total identified (%)				99.98	100	99.89	99.99	99.80	99.76	

[a] Kovats index determined experimentally on RTX-5 column relative to C8–C30 n-alkanes. [b] Published Kovats retention indices. Identification was based on comparison of the compounds mass spectral data (MS) and Kovats retention indices (RIs) with those of NIST Mass Spectral Library (2011), Wiley Registry of Mass Spectral Data 8th edition and literature.

Generally, thirty-nine compounds were tentatively identified including β-pinene, D-limonene, γ-terpinene, linalool, and dimethyl anthranilate as the main components. A great variability in the percentage of the chemical composition among different cultivars was observed. For example, β-pinene content was dominant in Sm cultivar (34.24 ± 3.50%) followed by Km cultivar (19.62 ± 0.37%), while in Wm, Am, and Bm it constitutes only 3.53 ± 0.68%, 2.48 ± 0.11%, and 1.80 ± 0.38% of the essential oil content, respectively. The highest D-limonene content was found in Wm cultivar (12.49 ± 0.21%) while it was totally absent in Sm cultivar. γ-Terpinene was present in all cultivars with amount varying from 48.56 ± 1.01% in Cm cultivar to 6.19 ± 0.31% in Sm cultivar. The highest amount of linalool was present in Km (22.2 ± 0.43%) and represented its major component, contrarily it was absent in Am and Bm. Additionally, certain compounds were present exclusively in certain cultivars in an appreciable amount, for example, m-mentha-6,8-diene was identified only in Sm and accounted for 4.12 ± 0.18% of its essential oil content, while β-thujene and thymol uniquely found in Wm, displayed 3.34 ± 0.10% and 10.98 ± 0.27% of its essential oil content, respectively.

Previous studies were concerned with the chemical composition of *C. reticulata* leaf essential oils. Lota et al. (2000) reported the variation in the content of γ-terpinene (0.2–61.3%), dimethyl anthranilate (trs–58%), sabinene (0.2–59.4%), linalool (0.2–54.3%), limonene (1.5–44.3%), *p*-cymene (tr–20.4%), β-ocimene (0.6–13.7%), β-pinene (0.1–10.7%), and terpinen-4-ol (0.1–10.6%) among forty-one mandarin cultivars [39]. Moreover, the chemical composition of the essential oils obtained from the leaves of six *C. reticulata* cultivars grown in Nigeria showed a discrepancy in the content of sabinene (1.4–39.7%), γ-terpinene (0.7–32.8%), *p*-cymene (0.3–27.4%), 3-carene (tr–11.6%), and β-ocimene (3.2–19.7%) [24]. These reports validate the difference in the essential oil content between the investigated mandarin cultivars as stated in our study. Another study performed by Karioti et al. (2007) on Nigerian *C. reticulata* displayed a predominance of γ-terpinene (53.0%) and linalool (16.1%) in the leaf essential oil [40], this amount of γ-terpinene is comparable to that found in Cm cultivar, while linalool percentage was near that found in Sm and Km cultivars.

In agreement with our results, Fleisher et al. (1990) reported dimethyl anthranilate as the major constituent of balady mandarin (Bm) which is accredited to 58.77 ± 2.04% of Bm oil content in our study [41]. Previous work on *C. reticulata* grown in Egypt, reported dimethyl anthranilate (65.3%) as a key component of the leaf essential oil followed by γ-terpinene (19.8%) and limonene (4.5%) [29]. These data are in line with the result of Am, Bm, and Wm essential oils. On the other hand, C. Blanco et al. (1995) reported linalool (52.66%), limonene (8.32%), and trans-β-ocimene (7.87%) as the major constituents of Colombian *C. reticulata* leaf oil, these components were found in our studied cultivar ranging from nil to 22.20% for linalool, from nil to 12.49% for limonene, and from 0.06 to 0.31% for trans-β-ocimene [42], which reveal that a difference in the geographical distribution between oils obtained from the same species marks a variation in the essential oil composition.

2.3. Chemometric Analysis Based on GC-MS Analysis

Due to the complexity and high dimensionality of GC-MS-based data comprising both qualitative and quantitative variances among different citrus cultivars, multivariate analysis was applied using principal component analysis (PCA) and hierarchal cluster analysis (HCA) to discriminate between closely related cultivars, as well as to detect any significant relationship between them [43]. A matrix of the total number of samples and their replicates (18 samples) multiplied by 39 variables (GC/MS peak area %) was constructed in MS Excel®, then subjected to multivariate analysis (PCA and HCA). Owing to the large number of variables, PCA was applied first to reduce the dimensionality of the multiple data set in addition to removing the redundancy in the variables, utilizing raw data (Peak area % for each compound as in Table 2). Figure 1a,b represents PCA score and loading plots based on GC-MS metabolic profiles of different citrus cultivars, respectively.

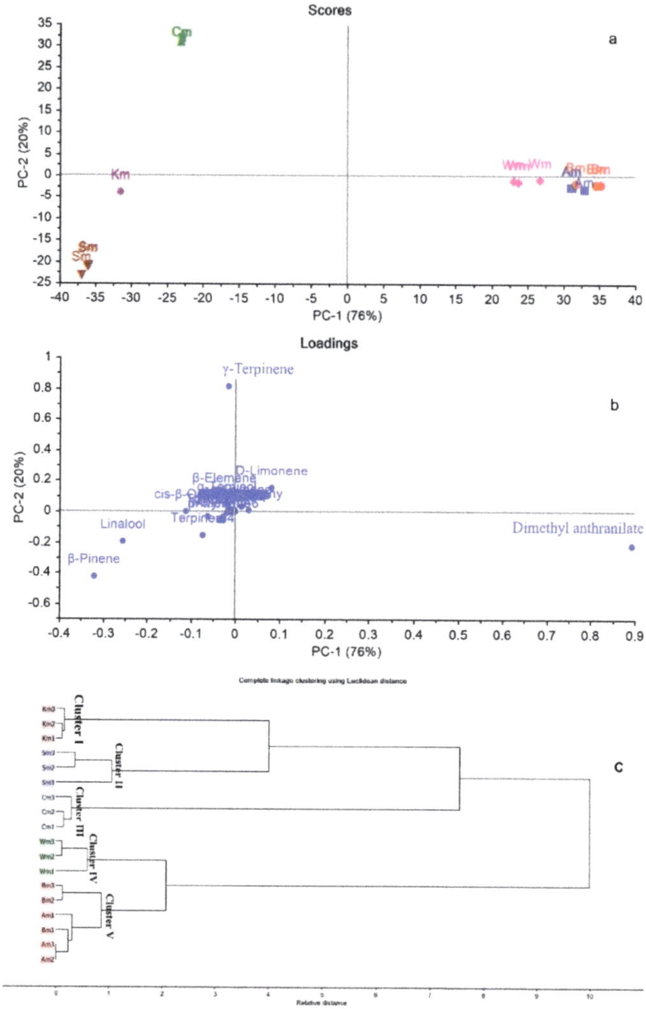

Figure 1. PCA score plot (**a**), loading plot (**b**), HCA dendogram (**c**) based on GC-MS metabolic profiles of different citrus cultivars based on the identification of volatile compounds displayed in Table 2. Refer to Table 1 for cultivar abbreviations.

PCA score plot (Figure 1a) explained about 96% of the variation in the dataset by the first two PCs, where PC1 accounting for 76% and PC2 for 20% of the variance. Different *C. reticulata* cultivars were assembled into four main groups on three different quadrants. Cultivars Cm, Sm, and Km were positioned on negative PC1, where Cm was positioned on the upper left quadrant. However, Sm and Km were placed on the lower one, well-discriminated from each other. Through clear investigation of the PCA score plot, it was observed that cultivars Am, Bm, and Wm were located on positive PC1, closely related to each other, where Am and Bm were superimposed on each other. The loading plot (Figure 1b) displayed the main discriminating markers responsible for PCA score plot pattern. Γ-Terpinene, linalool, and β-pinene were the key markers accountable for the segregation of Cm, Km, and Sm, respectively. However, dimethyl anthranilate was the compound responsible for the closeness of Am, Bm, and Wm cultivars.

Additionally, HCA was applied as an unsupervised pattern recognition method to confirm results obtained by PCA. The dendrograms displayed in Figure 1c, revealed segregation of different citrus cultivars into five main clusters. Cluster I, II, and III displayed Km, Sm, and Cm, respectively. HCA dendrograms revealed the closeness of Wm (Cluster IV) to Am and Bm that were grouped together in cluster V. HCA results endorsed that of PCA.

In an attempt to explore the ability of multivariate analysis to discriminate between closely related cultivars (Am, Bm, and Wm), PCA was applied on the GC-MS metabolic profiles of these three cultivars solely. Figure 2a,b displayed PCA score plot and loading plot, respectively. PCA score plot (Figure 2a) explained about 98% of the variation in the dataset by the first two PCs, where PC1 accounting for 95% and PC2 for 3% of the variance. Am, Bm, and Wm were completely segregated from each other in three different quadrants, where Am and Bm were positioned on PC1 on the right side of the plot confirming that they are closely related to each other in comparison with Wm that was located on negative PC1. Upon examination of the loading plot (Figure 2b), it was observed that dimethyl anthranilate was the major marker responsible for discriminating Am from other cultivars (Bm and Wm). However, γ-terpinene and β-caryophyllene were the main distinctive markers accountable for Bm cultivar segregation. Nevertheless, no distinctive marker was recognized in the loading plot for the separation of Wm. From this study, it was concluded that complete metabolic profile is mandatory for discrimination between closely related cultivars. For example, dimethyl anthranilate was the major component in the three cultivars (Am, Bm, and Wm); however, by applying multivariate analysis, it was observed that this marker could only discriminate Am from other cultivars. Similarly, regarding γ-terpinene is present approximately in the same percentage in the three cultivars; however, it discriminated Bm cultivar from other cultivars.

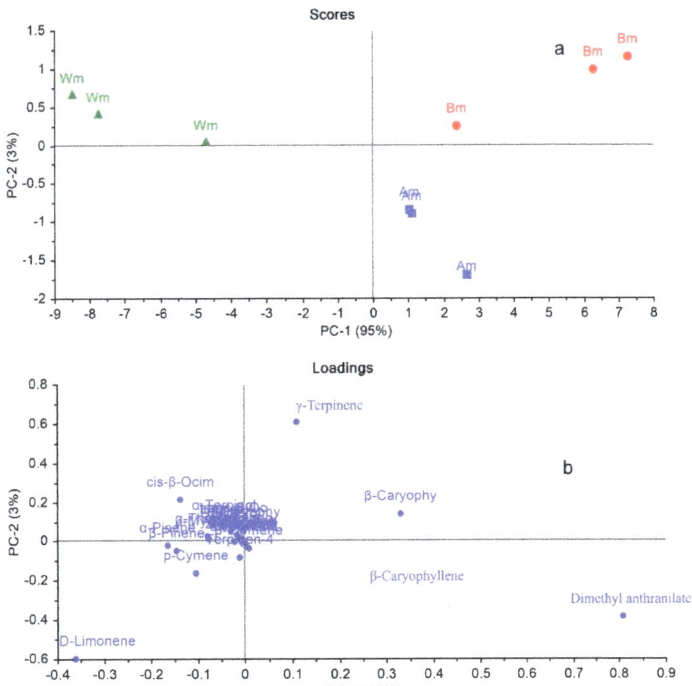

Figure 2. PCA score plot (**a**), loading plot (**b**) based on GC-MS metabolic profiles of Am, Bm, and Wm cultivars based on the identification of volatile compounds displayed in Table 2. Refer to Table 1 for cultivar abbreviations.

2.4. In Vitro Antiaging Activity

The antiaging capability of C. *reticulata* leaf cultivars essential oil was assessed via measuring their hyaluronidase, collagenase, and elastase enzyme inhibitory activity (Figure 3). Hyaluronic acid is one of the major building blocks of soft connective tissues that have a crucial role in maintaining the elasticity and moisture content of the skin, thus reducing wrinkles [44]. Hyaluronidase enzyme converts hyaluronic acid to small oligosaccharide moieties, consequently inhibitors of hyaluronidase are useful in hydrating the skin and delaying the aging process [45,46]. Among the tested cultivars Km and Cm were the most active with an IC_{50} value of 2.19 ± 0.10 and 3.22 ± 0.14 μg/mL, respectively, followed by Wm (IC_{50} = 8.43 ± 0.38 μg/mL) compared with the standard 6-*O*-palmitoyl-L-ascorbic acid exhibiting an IC_{50} of 1.56 ± 0.07 μg/mL. It is worth noting that Km cultivar showed no significant difference from the standard.

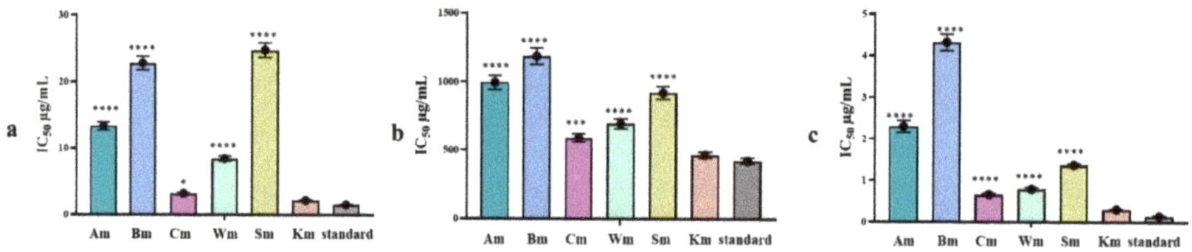

Figure 3. (**a**) Hyaluronidase, (**b**) collagenase, and (**c**) elastase inhibitory activity of C. *reticulata* leaf essential oil cultivar. The results are expressed as the mean ± SD, n = 3. Asterisks indicate significant differences from the standard drug (*, ***, ****, $p < 0.05$, $p < 0.0009$, $p < 0.0001$).

Collagen is considered the most abundant protein in the body and plays a crucial role in the structural support, elasticity, strength, and flexibility of the skin connective tissue. Collagenase an enzyme in the matrix metalloproteinase responsible for degrading collagen and thus inhibition of collagenase affects the elasticity of the skin and delays the wrinkling process. Km showed the highest inhibitory action on collagenase with an IC_{50} value of 465.9 ± 23.7 with no significant difference from the standard phenathroline used (IC_{50} = 423.2 ± 21.5 μg/mL). Whereas, elastin protein found in the connective tissue and catalyzed by elastase enzyme is the chief component of elastic fibers and preserves skin elasticity. Elastin degradation is accelerated by age and UV radiation due to the increase in the elastin action, resulting in skin aging [47]. In vitro elastase inhibitory activity revealed that Km, Cm, and Wm were the most active cultivars with IC_{50} of 0.31 ± 0.01, 0.66 ± 0.03, and 0.79 ± 0.04 μg/mL, respectively. However, Km only showed no significant difference from the standard FK 706 used with IC_{50} of 0.15 ± 0.01 μg/mL.

It is obvious from our results that Cm, Km, and Wm exert a promising inhibitory action against tested enzymes. These results may be explained by the difference in the chemical composition of Km and Cm and their segregation from other cultivars as displayed in the PCA score plot and HCA. The pronounced activity of Km might be accredited to the thymol (10.98 ± 0.27%) content uniquely found in this cultivar and linalool (22.20 ± 0.43%) present in the highest percentage in Km. The high γ-terpinene content found in Cm (48.56 ± 1.01 μg/mL) compared with other cultivars might be the reason beyond its enzyme inhibitory activity. Regarding Wm activity, we could not attribute its antiaging activity to the dimethyl anthranilate content only (49.06 ± 1.94 μg/mL), but also to the synergistic effect of the whole essential oil component, as Am and Bm cultivars showing nearly similar dimethyl anthranilate content displayed lower inhibitory activity on the studied enzymes.

2.5. In Silico Molecular Docking Studies on the Target Enzymes

The promising inhibitory activity of certain C. *reticulata* cultivars essential oil against hyaluronidase, collagenase, and elastase encouraged us to conduct a docking study of the major identified compounds with the enzymes of interest. This study aimed to discover the potential binding modes in which the essential oils exert its inhibitory action. The major compounds identified were docked into the 3D coordinates of hyaluronidase, collagenase, and elastase using the following PDB IDs: 1fcv, 456c, and 6qeo, respectively. The applied docking parameters were validated by re-docking each co-crystalized ligand into its corresponding active site. The calculated RMSD values between the docked pose and the co-crystalized pose were 0.63, 0.72, and 0.99 Å for hyaluronidase, collagenase, and elastase, respectively, ensuring the validity of the docking protocol. The re-docking of each co-crystalized ligand resulted in docking scores of −8.1, −9.6, and −10.4 Kcal/mole for hyaluronidase, collagenase, and elastase, respectively. The docking of the major compounds to the three enzymes resulted in good acceptable scores, comparable to those of the reference compounds. Table 3 summarizes the docking scores of essential oils major compounds against the three potential target enzymes and their corresponding docking scores.

Table 3. Docking score of major compounds of *C. reticulata* cultivars with hyaluronidase, collagenase, and elastase binding sites.

Compound Name	Docking Score with Hyaluronidase (1fcv) Kcal/mole	Docking Score with Collagenase (456c) Kcal/mole	Docking Score with Elastase (6eoq) Kcal/mole
α-Pinene	−7.1	−6.8	−5.9
D-Limonene	−5.8	−7.3	−6.8
cis-β-Ocimene	−6.6	−7.3	−7.4
γ-Terpinene	−7.1	−7.1	−6.9
Linalool	−8.1	−7.9	−9.1
Terpinen-4-ol	−7.3	−7.4	−6.7
Thymol	−9.8	−10.2	−12.3
Dimethyl anthranilate	−8.2	−8.6	−8.0

Interestingly, thymol, dimethyl anthranilate, and linalool and were the most active compounds on the three targets, achieving respective scores of −9.8, −8.2, and −8.1 Kcal/mole with hyaluronidase, −10.2, −8.6, and −7.9 Kcal/mole with collagenase, and −12.3, −8.0, and −9.1 Kcal/mole with elastase, respectively. Inspecting Figure 4, thymol interacted through hydrogen bonds with Asp111, Glu113, Tyr184, Tyr227, and Gln271 and hydrophobic interaction with Trp301, dimethyl anthranilate was found to interact with hyaluronidase through hydrogen bonds with Asp111 and Tyr227 and hydrophobic interactions with Trp267 and Trp301. Similarly, linalool interacted by hydrogen bonds with Asp111, Glu113, Tyr227, and Gln271 and hydrophobic interaction Trp301. As depicted by Figure 5 the top three compounds strongly interacted with the collagenase enzyme in which, thymol forms hydrogen bonds with Gly237, Ala238, and Phe241, besides hydrophobic interaction with Leu239 and Thr245, dimethyl anthranilate formed hydrogen bonds with Ala238, Leu239, Ile243, Tyr244, and Thr245 in addition to hydrophobic interaction with Thr245, while linalool formed six hydrogen bond interactions with Ala238, Phe241, Ile243, Tyr244, and Thr245. Moreover, Figure 6 shows that thymol interacted with Thr41, Cys42, and Cys191 through four hydrogen bond interactions, in addition to one hydrophobic interaction with His57, dimethyl anthranilate engaged in six hydrogen bond interactions with Cys42, His57, and Cys58 found in the active site of the elastase, while linalool interacted through hydrogen bond interactions with Thr41, Cys42, Gln192 and Ser195, in addition to hydrophobic interaction with His57. In conclusion, the three compounds namely, thymol, dimethyl anthranilate, and linalool had the best ability to strongly interact with the three enzymes hyaluronidase, collagenase, and elastase, achieving acceptable docking scores that sometimes exceeded those of the reference compounds. These acceptable scores were achieved through the establishment of many hydrophobic and hydrogen bond interactions. It is worth noting that thymol having the best docking scores in all tested enzyme was identified only in Km, this explains why it exhibited the highest in vitro enzyme inhibition action. Thus, the observed strong binding interactions validated the activities of the essential oils and suggested their possible mechanisms of action.

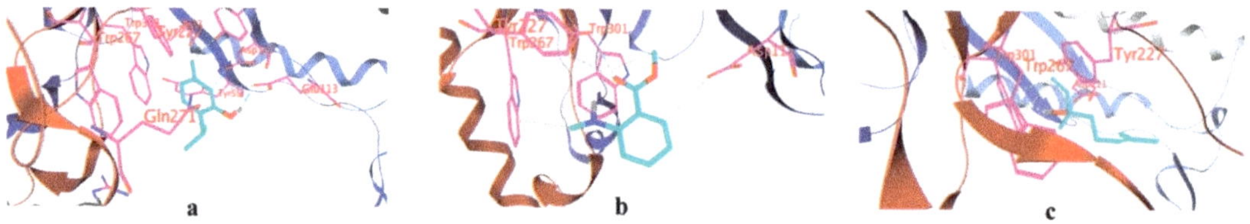

Figure 4. Binding modes of thymol (**a**), dimethyl anthranilate (**b**), and linalool (**c**) with hyaluronidase.

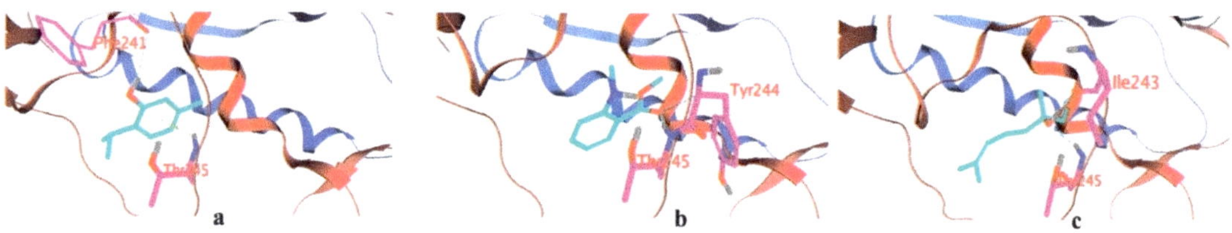

Figure 5. Binding modes of thymol (**a**), dimethyl anthranilate (**b**), and linalool (**c**) with collagenase.

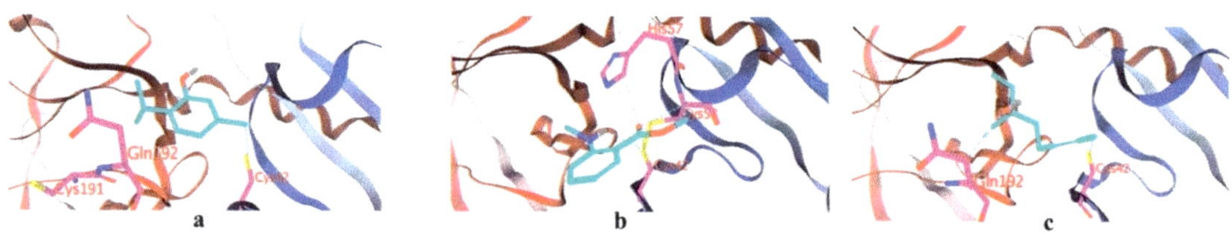

Figure 6. Binding modes of thymol (**a**), dimethyl anthranilate (**b**), and linalool (**c**) with elastase.

3. Materials and Methods

3.1. Plant Material

Leaves were randomly collected from different trees for each cultivar (6 cultivars of *Citrus reticulata*), cultivated under the same climatic conditions and during the fruit ripening stage from Citrus Department botanical garden (geographical coordinates: 30.020111741763642, 31.206797515343563), Horticulture Research Institute, Agriculture Research Center, Giza, Egypt (Table 1). All leaves were collected at the same phenological stage (spring growth cycle) in March 2021 (10 days after flowering, 10th week of the year). Plant materials were botanically identified by Prof. Gamal Farag Abdel Rahman, Head of Citrus Department. Voucher specimens of all collected samples were kept at the Pharmacognosy Department, Faculty of Pharmacy, Ain Shams University with codes (PHG-P-CR-391 to PHG-P-CR-396).

The type of the soil was light clay soil, (the temperature was about 25 °C with slight rain). The trees were 20 years old, where approximately 5 to 7 trees were used for leaves sampling for each cultivar. The distance of plants between the lines were approximately 1.5 m and 1 m on the line. Phosphate and organic fertilizers and agricultural sulfur were used. Phosphate fertilizers were added in the form of mono-superphosphate at a rate of 30 kg per feddan, during the months of December and January mixed with fully decomposed municipal fertilizers (15–20 m^3 per feddan) and 100 kg of agricultural sulfur until it decomposed before the spring season. Trees were irrigated by immersion every 30 to 45 days.

3.2. Hydro-Distillation of the Essential Oil and GC-MS Analysis

Entire fresh leaves (500 g) were submitted to hydro-distillation within 2 days from collection for 3 h using a Clevenger type apparatus. The oils obtained were recovered, weighed, dried over anhydrous sodium sulfate, and stored in amber and air-tight sealed vials at −20 °C until use. The yield (expressed in % (w/w)) was calculated based on the initial plant weight. The essential oils were analyzed one week after hydro distillation on GC-MS (Shimadzu GCMS-QP 2010, Koyoto, Japan) equipped with Rtx-5MS capillary column (30 m length × 0.25 mm i.d. × 0.25 μm film thickness) (Restek, Bellefonte, PA, USA). The oven temperature was kept at 45 °C for 2 min (isothermal) and programmed to 300 °C at 5 °C/min and kept constant at 300 °C for 10 min (isothermal); injector temperature was 250 °C. Helium was used as a carrier gas with the constant flow rate set at 1.41 mL/min. Diluted samples (1% v/v) were injected with a split ratio 15:1, and the injected volume was 1 μL. The MS operating parameters were as follows: interface temperature: 280 °C; ion source temperature: 200 °C; EI mode: 70 eV; scan range: 35–500 amu. Each sample was analyzed in triplicate [48]. Different compounds of the essential oils were comprehended with the aid of the NIST 05 database (NIST Mass Spectral Database, PC-Version 5.0, 2005, National Institute of Standardization and Technology, Gaithersburg, MD, USA). The Automated Mass Spectral Deconvolution and Identification System (AMDIS 2.64, NIST Gaithersburg, MD, USA) deconvoluted the measured mass spectra. The spectra of individual components were transferred to the NIST Mass Spectral Search Program MS Search 2.0 where they were matched against reference compounds of the NIST Mass Spectral Library 2005.

3.3. Multivariate Analysis

The data obtained from GC-MS were subjected to chemometric analysis, Principal Component Analysis (PCA) was applied as an initiative step in data investigation to afford an overview of all cultivar variability and to identify markers responsible for this variation. Hierarchal cluster analysis (HCA) was then utilized to allow clustering of different cultivars. The clustering patterns were built by applying the complete linkage way. This exhibition is more effective when the distance between samples (points) is computed by the Euclidean method [49]. PCA and HCA were accomplished using CAMO's Unscrambler® X 10.4 software (Computer-Aided Modeling, AS, Oslo, Norway).

3.4. In Vitro Antiaging Activity

3.4.1. Hyaluronidase Inhibition Assay

The assay was performed using hyaluronidase inhibitor screening assay kit QuantiChrom™ (BioAssay system, CA, USA) following the manufacturer's protocol. The assay was based on a turbidimetric reaction by measuring the amount of hyaluronic acid hydrolyzed. In brief, hyaluronidase from bovine testes (Sigma-Aldrich) was prepared freshly in 0.1 M acetate buffer. Serial dilutions of the essential oils were performed using DMSO. Forty microliters of hyaluronidase were transferred in each well of a 96-well plate then 20 μL of tested essential oil was added. The plate was incubated for 15 min at room temperature, then 10 μL substrate and 35 μL buffer were added to the plate, mixed, and incubated for 20 min at room temperature, the decrease in turbidity was measured spectrophotometry at 600 nm. 6-O-Palmitoyl-L-Ascorbic Acid (Sigma-Aldrich) was used as a standard hyaluronidase inhibitor.

3.4.2. Collagenase Inhibition Assay

The fluorometric collagenase inhibitor screening kit (Biovision, Catalog # K833-100, CA, USA) was used in the assay following the kit protocol. The kit utilizes Self-Quenched BODIPY conjugate of Type-B gelatin as a fluorogenic substrate to monitor the activity of collagenase. Serial dilutions of the essential oil and the standard (1,10)-Phenanthroline were mixed with 5 μL diluted collagenase and 44 μL collagenase buffer. Fluorescence was measured at 490/520 nm in a kinetic mode at 37 °C for 30–60 min.

3.4.3. Elastase Inhibition Assay

Elastase inhibitory activity was measured following the kit protocol (Molecular Probes', Catalog # E-12056 EnzChek®, Leiden, Netherlands). In brief, fifty microliters of the reaction buffer (pH 8 Tris-HCL buffer) were added into each well together with 50 µL of 100 µg/mL DQ™ elastin substrate working solution to provide a final substrate concentration of 25 µg/mL. Serial dilutions of the oils were added, mixed, and incubated for 30 min at room temperature. Fluorescence intensity was measured using a fluorescence microplate reader equipped with standard fluorescein filters. FK706 was used as a standard elastase inhibitor [50,51].

3.5. Statistical Analysis

All experiments were carried out three times in triplicate. Data are expressed as mean $\pm$ standard deviation. The IC_{50} values were calculated, and the results were presented using GraphPad Prism® software (Version 7, graph-Pad software Inc., San Diego, CA, USA).

3.6. In Silico Molecular Docking Study

The docking studies were conducted using Molecular Operating Environment (MOE 2019.02) Software [52,53]. The X-ray crystal structures of hyaluronidase, collagenase, and elastase were downloaded from the protein data bank using the following PDB IDs: 1fcv, 456c and 6qeo, respectively. Hydrogens and charges of the receptors were optimized using AMBER10: EHT embedded in MOE software. The binding site of the three enzymes was constructed where the corresponding co-crystalized ligand is bound. Major compounds identified in the essential oils were sketched using the 2D builder of MOE2019 and converted to 3D structures using the same software. Compounds were docked into the EGFR binding domain using triangular matcher and London dg as a placement and scoring methods, respectively. At last, 2D and 3D interaction diagrams were generated by MOE to analyze the docking results.

4. Conclusions

Metabolic profiling of the essential oils of *C. reticulata* leaf cultivars resulted in the identification of thirty-nine compounds including β-pinene, D-limonene, γ-terpinene, linalool, and dimethyl anthranilate as the main components. Qualitative and quantitative variabilities among the chemical composition of different cultivars was observed using PCA and HCA, which indicate that a complete metabolic profile is mandatory for discrimination between closely related cultivars. Cm, Km, and Wm exerted a promising inhibitory action against tested aging enzymes. In silico studies on the major compounds confirmed the activities of the essential oils and suggested their possible mechanisms of action. From our study we can conclude that certain cultivars of *Citrus reticulata* can be proposed as a promising candidate for incorporation in antiaging skin care products.

Author Contributions: Conceptualization, H.A.G. and N.M.F.; methodology, N.M.F., S.S.E. and H.A.G.; software, D.F.B., N.M.F. and S.S.E.; validation, H.A.G., R.T.M. and S.S.E.; formal analysis, N.M.F. and D.F.B.; investigation, N.M.F., S.S.E. and H.A.G.; resources, S.S.E., R.T.M. and D.F.B.; writing—original draft preparation, N.M.F. and H.A.G.; writing—review and editing, S.S.E., D.F.B. and R.T.M.; supervision, S.S.E. and H.A.G.; funding acquisition, D.F.B., R.T.M. and S.S.E. All authors have read and agreed to the published version of the manuscript.

Funding: This research work was funded by Institutional Fund Projects under grant no. (IFPIP:1004-249-1443). The authors gratefully acknowledge technical and financial support provided by the Ministry of Education and King Abdulaziz University, DSR, Jeddah, Saudi Arabia.

Institutional Review Board Statement: Not applicable.

Data Availability Statement: Data supporting the reported results can be found at, NIST Chemistry Webbook, https://webbook.nist.gov/chemistry/, (accessed on 10 September 2022).

Acknowledgments: The authors gratefully acknowledge technical and financial support provided by the Ministry of Education and King Abdulaziz University, DSR, Jeddah, Saudi Arabia.

Conflicts of Interest: The authors declare no conflict of interest.

Sample Availability: Samples of the essentials are available from the authors.

References

1. Dosoky, N.S.; Setzer, W.N. Biological Activities and Safety of *Citrus* spp. Essential Oils. *Int. J. Mol. Sci.* **2018**, *19*, 1966. [CrossRef] [PubMed]
2. Gioffrè, G.; Ursino, D.; Labate, M.L.C.; Giuffrè, A.M. The peel essential oil composition of bergamot fruit (*Citrus bergamia*, Risso) of Reggio Calabria (Italy): A review. *Emir. J. Food Agric.* **2020**, *32*, 835–845. [CrossRef]
3. Vekiari, S.A.; Protopapadakis, E.E.; Papadopoulou, P.; Papanicolaou, D.; Panou, C.; Vamvakias, M. Composition and seasonal variation of the essential oil from leaves and peel of a Cretan lemon variety. *J. Agric. Food Chem.* **2002**, *50*, 147–153. [CrossRef]
4. Maria, G.A.; Riccardo, N. *Citrus bergamia*, Risso: The peel, the juice and the seed oil of the bergamot fruit of Reggio Calabria (South Italy). *Emir. J. Food Agric.* **2020**, *32*, 522–532. [CrossRef]
5. Zouaghi, G.; Najar, A.; Aydi, A.; Claumann, C.A.; Zibetti, A.W.; Ben Mahmoud, K.; Jemmali, A.; Bleton, J.; Moussa, F.; Abderrabba, M. Essential oil components of Citrus cultivar 'MALTAISE DEMI SANGUINE' (*Citrus sinensis*) as affected by the effects of rootstocks and viroid infection. *Int. J. Food Prop.* **2019**, *22*, 438–448. [CrossRef]
6. Ferhat, M.-A.; Boukhatem, M.N.; Hazzit, M.; Meklati, B.Y.; Chemat, F. Cold pressing, hydrodistillation and microwave dry distillation of citrus essential oil from Algeria: A comparative study. *Electron. J. Biotechnol.* **2016**, *1*, 30–41.
7. Sulzbach, M.; Silva, M.A.S.d.; Gonzatto, M.P.; Marques, M.M.O.; Böettcher, G.N.; Silvestre, W.P.; Silva, J.C.R.L.; Pauletti, G.F.; Schwarz, S.F. Effect of distillation methods on the leaf essential oil of some Citrus cultivars. *J. Essent. Oil Res.* **2021**, *33*, 452–463. [CrossRef]
8. Nguyen, H.; Campi, E.; Jackson, W.; Patti, A. Effect of oxidative deterioration on flavour and aroma compounds of lemon oil. *Food Chem.* **2009**, *112*, 388–393. [CrossRef]
9. Steuer, B.; Schulz, H.; Läger, E. Classification and analysis of citrus oils by NIR spectroscopy. *Food Chem.* **2001**, *72*, 113–117. [CrossRef]
10. Bora, H.; Kamle, M.; Mahato, D.K.; Tiwari, P.; Kumar, P. Citrus Essential Oils (CEOs) and Their Applications in Food: An Overview. *Plants* **2020**, *9*, 357. [CrossRef]
11. Fisher, K.; Phillips, C. Potential antimicrobial uses of essential oils in food: Is citrus the answer? *Trends Food Sci. Technol.* **2008**, *19*, 156–164. [CrossRef]
12. Zhang, W.; Liu, D.; Fu, X.; Xiong, C.; Nie, Q. Peel Essential Oil Composition and Antibacterial Activities of *Citrus sinensis* L. Osbeck and *Citrus reticulata* Blanco. *Horticulturae* **2022**, *8*, 793. [CrossRef]
13. Yabalak, E.; Erdoğan Eliuz, E.A.; Nazlı, M.D. Evaluation of *Citrus reticulata* essential oil: Chemical composition and antibacterial effectiveness incorporated gelatin on *E. coli* and *S. aureus*. *Int. J. Environ. Health Res.* **2022**, *32*, 1261–1270. [CrossRef] [PubMed]
14. Viuda-Martos, M.; Ruiz-Navajas, Y.; Fernández-López, J.; Pérez-Álvarez, J. Antifungal activity of lemon (*Citrus lemon* L.), mandarin (*Citrus reticulata* L.), grapefruit (*Citrus paradisi* L.) and orange (*Citrus sinensis* L.) essential oils. *Food Control* **2008**, *19*, 1130–1138. [CrossRef]
15. Mahdi, A.A.; Al-Maqtari, Q.A.; Mohammed, J.K.; Al-Ansi, W.; Cui, H.; Lin, L. Enhancement of antioxidant activity, antifungal activity, and oxidation stability of *Citrus reticulata* essential oil nanocapsules by clove and cinnamon essential oils. *Food Biosci.* **2021**, *43*, 101226. [CrossRef]
16. Ravizza, R.; Gariboldi, M.B.; Molteni, R.; Monti, E. Linalool, a plant-derived monoterpene alcohol, reverses doxorubicin resistance in human breast adenocarcinoma cells. *Oncol. Rep.* **2008**, *20*, 625–630. [CrossRef]
17. Jing, L.; Zhang, Y.; Fan, S.; Gu, M.; Guan, Y.; Lu, X.; Huang, C.; Zhou, Z. Preventive and ameliorating effects of citrus D-limonene on dyslipidemia and hyperglycemia in mice with high-fat diet-induced obesity. *Eur. J. Pharmacol.* **2013**, *715*, 46–55. [CrossRef]
18. Moufida, S.d.; Marzouk, B. Biochemical characterization of blood orange, sweet orange, lemon, bergamot and bitter orange. *Phytochemistry* **2003**, *62*, 1283–1289. [CrossRef]
19. Legault, J.; Pichette, A. Potentiating effect of beta-caryophyllene on anticancer activity of alpha-humulene, isocaryophyllene and paclitaxel. *J. Pharm. Pharmacol.* **2007**, *59*, 1643–1647. [CrossRef]
20. Silveira, V.; Rubio, K.T.S.; Martucci, M.E.P. Anxiolytic effect of *Anthemis nobilis* L. (roman chamomile) and *Citrus reticulata* Blanco (tangerine) essential oils using the light-dark test in zebrafish (Danio rerio). *J. Ethnopharmacol.* **2022**, *298*, 115580. [CrossRef]
21. Suwannayod, S.; Sukontason, K.L.; Somboon, P.; Junkum, A.; Leksomboon, R.; Chaiwong, T.; Jones, M.K.; Sripa, B.; Balthaisong, S.; Phuyao, C.; et al. Activity of Kaffir lime (*Citrus hystrix*) essential oil against blow flies and house fly. *Southeast Asian J. Trop. Med. Public Health* **2018**, *49*, 32–45.
22. Chutia, M.; Deka Bhuyan, P.; Pathak, M.G.; Sarma, T.C.; Boruah, P. Antifungal activity and chemical composition of *Citrus reticulata* Blanco essential oil against phytopathogens from North East India. *LWT Food Sci. Technol.* **2009**, *42*, 777–780. [CrossRef]
23. Curk, F.; Ancillo, G.; Garcia-Lor, A.; Luro, F.; Perrier, X.; Jacquemoud-Collet, J.P.; Navarro, L.; Ollitrault, P. Next generation haplotyping to decipher nuclear genomic interspecific admixture in *Citrus* species: Analysis of chromosome 2. *BMC Genet.* **2014**, *15*, 152. [CrossRef] [PubMed]

24. Adeleke, K.; Lawal, O.A.; Abanikannda, O.; Olaniyan, A.; Setzer, W. Citrus Essential Oil of Nigeria Part IV: Volatile Constituents of Leaf Oils of Mandarins (*Citrus reticulata* Blanco) From Nigeria. *Rec. Nat. Prod.* **2010**, *4*, 156.
25. Yabesh, J.E.; Prabhu, S.; Vijayakumar, S. An ethnobotanical study of medicinal plants used by traditional healers in silent valley of Kerala, India. *J. Ethnopharmacol.* **2014**, *154*, 774–789. [CrossRef]
26. Yi, F.; Jin, R.; Sun, J.; Ma, B.; Bao, X. Evaluation of mechanical-pressed essential oil from Nanfeng mandarin (*Citrus reticulata* Blanco cv. Kinokuni) as a food preservative based on antimicrobial and antioxidant activities. *LWT* **2018**, *95*, 346–353. [CrossRef]
27. Tao, N.; Jia, L.; Zhou, H. Anti-fungal activity of *Citrus reticulata* Blanco essential oil against *Penicillium italicum* and *Penicillium digitatum*. *Food Chem.* **2014**, *153*, 265–271. [CrossRef]
28. D'Almeida, R.E.; Sued, N.; Arena, M.E. *Citrus paradisi* and *Citrus reticulata* essential oils interfere with *Pseudomonas aeruginosa* quorum sensing in vivo on *Caenorhabditis elegans*. *Phytomed. Plus* **2022**, *2*, 100160. [CrossRef]
29. Eldahshan, O. Comparison of Chemical and Antimicrobial Studies of Egyptian Mandarin Leaves and Green Branches Volatile Oil. *European. J. Med. Plants* **2015**, *5*, 248–254. [CrossRef]
30. Mehmood, F.; Khan, Z.-U.-D.; Shahzadi, P.; Yaseen, T.; Mughal, T.; Raza, S.; Qasim, M. A comparative study of in vitro total antioxidant, in vivo antidiabetic and antimicrobial activities of essential oils from leaves and rind of *Citrus reticulata* Blanco cv. Murcot (honey). *Pak. J. Bot.* **2013**, *45*, 1571–1576.
31. Zhou, X.M.; Zhao, Y.; He, C.C.; Li, J.X. Preventive effects of *Citrus reticulata* essential oil on bleomycin-induced pulmonary fibrosis in rats and the mechanism. *Zhong Xi Yi Jie He Xue Bao* **2012**, *10*, 200–209. [CrossRef] [PubMed]
32. Kamal, G.M.; Ashraf, M.; Hussain, A.; Shahzadi, A.; Chughtai, M.I. Antioxidant potential of peel essential oils of three Pakistani citrus species: *Citrus reticulata, Citrus sinensis* and *Citrus paradisii*. *Pak. J. Bot.* **2013**, *45*, 1449–1454.
33. Martins, M.H.G.; Fracarolli, L.; Vieira, T.M.; Dias, H.J.; Cruz, M.G.; Deus, C.C.H.; Nicolella, H.D.; Stefani, R.; Rodrigues, V.; Tavares, D.C.; et al. Schistosomicidal Effects of the Essential Oils of Citrus limonia and *Citrus reticulata* Against *Schistosoma mansoni*. *Chem. Biodivers.* **2017**, *14*, e1600194. [CrossRef]
34. Senol Deniz, F.S.; Orhan, I.E.; Duman, H. Profiling cosmeceutical effects of various herbal extracts through elastase, collagenase, tyrosinase inhibitory and antioxidant assays. *Phytochem. Lett.* **2021**, *45*, 171–183. [CrossRef]
35. Dalia, I.H.; Maged, E.M.; Assem, M.E. *Citrus reticulata* Blanco cv. Santra leaf and fruit peel: A common waste products, volatile oils composition and biological activities. *J. Med. Plants Res.* **2016**, *10*, 457–467. [CrossRef]
36. Denkova-Kostova, R.; Teneva, D.; Tomova, T.; Goranov, B.; Denkova, Z.; Shopska, V.; Slavchev, A.; Hristova-Ivanova, Y. Chemical composition, antioxidant and antimicrobial activity of essential oils from tangerine (*Citrus reticulata* L.), grapefruit (*Citrus paradisi* L.), lemon (*Citrus lemon* L.) and cinnamon (*Cinnamomum zeylanicum* Blume). *Z. Naturforsch. C* **2021**, *76*, 175–185. [CrossRef]
37. Chi, P.T.L.; Van Hung, P.; Le Thanh, H.; Phi, N.T.L.J.W.; Valorization, B. Valorization of citrus leaves: Chemical composition, antioxidant and antibacterial activities of essential oils. *Waste Biomass Valorization* **2020**, *11*, 4849–4857. [CrossRef]
38. Fanciullino, A.-L.; Tomi, F.; Luro, F.; Desjobert, J.M.; Casanova, J. Chemical variability of peel and leaf oils of mandarins. *Flavour Fragr. J.* **2006**, *21*, 359–367. [CrossRef]
39. Lota, M.-L.; de Rocca Serra, D.; Tomi, F.; Joseph, C. Chemical variability of peel and leaf essential oils of mandarins from *Citrus reticulata* Blanco. *Biochem. Syst. Ecol.* **2000**, *28*, 61–78. [CrossRef]
40. Karioti, A.; Skaltsa, H.; Gbolade, A.A. Constituents of the Distilled Essential Oils of *Citrus reticulata* and *C. paradisi* from Nigeria. *J. Essent. Oil Res.* **2007**, *19*, 520–522. [CrossRef]
41. Fleisher, Z.; Fleisher, A. Mandarin Leaf Oil (*Citrus reticulata* Blanco) Aromatic Plants of the Holy Land and the Sinai. Part III. *J. Essent. Oil Res.* **1990**, *2*, 331–334. [CrossRef]
42. Blanco Tirado, C.; Stashenko, E.E.; Combariza, M.Y.; Martinez, J.R. Comparative study of Colombian citrus oils by high-resolution gas chromatography and gas chromatography-mass spectrometry. *J. Chromatogr. A* **1995**, *697*, 501–513. [CrossRef]
43. Kammoun, A.K.; Altyar, A.E.; Gad, H.A. Comparative metabolic study of Citrus sinensis leaves cultivars based on GC–MS and their cytotoxic activity. *J. Pharm. Biomed. Anal.* **2021**, *198*, 113991. [CrossRef] [PubMed]
44. Liyanaarachchi, G.D.; Samarasekera, J.K.R.R.; Mahanama, K.R.R.; Hemalal, K.D. Tyrosinase, elastase, hyaluronidase, inhibitory and antioxidant activity of Sri Lankan medicinal plants for novel cosmeceuticals. *Ind. Crops Prod.* **2018**, *111*, 597–605. [CrossRef]
45. Miri, A.K.; Heris, H.K.; Mongeau, L.; Javid, F. Nanoscale viscoelasticity of extracellular matrix proteins in soft tissues: A multiscale approach. *J. Mech. Behav. Biomed. Mater.* **2014**, *30*, 196–204. [CrossRef] [PubMed]
46. Lee, J.; Moon, S.; Hong, Y.; Ahn, D.U.; Paik, H.-D. Anti-elastase and anti-hyaluronidase activity of phosvitin isolated from hen egg yolk. *Br. Poult. Sci.* **2020**, *61*, 17–21. [CrossRef]
47. Antonicelli, F.; Bellon, G.; Debelle, L.; Hornebeck, W. Elastin-elastases and inflamm-aging. *J. Dev. Biol.* **2007**, *79*, 99–155.
48. Gad, H.; Al-Sayed, E.; Ayoub, I. Phytochemical discrimination of Pinus species based on GC–MS and ATR-IR analyses and their impact on Helicobacter pylori. *Phytochem. Anal.* **2021**, *32*, 820–835. [CrossRef]
49. Brereton, R.G. (Ed.) *Chemometrics, Data Analysis for the Laboratory and Chemical Plant*; John Wiley & Sons, Ltd.: Hoboken, NJ, USA, 2003.
50. Stein, R.L.; Trainor, D.A. Mechanism of inactivation of human leukocyte elastase by a chloromethyl ketone: Kinetic and solvent isotope effect studies. *J. Biochem.* **1986**, *25*, 5414–5419. [CrossRef]
51. Powers, J.C.; Gupton, B.F.; Harley, A.D.; Nishino, N.; Whitley, R.J. Specificity of porcine pancreatic elastase, human leukocyte elastase and cathepsin G Inhibition with peptide chloromethyl ketones. *Acta Biochim. Biophys. Enzymol.* **1977**, *485*, 156–166. [CrossRef]

52. Scholz, C.; Knorr, S.; Hamacher, K.; Schmidt, B. DOCKTITE: A Highly Versatile Step-by-Step Workflow for Covalent Docking and Virtual Screening in the Molecular Operating Environment. *J. Chem. Inf. Model.* **2015**, *55*, 398–406. [CrossRef] [PubMed]
53. Vilar, S.; Cozza, G.; Moro, S. Medicinal chemistry and the molecular operating environment (MOE): Application of QSAR and molecular docking to drug discovery. *Curr. Top. Med. Chem.* **2008**, *8*, 1555–1572. [CrossRef] [PubMed]

plants

MDPI

Article

Comparative Metabolic Study of *Tamarindus indica* L.'s Various Organs Based on GC/MS Analysis, *In Silico* and *In Vitro* Anti-Inflammatory and Wound Healing Activities

Shaza H. Aly [1], Mahmoud A. El-Hassab [2], Sameh S. Elhady [3,4] and Haidy A. Gad [5,6,*]

1 Department of Pharmacognosy, Faculty of Pharmacy, Badr University in Cairo, Cairo 11829, Egypt
2 Department of Medicinal Chemistry, Faculty of Pharmacy, King Salman International University (KSIU), South Sinai 46612, Egypt
3 Department of Natural Products, Faculty of Pharmacy, King Abdulaziz University, Jeddah 21589, Saudi Arabia
4 Center for Artificial Intelligence in Precision Medicines, King Abdulaziz University, Jeddah 21589, Saudi Arabia
5 Department of Pharmacognosy, Faculty of Pharmacy, Ain Shams University, Cairo 11566, Egypt
6 Department of Pharmacognosy, Faculty of Pharmacy, King Salman International University (KSIU), South Sinai 46612, Egypt
* Correspondence: haidygad@pharma.asu.edu.eg

Abstract: The chemical composition of the *n*-hexane extract of *Tamarindus indica's* various organs—bark, leaves, seeds, and fruits (TIB, TIL, TIS, TIF)—was investigated using gas chromatography-mass spectrometry (GC/MS) analysis. A total of 113 metabolites were identified, accounting for 93.07, 83.17, 84.05, and 85.08 % of the total identified components in TIB, TIL, TIS, and TIF, respectively. Lupeol was the most predominant component in TIB and TIL, accounting for 23.61 and 22.78%, respectively. However, *n*-Docosanoic acid (10.49%) and methyl tricosanoate (7.09%) were present in a high percentage in TIS. However, α-terpinyl acetate (7.36%) and α-muurolene (7.52%) were the major components of TIF *n*-hexane extract. By applying a principal component analysis (PCA) and hierarchal cluster analysis (HCA) to GC/MS-based metabolites, a clear differentiation of *Tamarindus indica* organs was achieved. The anti-inflammatory activity was evaluated in vitro on lipopolysaccharide (LPS)-induced RAW 264.7 macrophages. In addition, the wound healing potential for the *n*-hexane extract of various plant organs was assessed using the in-vitro wound scratch assay using Human Skin Fibroblast cells. The tested extracts showed considerable anti-inflammatory and wound-healing activities. At a concentration of 10 μg/mL, TIL showed the highest nitric oxide (NO) inhibition by 53.97 ± 5.89%. Regarding the wound healing potential, after 24 h, TIB, TIL, TIS, and TIF *n*-hexane extracts at 10 g/mL reduced the wound width to 1.09 ± 0.04, 1.12 ± 0.18, 1.09 ± 0.28, and 1.41 ± 0.35 mm, respectively, as compared to the control cells (1.37 ± 0.15 mm). These findings showed that the *n*-hexane extract of *T. indica* enhanced wound healing by promoting fibroblast migration. Additionally, a docking study was conducted to assess the major identified phytoconstituents' affinity for binding to glycogen synthase kinase 3-β (GSK3-β), matrix metalloproteinases-8 (MMP-8), and nitric oxide synthase (iNOS). Lupeol showed the most favourable binding affinity to GSK3-β and iNOS, equal to −12.5 and −13.7 Kcal/mol, respectively, while methyl tricosanoate showed the highest binding affinity with MMP-8 equal to −13.1 Kcal/mol. Accordingly, the *n*-hexane extract of *T. indica's* various organs can be considered a good candidate for the management of wound healing and inflammatory conditions.

Keywords: *Tamarindus indica* L.; GC/MS; anti-inflammatory; wound healing; molecular docking; chemometric analysis; drug discovery; public health

Citation: Aly, S.H.; El-Hassab, M.A.; Elhady, S.S.; Gad, H.A. Comparative Metabolic Study of *Tamarindus indica* L.'s Various Organs Based on GC/MS Analysis, *In Silico* and *In Vitro* Anti-Inflammatory and Wound Healing Activities. *Plants* **2023**, *12*, 87. https://doi.org/10.3390/plants12010087

Academic Editors: Hazem Salaheldin Elshafie, Ippolito Camele and Adriano Sofo

Received: 4 December 2022
Revised: 17 December 2022
Accepted: 21 December 2022
Published: 23 December 2022

1. Introduction

Tamarindus is a monospecific genus belonging to the Fabaceae family and usually has been identified under the name Tamarind [1]. *Tamarindus indica* Linn. is a tropical fruit tree that can be found growing wild in many tropical and subtropical areas [2]. Tamarind is a semi-deciduous tree, slow growing but long-lived, that can reach a height of up to 30 m. The leaves are alternate and paripinnate, and its fruit is a legume appearing as indehiscent pods with 1–10 seeds, which represent the main criteria of plants belonging to the Fabaceae family [3–6]. The fruit was especially well-known and commonly used by the ancient Egyptians and widely consumed and traded in Africa [3]. The fruit has usually been used in food dishes through different applications as a condiment and spice and could be eaten fresh or raw. Tamarind is sold as an entire pod, paste, or concentrate in speciality food stores all throughout the world [7,8]. Its fruit has been reported in the literature for its pharmacological antioxidant, anti-inflammatory, antidiabetic, anti-Alzheimer, and antihyperlipidemic properties [9–11]. Tamarind has been commonly used in folk medicine for the treatment of diarrhoea and dysentery, helminth infections, fever, and abdominal pain. It is also used as an anti-malarial and laxative and has a role in wound healing [3]. Traditional Indian and African medicine has made substantial use of different tamarind products, including its leaves, fruit, and seeds [12]. Across many areas of central West Africa, tamarind leaves and bark are used traditionally to alleviate wounds [3,13]. Leaves of Tamarind have nutritional value due to their richness in vitamins and minerals [12]. Its leaves have exhibited potential antioxidant, antibacterial and antifungal, anti-inflammatory, and analgesic activities [14–17]. Additionally, the leaves of Tamarind have an anti-inflammatory role in Indian folk medicine [18].

Regarding the bark, it is effective as an astringent and tonic agent in lotions or poultices to relieve sores, ulcers, boils, and rashes, owing to its richness with tannins [12]. In addition, the bark showed in vivo antihyperglycemic, anti-inflammatory, and analgesic activities [17,19]. Moreover, leaves and bark have been reported for their anthelmintic activity [20].

Wound healing is a normal complex repair process that requires synchronisation between various biological and immunological systems, including many phases of inflammation, proliferation, and contraction of the wound, with the formation of granulation tissues and remodelling [21–23]. Plants and plant products, such as *Aloe vera*, *Sophora flavescens*, *Punica granatum* L., *Mimosa pudica* L., and *Hibiscus rosa sinensis* L., have been used in traditional medicine to cure and prevent diseases, especially wounds [24–26].

Recently, there has been a great demand for the development of natural products to cure different conditions owing to their safety, availability, versatile biological activities, and unique secondary metabolites [27–31]. Consequently, the exploration of medicinal plants as new candidates for wound healing would be valuable and beneficial, especially medicinal plants that are characterised by biocompatibility, wound healing, and anti-inflammatory properties [22,23,32,33]. Many reports have revealed that using *n*-hexane as an extracting solvent targets non-polar compounds [21,34,35], and many studies have proven the role of *n*-hexane extract from various plants in anti-inflammatory and wound healing conditions [36–39].

Thus, the purpose of this present study was to compare and determine the chemical composition of the *n*-hexane extract of *T. indica* L.'s various organs using GC/MS analysis. Further, the principal component analysis (PCA) and hierarchical cluster analysis (HCA) were applied to discriminate among various plant organs based on their chemical profiles, which could be used as a tool for the detection of similarities and differences among the four organs. Furthermore, the *n*-hexane extract obtained from *T. indica* L.'s various organs were investigated for their anti-inflammatory and wound healing properties for the first time along with a molecular docking study to correlate the chemical composition with the mentioned biological activity. To the best of our knowledge, this is the first study on the wound-healing properties for the various organs of *T. indica* L. *n*-hexane extract.

2. Results and Discussion

2.1. GC/MS Analysis of the N-Hexane Extract of *Tamarindus indica* L. Various Organs

The chemical composition of the *n*-hexane extract of *T. indica*, bark, leaves, seeds, and fruit (TIB, TIL, TIS, and TIL) were investigated using GC/MS analyses (Table 1). The yields of extraction of TIB, TIL, TIS, and TIL using *n*-hexane were 4.20, 6.10, 2.07, and 3.12% *w*/*w*, respectively. The samples encompassed 113 constituents that accounted for 93.07, 83.17, 84.05, and 85.08 % of the total identified components in bark, leaves, seeds, and fruits of *T. indica*, respectively. The major constituents identified in the *n*-hexane extract of TIB were lupeol (23.61%), lupeol acetate (23.02%), lupenone (9.36%), *n*-tetratriacontane (8.16%), 24-methylenecycloartanol (5.87%), 1-heptatriacontanol (4.25%), and γ-sitosterol (4.17%). Regarding the *n*-hexane extract of TIL, lupeol (22.78%), lupenone (13.08%), γ-sitosterol (8.83%), (17*E*)-Cholesta-17,24-diene-3,6-diol (8.23%), *n*-nonacosane (4.13%), Betulinaldehyde (3.70%), and β-Amyrone (2.05%) were characterised as the chief metabolites. In this case, lupenone and lupeol were present in higher percentages as compared to other organs. Moreover, *n*-Docosanoic acid (10.49%), methyl tricosanoate (7.09%), tetracosanal (5.42%), 2-methylhexacosane (4.31%), 11-methylpentacosane (4.22%), and heneicosyl acetate (2.94%) displayed the highest percentages in the *n*-hexane extract of TIS. Additionally, among the major constituents, especially in TIS, are *n*-octacosane (4.63%), *n*-heptacosane (4.17%), *cis*-13,16-docosadienoic acid (3.70%), *n*-hexacosane (2.47%), *n*- pentacosane (2.46%), and methyl tetracosanoate (2.17%). Furthermore, the *n*-hexane extract of TIF showed a predominantly high percentage of α-terpinyl acetate (7.36%), α-muurolene (7.52%), 2-methyl hexadecane (5.70%), methyl pentadecanoate (4.53%), α-eicosene (3.44%), and linolenic acid (3.26%) in addition to chief constituents *n*-heptacosane (6.67%), *n*-nonacosane (5.42%), squalene (4.17%), *cis*-13,16-docosadienoic acid (3.76%), lupeol (3.76%), *n*-tetratriacontane (3.38%), and hentriacontane (3.32%). It was notable that γ-sitosterol is the chief component in the four organs ranging from (3.79%) in TIS to (11.28%) in TIF. It was notable that squalene, n-nonacosane, n-hentriacontane, 5α-stigmast-22-en-3β-ol, and γ-sitosterol were detected in the four organs. The major compounds present in the four various organs of *T. indica* are represented in (Figure 1).

Table 1. Chemical profile of *n*-hexane extract from various organs of *T. indica*: TIB (bark), TIL (leaves), TIS (seeds), and TIF (fruits) using GC/MS analysis.

No.	R_t (min)	Compound Name	$RI_{Exp.}$ [a]	RI_{Lit} [b]	Molecular Formula	Peak Area (%)			
						TIB	TIL	TIS	TIF
1	6.94	α-pinene	930	930	$C_{10}H_{16}$	-	-	-	0.38
2	8.16	α-Sabinene	971	971	$C_{10}H_{16}$	-	-	-	0.57
3	9.90	D-Limonene	1028	1028	$C_{10}H_{16}$	-	-	-	1.35
4	9.97	1,8-Cineole	1030	1030	$C_{10}H_{18}O$	-	-	-	0.61
5	13.93	Isoborneol	1156	1156	$C_{10}H_{18}O$	-	-	-	0.87
6	19.00	α-Terpinyl acetate	1328	1327	$C_{12}H_{20}O_2$	0.03	-	-	7.36
7	23.59	α-Muurolene	1494	1494	$C_{15}H_{24}$	0.04	-	-	7.52
8	27.69	2-Methylhexadecane	1668	1666	$C_{17}H_{36}$	0.03	-	-	5.70
9	31.07	Hexadecanal; Palmitaldehyde	1810	1811	$C_{16}H_{32}O$	-	-	-	0.36
10	31.25	Methyl pentadecanoate	1819	1820	$C_{16}H_{32}O_2$	-	-	-	4.53
11	31.52	Neophytadiene	1833	1837	$C_{20}H_{38}$	-	0.2	-	-
12	31.67	Hexahydrofarnesyl acetone	1841	1842	$C_{18}H_{36}O$	-	0.09	-	-
13	32.43	Heptadecan-2-one	1880	1892	$C_{17}H_{34}O$	0.03	0.07	-	-
14	33.37	Methyl palmitate	1928	1928	$C_{17}H_{34}O_2$	0.11	-	-	-
15	34.42	α-Eicosene	1980	1986	$C_{20}H_{40}$	-	-	-	3.44
16	35.62	Octadecanal	2040	2034	$C_{18}H_{36}O$	-	-	0.58	-
17	36.73	n-Heneicosane	2098	2100	$C_{21}H_{44}$	0.05	-	-	-
18	36.85	Methyl linolenate	2104	2108	$C_{19}H_{32}O_2$	0.07	-	-	-
19	37.06	Phytol	2116	2116	$C_{20}H_{40}O$	-	1.97	-	-
20	37.32	Linolenic acid	2130	2134	$C_{18}H_{30}O_2$	-	-	1.23	3.26
21	37.88	Ethyl linolate	2160	2164	$C_{20}H_{36}O_2$	0.08	-	-	-
22	38.48	Stearic acid=n-Octadecanoic acid	2192	2180	$C_{18}H_{36}O_2$	-	-	0.64	-
23	38.59	n-Docosane	2198	2200	$C_{22}H_{46}$	0.07	-	0.82	
24	38.89	Phytol acetate	2215	2218	$C_{22}H_{42}O_2$	-	-	0.67	-
25	39.17	Methyl nonadecanoate	2231	2230	$C_{20}H_{40}O_2$	-	-	0.59	-
26	40.12	Heneicosan-3-one	2285	2283	$C_{21}H_{42}O$	-	-	0.54	-
27	40.20	1-Eicosanol	2289	2292	$C_{20}H_{42}O$	0.09	-	-	-
28	40.36	n- Tricosane	2298	2300	$C_{23}H_{48}$	0.17	0.04	-	-
29	40.65	Methyl cis-11-eicosenoate	2315	2302	$C_{21}H_{40}O_2$	-	-	1.83	-

Table 1. *Cont.*

No.	R_t (min)	Compound Name	$RI_{Exp.}$ [a]	RI_{Lit} [b]	Molecular Formula	Peak Area (%)			
						TIB	TIL	TIS	TIF
30	41.15	Methyl eicosanoate	2344	2339	$C_{21}H_{42}O_2$	-	-	0.77	-
31	41.41	4,8,12,16-Tetramethylheptadecan-4-olide	2360	2364	$C_{21}H_{40}O_2$	-	0.07	0.40	-
32	41.57	n-Eicosanoic acid	2369	2380	$C_{20}H_{40}O_2$	-	-	1.01	-
33	42.07	2,2′-Methylene-bis-(6-tert butyl-4-methylphenol)	2398	2398	$C_{23}H_{32}O_2$	0.10	0.06	-	-
34	42.24	n-Tetracosane	2409	2400	$C_{24}H_{50}$	-	-	0.53	-
35	42.48	Docosanal	2424	2426	$C_{22}H_{44}O$	-	-	0.64	1.48
36	42.61	Methyl heneicosanoate	2431	2430	$C_{22}H_{44}O_2$	0.07	-	-	-
37	42.85	1-Docosanol	2446	2456	$C_{22}H_{46}O$	-	-	0.54	-
38	43.1	2-Methyltetracosane	2461	2465	$C_{25}H_{52}$	-	-	1.28	-
39	43.48	1-Pentatricosene	2484	2485	$C_{35}H_{70}$	-	-	1.77	-
40	43.62	Cyclogallipharaol	2493	2499	$C_{21}H_{36}O$	0.35	-	0.81	0.72
41	43.70	n- Pentacosane	2498	2500	$C_{25}H_{52}$	0.24	0.23	2.46	-
42	43.87	Heneicosyl acetate	2509	2509	$C_{23}H_{46}O_2$	-	-	2.94	-
43	43.96	Palmitic acid β-monoglyceride	2514	2519	$C_{19}H_{38}O_4$	0.07	-	-	-
44	44.230	Methyl docosanoate	2531	2531	$C_{23}H_{46}O_2$	0.08	-	-	-
45	44.305	11-Methylpentacosane	2536	2529	$C_{26}H_{54}$	-	-	4.22	-
46	44.675	cis-13,16-Docasadienoic acid	2560	2566	$C_{22}H_{40}O_2$	-	-	3.70	3.76
47	44.84	n-Docosanoic acid	2570	2569	$C_{22}H_{44}O_2$	-	-	10.49	-
48	45.18	Ethyl docosanoate	2592	2593	$C_{24}H_{48}O_2$			1.44	
49	45.275	n-Hexacosane	2598	2600	$C_{26}H_{54}$	0.11	0.20	2.47	-
50	45.43	Methyl tricosanoate	2608	2615	$C_{24}H_{48}O_2$	-	-	7.09	-
51	45.62	Erucylamide	2621	2625	$C_{22}H_{43}NO$	-	-	1.32	-
52	45.83	Tetracosanal	2635	2632	$C_{24}H_{48}O$	-	-	5.42	-
53	46.19	2-Methylhexacosane	2658	2662	$C_{27}H_{56}$	-	-	4.31	-
54	46.793	1-Heptacosene	2698	2694	$C_{27}H_{54}$	0.58	-	-	-
55	46.805	n-Heptacosane	2699	2700	$C_{27}H_{56}$	-	1.37	4.17	6.67
56	47.310	Methyl tetracosanoate	2733	2731	$C_{25}H_{50}O_2$	0.14	-	2.17	-
57	47.77	2-Methylheptacosane	2765	2761	$C_{28}H_{58}$	-	-	0.59	-
58	48.02	16-Acetoxycarterochaetol	2782	2787	$C_{22}H_{36}O_3$	-	-	0.26	-
59	48.260	n-Octacosane	2798	2800	$C_{28}H_{58}$	0.09	0.44	4.63	-
60	48.800	Squalene	2836	2835	$C_{30}H_{50}$	0.24	1.29	1.82	4.17
61	48.865	n-Hexacosanal	2841	2833	$C_{26}H_{52}O$	0.04	0.09	-	-
62	48.94	n-Hexacosanol	2846	2848	$C_{26}H_{54}O$	-	-	0.25	-
63	49.26	2-Methyloctacosane	2869	2860	$C_{29}H_{60}$	-	-	1.32	-
64	49.44	1-Nonacosene	2882	2884	$C_{29}H_{58}$	-	-	0.67	-
65	49.730	n-Nonacosane	2902	2900	$C_{29}H_{60}$	1.26	4.13	1.69	5.42
66	50.175	15-Methylnonacosane	2935	2931	$C_{30}H_{62}$	0.03	-	-	-
67	50.285	Methyl hexacosanoate	2943	2940	$C_{27}H_{54}O_2$	-	0.17	-	-
68	50.515	2-Methylnonacosane	2960	2960	$C_{30}H_{62}$	0.04	-	-	-
69	51.040	n-Triacontane	2998	3000	$C_{30}H_{62}$	0.04	0.37	-	-
70	51.130	Benzyl icosanoate	3005	3003	$C_{27}H_{46}O_2$	0.02	-	-	-
71	51.210	1-Heptacosanol	3011	3016	$C_{27}H_{56}O$	-	0.06	-	-
72	51.655	n-Octacosanal	3045	3039	$C_{28}H_{56}O$	-	0.29	-	0.44
73	51.850	2-Methyltriacontane	3059	3060	$C_{31}H_{64}$	-	0.09	-	-
74	52.370	n-Hentriacontane	3099	3100	$C_{31}H_{64}$	0.08	1.73	0.61	3.32
75	52.475	Octacosanol	3107	3110	$C_{28}H_{58}O$	1.02	-	-	1.89
76	52.72	13-Methylhentriacontane	3126	3130	$C_{32}H_{66}$	-	0.15	0.85	-
77	52.835	Campesterol	3134	3131	$C_{28}H_{48}O$	-	0.05	-	-
78	53.015	Cholest-3,5-diene	3148	-	$C_{27}H_{44}$	0.07	0.09	-	-
79	53.085	*α*-Tocopherol	3154	3149	$C_{29}H_{50}O_2$	0.03	1.01	-	-
80	53.155	Ergosta-5,8,22-trien-3-ol, (3*β*,22*E*)-	3159	3158	$C_{28}H_{44}O$	-	0.40	-	-
81	53.340	Stigmasterol	3173	3170	$C_{29}H_{48}O$	0.11	0.25	-	-
82	53.475	*β*-Sitosterol	3183	3197	$C_{29}H_{50}O$	-	0.10	-	-
83	53.710	n-Dotriacontane	3201	3200	$C_{32}H_{66}$	0.26	0.23	-	-
84	53.925	Cholest-5-ene, 3*β*-methoxy	3216	3216	$C_{28}H_{48}O$	-	0.08	-	-
85	54.375	Triacontanal	3248	3251	$C_{30}H_{60}O$	-	0.47	-	0.69
86	54.575	5*α*-Stigmast-22-en-3β-ol	3261	3253	$C_{29}H_{50}O$	1.09	0.84	0.72	1.39
87	55.070	Chondrillasterol	3296	3295	$C_{29}H_{48}O$	1.42	0.07	-	-
88	55.130	n-Tritriacontane	3300	3300	$C_{33}H_{68}$	-	0.41	-	-
89	55.230	Lanosterol	3306	3302	$C_{30}H_{50}O$	0.25	0.13	-	-
90	55.310	1-Triacontanol	3311	3306	$C_{33}H_{68}O$	0.38	-	-	-
91	55.440	5*α*-Stigmastan-3*β*-ol	3320	3325	$C_{29}H_{52}O$	0.04	0.14	-	-
92	55.655	*β*-Amyrin	3332	3337	$C_{30}H_{50}O$	0.11	-	-	-
93	56.015	*γ*-Sitosterol	3353	3351	$C_{29}H_{50}O$	4.17	8.83	3.79	11.28
94	56.200	2-Methyldotriacontane	3364	3259	$C_{33}H_{68}$	-	0.52	-	0.76
95	56.335	*β*-Amyrone	3372	3327	$C_{30}H_{50}O$	1.78	2.05	-	-
96	56.445	*α*-Amyrin	3379	3376	$C_{30}H_{50}O$	-	0.33	-	-
97	56.740	n-Tetratriacontane	3397	3403	$C_{34}H_{70}$	8.16	1.21	-	3.38
98	57.020	*β*-Amyrin acetate	3446	3437	$C_{32}H_{52}O_2$	-	0.08	-	-
99	57.195	Lupenone	3488	3384	$C_{30}H_{48}O$	9.36	13.08	-	-
100	57.550	Lupeol acetate	3518	3525	$C_{32}H_{52}O_2$	23.02	1.68	-	-
101	57.68	Lupeol	3527	3500	$C_{30}H_{50}O$	23.61	22.78	-	3.76
102	58.015	(17*E*)-Cholesta-17,24-diene-3,6-diol	3550	-	$C_{27}H_{44}O_2$	0.29	8.23	-	-
103	58.170	Hexadecanoic acid, 3,7,11,15-tetramethyl-2-hexadecenyl ester	3561	3567	$C_{36}H_{70}O_2$	0.04	-	-	-
104	58.295	24-Methylenecycloartan-3-one	3570	-	$C_{31}H_{50}O$	0.45	-	-	-

Table 1. *Cont.*

No.	R_t (min)	Compound Name	$RI_{Exp.}$ [a]	RI_{Lit} [b]	Molecular Formula	Peak Area (%)			
						TIB	TIL	TIS	TIF
105	58.605	24-Methylenecycloartanol	3591	-	$C_{31}H_{52}O$	5.87	0.18	-	-
106	58.890	3*β*-Hydroxystigmast-5-en-7-one	3611	3609	$C_{29}H_{48}O_2$	-	1.03	-	-
107	59.095	Betulinaldehyde	3625	3628	$C_{30}H_{48}O_2$	-	3.70	-	-
108	59.440	Germanicol	3649	-	$C_{30}H_{50}O$	-	0.80	-	-
109	61.000	Betulin	3757	3760	$C_{30}H_{50}O_2$	1.78	-	-	-
110	62.04	Ursane-3,12-diol	3829	-	$C_{30}H_{52}O_2$	-	0.59	-	-
111	62.11	1-Heptatriacontanol	3834	-	$C_{37}H_{76}O$	4.25	-	-	-
112	62.39	Stigmastane-3,6-dione	3853	3601	$C_{29}H_{48}O_2$	1.16	-	-	-
113	62.79	9,19-Cyclolanost-23-ene-3,25-diol, (3*β*,23*E*)-	3881	-	$C_{32}H_{52}O_3$	-	0.70	-	-
		Monoterpene Hydrocarbon				-	-	-	2.30
		Oxygenated Monoterpene				0.03	-	-	8.84
		Sesquiterpene Hydrocarbon				0.04	-	-	7.52
		Oxygenated Sesquiterpene				-	0.09	-	-
		Diterpenoids				-	2.24	1.33	-
		Triterpenoids				61.06	47.41	1.82	7.93
		Steroids				13.47	11.86	4.51	12.67
		Fatty acids and fatty acids derivatives				0.97	8.40	35.22	11.55
		Straight-chain Hydrocarbons and derivatives				12.77	12.1	40.36	33.55
		Others				4.73	1.07	0.81	0.72
		Total identified compounds %				93.07	83.17	84.05	85.08

Compounds listed in order of their elution in RTX-5 GC column. Identification was based on comparison of the compound mass spectral data (MS) and retention indices (RI) with those of NIST Mass Spectral Library (2011), Wiley Registry of Mass Spectral Data 8th edition and literature. [a] Retention index calculated experimentally in RTX-5 column relative to n-alkane series (C8–C28). [b] Published retention indices.

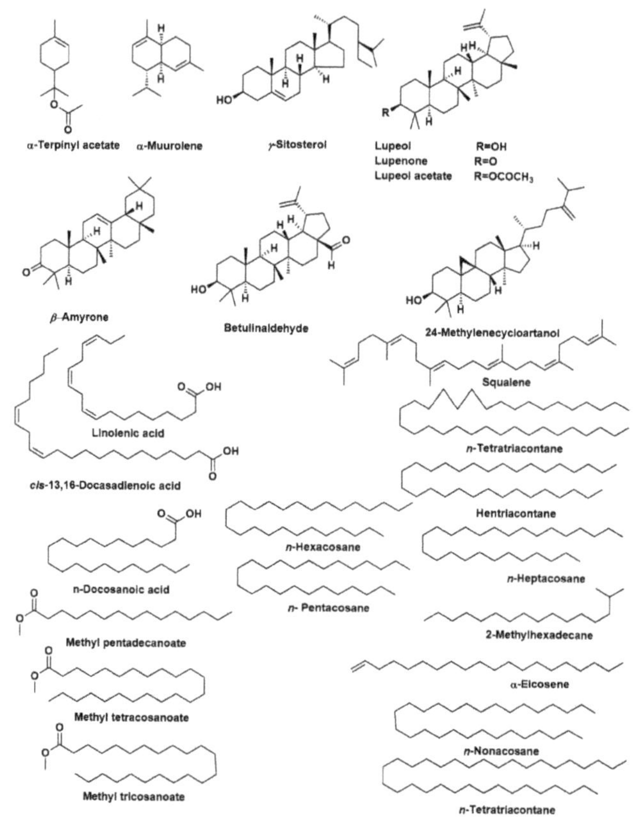

Figure 1. The major compounds identified in bark, leaves, seeds, and fruits of *Tamarindus indica*.

Triterpenoids and steroids are the predominant classes in the *T. indica* bark *n*-hexane extract, accounting for 61.06 % and 13.47%, respectively. Additionally, the *n*-hexane extract of *T. indica* leaves showed a high percentage of triterpenoids and steroids, representing 47.41% and 11.86% of the identified compounds, respectively, along with fatty acids and their derivatives accounting for 8.40 %. On the other hand, the most predominant classes of metabolites in the *n*-hexane extract of *T. indica* seeds were fatty acids and their derivatives and the straight-chain hydrocarbons, accounting for 35.22% and 40.36%, respectively. Meanwhile, the straight-chain hydrocarbons were the dominant class in the *n*-hexane extract of *T. indica* fruits, representing 33.55% of the total identified compounds, followed by steroids and fatty acids and their derivatives, accounting for 12.67% and 11.55%, respectively. The illustration of the metabolite distribution in the various organs of *T. indica* is represented in (Figure 2).

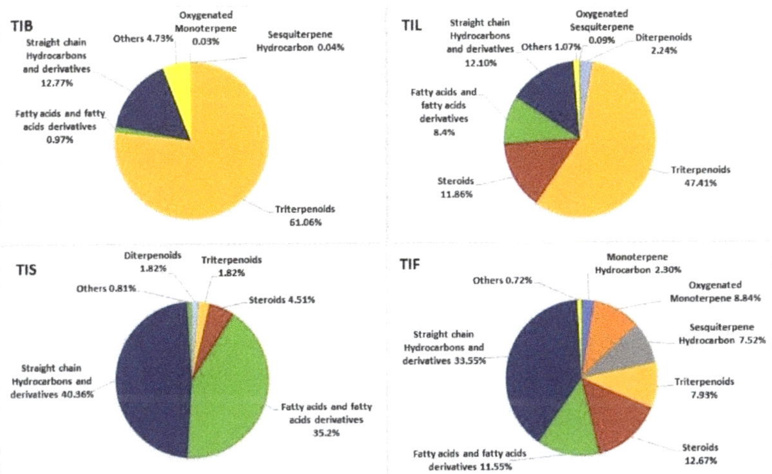

Figure 2. Pie charts demonstrate distribution of metabolite classes in percentages within various organs: TIB (bark), TIL (leaves), TIS (seeds), and TIF (fruits) of *Tamarindus indica*.

The fatty acids previously identified in the seeds of Tamarind from Sudan by Ibrahim et al. [40] were characterised as palmitic acid, oleic acid, and eicosanoic acid in addition to the identification of β-amyrin, β-sitosterol, and campesterol. *Cis*-vaccenic acid, 2-methyltetracosane, β-sitosterol, 9,12-octadecadienoic acid (*Z*, *Z*)-, and n-hexadecanoic acid were reported as the major metabolites identified in the oil of *T. indica* seeds collected from Nigeria. Another study by Carasek and Pawliszyn [41] reported that the volatile compounds in the tamarind fruit collected from Brazil included phenylacetaldehyde and furfural. Aldehyde and ester compounds were mostly identified by using Solid-Phase Microextraction (SPME) Fibres. It is important to note that the plant extract components can be influenced by variables including geographic, climatic, seasonal, and experimental settings. However, the aim of our study was to explore the difference in discriminating among the various organs of *T. indica*, not to study the effects of the above-mentioned variables from Gad et al. [42].

2.2. Chemometric Analysis Based on GC/MS Analysis

GC/MS-based metabolites showed both qualitative and quantitative variation among different *Tamarindus indica* organs. Consequently, chemometric analysis, representing both principal component analysis (PCA) and hierarchal cluster analysis (HCA), was applied to the relative peak areas of all the identified compounds of different *T. indica* organs to explore the similarities and differences among them. The PCA score plot and loading plot are shown in Figure 3a,b. In the PCA score plot, the percentage of the first two principal components explained 83% of the variance in the data, where PC1 accounted for 63%, while PC2 explained only 20% of the data discrepancies. From the score plot, various *T. indica* organs were divided into four main groups, and each part was scattered in a separate quadrant. TIL and TIB were positioned on the positive side of PC1, whereas, TIF and TIS were located on the negative side of PC1. An in-depth inspection of the loading plot (Figure 3b) revealed that lupeol acetate was the major discriminating marker for TIB. Although lupeol was the major identified component in both TIB and TIL, it was ©dentified as the main segregating marker for TIL in addition to lupenone.

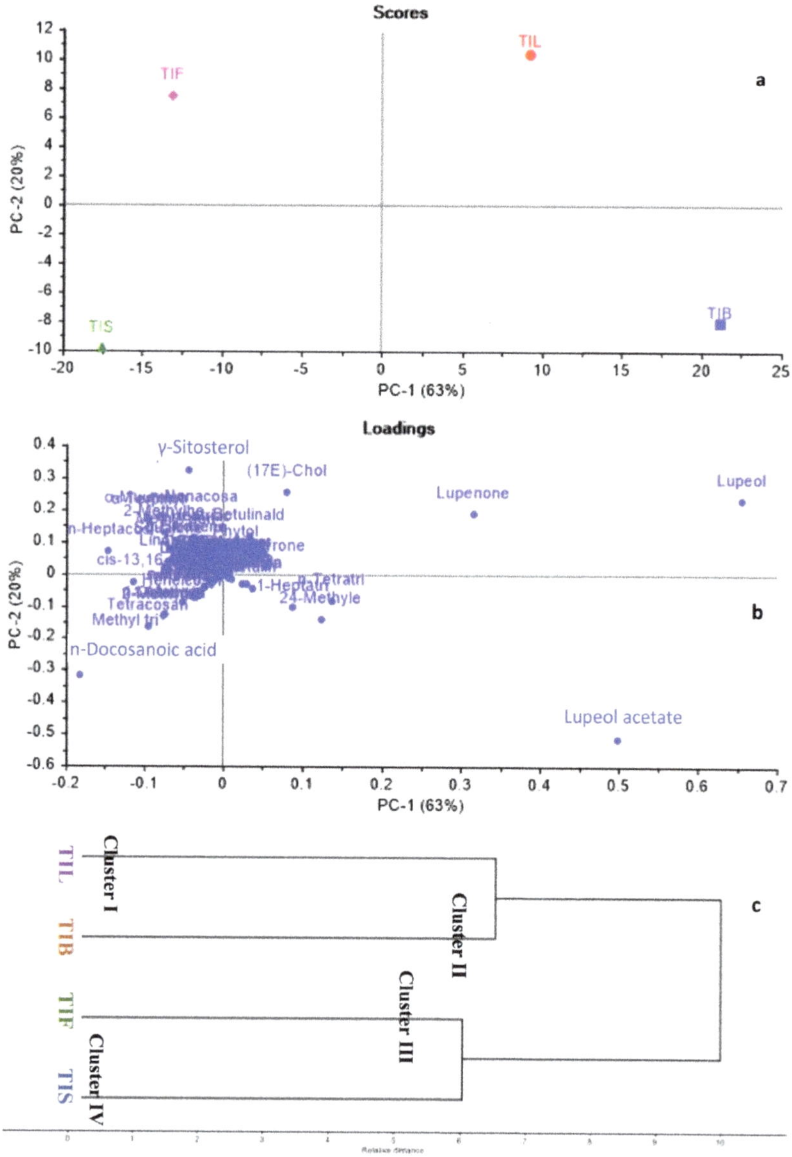

Figure 3. Principal component analysis score plot (**a**), loading plot (**b**), HCA dendrogram (**c**) based on GC-MS metabolites of various *T. indica* organs as identified in Table 1.

Regarding TIF and TIS, γ-Sitosterol and n-Docosanoic acid were the chef compounds responsible for their separation in separate quadrants. By applying HCA, the dendrogram (Figure 3c) categorised *T. indica* organs into four main clusters. Cluster I, II, III, and IV displayed TIL, TIB, TIF, and TIS, respectively. The dendrogram showed that TIL and TIB were closely related to each other's when compared to TIF and TIS. Both PCA and HCA successfully discriminated among various *T. indica* organs proving the effective application of chemometric analysis in combination with GC/MS-based metabolites for the discrimination among various plant organs.

2.3. *Anti-Inflammatory Activity of Tamarindus indica* L. *Various Organs*

The anti-inflammatory effects of the *n*-hexane extract of *T. indica* bark, leaves, seeds, and fruits (TIB, TIL, TIS, and TIF) on lipopolysaccharide (LPS)-induced RAW 264.7 macrophages were represented in (Table 2). The various organs of *T. indica* showed promising inhibition of LPS-induced nitric oxide (NO) release in RAW 264.7 macrophages. At a concentration of 10 µg/mL, TIB, TIL, TIS, and TIF *n*-hexane extracts showed %NO inhibition by 7.53 ± 1.69, 53.97 ± 5.89, 19.54 ± 1.19, and 26.66 ± 3.44%, respectively. Further, at a concentration of 100 µg/mL, TIB, TIL, TIS, and TIF *n*-hexane extracts showed %NO inhibition by 51.44 ± 1.17, 98.00 ± 1.90, 85.47 ± 0.22, and 83.69 ± 2.39%, respectively as compared to positive control L-N^G-nitro arginine methyl ester (L-NAME) (1 mM) that showed 84.64 ± 1.04% NO inhibition.

Table 2. The anti-inflammatory effects of the *n*-hexane extract from various organs of *T. indica*: TIB (bark), TIL (leaves), TIS (seeds), and TIF (fruits) on lipopolysaccharide (LPS)-induced RAW 264.7 macrophages.

Sample	%NO Inhibition	
	10 µg/mL	100 µg/mL
TIB	7.53 ±1.69 [b]	51.44 ± 1.17 [b]
TIL	53.97 ± 5.89 [b]	98.00 ± 1.90 [b]
TIS	19.54 ± 1.19 [b]	85.47 ± 0.22 [a]
TIF	26.66 ± 3.44 [b]	83.69 ± 2.39 [a]
L-NAME (1 mM)	84.64 ± 1.04 [a]	

Means bearing same scripts ([a] or [b]) are not significantly different from control at $p < 0.05$, Mean ± Standard error.

Previous studies revealed that the fruit pulp was applied to inflammations and traditionally used as a gargle for sore throats. Further, the bark decoction was used in cases of eye inflammation [12]. Rimbau et al. revealed the in vivo anti-inflammatory properties of different extracts of tamarind fruit pulp in mice ear oedema induced by arachidonic acid, and rats sub plantar oedema induced by carrageenan [43]. Moreover, the petroleum ether and ethyl acetate extracts of seeds of *T. indica* showed a significant anti-inflammatory and analgesic potential using carrageenan-induced paw oedema and cotton pellet-induced granuloma models in rats [44]. Among the major compounds in the *n*-hexane extract of *T. indica's* various organs were lupenone and sitosterol, which were reported for their anti-inflammatory activity [45]. Additionally, among the major identified compounds, lupeol and lupeol acetate were reported for their anti-inflammatory properties through regulating TNF-α, IL-2, and IL-β specific mRNA, reducing PGE-2 synthesis from macrophage, neutrophil migration, and the number of iNOS cells [46].

2.4. *Wound Healing Activity of Tamarindus indica* L. *Various Organs*

2.4.1. Cytotoxicity Assay

Evaluation of the cytotoxicity of the *n*-hexane extract of *T. indica* bark, leaves, seeds, and fruits (TIB, TIL, TIS, and TIL) on the HSF cells was necessary to assess the safety of the treatment dose. The cytotoxicity on HSF cells of the *n*-hexane extract of *T. indica's* various organs was evaluated using the SRB assay [47]. In our investigation, TIF and TIS (up to 100 µg/mL) had no cytotoxic effect on the HSF cells. While TIB and TIL at a concentration of 10 μg/mL showed HSF cell viability up to 91.04 ± 2.09 and 95.38 ± 0.86%, respectively (Figure 4). Therefore, a concentration of 10 μg/mL was safe and used further as the treatment dose in the scratch wound assay.

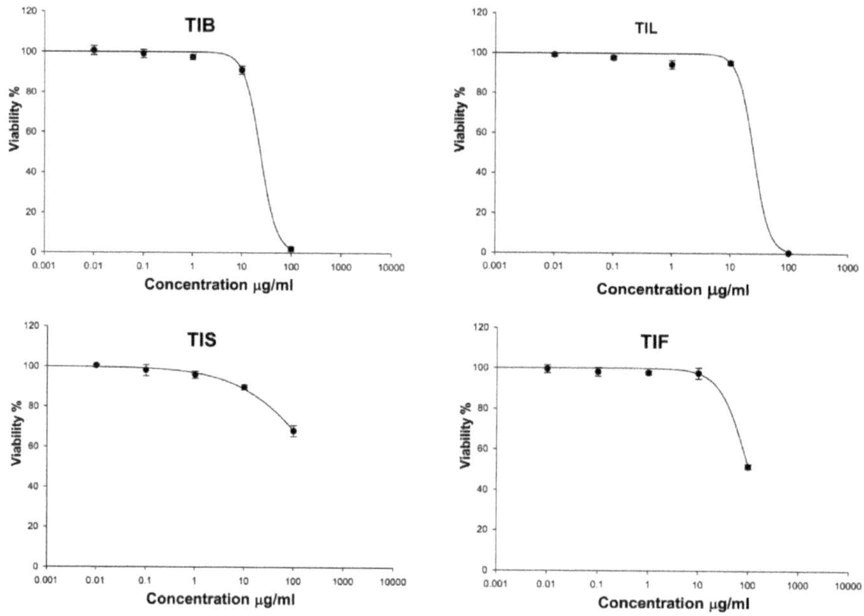

Figure 4. The effect of *n*-hexane extract from various organs of *T. indica*: TIB (bark), TIL (leaves), TIS (seeds), and TIF (fruits) on Human Skin Fibroblast cells (HSF) viability.

2.4.2. Scratch Wound Assay

In the present study, an in vitro scratch assay using HSF cells was used to assess TIB, TIL, TIS and TIF wound-healing activity. It was evaluated by changes in wound width by measuring the average distance between the borders of the scratches (Figures 5 and 6). The tested plant extracts at a concentration of 10 μg/mL decreased the wound width significantly compared with the control cells (Table 3).

Table 3. Wound width of the scratched Human Skin Fibroblast cells (HSF) incubated in the absence of the plant extract (negative control) and the presence of *n*-hexane extract from *T. indica*: TIB (bark), TIL (leaves), TIS (seeds), and TIF (fruits) (10 μg/mL).

Time (h)	Wound Width (mm)				
	TIB (10 μg/mL)	TIL (10 μg/mL)	TIS (10 μg/mL)	TIF (10 μg/mL)	Control
0	2.68 ± 0.02 [a]	2.75 ± 0.02 [a]	2.72 ± 0 [a]	2.74 ± 0.04 [a]	2.73 ± 0.03 [a]
24	1.09 ± 0.04 [a]	1.12 ± 0.18 [a]	1.09 ± 0.28 [a]	1.41 ± 0.35 [a]	1.37 ± 0.15 [a]
48	0 [a]	0.09 ± 0.16 [a]	0.13 ± 0.12 [a]	0.27 ± 0.12 [a]	0 [a]
72	0	0	0	0	0

Means bearing same script ([a]) are not significantly different from control at $p < 0.05$, Mean ± Standard error.

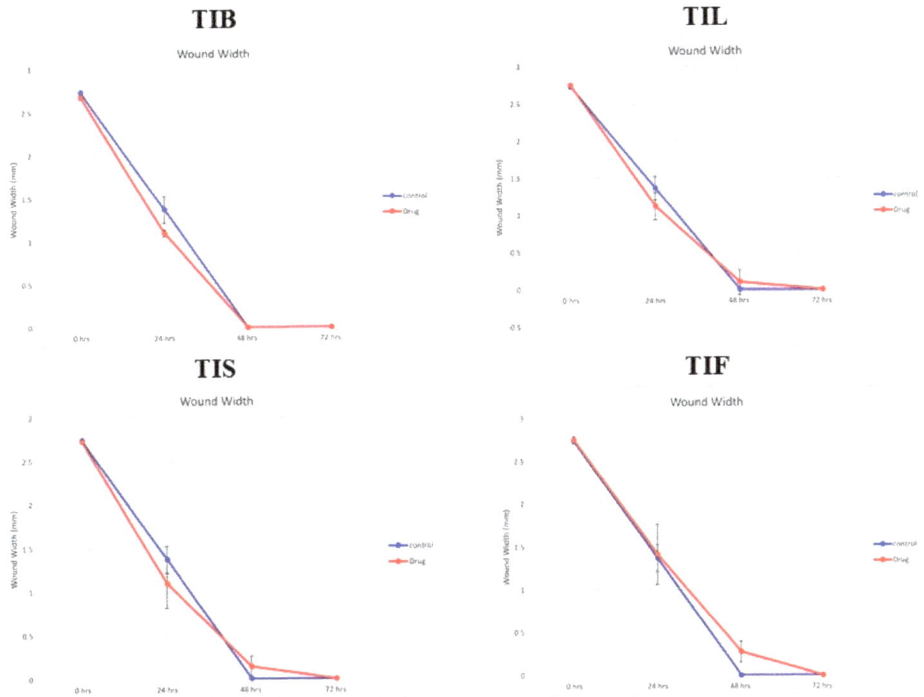

Figure 5. The wound width changes in the absence or presence of 10 µg/mL of *n*-hexane extract from various organs of *T. indica*: TIB (bark), TIL (leaves), TIS (seeds), and TIF (fruits).

After 24 h, the highest wound healing potential was recorded by TIB and TIS *n*-hexane extracts with wound widths equal to 1.09 ± 0.04 and 1.09 ± 0.28 mm, respectively. This result was followed by TIL and TIF with wound widths equal to 1.12 ± 0.18 and 1.41 ± 0.35 mm, respectively, as compared to the wound width in the control cells that equalled 1.37 ± 0.15 mm. As the wound width reduced as cell migration was enhanced, our findings revealed that both TIB and TIL *n*-hexane extracts exhibited almost complete cell migration after 48 h of observation. Meanwhile, TIS and TIF *n*-hexane extracts had wound-healing effects on HSF cells, but the time required to conduct wound closure was longer than for the negative control. This revealed that the *n*-hexane extract of *T. indica's* various organs exhibited wound-healing potential through fibroblast migration enhancement.

Previous reports revealed the traditional utilisation of tamarind in eye surgery for conjunctival cell adhesion and corneal wound healing [12]. In a study by Adeniyi et al., the wound-healing activity of *T. indica* leaves and pulp was investigated in the African catfish, and the results revealed the enhancement of the wound healing significantly in the fish-fed tamarind-fortified diets as a consequence of the elevation of antioxidant enzymes [48]. Attah et al. investigated the wound-healing potential of *T. indica* fruit paste in adult rabbits, and the results revealed wound closure acceleration and increasing epithelial migration and re-epithelialisation [49].

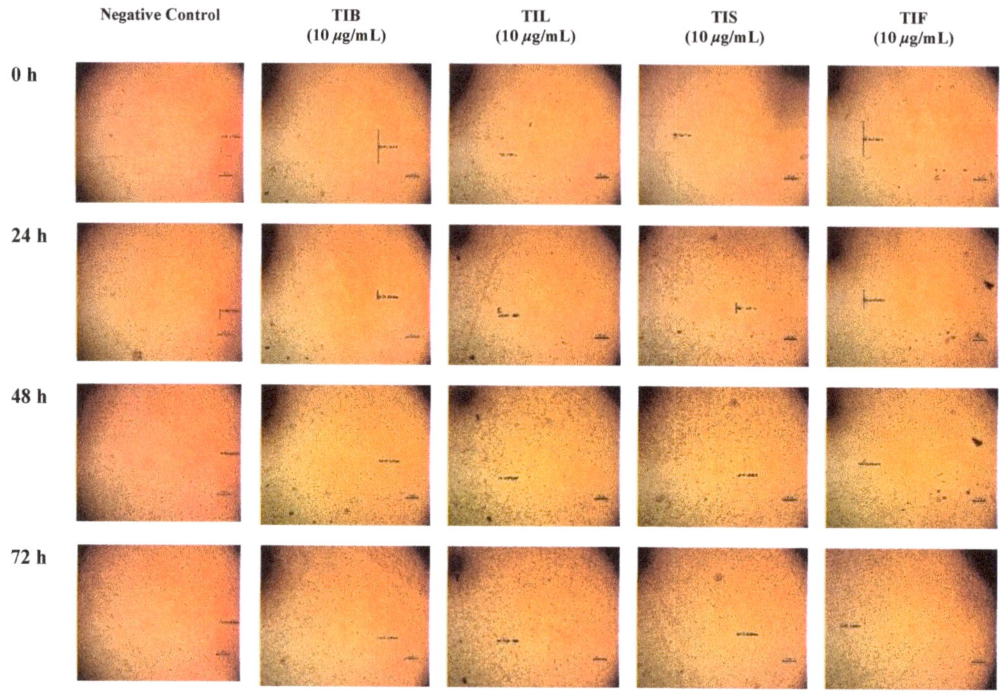

Figure 6. Microscopic images of (HSF) incubated in absence of the plant extract (negative control) and depicting the influence of 10 μg/mL of *n*-hexane extract from various organs of *T. indica*: TIB (bark), TIL (leaves), TIS (seeds), and TIF (fruits). Images were captured at 0, 24, 48 and 72 h. The boundaries of the scratched wounds are marked by dark lines.

The wound-healing potential of *n*-hexane extracts of various organs of *T. indica* can be linked to their chemical components. For example, γ-sitosterol is one of the most famous phytosterols with reported biological activity. Phytosterols were shown to inhibit MMP-1, reduce collagen breakdown, and promote the synthesis of collagen in human keratinocytes [50]. Moreover, they promote keratinocyte migration through the reduction of oxidative stress that effectively accelerates the healing process [51]. In previous studies, lupeol and lupeol acetate, which are the major identified components in the present study, were reported for their wound-healing properties [52–58]. Lupeol prevents collagen I depletion and restores levels of hydroxyproline, hexosamine, hexuronic acid, and matrix glycosaminoglycans, together with modulating collagen I expression in human fibroblasts during the proliferation phase. Further, it enhances fibroblast proliferation, angiogenesis, and growth factors in wound healing [52–54]. Another study reported the potential of lupeol for wound healing in streptozotocin-induced hyperglycaemic rats [56]. Another study by Malinowska et al. revealed that lupeol acetate was one of the most effective lupeol derivatives in the stimulation of the human skin cell proliferation process [58]. Bopage et al. reported that the presence of a 3-OH group in the lupeol structure is one of the essential features in the lupane skeleton for wound healing activity as compared to lupenone and lupeol acetate [59].

Linolenic acid and squalene are among the major identified constituents in the *n*-hexane extract of *T. indica* fruits; linolenic acid plays an important role in the wound healing process across its antioxidative and anti-inflammatory properties as well as by encouraging cell proliferation, raising collagen synthesis, promoting dermal reconstruction, and restoring the function of the skin's lipid barrier [60–63]. Further, squalene is reported to have anti-inflammatory and protective actions against skin damage. It can accelerate the healing of wounds by stimulating the macrophage response to inflammation. Squalene could be helpful during the resolution phase of wound healing [64].

The amounts of lupeol, lupeol acetate, lupenone, and γ-sitosterol show a correlation with the in vitro wound healing potential of the n-hexane extract of various organs of *T. indica*, and TIB was found to contain the highest amounts of these compounds according to the GC/MS analysis. The present study and the data in the literature show that terpenoids, sterols, and fatty acids play important roles in the wound-healing potential of *T. indica*. So, *T. indica* is suggested as a potent natural wound-healing product.

2.5. In Silico Molecular Docking Studies

This part was conducted to investigate the possible mechanism of action in which the identified major compounds exert their biological effect. Accordingly, the 3D structures of glycogen synthase kinase 3-β (GSK3-β), matrix metalloproteinases-8 (MMP-8), and nitric oxide synthase (iNOS) were downloaded from the protein data bank using the following IDs: 3F88, 5H8X, and 3N2R, respectively. After that, the twenty major compounds were docked into the active site vicinity of the three enzymes. Interestingly, all the compounds achieved acceptable binding scores upon docking with the three targets (Table 4).

Table 4. The docking scores achieved by the major identified compounds against the three enzymes.

Compound Name	Docking Scores (Kcal/mol)		
	Glycogen Synthase Kinase 3-β (GSK3-β) 3F88	**Nitric Oxide Reductase (iNOS) 3N2R**	**Matrix Metalloproteinases 8 (MMP-8) 5H8X**
Lupeol	−12.5	−13.7	−9.8
n-docosanoic acid	−11.7	−12.6	−11.9
methyl tricosanoate	−11.2	−11.8	−13.1
α-terpinyl acetate	−11.8	−10.1	−12.5
α-muurolene	−10.4	−11.8	−10.1
Gamma- sitosterol	−10.2	−11.9	−8.2
Lupenone	−10.1	−10.3	−7.6
Lupeol acetate	−11.3	−9.5	−7.7
n-tetratriacontane	−9.8	−7.6	−7.5
Betulinaldehyde	−8.7	−8.1	−7.4
β-Amyrone	−9.2	−6.5	−7.1
24-methylenecycloartanol	−8.3	−7.2	−6.8
methyl tricosanoate	−7.8	−7.4	−7.5
n-hexacosane	−8.2	−7.8	−7.3
cis-13,16-docasadienoic acid	−7.5	−8.3	−6.5
n-tetratriacontane	−7.9	−9.1	−5.9
methyl pentadecanoate	−7.3	−7.8	−7.6
Squalene	−6.8	−7.8	−7.2
Linolenic acid	−8.1	−8.5	−7.9
n- pentacosane	−7.3	−7.7	−7.4

To this end, it was expected that the identified major compounds would exert synergetic effects. In the docking of GSK3-β-lupeol, n-docosanoic acid, methyl tricosanoate, α-terpinyl acetate, and lupeol acetate achieved best docking scores of −12.5, −11.7, −11.2, −11.8, and −11.3 Kcal/Mol, respectively. As Figure 7 reveals, lupeol interacted with Val135, Tyr140, Arg141, Gln185, and Cys199; n-docosanoic acid interacted with Gly63, Lys183, and Cys199; methyl tricosanoate interacted with GSK3-β through binding with Phe67, Arg141, and Cys199; α-terpinyl acetate interacted with Lys85, Asp133, and Cys199; and lupeol acetate interacted with Lys85, Gln185, and Cys199. In the docking of matrix metalloproteinases-8 (MMP-8), lupeol, n-docosanoic acid, methyl tricosanoate, α-terpinyl acetate, and α-muurolene achieved the best docking scores of −9.8, −11.9, −13.1, −12.5, and −10.1 Kcal/Mol, respectively. As depicted in Figure 8, lupeol bound to MMP-8 through interactions with Ala161, Gly158, and Asn218; n-docosanoic acid interacted with Ala161, Glu198, His207, Pro217, and Asn218; methyl tricosanoate interacted with Asn85, Ala161, Ala163, Gln165, His 197, Glu198, His201, and Pro217; α-terpinyl acetate interacted with Ala161, Val194, His 197, Glu198, His207, Pro217, and Asn218; and α-muurolene interacted with MMP-8 through binding with His197, Leu214, and Tyr216. In the docking of Nitric oxide reductase (iNOS), lupeol, n-docosanoic acid, methyl tricosanoate, α-muurolene, and gamma-sitosterol achieved the best docking scores of −13.7, −12.6, −11.8, −11.8, and −11.9 Kcal/Mol, respectively. Looking at Figure 9, we see that lupeol was able to interact with the residues of iNOS through binding with Gln478, Tyr588, Glu592, Asp597, Trp678, and Val680; n-docosanoic acid interacted with Thr324, Arg414, Cys415, Gln478, Glu592, Trp678, and Asn697; methyl tricosanoate interacted with Trp409, Arg414, Cys415, Gly417, Met570, Phe584, and Glu592; α-muurolene interacted with Met336, Trp678, and Val680; and gamma-sitosterol interacted with Cys415, Val677, and Trp678. In conclusion, the docking results supported and justified the biological results giving rise to a synergetic effect for all the major components of the extract.

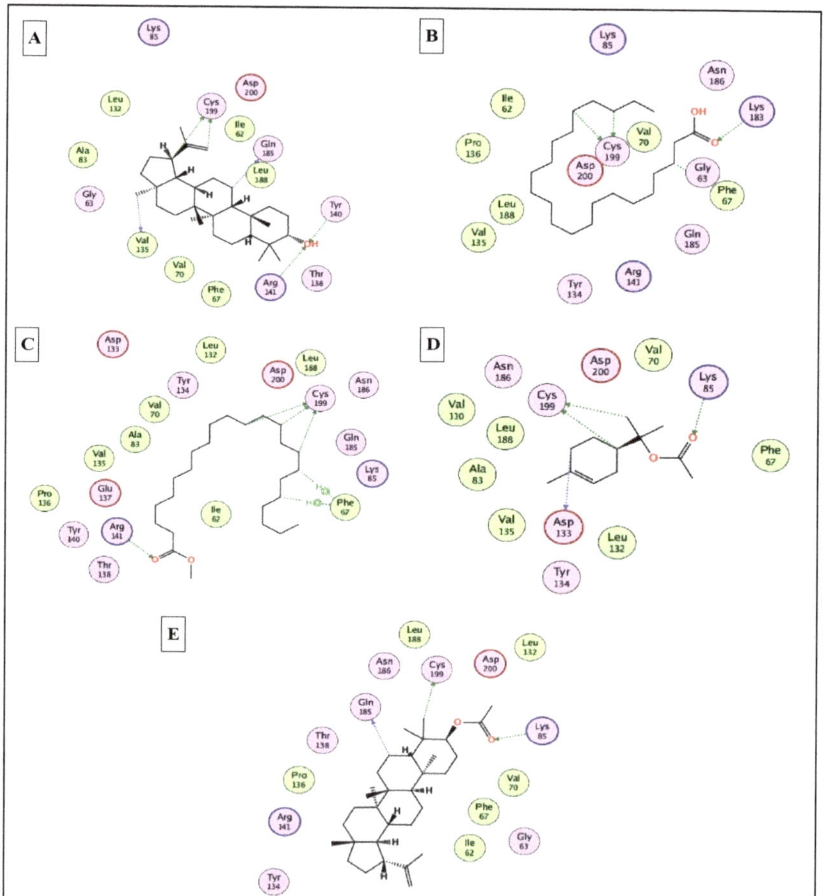

Figure 7. 2D binding modes of lupeol (**A**), n- docosanoic acid (**B**), methyl tricosanoate (**C**), α-terpinyl acetate (**D**), and lupeol acetate (**E**) to the active binding sites of glycogen synthase kinase 3-β (GSK3-β).

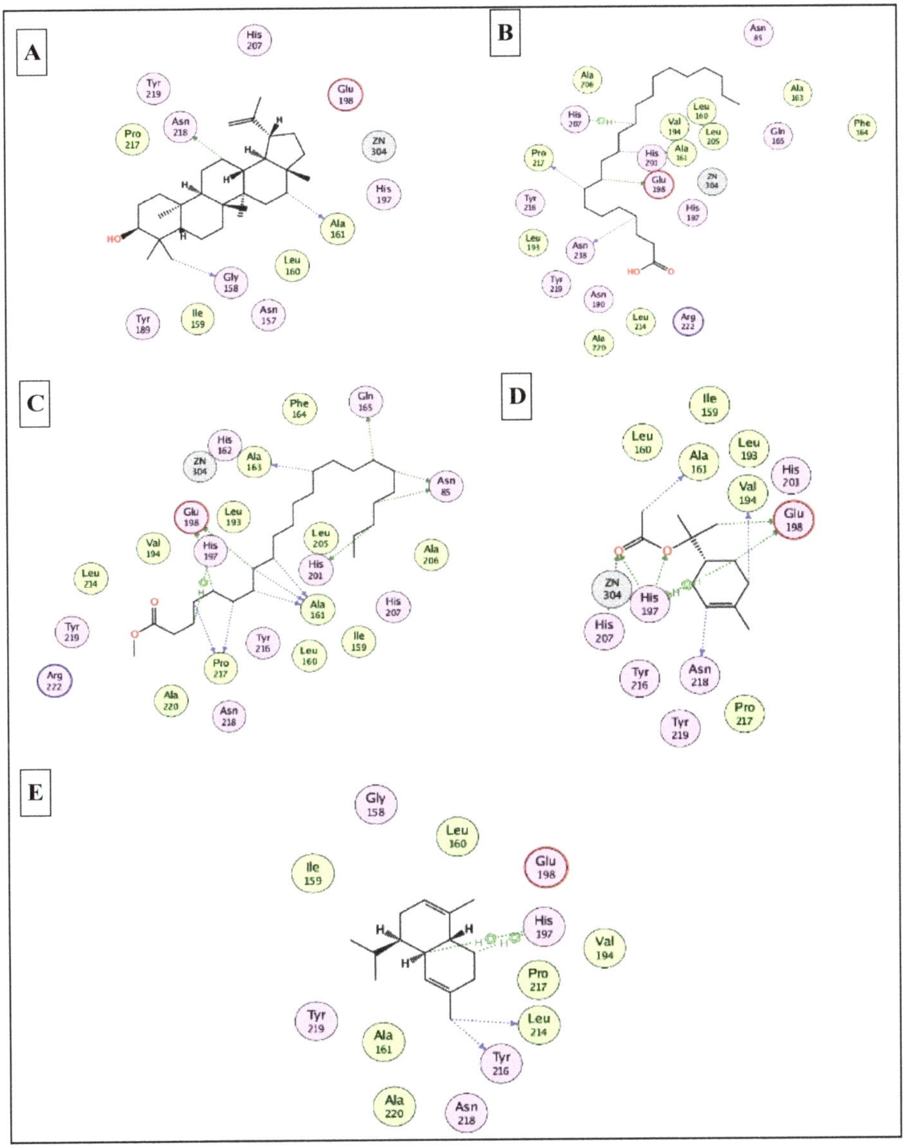

Figure 8. 2D binding modes of lupeol (**A**), n- docosanoic acid (**B**), methyl tricosanoate (**C**), α-terpinyl acetate (**D**), and α-muurolene (**E**) to the active binding sites of matrix metalloproteinases-8 (MMP-8).

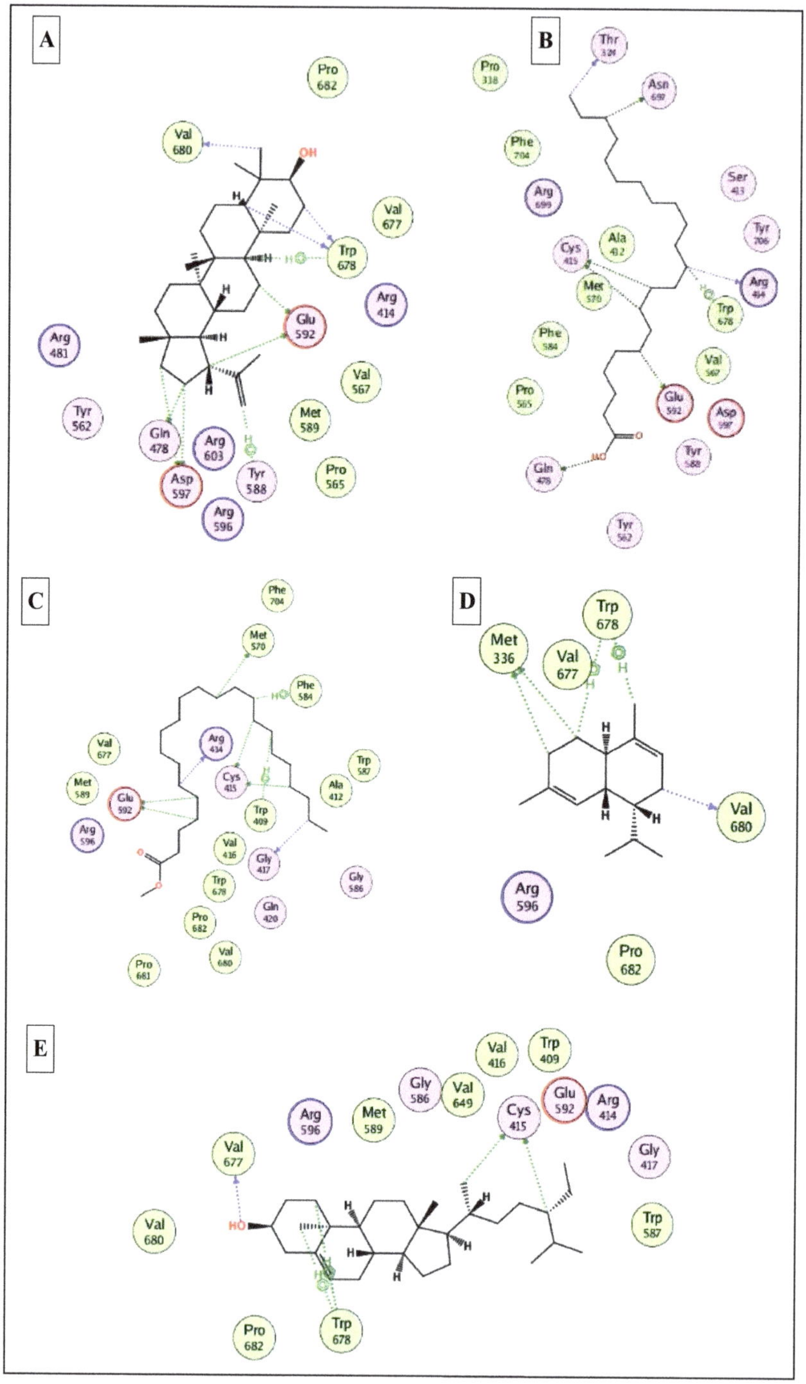

Figure 9. 2D binding modes of (**A**), n- docosanoic acid (**B**), methyl tricosanoate (**C**), α-muurolene (**D**), and gamma-sitosterol (**E**) to the active binding sites of nitric oxide synthase (iNOS).

3. Materials and Methods

3.1. Plant Material

Tamarindus indica L.'s (Fabaceae) various organs (Bark, leaves, seeds, and fruits) were collected from the Zoo Botanical Garden, Giza, Egypt, in October 2021. The various plant organs were kindly identified and authenticated by agricultural engineer Eng. Terase Labib, Consultant of Plant Taxonomy at the Ministry of Agriculture and El-Orman Botanical Garden, Giza, Egypt. Voucher specimens of the various plant organs with codes BUC-PHG-TIL-6, BUC- PHG-TIS-7, BUC- PHG-TIF-8, and BUC-PHG-TIB-9 were kept at the Department of Pharmacognosy, Badr University in Cairo, Cairo, Egypt.

3.2. Preparation of the N-Hexane Extracts of Various Organs

The dried samples of the barks, leaves, seeds, and fruits of *T. indica* (100 g each) were powdered and extracted by cold maceration method with *n*-hexane (500 mL × 3) separately, followed by filtration for 3 days. The filtrate of each plant material was completely evaporated *in vacuo* at 45 °C until dry to obtain the dried residue of the *n*-hexane extract (420 mg, 610 mg, 207 mg, and 312 mg) for barks, leaves, seeds, and fruits, respectively. The dried residue of each plant part was named as follows; *T. indica* bark (TIB), *T. indica* leaves (TIL), *T. indica* seeds (TIS), and *T. indica* fruits (TIF). All extracts were stored in a tight container at 4 °C in the refrigerator for further analysis [65].

3.3. GC/MS Analysis

Gas chromatography coupled with mass spectrometry (GC/MS) analyses were performed on a Shimadzu GCMS-QP 2010 (Shimadzu Corporation, Koyoto, Japan), provided using an Rtx-5MS (30 m × 0.25 mm i.d. × 0.25 µm film thickness) capillary column (Restek, Bellefonte, PA, USA) and attached to a Shimadzu mass spectrometer. The column temperature was initially set at 50 °C for 3 min. Then, the temperature was gradually increased from 50 to 300 °C at a rate of 5 °C/min and then isothermally maintained at 300 °C for 10 min. The temperature of the injector was kept at 280 °C. Helium was used as a carrier gas at a flow rate of 1.37 mL/min. The ion source and the interface were at temperatures of 280 and 220 °C, respectively. An injection of 1 µL of 1% *v*/*v* of diluted sample was achieved via a split mode adopting a split ratio of 15:1. Recording of the mass spectrum was performed in EI mode of 70 eV from m/z 35 to 500. Compound quantitation was performed based on the normalisation method, employing the reading of three chromatographic runs.

3.4. Identification of the N-Hexane Extract Components

The components of the *n*-hexane extracts were characterised by comparing their GC/MS spectra, fragmentation patterns, and retention indices with those reported in the literature data [66–72]. The retention indices were calculated relative to a homologous series of *n*-alkanes (C8–C28) injected under the same conditions.

3.5. Chemometric Analysis

The data obtained from GC-MS were subjected to multivariate analysis. Principal component analysis (PCA) was performed as the first step in data analysis to provide an overview of all observations and samples and to identify and evaluate groupings, trends, and strong outliers [73]. Hierarchal cluster analysis (HCA) was used to allow the clustering of samples. The clustering patterns were constructed by applying the complete linkage method. This presentation was more efficient when the distance between samples (points) was computed using the Euclidean method. Both PCA and HCA were achieved by utilising Unscrambler®X 10.4 from CAMO (Computer Aided Modeling, Viken, Norway) [74].

3.6. Anti-Inflammatory Activity

Murine macrophage RAW264.7 cells (ATCC®) were maintained in a complete Dulbecco's Modified Eagle's Medium (DMEM, Corning, NY, USA) supplemented with 10% foetal bovine serum, penicillin (100 U/mL), streptomycin sulphate (100 µg/mL), and 2 mM

L-glutamine in a humidified 5% CO_2 incubator. For passaging and treatment, cells were washed with phosphate-buffered saline and scrapped off the flasks using sterile scrappers (SPL, Spain). RAW 264.7 cell stock (0.5 × 106 cells/mL) was seeded into 96-well microwell plates and incubated overnight. The next day, the non-induced triplicate wells received a medium with the sample vehicle (DMSO, 0.1% *v*/*v*). The inflammation group of triplicate wells received the inducer of inflammation [lipopolysaccharide (LPS) as 100 ng/mL] in complete culture media containing 0.1% DMSO, *v*/*v*. Sample groups of triplicate wells received two screening amounts (10 and 50 μg/mL) of the sample dissolved in DMSO and diluted into culture media containing LPS (Final concentration of DMSO = 0.1%, by volume). L-NAME (L-N^G-nitro arginine methyl ester) (1 mM) was used as a standard NOS inhibitor. After 24 h of incubation, a Griess assay [75] was used to determine NO in all wells. Equal volumes of culture supernatants and Griess reagent were mixed and incubated at room temperature for 10 min to form the coloured diazonium salt and read at an absorbance of 520 nm. The NO Inhibition % of the test extract was calculated relative to the LPS-induced inflammation group, normalised to cell viability determined with the Alamar Blue™ reduction assay [76].

3.7. *Wound Healing Activity*

3.7.1. Cytotoxicity Assay

The cytotoxicity of the tested plant extract against the HSF (Human Skin Fibroblast) cell line was assessed prior to the wound-healing assay using the sulforhodamine B assay (SRB) [47]. The HSF cell line was obtained from Nawah Scientific Inc. (Mokatam, Cairo, Egypt). Cells were maintained in a DMEM medium supplemented with 100 mg/mL of streptomycin, 100 units/mL of penicillin, and 10% of heat-inactivated foetal bovine serum in a humidified 5% (*v*/*v*) CO_2 atmosphere at 37 °C. Aliquots of 100 μL of cell suspension (5×10^3 cells) were placed in 96-well plates and incubated in a complete medium for 24 h. Cells were treated with another aliquot of 100 μL medium containing DRSE at various concentrations (0.01, 0.1, 1, 10, and 100 μg/mL). After 72 h of drug exposure, cells were fixed by replacing the medium with 150 μL of 10% TCA and incubated at 4 °C for 1 h. The TCA solution was removed, and the cells were washed five times with distilled water. Aliquots of 70 μL SRB solution (0.4% *w*/*v*) were added and incubated in a dark place at room temperature for 10 min. Plates were washed three times with 1% acetic acid and allowed to air-dry overnight. Then, 150 μL of TRIS (10 mM) was added to dissolve the protein-bound SRB stain; the absorbance was measured at 540 nm using a BMGLABTECH®-FLUO star Omega microplate reader (Ortenberg, Germany).

3.7.2. Scratch Wound Assay

The wound-healing activity of TIB, TIL, TIS, and TIF *n*-hexane extract was evaluated using in vitro cell migration studies on HSF cells. Cells were plated at a density of 2×10^5/well onto a coated 12-well plate for the scratch wound assay and cultured overnight in 5% FBS-DMEM at 37 °C and 5% CO_2. On the next day, horizontal scratches were introduced into the confluent monolayer; the plate was washed thoroughly with PBS, control wells were replenished with the fresh medium, and drug wells were treated with fresh media containing the drug. Images were taken using an inverted microscope at the indicated time intervals. The plate was incubated at 37 °C and 5% CO_2 between time points. The experiment was done in triplicate. The acquired images are displayed in Figure 6 and were analysed using MII ImageView software version 3.7. Wound width was calculated as the average distance between the edges of the scratches; the wound width decreases as cell migration is induced. The results are displayed as mean ± standard deviation (Table 3) [77–79].

3.8. Statistical Analysis

Statistical analyses were done using a one-way analysis of variance (ANOVA) followed by the Tukey–Kramer multiple comparison test ($p < 0.05$). Statistical analyses were performed using GraphPad Prism 6.01 (GraphPad Inc., La Jolla, CA, USA).

3.9. In Silico Molecular Docking Studies

The X-ray 3D structures of glycogen synthase kinase 3-β (GSK3-β), matrix metalloproteinases-8 (MMP-8), and nitric oxide synthase (iNOS) were downloaded from the protein data bank www.pdb.org (accessed on 15 August 2022) using the following IDs: 3F88, 5H8X, and 3N2R, respectively [80–82]. All the docking studies were conducted using MOE 2019 [83], which was also used to generate the 2D interaction diagrams between the docked ligands and their potential targets. The identified major compounds were prepared using the default parameters and saved in a single MDB file. The active site for each target was determined from the binding of the corresponding co-crystalised ligand. Finally, the docking was finalised by docking the MDB file containing all the major compounds into the active site of the three enzymes.

4. Conclusions

The present study investigated the secondary metabolites in the *n*-hexane extract of *T. indica's* various organs and their in vitro anti-inflammatory and wound healing properties using the scratch assay with HSF cells. The GC/MS analysis revealed that triterpenoids and steroids were the predominant classes in the *T. indica* bark and leaf n-hexane extracts. Additionally, the seed and fruit *n*-hexane extracts showed a high percentage of fatty acids and higher hydrocarbons. PCA and HCA successfully discriminated various *T. indica* organs based on their GC/MS metabolites. The tested extracts showed promising anti-inflammatory and wound-healing properties. Additionally, the major characterised phytoconstituents achieved promising docking scores in the active sites of glycogen synthase kinase 3-β (GSK3-β), matrix metalloproteinases-8 (MMP-8), and nitric oxide synthase (iNOS), which give the *n*-hexane extract of the various *T. indica* organs a chance to be incorporated in the pharmaceutical products for wound healing after further in vivo and clinical trials.

Author Contributions: Conceptualization, S.H.A. and H.A.G.; methodology, S.H.A. and H.A.G.; software, H.A.G., M.A.E.-H. and S.S.E.; validation, S.H.A. and S.S.E.; formal analysis, S.H.A. and M.A.E.-H.; investigation, S.H.A., S.S.E. and H.A.G.; resources, S.S.E. and M.A.E.-H.; writing—original draft preparation, S.H.A. and H.A.G.; writing—review and editing, S.S.E. and M.A.E.-H.; supervision, S.S.E. and H.A.G.; funding acquisition, S.S.E. All authors have read and agreed to the published version of the manuscript.

Funding: This research work was funded by Institutional Fund Projects under grant no. (IFPIP:1919-166-1443). The authors gratefully acknowledge the technical and financial support provided by the Ministry of Education and King Abdulaziz University, DSR, Jeddah, Saudi Arabia.

Institutional Review Board Statement: Not applicable.

Informed Consent Statement: Not applicable.

Data Availability Statement: Data supporting the reported results can be found at NIST Chemistry Webbook, https://webbook.nist.gov/chemistry/ (accessed on 10 October 2022).

Acknowledgments: This research work was funded by Institutional Fund Projects under grant no. (IFPIP:1919-166-1443). The authors gratefully acknowledge the technical and financial support provided by the Ministry of Education and King Abdulaziz University, DSR, Jeddah, Saudi Arabia.

Conflicts of Interest: The authors declare no conflict of interest.

Sample Availability: Samples of the essentials are available from the authors.

References

1. Elmaidomy, A.H.; Abdelmohsen, U.R.; Alsenani, F.; Aly, H.F.; Shams, S.G.E.; Younis, E.A.; Ahmed, K.A.; Sayed, A.M.; Owis, A.I.; Afifi, N.; et al. The Anti-Alzheimer Potential of Tamarindus Indica: An in Vivo Investigation Supported by in Vitro and in Silico Approaches. *RSC Adv.* **2022**, *12*, 11769–11785. [CrossRef] [PubMed]
2. Chong, U.R.W.; Abdul-Rahman, P.S.; Abdul-Aziz, A.; Hashim, O.H.; Mat Junit, S. *Tamarindus indica* Extract Alters Release of Alpha Enolase, Apolipoprotein A-I, Transthyretin and Rab GDP Dissociation Inhibitor Beta from HepG2 Cells. *PLoS ONE* **2012**, *7*, e39476. [CrossRef] [PubMed]
3. Havinga, R.M.; Hartl, A.; Putscher, J.; Prehsler, S.; Buchmann, C.; Vogl, C.R. *Tamarindus indica* L. (Fabaceae): Patterns of Use in Traditional African Medicine. *J. Ethnopharmacol.* **2010**, *127*, 573–588. [CrossRef] [PubMed]
4. Lim, C.Y.; Mat Junit, S.; Abdulla, M.A.; Abdul Aziz, A. In Vivo Biochemical and Gene Expression Analyses of the Antioxidant Activities and Hypocholesterolaemic Properties of *Tamarindus indica* Fruit Pulp Extract. *PLoS ONE* **2013**, *8*, e70058. [CrossRef]
5. Aly, S.H.; Elissawy, A.M.; Eldahshan, O.A.; Elshanawany, M.A.; Efferth, T.; Singab, A.N.B. The Pharmacology of the Genus *Sophora* (Fabaceae): An Updated Review. *Phytomedicine* **2019**, *64*, 153070. [CrossRef] [PubMed]
6. Eldahshan, O.; Aly, S.; Elissawy, A.; Elshanawany, M.; Singab, A.N. Morphological and Genetic Characteristics of *Sophora secundiflora* and *Sophora tomentosa* (Fabaceae) Cultivated in Egypt. *Taeckholmia* **2019**, *39*, 103–129. [CrossRef]
7. Abdelrahman, G.H.; Mariod, A.A. *Wild Fruits: Composition, Nutritional Value and Products*; Springer: Berlin/Heidelberg, Germany, 2019; ISBN 9783030318857.
8. Kuru, P. Tamarindus Indica and Its Health Related Effects. *Asian Pac. J. Trop. Biomed.* **2014**, *4*, 676–681. [CrossRef]
9. Martinello, F.; Soares, S.M.; Franco, J.J.; Santos, A.C.; Sugohara, A.; Garcia, S.B.; Curti, C.; Uyemura, S.A. Hypolipemic and Antioxidant Activities from *Tamarindus indica* L. Pulp Fruit Extract in Hypercholesterolemic Hamsters. *Food Chem. Toxicol. Int. J. Publ. Br. Ind. Biol. Res. Assoc.* **2006**, *44*, 810–818. [CrossRef] [PubMed]
10. Bhadoriya, S.S.; Mishra, V.; Raut, S.; Ganeshpurkar, A.; Jain, S.K. Anti-Inflammatory and Antinociceptive Activities of a Hydroethanolic Extract of *Tamarindus indica* Leaves. *Sci. Pharm.* **2012**, *80*, 685–700. [CrossRef]
11. Maiti, R.; Jana, D.; Das, U.K.; Ghosh, D. Antidiabetic Effect of Aqueous Extract of Seed of *Tamarindus indica* in Streptozotocin-Induced Diabetic Rats. *J. Ethnopharmacol.* **2004**, *92*, 85–91. [CrossRef]
12. De Caluwé, E.; Halamová, K.; Van Damme, P. Tamarind (*Tamarindus indica* L.): A Review of Traditional Uses, Phytochemistry and Pharmacology. *ACS Symp. Ser.* **2009**, *1021*, 85–110. [CrossRef]
13. Naeem, N.; Nadeem, F.; Azeem, M.W.; Dharmadasa, R.M. *Tamarindus indica*—A Review of Explored Potentials. *Int. J. Chem. Biochem. Sci.* **2017**, *12*, 98–106.
14. Escalona-Arranz, J.; Péres-Roses, R.; Urdaneta-Laffita, I.; Camacho-Pozo, M.; Rodríguez-Amado, J.; Licea-Jiménez, I. Antimicrobial Activity of Extracts from *Tamarindus indica* L. Leaves. *Pharmacogn. Mag.* **2010**, *6*, 242–247. [CrossRef] [PubMed]
15. Al-Fatimi, M.; Wurster, M.; Schröder, G.; Lindequist, U. Antioxidant, Antimicrobial and Cytotoxic Activities of Selected Medicinal Plants from Yemen. *J. Ethnopharmacol.* **2007**, *111*, 657–666. [CrossRef] [PubMed]
16. Meléndez, P.A.; Capriles, V.A. Antibacterial Properties of Tropical Plants from Puerto Rico. *Phytomedicine* **2006**, *13*, 272–276. [CrossRef]
17. Komakech, R.; Kim, Y.; Matsabisa, G.M.; Kang, Y. Anti-Inflammatory and Analgesic Potential of *Tamarindus indica* Linn. (Fabaceae): A Narrative Review. *Integr. Med. Res.* **2019**, *8*, 181–186. [CrossRef]
18. Meher, B.; Dash, D.K.; Kumar Dash, D.; Roy, A. A Review on: Phytochemistry, Pharmacology and Traditional Uses of *Tamarindus indica* L. *World J. Pharm. Pharm. Sci.* **2014**, *3*, 229.
19. Yerima, M.; Anuka, J.A.; Salawu, O.A.; Abdu-Aguye, I. Antihyperglycaemic Activity of the Stem-Bark Extract of *Tamarindus indica* L. on Experimentally Induced Hyperglycaemic and Normoglycaemic Wistar Rats. *Pak. J. Biol. Sci. PJBS* **2014**, *17*, 414–418. [CrossRef]
20. Das, S.S.; Dey, M.; Ghosh, A.K. Determination of Anthelmintic Activity of the Leaf and Bark Extract of *Tamarindus indica* Linn. *Indian J. Pharm. Sci.* **2011**, *73*, 104–107. [CrossRef]
21. Okur, M.E.; Karadağ, A.E.; Okur, N.Ü.; Özhan, Y.; Sipahi, H.; Ayla, Ş.; Daylan, B.; Demirci, B.; Demirci, F. In Vivo Wound Healing and in Vitro Anti-Inflammatory Activity Evaluation of *Phlomis russeliana* Extract Gel Formulations. *Molecules* **2020**, *25*, 2695. [CrossRef]
22. Bakr, R.O.; Amer, R.I.; Attia, D.; Abdelhafez, M.M.; Al-Mokaddem, A.K.; El Gendy, A.N.; El-Fishawy, A.M.; Fayed, M.A.A.; Gad, S.S. In-Vivo Wound Healing Activity of a Novel Composite Sponge Loaded with Mucilage and Lipoidal Matter of Hibiscus Species. *Biomed. Pharmacother.* **2021**, *135*, 111225. [CrossRef] [PubMed]
23. Imran, M.; Sharma, J.N.; Kamal, M.; Asif, M. Standardization and Wound-Healing Activity of Petroleum, Ethanolic and Aqueous Extracts of *Ficus benghalensis* Leaves. *Pharm. Chem. J.* **2021**, *54*, 1057–1062. [CrossRef]
24. Agyare, C.; Boakye, Y.D.; Bekoe, E.O.; Hensel, A.; Dapaah, S.O.; Appiah, T. Review: African Medicinal Plants with Wound Healing Properties. *J. Ethnopharmacol.* **2016**, *177*, 85–100. [CrossRef] [PubMed]
25. Haque, S.D.; Saha, S.K.; Salma, U.; Nishi, M.K.; Rahaman, M.S. Antibacterial Effect of Aloe Vera (*Aloe barbadensis*) Leaf Gel against *Staphylococcus aureus*, *Pseudomonas aeruginosa*, *Escherichia coli* and *Klebsiella pneumoniae*. *Mymensingh Med. J.* **2019**, *28*, 490–496. [CrossRef] [PubMed]
26. Shedoeva, A.; Leavesley, D.; Upton, Z.; Fan, C. Wound Healing and the Use of Medicinal Plants. *Evid.-Based Complement. Altern. Med.* **2019**, *2019*, 2684108. [CrossRef] [PubMed]

27. Ads, E.N.; Hassan, S.I.; Rajendrasozhan, S.; Hetta, M.H.; Aly, S.H.; Ali, M.A. Isolation, Structure Elucidation and Antimicrobial Evaluation of Natural Pentacyclic Triterpenoids and Phytochemical Investigation of Different Fractions of *Ziziphus spina-christi* (L.) Stem Bark Using LCHRMS Analysis. *Molecules* **2022**, *27*, 1805. [CrossRef]
28. El-Nashar, H.A.S.; Aly, S.H.; Ahmadi, A.; El-Shazly, M. The Impact of Polyphenolics in the Management of Breast Cancer: Mechanistic Aspects and Recent Patents. *Recent Pat. Anticancer Drug Discov.* **2021**, *17*, 358–379. [CrossRef]
29. Aly, S.H.; Elgindi, M.R.; Singab, A.E.N.B.; Mahmoud, I.I. *Hyophorbe verschaffeltii* DNA Profiling, Chemical Composition of the Lipophilic Fraction, Antimicrobial, Anti-Inflammatory and Cytotoxic Activities. *Res. J. Pharm. Biol. Chem. Sci.* **2016**, *7*, 120–130.
30. Aly, S.H.; Elissawy, A.M.; Fayez, A.M.; Eldahshan, O.A.; Elshanawany, M.A.; Singab, A.N.B. Neuroprotective Effects of *Sophora secundiflora*, *Sophora tomentosa* Leaves and Formononetin on Scopolamine-Induced Dementia. *Nat. Prod. Res.* **2021**, *35*, 5848–5852. [CrossRef]
31. Saber, F.R.; Aly, S.H.; Khallaf, M.A.; El-Nashar, H.A.S.; Fahmy, N.M.; El-Shazly, M.; Radha, R.; Prakash, S.; Kumar, M.; Taha, D.; et al. *Hyphaene thebaica* (Areceaeae) as a Promising Functional Food: Extraction, Analytical Techniques, Bioactivity, Food, and Industrial Applications. *Food Anal. Methods* **2022**, 1–21. [CrossRef]
32. Raslan, M.A.; Afifi, A.H. In Vitro Wound Healing Properties, Antioxidant Activities, HPLC-ESI-MS/MS Profile and Phytoconstituents of the Stem Aqueous Methanolic Extract of *Dracaena reflexa* Lam. *Biomed. Chromatogr.* **2022**, *36*, e5352. [CrossRef] [PubMed]
33. Labib, R.M.; Ayoub, I.M.; Michel, H.E.; Mehanny, M.; Kamil, V.; Hany, M.; Magdy, M.; Moataz, A.; Maged, B.; Mohamed, A. Appraisal on the Wound Healing Potential of *Melaleuca alternifolia* and *Rosmarinus officinalis* L. Essential Oil-Loaded Chitosan Topical Preparations. *PLoS ONE* **2019**, *14*, e0219561. [CrossRef] [PubMed]
34. El-Nashar, H.A.S.; Mostafa, N.M.; Eldahshan, O.A.; Singab, A.N.B. Chemical Composition, Cytotoxic and Anti-Arthritic Activities of Hexane Extracts of Certain *Schinus* Species. *J. Pharm. Pharmacol.* **2021**, *9*, 378–386. [CrossRef]
35. Aly, S.H.; Eldahshan, O.A.; Al-Rashood, S.T.; Binjubair, F.A.; El-Hassab, M.A.; Eldehna, W.M.; Acqua, S.D.; Zengin, G. Chemical Constituents, Antioxidant, and Enzyme Inhibitory Activities Supported by In-Silico Study of n-Hexane Extract and Essential Oil of Guava Leaves. *Molecules* **2022**, *27*, 8979. [CrossRef]
36. Ali, A.; Garg, P.; Goyal, R.; Kaur, G.; Li, X.; Negi, P.; Valis, M.; Kuca, K.; Kulshrestha, S. A Novel Herbal Hydrogel Formulation of *Moringa oleifera* for Wound Healing. *Plants* **2020**, *10*, 25. [CrossRef]
37. Ayoub, I.M.; Korinek, M.; El-Shazly, M.; Wetterauer, B.; El-Beshbishy, H.A. Activity of *Chasmanthe aethiopica* Leaf Extract and Its Profiling Using LC/MS and GLC/MS. *Plants* **2021**, *10*, 1118. [CrossRef]
38. Sakib, S.A.; Tareq, A.M.; Islam, A.; Rakib, A.; Islam, M.N.; Uddin, M.A.; Rahman, M.M.; Seidel, V.; Emran, T. Bin Anti-Inflammatory, Thrombolytic and Hair-Growth Promoting Activity of the n-Hexane Fraction of the Methanol Extract of *Leea indica* Leaves. *Plants* **2021**, *10*, 1081. [CrossRef]
39. Carullo, G.; Sciubba, F.; Governa, P.; Mazzotta, S.; Frattaruolo, L.; Grillo, G.; Cappello, A.R.; Cravotto, G.; Di Cocco, M.E.; Aiello, F. Mantonico and Pecorello Grape Seed Extracts: Chemical Characterization and Evaluation of In Vitro Wound-Healing and Anti-Inflammatory Activities. *Pharmaceuticals* **2020**, *13*, 97. [CrossRef]
40. Ibrahim, N.A.; El-Gengaihi, S.; El-Hamidi, A.; Bashandy, S.A.E. Chemical and Biological Evaluation of *Tamarindus indica* L. Growing in Sudan. *Acta Hortic.* **1995**, *390*, 51–58. [CrossRef]
41. Carasek, E.; Pawliszyn, J. Screening of Tropical Fruit Volatile Compounds Using Solid-Phase Microextraction (SPME) Fibers and Internally Cooled SPME Fiber. *J. Agric. Food Chem.* **2006**, *54*, 8688–8696. [CrossRef]
42. Gad, H.A.; Mukhammadiev, E.A.; Zengen, G.; Musayeib, N.M.A.; Hussain, H.; Bin Ware, I.; Ashour, M.L.; Mamadalieva, N.Z. Chemometric Analysis Based on GC-MS Chemical Profiles of Three Stachys Species from Uzbekistan and Their Biological Activity. *Plants* **2022**, *11*, 1215. [CrossRef] [PubMed]
43. Rimbau, V.; Cerdan, C.; Vila, R.; Iglesias, J. Antiinflammatory Activity of Some Extracts from Plants Used in the Traditional Medicine of North-African Countries (II). *Phyther. Res.* **1999**, *13*, 128–132. [CrossRef]
44. Hivrale, M.G.; Bandawane, D.D.; Mali, A.A. Anti-Inflammatory and Analgesic Activities of Petroleum Ether and Ethyl Acetate Fractions of *Tamarindus indica* Seeds. *Orient. Pharm. Exp. Med.* **2013**, *13*, 319–326. [CrossRef]
45. Dias, A.M.A.; Rey-Rico, A.; Oliveira, R.A.; Marceneiro, S.; Alvarez-Lorenzo, C.; Concheiro, A.; Júnior, R.N.C.; Braga, M.E.M.; De Sousa, H.C. Wound Dressings Loaded with an Anti-Inflammatory Jucá (*Libidibia ferrea*) Extract Using Supercritical Carbon Dioxide Technology. *J. Supercrit. Fluids* **2013**, *74*, 34–45. [CrossRef]
46. Lucetti, D.L.; Lucetti, E.C.; Bandeira, M.A.M.; Veras, H.N.; Silva, A.H.; Leal, L.K.A.; Lopes, A.A.; Alves, V.C.; Silva, G.S.; Brito, G.A.; et al. Anti-Inflammatory Effects and Possible Mechanism of Action of Lupeol Acetate Isolated from *Himatanthus drasticus* (Mart.) Plumel. *J. Inflamm.* **2010**, *7*, 60. [CrossRef]
47. Skehan, P.; Storeng, R.; Scudiero, D.; Monks, A.; McMahon, J.; Vistica, D.; Warren, J.T.; Bokesch, H.; Kenney, S.; Boyd, M.R. New Colorimetric Cytotoxicity Assay for Anticancer—Drug Screening. *J. Natl. Cancer Inst.* **1990**, *82*, 1107–1112. [CrossRef]
48. Adeniyi, O.V.; Olaifa, F.E.; Emikpe, B.O.; Oyagbemi, A.A. Experimental Evaluation of the Wound-Healing and Antioxidant Activities of Tamarind (*Tamarindus indica*) Pulp and Leaf Meal in the African Catfish (*Clarias gariepinus*). *Acta Vet. Eurasia* **2018**, *44*, 63–72. [CrossRef]
49. Attah, M.O.; Ishaya, H.B.; Chiroma, M.S.; Amaza, D.; Balogun, S.U.; Jacks, T. Effect of *Tamarindus indica* (Linn) on the Rate of Wound Healing in Adult Rabbits. *IOSR J. Dent. Med. Sci.* **2015**, *14*, 2279–2861. [CrossRef]

50. Poljšak, N.; Kočevar Glavač, N. *Tilia* Sp. Seed Oil—Composition, Antioxidant Activity and Potential Use. *Appl. Sci.* **2021**, *11*, 4932. [CrossRef]
51. Hernandez, G.R.; Hernandez Garcia, D.Y.; Sanchez, M.L. Healing Cream *Tournefortia hirsutissima* L. *Med. Aromat. Plants* **2017**, *6*, 4–6. [CrossRef]
52. Hata, K.; Hori, K.; Takahashi, S. Role of P38 MAPK in Lupeol-Induced B16 2F2 Mouse Melanoma Cell Differentiation. *J. Biochem.* **2003**, *134*, 441–445. [CrossRef] [PubMed]
53. Geetha, T.; Varalakshmi, P. Anti-Inflammatory Activity of Lupeol and Lupeol Linoleate in Rats. *J. Ethnopharmacol.* **2001**, *76*, 77–80. [CrossRef] [PubMed]
54. Pereira Beserra, F.; Xue, M.; Maia, G.L.d.A.; Leite Rozza, A.; Helena Pellizzon, C.; Jackson, C.J. Lupeol, a Pentacyclic Triterpene, Promotes Migration, Wound Closure, and Contractile Effect In Vitro: Possible Involvement of PI3K/Akt and P38/ERK/MAPK Pathways. *Molecules* **2018**, *23*, 2819. [CrossRef] [PubMed]
55. Patel, S.; Srivastava, S.; Singh, M.R.; Singh, D. Preparation and Optimization of Chitosan-Gelatin Films for Sustained Delivery of Lupeol for Wound Healing. *Int. J. Biol. Macromol.* **2018**, *107*, 1888–1897. [CrossRef] [PubMed]
56. Beserra, F.P.; Vieira, A.J.; Gushiken, L.F.S.; De Souza, E.O.; Hussni, M.F.; Hussni, C.A.; Takahira, R.K.; Nóbrega, R.H.; Martinez, E.R.M.; Jackson, C.J.; et al. Lupeol, a Dietary Triterpene, Enhances Wound Healing in Streptozotocin-Induced Hyperglycemic Rats with Modulatory Effects on Inflammation, Oxidative Stress, and Angiogenesis. *Oxidative Med. Cell. Longev.* **2019**, *2019*, 3182627. [CrossRef] [PubMed]
57. Bopage, N.S.; Kamal Bandara Gunaherath, G.M.; Jayawardena, K.H.; Wijeyaratne, S.C.; Abeysekera, A.M.; Somaratne, S. Dual Function of Active Constituents from Bark of *Ficus Racemosa* L. in Wound Healing. *BMC Complement. Altern. Med.* **2018**, *18*, 29. [CrossRef] [PubMed]
58. Malinowska, M.; Miroslaw, B.; Sikora, E.; Ogonowski, J.; Wojtkiewicz, A.M.; Szaleniec, M.; Pasikowska-Piwko, M.; Eris, I. New Lupeol Esters as Active Substances in the Treatment of Skin Damage. *PLoS ONE* **2019**, *14*, e0214216. [CrossRef]
59. Bopage, N.S.; Jayawardena, K.H.; Wijeyaratne, C.; Abeysekera, A.M.; Gunaherath, G.M.K.B. Wound Healing Activity of Some Lupeol Derivatives Using. *Life Sci.* **2016**, 3–5.
60. Poljšak, N.; Kreft, S.; Kočevar Glavač, N. Vegetable Butters and Oils in Skin Wound Healing: Scientific Evidence for New Opportunities in Dermatology. *Phyther. Res.* **2020**, *34*, 254–269. [CrossRef]
61. Lewinska, A.; Zebrowski, J.; Duda, M.; Gorka, A.; Wnuk, M. Fatty Acid Profile and Biological Activities of Linseed and Rapeseed Oils. *Molecules* **2015**, *20*, 22872–22880. [CrossRef]
62. Xu, A.L.; Xue, Y.-Y.; Tao, W.-T.; Wang, S.-Q.; Xu, H.-Q. Oleanolic Acid Combined with Olaparib Enhances Radiosensitization in Triple Negative Breast Cancer and Hypoxia Imaging with (18)F-FETNIM Micro PET/CT. *Biomed. Pharmacother.* **2022**, *150*, 113007. [CrossRef] [PubMed]
63. Lin, T.-K.; Zhong, L.; Santiago, J.L. Anti-Inflammatory and Skin Barrier Repair Effects of Topical Application of Some Plant Oils. *Int. J. Mol. Sci.* **2017**, *19*, 70. [CrossRef] [PubMed]
64. Sánchez-Quesada, C.; López-Biedma, A.; Toledo, E.; Gaforio, J.J. Squalene Stimulates a Key Innate Immune Cell to Foster Wound Healing and Tissue Repair. *Evid.-Based Complement. Altern. Med.* **2018**, *2018*, 9473094. [CrossRef]
65. Abd El-Ghffar, E.A.; El-Nashar, H.A.S.; Eldahshan, O.A.; Singab, A.N.B. GC-MS Analysis and Hepatoprotective Activity of the n-Hexane Extract of *Acrocarpus fraxinifolius* Leaves against Paracetamol-Induced Hepatotoxicity in Male Albino Rats. *Pharm. Biol.* **2017**, *55*, 444–449. [CrossRef]
66. Rostad, C.E.; Pereira, W.E. Kovats and Lee Retention Indices Determined by Gas Chromatography/Mass Spectrometry for Organic Compounds of Environmental Interest. *J. High Resolut. Chromatogr.* **1986**, *9*, 328–334. [CrossRef]
67. Ivanov, I.; Dincheva, I.; Badjakov, I.; Petkova, N.; Denev, P.; Pavlov, A. GC-MS Analysis of Unpolar Fraction from *Ficus carica* L. (Fig) Leaves. *Int. Food Res. J.* **2018**, *25*, 282–286.
68. de Carvalho, F.M.d.A.; Schneider, J.K.; de Jesus, C.V.F.; de Andrade, L.N.; Amaral, R.G.; David, J.M.; Krause, L.C.; Severino, P.; Soares, C.M.F.; Bastos, E.C.; et al. Brazilian Red Propolis: Extracts Production, Physicochemical Characterization, and Cytotoxicity Profile for Antitumor Activity. *Biomolecules* **2020**, *10*, 726. [CrossRef]
69. El-Nashar, H.A.S.; Eldehna, W.M.; Al-Rashood, S.T.; Alharbi, A.; Eskandrani, R.O.; Aly, S.H. GC/MS Analysis of Essential Oil and Enzyme Inhibitory Activities of *Syzygium cumini* (Pamposia) Grown in Docking Studies. *Molecules* **2021**, *26*, 6984. [CrossRef]
70. Aly, S.H.; Elissawy, A.M.; Eldahshan, O.A.; Elshanawany, M.A.; Singab, A.N.B. Phytochemical Investigation Using GC/MS Analysis and Evaluation of Antimicrobial and Cytotoxic Activities of the Lipoidal Matter of Leaves of *Sophora secundiflora* and *Sophora tomentosa*. *Arch. Pharm. Sci. Ain Shams Univ.* **2020**, *4*, 207–214. [CrossRef]
71. Aly, S.H.; Elissawy, A.M.; Eldahshan, O.A.; Elshanawany, M.A.; Singab, A.N.B. Variability of the Chemical Composition of the Essential Oils of Flowers and the Alkaloid Contents of Leaves of *Sophora secundiflora* and *Sophora tomentosa*. *J. Essent. Oil-Bear. Plants* **2020**, *23*, 442–452. [CrossRef]
72. Gad, H.A.; Ayoub, I.M.; Wink, M. Phytochemical Profiling and Seasonal Variation of Essential Oils of Three *Callistemon* Species Cultivated in Egypt. *PLoS One* **2019**, *14*, e0219571. [CrossRef] [PubMed]
73. Gad, H.A.; Mamadalieva, R.Z.; Khalil, N.; Zengin, G.; Najar, B.; Khojimatov, O.K.; Al Musayeib, N.M.; Ashour, M.L.; Mamadalieva, N.Z. GC-MS Chemical Profiling, Biological Investigation of Three *Salvia* Species Growing in Uzbekistan. *Molecules* **2022**, *27*, 5365. [CrossRef] [PubMed]
74. Brereton, R.G. *Applied Chemometrics for Scientists*; John Wiley & Sons: Chichester, UK, 2007; pp. 1–379.

75. Yoo, M.-S.; Shin, J.-S.; Choi, H.-E.; Cho, Y.-W.; Bang, M.-H.; Baek, N.-I.; Lee, K.-T. Fucosterol Isolated from *Undaria pinnatifida* Inhibits Lipopolysaccharide-Induced Production of Nitric Oxide and Pro-Inflammatory Cytokines via the Inactivation of Nuclear Factor-KB and P38 Mitogen-Activated Protein Kinase in RAW264.7 Macrophages. *Food Chem.* **2012**, *135*, 967–975. [CrossRef] [PubMed]
76. Oliveira, T.; Figueiredo, C.A.; Brito, C.; Stavroullakis, A.; Prakki, A.; Da Silva Velozo, E.; Nogueira-Filho, G. Effect of *Allium cepa* L. on Lipopolysaccharide-Stimulated Osteoclast Precursor Cell Viability, Count, and Morphology Using 4′,6-Diamidino-2-Phenylindole-Staining. *Int. J. Cell Biol.* **2014**, *2014*, 535789. [CrossRef] [PubMed]
77. Main, K.A.; Mikelis, C.M.; Doçi, C.L. In Vitro Wound Healing Assays to Investigate Epidermal Migration. In *Methods in Molecular Biology*; Springer: New York, NY, USA, 2019; Volume 2109, pp. 147–154.
78. Martinotti, S.; Ranzato, E. Scratch Wound Healing Assay. In *Methods in Molecular Biology*; Springer: Berlin/Heidelberg, Germany, 2019; Volume 2109, pp. 225–229.
79. Jonkman, J.E.N.; Cathcart, J.A.; Xu, F.; Bartolini, M.E.; Amon, J.E.; Stevens, K.M.; Colarusso, P. An Introduction to the Wound Healing Assay Using Live-Cell Microscopy. *Cell Adhes. Migr.* **2014**, *8*, 440–451. [CrossRef]
80. Saitoh, M.; Kunitomo, J.; Kimura, E.; Hayase, Y.; Kobayashi, H.; Uchiyama, N.; Kawamoto, T.; Tanaka, T.; Mol, C.D.; Dougan, D.R.; et al. Design, Synthesis and Structure-Activity Relationships of 1,3,4-Oxadiazole Derivatives as Novel Inhibitors of Glycogen Synthase Kinase-3β. *Bioorg. Med. Chem.* **2009**, *17*, 2017–2029. [CrossRef]
81. Tauro, M.; Laghezza, A.; Loiodice, F.; Piemontese, L.; Caradonna, A.; Capelli, D.; Montanari, R.; Pochetti, G.; Di Pizio, A.; Agamennone, M.; et al. Catechol-Based Matrix Metalloproteinase Inhibitors with Additional Antioxidative Activity. *J. Enzym. Inhib. Med. Chem.* **2016**, *31*, 25–37. [CrossRef]
82. Xue, F.; Huang, J.; Ji, H.; Fang, J.; Li, H.; Martásek, P.; Roman, L.J.; Poulos, T.L.; Silverman, R.B. Structure-Based Design, Synthesis, and Biological Evaluation of Lipophilic-Tailed Monocationic Inhibitors of Neuronal Nitric Oxide Synthase. *Bioorg. Med. Chem.* **2010**, *18*, 6526–6537. [CrossRef]
83. Vilar, S.; Cozza, G.; Moro, S. Medicinal Chemistry and the Molecular Operating Environment (MOE): Application of QSAR and Molecular Docking to Drug Discovery. *Curr. Top. Med. Chem.* **2008**, *8*, 1555–1572. [CrossRef]

plants

MDPI

Article

Volatile Oils Discrepancy between Male and Female *Ochradenus arabicus* and Their Allelopathic Activity on *Dactyloctenium aegyptium*

Ahmed M. Abd-ElGawad [1,2,*], Abdulaziz M. Assaeed [1], Abd El-Nasser G. El Gendy [3], Basharat A. Dar [1] and Abdelsamed I. Elshamy [4]

1 Plant Production Department, College of Food & Agriculture Sciences, King Saud University, P.O. Box 2460, Riyadh 11451, Saudi Arabia
2 Department of Botany, Faculty of Science, Mansoura University, Mansoura 35516, Egypt
3 Medicinal and Aromatic Plants Research Department, National Research Centre, Cairo 11865, Egypt
4 Department of Natural Compounds Chemistry, National Research Centre, 33 El Bohouth St., Dokki, Giza 12622, Egypt
* Correspondence: aibrahim2@ksu.edu.sa or dgawad84@mans.edu.eg; Tel.: +966-562680864

Abstract: Volatile oils (VOs) composition of plants is affected by several exogenous and endogenous factors. Male and female plants of the dioecious species exhibit variation in the bioactive constituents' allocation. The chemical variation in the VOs between male and female plants is not well studied. In the present study, the chemical characterization of the VOs extracted from aerial parts of male and female ecospecies of *Ochradenus arabicus* was documented. Additionally, the extracted VOs were tested for their allelopathic activity against the weed *Dactyloctenium aegyptium*. Via GC-MS analysis, a total of 53 compounds were identified in both male and female plants. Among them, 49 compounds were identified from male plants, and 47 compounds were characterized in female plants. Isothiocyanates (47.50% in male and 84.32% in female) and terpenes (48.05% in male and 13.22% in female) were the main components of VOs, in addition to traces of carotenoid-derived compounds and hydrocarbons. The major identified compounds of male and female plants are *m*-tolyl isothiocyanate, benzyl isothiocyanate, butyl isothiocyanate, isobutyl isothiocyanate, carvone, and α-bisabolol, where they showed variation in the concentration between male and female plants. The *O. arabicus* VOs of the male plants attained IC_{50} values of 51.1, 58.1, and 41.9 $\mu L\ L^{-1}$ for the seed germination, seedling shoot growth, and seedling root growth of the weed (*D. aegyptium*), respectively, while the females showed IC_{50} values of 56.7, 63.9, and 40.7 $\mu L\ L^{-1}$, respectively. The present data revealed that VOs composition and bioactivity varied significantly with respect to the plant gender, either qualitatively or quantitatively.

Keywords: *Ochradenus arabicus*; essential oils; phytotoxicity; weed control; isothiocyanates

Citation: Abd-ElGawad, A.M.; Assaeed, A.M.; El Gendy, A.E.-N.G.; Dar, B.A.; Elshamy, A.I. Volatile Oils Discrepancy between Male and Female *Ochradenus arabicus* and Their Allelopathic Activity on *Dactyloctenium aegyptium*. *Plants* **2023**, *12*, 110. https://doi.org/10.3390/plants12010110

Academic Editors: Adriano Sofo, Ippolito Camele and Hazem Salaheldin Elshafie

Received: 15 November 2022
Revised: 21 December 2022
Accepted: 21 December 2022
Published: 26 December 2022

1. Introduction

The plant kingdom encompasses about 320,000 known species that are very rich resources of metabolites, which play crucial roles in the growth, reproduction, and defense mechanism of plants [1]. A myriad of metabolites with potent bioactivities was reported from the plants including essential and/or volatile oils [2]. The volatile oils (VOs) derived from the plants are mixtures of volatile low molecular weights constituents extracted by several extraction techniques. These compounds are categorized under various classes such as mono-, sesqui-, and di-terpenes carotenoid-derived compounds, apo-carotenoid-derived compounds, phenylpropanoids, and other hydrocarbons [3,4]. The VOs were reported to exert several biological and pharmaceutical potentialities such as antiviral [5], antimicrobial [6], anticancer [7,8], anti-inflammatory [9,10], antipyretic [9,11], and antiulcer [12] effects.

The VOs of plants are affected by several external factors including environmental factors, such as temperature, light, moisture, atmospheric oxygen, precipitation, soil characteristics [13,14], geographic variations [15], seasonal variation, and climatic factors [16]. In addition, VOs can be affected by endogenous factors such as developmental stages [17], genetic variability [18,19], and variety [20]. Moreover, plant sex has been reported to affect the chemical composition of the VOs either quantitively or qualitatively in many plants, such as *Laurus nobilis* L. [21] and *Juniperus communis* L. [22].

The plants belonging to the *Ochradenus* genus (Family: Resedaceae), including around eight species, are widely distributed in Southwest Asia, North Africa, and the Arabian Peninsula [23]. *Ochradenus arabicus* Chaudhary, Hillc. & A.G.Mill. is a compact twiggy dioecious shrub (up to 75 cm tall) with small yellow flowers as well as yellow, papery fruits [24]. The male and female reproductive organs are present in separate individuals. *O. arabicus* is a plant endemic to the Arabian Peninsula, where it is reported only in the flora of Oman, Saudi Arabia, Yemen (Soqotra), and the United Arab Emirates. This shrub is one of the most common medicinal plants with several significant bioactivities such as antioxidant, anticancer, antimicrobial, antidiabetic, anti-indole acetic acid genotoxicity, and allopathic [25–27]. The documented phytochemical studies of the various extracts of *O. arabicus* revealed the identification of phenolic compounds including several flavonoids [25,26]. Recently, the aroma profiling, antioxidant, antimicrobial, and antidiabetic effects of VOs derived from the different organs of the *O. arabicus* collected from Oman were documented [28]. Although *O. arabicus* is a dioecious plant, in this study, the authors did not explain if they targeted the male or female plant of *O. arabicus*. Additionally, we hypothesized that the chemical composition of the VOs of *O. arabicus* would be varied according to the sex of the plant. Thereby, the objectives of the current work were (1) to assess the variation in the chemical composition of the VOs derived from the aerial parts of the male and female plants of *O. arabicus* collected from Saudi Arabia, and (2) to evaluate the allelopathic activity of the extracted VOs from male and female plants.

2. Results and Discussion

2.1. Male and Female *Ochradenus arabicus* VOs Chemical Profiling

The air-dried samples of male and female *O. arabicus* were subjected to hydrodistillation for three hours over the Clevenger apparatus, where they produced 0.08% and 0.07% (v/w) of the pale-yellow VOs, respectively. The analysis of the VOs samples was performed by the GC-MS (Figure 1). The chemical profiles were summarized in Table 1 including the Rt (retention times), literature, and experimental KI (retention indices), along with relative concentrations. The data revealed that VOs' profiles of male and female *O. arabicus* included nine classes of components (Figure 2). The female plants were rich in isothiocyanates (84.32%) compared to the male plants (46.50%). However, male plants were richer in terpenoid compounds (47.27), compared to female plants (12.68%). The majority of the terpenoid classes are oxygenated monoterpenes and oxygenated sesquiterpenes in both male and female *O. arabicus* (Figure 2). Generally, the oxygenated compounds were higher in male (43.27%) than female plants (11.20%), while the non-oxygenated compounds were identified in low concentration in both male (4.39%) and female plants (2.31%). The abundance of the oxygenated compounds was already reported in the previous analysis of EO derived from different organs of *O. arabicus* collected from Oman but without specification of the gender [28].

Overall, 53 compounds were identified in the VOs of both male and female *O. arabicus*. Forty-nine compounds (94.72% of the total mass) were identified in the VO of the male plant, while 47 compounds, with a total relative concentration of 97.78%, were assigned from the female plant. The isothiocyanates were assigned as the major constituents of both genders, while the female was richer than the male plants. Four isothiocyanates were characterized with high relative concentrations in male and female ecospecies including *m*-tolyl isothiocyanate (35.3 and 55.41%), benzyl isothiocyanate (4.88 and 14.08%), butyl isothiocyanate (4.77 and 6.84%), and isobutyl isothiocyanate (1.55 and 7.99%) (Figure 3).

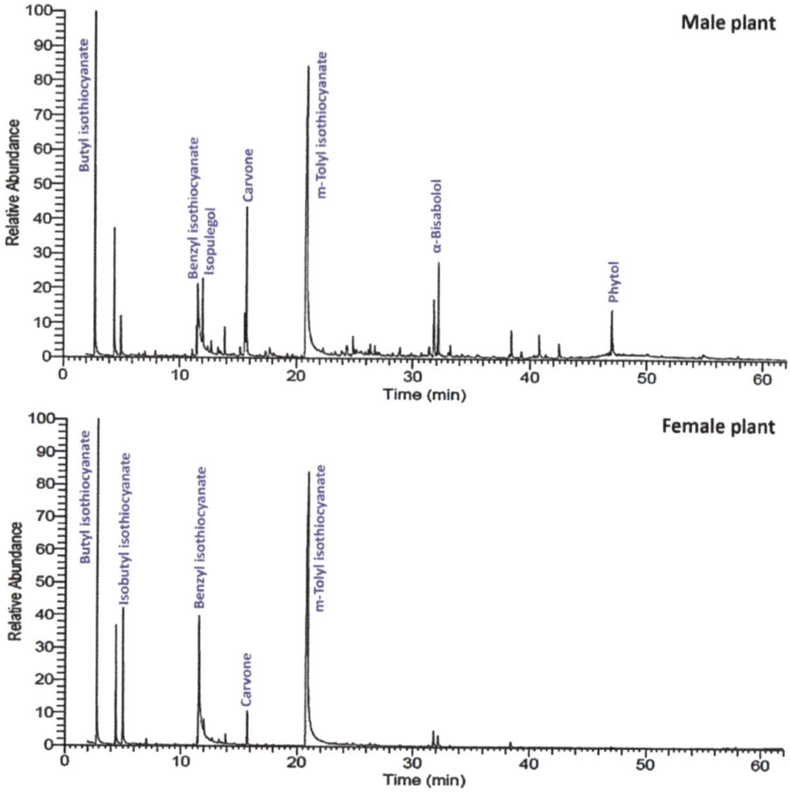

Figure 1. GC-MS chromatograms of male and female *Ochradenus arabicus* volatile oils.

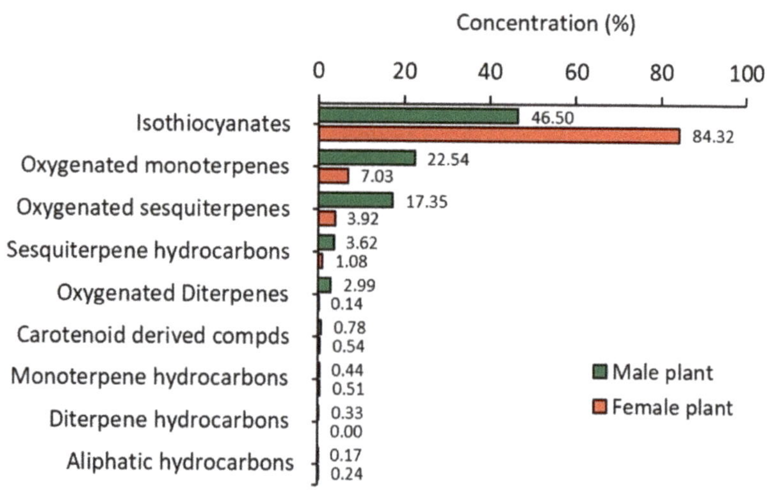

Figure 2. Various classes of the identified chemical compounds in male and female plants of *Ochradenus arabicus*.

Table 1. Volatile oil components of male and female plants of *Ochradenus arabicus*.

No	Compound Name	Rt [1]	Conc. % [2]		KI	
			Male	Female	Lit. [3]	Exp. [4]
	Monoterpene hydrocarbons					
1	α-Terpinene	6.55	0.15 ± 0.01	0.09 ± 0.01	1014	1015
2	γ-Terpinene	7.92	0.29 ± 0.02	0.42 ± 0.02	1054	1053
	Oxygenated Monoterpenes					
3	Eucalyptol	6.87	0.11 ± 0.01	0.05 ± 0.00	1026	1024
4	Linalool	7.05	0.18 ± 0.01	0.04 ± 0.00	1096	1097
5	1-Terpineol	9.51	0.00	0.13 ± 0.01	1133	1134
6	cis-Verbenol	10.47	0.11 ± 0.01	0.14 ± 0.01	1137	1135
7	trans-Pinocarveol	11.11	0.35 ± 0.02	0.13 ± 0.01	1139	1141
8	Camphor	11.21	0.00	0.03 ± 0.00	1141	1143
9	Menthone	11.44	1.42 ± 0.06	0.56 ± 0.02	1148	1147
10	Isopulegol	12.00	3.49 ± 0.09	1.10 ± 0.05	1149	1150
11	4-Terpineol	12.73	0.64 ± 0.03	0.00	1177	1175
12	α-Terpineol	13.44	0.58 ± 0.02	0.42 ± 0.01	1188	1190
13	trans-Carveol	13.89	1.52 ± 0.06	0.95 ± 0.03	1215	1213
14	Pulegone	15.21	0.82 ± 0.03	0.00	1233	1231
15	*trans*-chrysanthenyl acetate	15.57	3.02 ± 0.08	0.11 ± 0.01	1235	1236
16	Carvone	15.74	7.80 ± 0.16	3.01 ± 0.12	1239	1241
17	Bornyl acetate	16.90	0.11 ± 0.01	0.04 ± 0.01	1254	1257
18	Thymol	17.40	0.51 ± 0.02	0.19 ± 0.01	1290	1291
19	2-Adamantanone	17.73	0.84 ± 0.03	0.04 ± 0.01	1311	1314
20	trans-sabinenehydrate acetate	24.39	1.04 ± 0.05	0.09 ± 0.01	1577	1574
	Sesquiterpene hydrocarbons					
21	α-Cubebene	19.25	0.23 ± 0.01	0.10 ± 0.01	1351	1353
22	α-Ylangene	20.47	0.12 ± 0.01	0.07 ± 0.00	1373	1371
23	α-Duprezianene	23.34	0.26 ± 0.01	0.14 ± 0.01	1387	1384
24	Davana ether-1	24.90	1.16 ± 0.09	0.23 ± 0.01	1433	1430
25	Spirolepechinene	25.22	0.36 ± 0.02	0.08 ± 0.00	1451	1449
26	Dihydro-β-agarofuran	25.99	0.22 ± 0.01	0.06 ± 0.00	1503	1505
27	γ-Cadinene	26.22	0.32 ± 0.02	0.09 ± 0.01	1513	1515
28	α-Cadinene	26.39	0.73 ± 0.02	0.26 ± 0.01	1537	1539
29	α-Cadinene	27.14	0.22 ± 0.01	0.05 ± 0.00	1538	1535
	Oxygenated Sesquiterpenes					
30	Widdrol hydroxyether	19.94	0.00	0.06 ± 0.00	1479	1480
31	6-epi-shyobunol	26.97	0.79 ± 0.02	0.21 ± 0.01	1517	1516
32	*E*-Nerolidol	28.27	0.87 ± 0.02	0.00	1563	1560
33	Spathulenol	28.81	0.28 ± 0.01	0.04 ± 0.00	1578	1590
34	Caryophyllene oxide	28.93	0.56 ± 0.02	0.11 ± 0.00	1583	1581
35	Davanone	29.65	0.41 ± 0.01	0.13 ± 0.01	1587	1585
36	Cubenol	31.39	1.17 ± 0.06	0.23 ± 0.01	1646	1645
37	Calarene epoxide	31.81	3.79 ± 0.08	1.68 ± 0.08	1671	1670
38	α-Bisabolol	32.21	5.77 ± 0.11	1.11 ± 0.07	1685	1687
39	epi-Nootkatol	32.48	0.09 ± 0.01	0.00	1699	1601
40	Juniper camphor	33.07	0.37 ± 0.01	0.00	1700	1703
41	Drimenol	33.24	0.67 ± 0.03	0.13 ± 0.01	1767	1768
42	Hexahydrofarnesyl acetone	38.42	1.58 ± 0.07	0.54 ± 0.02	1845	1843
43	Farnesyl acetone C	40.76	1.29 ± 0.06	0.11 ± 0.01	1921	1924
	Diterpene hydrocarbons					
44	Cembrene	36.89	0.33 ± 0.02	0.00	1937	1939
	Oxygenated Diterpenes					
45	Phytol	46.98	2.99 ± 0.08	0.14 ± 0.01	1942	1945
	Carotenoid derived compounds					
46	Theaspirane B	18.08	0.21 ± 0.01	0.07 ± 0.00	1302	1300
47	α-Ionone	25.04	0.28 ± 0.02	0.04 ± 0.00	1430	1432
	Isothiocyanates					
48	Butyl isothiocyanate	4.43	4.77 ± 0.10	6.84 ± 0.13	943	941
49	Isobutyl isothiocyanate	5.01	1.55 ± 0.05	7.99 ± 0.17	978	976
50	Benzyl isothiocyanate	11.56	4.88 ± 0.07	14.08 ± 0.26	1367	1369
51	*m*-Tolyl isothiocyanate	20.93	35.30 ± 0.33	55.41 ± 0.46	1970	1972
	Aliphatic hydrocarbons					
52	*n*-Docosane	52.46	0.00	0.08 ± 0.00	2200	2200
53	*n*-Tricosane	57.84	0.17 ± 0.01	0.16 ± 0.01	2300	2300
	Total		94.72	97.78		

[1] Rt: Retention time; [2] values are average ± SD, [3] KI_{exp}: experimental Kovats retention index; [4] KI_{lit}: Kovats retention index on DB-5 column with reference to n-alkanes.

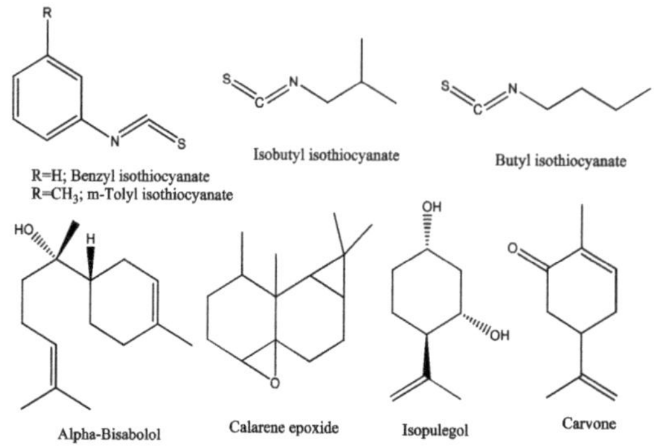

Figure 3. Chemical structures of the main identified compounds in the volatile oils of *Ochradenus arabicus*.

The abundance of isothiocyanates in the present study is not in agreement with previously documented data on the VOs of the stems, leaves, and flowers of *O. arabicus* collected from Oman [28]. This wide variation between our data and that identified in the Omani *O. arabicus* can be ascribed to the extraction technique, where Ullah, et al. [28] made the extraction by heating the Clevenger machine for a long time till no further oil was extracted. Chen and Ho [29] described that isothiocyanate is easily decomposed with refluxing for one hour at 100 °C. Additionally, De Nicola, et al. [30] deduced that the benzylic-Isothiocyanates are unstable and easily converted to other derivatives after refluxing at 90 °C. Thereby, the absence of isothiocyanates in the data of the Omani *O. arabicus* could be attributed to the degradation of the isothiocyanate due to the extraction process.

The isothiocyanates were characterized as main volatiles in the VOs of several plants such as *Wasabia japonica* (Miq.) Koidz. [31]; *Brassica oleracea* L.; *B. rapa* L.; *Armoracia lapathifolia* G. Gaertn., B. Mey. & Scherb.; *Eutrema japonicum* (Miq.) Koidz.; and *Carica papaya* L. [32].

Terpenoids represented the main constituents of the male *O. arabicus*, with a higher relative concentration than the female plant. Six classes of terpenes were characterized from the male plant comprising monoterpene hydrocarbons (0.44%), oxygenated monoterpenes (22.54%), sesquiterpene hydrocarbons (3.62%), oxygenated sesquiterpene (17.35%), diterpene hydrocarbons (0.33%), and oxygenated diterpenes (2.99%). Five terpene classes were assigned from the female plant including monoterpene hydrocarbons (0.51%), oxygenated monoterpenes (7.03%), sesquiterpene hydrocarbons (1.08%), oxygenated sesquiterpenes (3.92%), and oxygenated diterpenes (0.14%) (Figure 2). The present data are in harmony with the data published for Omani *O. arabicus* [28].

Carvone (7.80%), α-bisabolol (5.77%), calarene epoxide (3.79%), isopulegol (3.49%) (Figure 3), *trans*-chrysanthenyl acetate (3.02), and phytol (2.99%) were found as the fundamental components of the VO of the male plant. Furthermore, carvone (3.01%), calarene epoxide (1.68%), and α-bisabolol (1.11%) were the major constituents of the female plants' VO. The diversity of terpenoids in the current findings was in harmony with previously reported data of VO derived from *O. arabicus* collected from Oman [28]. However, the composition of the VO profile is different, i.e., the major compounds of the current study varied from those reported for the Omani ecospecies. Carvone as the main compound of the current study was reported as the main constituent of several plants such as *Tanacetum balsamita* L. [33], *Mentha longifolia* (L.) Huds., and *M. spicata* L. [34]. Furthermore, the major compound in the present study, α-bisabolol, was a common major compound in EOs of numerous plants such as *Matricaria chamomilla* L., *Salvia runcinata* L.f., *Smyrniopsis aucheri* Boiss., *Eremanthus erythropappus* (DC.) MacLeish, and other *Vanillosmopsis* Sch.Bip. species [35].

Finally, traces of the carotenoid-derived compounds, represented by two compounds, theaspirane B and α-ionone, were identified in the VOs of both male and female *O. arabicus*.

In addition, one hydrocarbon compound, *n*-tricosane (0.17%), was identified in VOs of the male plants while two hydrocarbons, *n*-docosane (0.08) and *n*-tricosane (0.16%), were assigned from female plants. The tricosane was reported in trace amounts (0.53%) in the leaf VOs of the *O. arabicus* collected from Oman [28].

The significant variation in the chemical constituents between the current results and previous data might be ascribed to variations in organs, genotypes, ages, climate, weather, humidity, and environmental conditions [36–38]. Moreover, our data revealed the role of the gender of the plant on the phytochemical compositions including the VOs [39,40]. The richness of the compounds in the male plants compared to the female ones could be attributed to the fact that female plants invest less in chemical defense and more into biomass production than male plants [41], where the VOs can be considered a good indicator for the degree of chemical defense in plants due to its distinct variability with the biotic and abiotic stresses [22,42].

2.2. Chemometric Analysis of the VOs from Male and Female O. arabicus

In order to show the variation between male and female plants of *O. arabicus*, the data of the concentration of all identified compounds in the VOs were subjected to the Principal Components Analysis (PCA), where it revealed a slight variation in the composition (Figure 4). The main components that showed clear segregation were isobutyl isothiocyanate, benzyl isothiocyanate, m-tolyl isothiocyanate, butyl isothiocyanate, farnesyl acetone C, isopulegol, phytol, α -bisabolol, and carvone (Figure 4a). The male plants showed a close correlation with m-tolyl isothiocyanate and calarene epoxide, while female plants showed a close correlation with benzyl isothiocyanate and m-tolyl isothiocyanate (Figure 4c). This variation between male and female plants supports the issue that a plant's gender affects the chemical composition of the secondary metabolites [21,40].

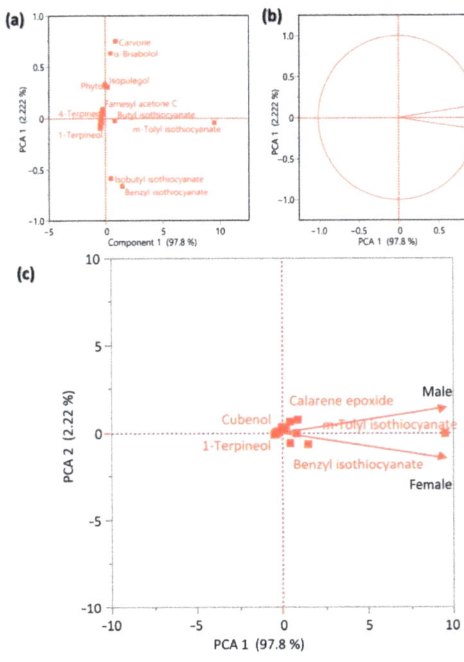

Figure 4. The Principal Components Analysis (PCA) of the identified volatile compounds in male and female *O. arabicus*. (**a**) The observation in the PCA space, (**b**) correlation circle (variables chart), and (**c**) biplot.

2.3. Allelopathic Activity of Male and Female O. arabicus VOs on the Weed D. aegyptium

The VOs extracted from both male and female plants of *O. arabicus* showed significant allelopathic activity against the weed *Dactyloctenium aegyptium* (L.) Willd in a dose-dependent trend (Figure 5). The highest concentration of the VOs (100 μL L^{-1}) showed inhibition of *D. aegyptium* germination by 91.18% and 76.47% for male and female plants, respectively (Figure 5a). The root growth of *D. aegyptium* seedlings was more affected by the VOs compared to the shoot; this could be ascribed to direct contact with the VOs in the medium as well as the permeability of root cells [38,43,44]. The shoot growth of *D. aegyptium* was inhibited by 88.84% and 70.62% after treatment at a concentration of 100 μL L^{-1} of the VOs (Figure 5b). However, at a concentration of 100 μL L^{-1} of the VOs application, the seedling root growth was totally inhibited for both male and female plant extracts, while at a concentration of 75 μL L^{-1}, the root growth was decreased by 96.50% and 92.42% for male and female plants, respectively (Figure 5c).

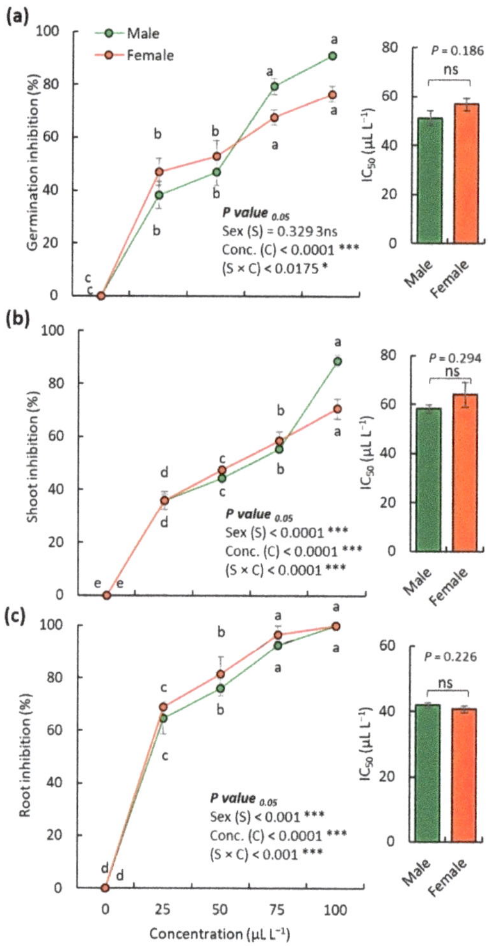

Figure 5. Allelopathic activity of various concentrations of the extracted volatile compounds of male and female *O. arabicus* against the seed germination (**a**), shoot growth (**b**), and root growth (**c**) of *D. aegyptium*. * $p < 0.05$, *** $p < 0.001$, and "ns" for $p > 0.05$.

Based on the IC_{50} values, no significant variation in either the germination or seedling growth of *D. aegyptium* was observed between the male and female plants of *O. arabicus* (Figure 5). The *O. arabicus* VOs of the male plants attained IC_{50} values of 51.10, 58.05, and 63.91 μL L^{-1} for the seed germination, seedling shoot growth, and seedling root growth, respectively, while the female plants showed IC_{50} values of 56.75, 41.93, and 40.71 μL L^{-1}, respectively. These data showed that the variation in the VOs composition (either quality or quantity) has a consequential effect on its biological activities [45]. In this context, the male plants of *Baccharis dracunculifolia* DC. have been reported to produce higher levels of essential oil and phenolic compounds compared to female plants, which leads to greater antioxidant capacity [46]. Moreover, the bioactive compounds (total phenolics, flavonoids, and tannins) and antioxidant activities of *Pistacia atlantica* Desf. were more influenced by growing region than by gender [47].

The major compound, m-tolyl isothiocyanate, in the present study has been reported to possess allelopathic activity against various crops such as wheat, lettuce, cowpea, and barnyard grass [48]. In addition, isothiocyanate is reported as a potential inhibitor of germination and growth in many weedy species such as *Cyperus rotundus* L. and *C. esculentus* L. [49]. Moreover, it seems that plants produce isothiocyanate compounds as a defense strategy, where these compounds showed inhibitory activity against microbial plant pathogens [50] and insects [51]. In the same context, the oxygenated monoterpene carvone (a major compound in the present study) has been described to inhibit the growth of weeds [52].

The slight variation in the allelopathic activities between male and female plants of *O. arabicus* VOs in the present study could be attributed to the variation in chemical composition [45]. It is worth mentioning here that oxygenated compounds have more biological activities compared to non-oxygenated compounds, where active groups/sites in the oxygenated compounds showed a more interactive effect [38,53]. In the present study, the oxygenated compounds in the male plants (43.27%) were higher than in female plants (11.20%), which could explain the higher allelopathic activity of the male *O. arabicus* VOs.

Comparing male and female plants of *O. arabicus*, the female plants have a higher content of isothiocyanates (76.33%) and lower content of oxygenated compounds, compared to the male plants, which have a higher content of oxygenated compounds and lower content of isothiocyanates (44.95%). This situation could explain the comparable allelopathic activity against *D. aegyptium*.

3. Materials and Methods

3.1. Plant Collection and Preparation

Three samples of the aerial parts (aboveground parts) of either male or female plants of *O. arabicus* were separately collected in paper bags from different individuals (n = 10) growing in sandy habitats at Thadiq, 130 km northern Riyadh City (25°12′55.3″ N, 45°54′53.6″ E). The plant specimen was identified by Prof. Dr. Abdulaziz Assaeed (an author) according to flora books [24,54]. A voucher sample was prepared and deposited in the herbarium of the Plant Production Department, College of Food and Agricultural Sciences, King Saud University with ID: KSU-AGRIC-181501001 (Figure 6). The samples (about 2 kg) were air dried in a shaded place at room temperature (25 ± 3 °C) for one week, till complete dryness; crushed into powder with a grinder; then packaged in paper bags and stored in the fridge at 4 °C till further analyses.

3.2. Extraction of the VOs, GC–MS Analysis, and Components Identification

The VOs of the male and female plants of *O. arabicus* were extracted separately from 200 g of the air-dried plant materials. In brief, the two samples were separately subjected to a hydrodistillation process for 3 h via the Clevenger apparatus. The separation of the VOs was performed by *n*-hexane and then dried with 0.5 g anhydrous sodium sulphate. These extractions were applied to the three collected samples of either male or female plants. All the extracted VOs samples were deposited at 4 °C in glass vials till the Gas

Chromatography-Mass Spectrometry (GC-MS) Analysis as well as the biological assays were performed.

Figure 6. *Ochradenus arabicus* Chaudhary, Hillc. & A.G.Mill. female shrub. (**a**) Overview of the female shrub, (**b**) close view of fruiting branches, and (**c**) male plant. Photos by Dr. Abulaziz Assaeed (an author).

The GC-MS analysis of all extracted VOs samples was performed according to previously documented conditions [9,11]. The chemical components' identification and authentication were carried out depending upon AMDIS software (Automated Mass spectral Deconvolution and Identification), NIST database, Wiley spectral collection, and *n*-alkanes (C_8–C_{22}) retention indices.

3.3. Allelopathic Activity Bioassay

To determine the allelopathic activity of the extracted VOs, various concentrations of the VOs were prepared and tested on the germination and seedling growth of the weed *D. aegyptium*. In brief, concentrations of 0, 25, 50, 75, and 100 $\mu L\ L^{-1}$ of the VOs were prepared using 1% of Tween 80® (Sigma-Aldrich, Darmstadt, Germany). The seeds of the targeted weed (*D. aegyptium*) were collected from the infested field and sterilized with 1% sodium hypochlorite for 3 min, followed by washing with distilled water three times, and they were then dried in air and stored in glass vials till further analysis. In Petri plates, 20 seeds of *D. aegyptium* were lined over filter paper (Whatman Grade 1), which was moistened with 5 mL of each concentration. To avoid leakage of the VOs, the Petri plates were sealed with a tape of Parafilm® (Sigma, St. Louis, MO, USA). A control of Tween 80® (1%) was performed with the same procedures as the treatments. A total of 90 plates [5 treatments (4 concentrations + 1 control) × 3 replications × 3 experiment times × gender (male and female plants)] were prepared and then incubated in a growth chamber adjusted at 25 ± 2 °C with a light cycle of 12 h light/12 h dark. The germination of seeds was observed and counted daily, where the seed was counted as germinated when the radicle sprouted with a 2 mm length. After 10 days of treatment, the number of germinated seeds as well as the lengths of seedling radicles and shoots were measured in mm. The inhibition of seed germination or seedling growth was calculated upon the following equation:

$$A = 100 \times \left\{ \frac{No./Length_{control} - No./treatment_{control}}{No./Length_{control}} \right\}$$

The experiment was repeated three times with three replications for each treatment and control, and the average ± standard errors were calculated by MS-EXCEL 2019.

3.4. Statistical Analaysis

The data of the allelopathic activity experiment in triplicates were subjected to two-way ANOVA with gender as the first factor and concentration of the extract as the second factor at a probability level of 0.05. The analysis was followed by Tukey's HSD test using the CoStat software program, version 6.311 (CoHort Software, Monterey, CA, USA). In addition, to test the significant variation between male and female plants, the IC50 values were subjected to a two-tailed *t*-test. On the other hand, the dataset of the concentration (%) of all identified compounds in both VOs of male and female plants of *O. arabicus* was prepared and subjected to principal component analysis (PCA) using JMP® Pro 16.0.0, SAS Institute Inc., Cary, NC, USA.

4. Conclusions

The present study revealed for the first time substantial variations in the chemical profile of the VOs between male and female plants of *O. arabicus*, either in quantity or quality of the chemical compounds. The *m*-tolyl isothiocyanate, benzyl isothiocyanate, butyl isothiocyanate, isobutyl isothiocyanate, carvone, and α-bisabolol were the major constituents in both genders taking into consideration the differences in their relative concentrations. These data varied from those reported for Omani *O. arabicus*, although the gender was not clarified in that study. These findings support that plant gender has a significant effect on secondary metabolites in plants, coupled with environmental, climatic, and genetic factors. The extracted VOs from the two genders were found to exhibit significant allelopathic effects via the suppression of seed germination and shoot and root growth of the weed *D. aegyptium*. Furthermore, a slight difference in allelopathic activity was determined in the present study between male and female plants. This activity could be ascribed to the higher content of isothiocyanates in female plants compared to male, while male plants attained high content of oxygenated terpenes, particularly carvone and α-bisabolol, compared to female.

Author Contributions: Conceptualization, A.M.A.-E., A.M.A. and A.I.E.; formal analysis, A.M.A.-E., A.I.E., B.A.D. and A.E.-N.G.E.G.; investigation, A.M.A.-E., A.M.A. and A.I.E.; writing—original draft preparation, A.M.A.-E. and A.I.E.; writing—review and editing, A.M.A.-E., A.M.A., A.I.E., B.A.D. and A.E.-N.G.E.G. All authors have read and agreed to the published version of the manuscript.

Funding: Deputyship for Research & Innovation, Ministry of Education in Saudi Arabia funded this research work through the project no. (IFKSURG-2-49).

Data Availability Statement: Not applicable.

Acknowledgments: Authors thank the Deputyship for Research & Innovation, Ministry of Education in Saudi Arabia for funding this research work through the project no. (IFKSURG-2-49).

Conflicts of Interest: The authors declare no conflict of interest.

References

1. Reece, J.B.; Urry, L.A.; Cain, M.L.; Wasserman, S.A.; Minorsky, P.V.; Jackson, R.B.; Rawle, F.E.; Durnford, D.G.; Moyes, C.D.; Scott, K. *Campbell Biology, Third Canadian Edition*; Pearson Education Canada: North York, ON, Canada, 2020.
2. El-Shemy, H. *Essential Oils: Oils of Nature*; IntechOpen: London, UK, 2020.
3. Hanif, M.A.; Nisar, S.; Khan, G.S.; Mushtaq, Z.; Zubair, M. Essential Oils. In *Essential Oil Research: Trends in Biosynthesis, Analytics, Industrial Applications and Biotechnological Production*; Malik, S., Ed.; Springer International Publishing: Cham, Switzerland, 2019; pp. 3–17.
4. Rhind, J.; Pirie, D. *Essential Oils: A Handbook for Aromatherapy Practice*; Singing Dragon: London, UK, 2012.
5. El Gendy, A.E.-N.G.; Essa, A.F.; El-Rashedy, A.A.; Elgamal, A.M.; Khalaf, D.D.; Hassan, E.M.; Abd-ElGawad, A.M.; Elgorban, A.M.; Zaghloul, N.S.; Alamery, S.F. Antiviral potentialities of chemical characterized essential oils of *Acacia nilotica* bark and fruits against hepatitis A and herpes simplex viruses: *In vitro, in silico*, and molecular dynamics studies. *Plants* **2022**, *11*, 2889. [CrossRef] [PubMed]
6. Abd-ElGawad, A.M.; El-Amier, Y.A.; Bonanomi, G.; Gendy, A.E.-N.G.E.; Elgorban, A.M.; Alamery, S.F.; Elshamy, A.I. Chemical composition of *Kickxia aegyptiaca* essential oil and its potential antioxidant and antimicrobial activities. *Plants* **2022**, *11*, 594. [CrossRef] [PubMed]
7. Angelini, P.; Tirillini, B.; Akhtar, M.S.; Dimitriu, L.; Bricchi, E.; Bertuzzi, G.; Venanzoni, R. Essential Oil with Anticancer Activity: An Overview. In *Anticancer Plants: Natural Products and Biotechnological Implements*; Akhtar, M.S., Swamy, M.K., Eds.; Springer: Singapore, 2018; Volume 2, pp. 207–231.
8. Elgamal, A.M.; Ahmed, R.F.; Abd-ElGawad, A.M.; El Gendy, A.E.-N.G.; Elshamy, A.I.; Nassar, M.I. Chemical profiles, anticancer, and anti-aging activities of essential oils of *Pluchea dioscoridis* (L.) DC. and *Erigeron bonariensis* L. *Plants* **2021**, *10*, 667. [CrossRef] [PubMed]
9. Elshamy, A.I.; Ammar, N.M.; Hassan, H.A.; Al-Rowaily, S.L.; Ragab, T.I.; El Gendy, A.E.-N.G.; Abd-ElGawad, A.M. Essential oil and its nanoemulsion of *Araucaria heterophylla* resin: Chemical characterization, anti-inflammatory, and antipyretic activities. *Ind. Crops Prod.* **2020**, *148*, 112272. [CrossRef]
10. Abd-ElGawad, A.M.; Elgamal, A.M.; Ei-Amier, Y.A.; Mohamed, T.A.; El Gendy, A.E.-N.G.; Elshamy, A.I. Chemical composition, allelopathic, antioxidant, and anti-inflammatory activities of sesquiterpenes rich essential oil of *Cleome amblyocarpa* barratte & murb. *Plants* **2021**, *10*, 1294.
11. Abdelhameed, M.F.; Asaad, G.F.; Ragab, T.I.; Ahmed, R.F.; El Gendy, A.E.-N.G.; El-Rahman, A.; Sahar, S.; Elgamal, A.M.; Elshamy, A.I. Oral and topical anti-inflammatory and antipyretic potentialities of *Araucaria bidiwillii* shoot essential oil and its nanoemulsion in relation to chemical composition. *Molecules* **2021**, *26*, 5833. [CrossRef]
12. Ammar, N.M.; Hassan, H.A.; Ahmed, R.F.; El-Gendy, A.E.-N.G.; Abd-ElGawad, A.M.; Farrag, A.R.H.; Farag, M.A.; Elshamy, A.I.; Afifi, S.M. Gastro-protective effect of *Artemisia sieberi* essential oil against ethanol-induced ulcer in rats as revealed via biochemical, histopathological and metabolomics analysis. *Biomarkers* **2022**, *27*, 247–257. [CrossRef]
13. Boira, H.; Blanquer, A. Environmental factors affecting chemical variability of essential oils in *Thymus piperella* L. *Biochem. Syst. Ecol.* **1998**, *26*, 811–822. [CrossRef]
14. Abd-ElGawad, A.M.; Elshamy, A.I.; El-Amier, Y.A.; El Gendy, A.E.-N.G.; Al-Barati, S.A.; Dar, B.A.; Al-Rowaily, S.L.; Assaeed, A.M. Chemical composition variations, allelopathic, and antioxidant activities of *Symphyotrichum squamatum* (Spreng.) Nesom essential oils growing in heterogeneous habitats. *Arab. J. Chem.* **2020**, *13*, 4237–4245. [CrossRef]
15. Figueiredo, A.C.; Barroso, J.G.; Pedro, L.G.; Scheffer, J.J. Factors affecting secondary metabolite production in plants: Volatile components and essential oils. *Flavour Fragr. J.* **2008**, *23*, 213–226. [CrossRef]
16. Mehalaine, S.; Chenchouni, H. Quantifying how climatic factors influence essential oil yield in wild-growing plants. *Arab. J. Geosci.* **2021**, *14*, 1257. [CrossRef]
17. Li, Y.; Kong, D.; Fu, Y.; Sussman, M.R.; Wu, H. The effect of developmental and environmental factors on secondary metabolites in medicinal plants. *Plant Physiol. Biochem.* **2020**, *148*, 80–89. [CrossRef] [PubMed]
18. Barra, A. Factors affecting chemical variability of essential oils: A review of recent developments. *Nat. Prod. Commun.* **2009**, *4*, 1147–1154. [CrossRef] [PubMed]
19. Gupta, A.; Mishra, R.; Singh, A.; Srivastava, A.; Lal, R. Genetic variability and correlations of essential oil yield with agro-economic traits in *Mentha* species and identification of promising cultivars. *Ind. Crops Prod.* **2017**, *95*, 726–732. [CrossRef]
20. Hassan, E.M.; El Gendy, A.E.-N.G.; Abd-ElGawad, A.M.; Elshamy, A.I.; Farag, M.A.; Alamery, S.F.; Omer, E.A. Comparative chemical profiles of the essential oils from different varieties of *Psidium guajava* L. *Molecules* **2020**, *26*, 119. [CrossRef] [PubMed]
21. Demirbolat, I.; Karik, Ü.; Erçin, E.; Kartal, M. Gender dependent differences in composition, antioxidant and antimicrobial activities of wild and cultivated *Laurus nobilis* L. Leaf and flower essential oils from Aegean region of Turkey. *J. Essent. Oil Bear. Plants* **2020**, *23*, 1084–1094. [CrossRef]
22. Markó, G.; Németh, I.; Gyuricza, V.; Altbäcker, V. Sex-specific differences in *Juniperus communis*: Essential oil yield, growth-defence conflict and population sex ratio. *AoB PLANTS* **2021**, *13*, plab021. [CrossRef] [PubMed]
23. Miller, A.G. Revision of *Ochradenus*. *Notes-R. Bot. Gard. Edinb.* **1984**, *41*, 491–504.
24. Collenette, S. *Wildflowers of Saudi Arabia*; National Commission for Wildlife Conservation and Development (NCWCD): Riyadh, Saudi Arabia, 1999.

25. Hussain, J.; Rehman, N.U.; Khan, A.L.; Ali, L.; Kim, J.-S.; Zakarova, A.; Al-Harrasi, A.; Shinwari, Z.K. Phytochemical and biological assessment of medicinally important plant *Ochradenus arabicus*. *Pak. J. Bot.* **2014**, *46*, 2027–2034.
26. Ali, M.A.; Farah, M.A.; Al-Hemaid, F.M.; Abou-Tarboush, F.M.; Al-Anazi, K.M.; Wabaidur, S.; Alothman, Z.; Lee, J. Assessment of biological activity and UPLC–MS based chromatographic profiling of ethanolic extract of *Ochradenus arabicus*. *Saudi J. Biol. Sci.* **2016**, *23*, 229–236. [CrossRef]
27. Alanazi, K.M.; Al-Kawmani, A.A.; Farah, M.A.; Hailan, W.A.; Alsalme, A.; Al-Zaqri, N.; Ali, M.A.; Almansob, A.M. Amelioration of indole acetic acid-induced cytotoxicity in mice using zinc nanoparticles biosynthesized with *Ochradenus arabicus* leaf extract. *Saudi J. Biol. Sci.* **2021**, *28*, 7190–7201. [CrossRef]
28. Ullah, O.; Shah, M.; Rehman, N.U.; Ullah, S.; Al-Sabahi, J.N.; Alam, T.; Khan, A.; Khan, N.A.; Rafiq, N.; Bilal, S.; et al. Aroma profile and biological effects of *Ochradenus arabicus* essential oils: A comparative study of stem, flowers, and leaves. *Molecules* **2022**, *27*, 5197. [CrossRef] [PubMed]
29. Chen, C.-W.; Ho, C.-T. Thermal degradation of allyl isothiocyanate in aqueous solution. *J. Agric. Food Chem.* **1998**, *46*, 220–223. [CrossRef] [PubMed]
30. De Nicola, G.R.; Montaut, S.; Rollin, P.; Nyegue, M.; Menut, C.; Iori, R.; Tatibouët, A. Stability of benzylic-type isothiocyanates in hydrodistillation-mimicking conditions. *J. Agric. Food Chem.* **2013**, *61*, 137–142. [CrossRef] [PubMed]
31. Kumagai, H.; Kashima, N.; Seki, T.; Sakurai, H.; Ishii, K.; Ariga, T. Analysis of volatile components in essential oil of upland wasabi and their inhibitory effects on platelet aggregation. *Biosci. Botechnology Biochem.* **1994**, *58*, 2131–2135. [CrossRef]
32. Kyriakou, S.; Trafalis, D.T.; Deligiorgi, M.V.; Franco, R.; Pappa, A.; Panayiotidis, M.I. Assessment of methodological pipelines for the determination of isothiocyanates derived from natural sources. *Antioxidants* **2022**, *11*, 642. [CrossRef]
33. Başer, K.H.C.; Demirci, B.; Tabanca, N.; Özek, T.; Gören, N. Composition of the essential oils of *Tanacetum armenum* (DC.) Schultz Bip., *T. balsamita* L., *T. chiliophyllum* (Fisch. & Mey.) Schultz Bip. var. *chiliophyllum* and *T. haradjani* (Rech. fil.) Grierson and the enantiomeric distribution of camphor and carvone. *Flavour Fragr. J.* **2001**, *16*, 195–200.
34. Younis, Y.M.; Beshir, S.M. Carvone-rich essential oils from *Mentha longifolia* (L.) Huds. ssp. *schimperi* Briq. and *Mentha spicata* L. grown in Sudan. *J. Essent. Oil Res.* **2004**, *16*, 539–541. [CrossRef]
35. Kamatou, G.P.; Viljoen, A.M. A review of the application and pharmacological properties of α-bisabolol and α-bisabolol-rich oils. *J. Am. Oil Chem. Soc.* **2010**, *87*, 1–7. [CrossRef]
36. Al-Rowaily, S.L.; Abd-ElGawad, A.M.; Assaeed, A.M.; Elgamal, A.M.; Gendy, A.E.-N.G.E.; Mohamed, T.A.; Dar, B.A.; Mohamed, T.K.; Elshamy, A.I. Essential oil of *Calotropis procera*: Comparative chemical profiles, antimicrobial activity, and allelopathic potential on weeds. *Molecules* **2020**, *25*, 5203. [CrossRef]
37. Tsusaka, T.; Makino, B.; Ohsawa, R.; Ezura, H. Genetic and environmental factors influencing the contents of essential oil compounds in *Atractylodes lancea*. *PLoS ONE* **2019**, *14*, e0217522. [CrossRef] [PubMed]
38. Abd-ElGawad, A.M.; El Gendy, A.E.-N.G.; Assaeed, A.M.; Al-Rowaily, S.L.; Alharthi, A.S.; Mohamed, T.A.; Nassar, M.I.; Dewir, Y.H.; Elshamy, A.I. Phytotoxic effects of plant essential oils: A systematic review and structure-activity relationship based on chemometric analyses. *Plants* **2020**, *10*, 36. [CrossRef] [PubMed]
39. Faughnan, M.S.; Hawdon, A.; Ah-Singh, E.; Brown, J.; Millward, D.; Cassidy, A. Urinary isoflavone kinetics: The effect of age, gender, food matrix and chemical composition. *Br. J. Nutr.* **2004**, *91*, 567–574. [CrossRef] [PubMed]
40. Struiving, S.; Hacke, A.C.M.; Simionatto, E.L.; Scharf, D.R.; Klimaczewski, C.V.; Besten, M.A.; Heiden, G.; Boligon, A.A.; Rocha, J.B.T.; Vellosa, J.C.R. Effects of gender and geographical origin on the chemical composition and antiradical activity of *Baccharis myriocephala* and *Baccharis trimera*. *Foods* **2020**, *9*, 1433. [CrossRef]
41. Litto, M.; Scopece, G.; Fineschi, S.; Schiestl, F.P.; Cozzolino, S. Herbivory affects male and female reproductive success differently in dioecious *Silene latifolia*. *Entomol. Exp. Appl.* **2015**, *157*, 60–67. [CrossRef]
42. Phillips, M.A.; Croteau, R.B. Resin-based defenses in conifers. *Trends Plant Sci.* **1999**, *4*, 184–190. [CrossRef]
43. Abd El-Gawad, A.M. Chemical constituents, antioxidant and potential allelopathic effect of the essential oil from the aerial parts of *Cullen plicata*. *Ind. Crops Prod.* **2016**, *80*, 36–41. [CrossRef]
44. Elshamy, A.; Abd-ElGawad, A.; Mohamed, T.; El Gendy, A.E.N.; Abd El Aty, A.A.; Saleh, I.; Moustafa, M.F.; Hussien, T.A.; Pare, P.W.; Hegazy, M.E.F. Extraction development for antimicrobial and phytotoxic essential oils from Asteraceae species: *Achillea fragrantissima*, *Artemisia judaica* and *Tanacetum sinaicum*. *Flavour Fragr. J.* **2021**, *36*, 352–364. [CrossRef]
45. Ketnawa, S.; Reginio, F.C., Jr.; Thuengtung, S.; Ogawa, Y. Changes in bioactive compounds and antioxidant activity of plant-based foods by gastrointestinal digestion: A review. *Crit. Rev. Food Sci.* **2022**, *62*, 4684–4705. [CrossRef]
46. Tomazzoli, M.M. *Phytochemical Evaluation, Antioxidant Activity and Propagation Methods of Baccharis dracunculifolia DC: The Main Botanical Source of Brazilian Green Propolis*; Universidade Federal Do Paraná: Curitiba, Brazil, 2020.
47. Ben Ahmed, Z.; Yousfi, M.; Viaene, J.; Dejaegher, B.; Demeyer, K.; Mangelings, D.; Vander Heyden, Y. Seasonal, gender and regional variations in total phenolic, flavonoid, and condensed tannins contents and in antioxidant properties from *Pistacia atlantica* ssp. leaves. *Pharm. Biol.* **2017**, *55*, 1185–1194. [CrossRef]
48. Bialy, Z.; Oleszek, W.; Lewis, J.; Fenwick, G. Allelopathic potential of glucosinolates (mustard oil glycosides) and their degradation products against wheat. *Plant Soil* **1990**, *129*, 277–281. [CrossRef]
49. Norsworthy, J.K.; Malik, M.S.; Jha, P.; Oliveira, M.J. Effects of isothiocyanates on purple (*Cyperus rotundus* L.) and yellow nutsedge (*Cyperus esculentus* L.). *Weed Biol. Manag.* **2006**, *6*, 131–138. [CrossRef]

50. Smolinska, U.; Morra, M.; Knudsen, G.; James, R. Isothiocyanates produced by Brassicaceae species as inhibitors of *Fusarium oxysporum*. *Plant Dis.* **2003**, *87*, 407–412. [CrossRef] [PubMed]
51. Du, Y.; Grodowitz, M.J.; Chen, J. Insecticidal and enzyme inhibitory activities of isothiocyanates against red imported fire ants, *Solenopsis invicta*. *Biomolecules* **2020**, *10*, 716. [CrossRef]
52. Razavi, S.M.; Badihi, M.; Nasrollahi, P. Inhibitory potential of (-)-carvone and carvone-PLGA composite on plant pathogens and common weeds. *Arch. Phytopathol. Plant Prot.* **2022**, *55*, 926–936. [CrossRef]
53. Abd-ElGawad, A.M.; El Gendy, A.E.-N.G.; Assaeed, A.M.; Al-Rowaily, S.L.; Omer, E.A.; Dar, B.A.; Al-Taisan, W.'a.A.; Elshamy, A.I. Essential oil enriched with oxygenated constituents from invasive plant *Argemone ochroleuca* exhibited potent phytotoxic effects. *Plants* **2020**, *9*, 998. [CrossRef]
54. Chaudhary, S.A. *Flora of the Kingdom of Saudi Arabia*; Ministry of Agriculture and Water: Riyadh, Saudi Arabia, 1999; Volume 1.

plants

MDPI

Article

Composition, Antibacterial Efficacy, and Anticancer Activity of Essential Oil Extracted from *Psidium guajava* (L.) Leaves

Aftab Alam [1,*], Talha Jawaid [2], Saud M. Alsanad [2], Mehnaz Kamal [3] and Mohamed F. Balaha [4,5]

1 Department of Pharmacognosy, College of Pharmacy, Prince Sattam Bin Abdulaziz University, Al-Kharj 11942, Saudi Arabia
2 Department of Pharmacology, College of Medicine, Imam Mohammad Ibn Saud Islamic University (IMSIU), Riyadh 13317, Saudi Arabia
3 Department of Pharmaceutical Chemistry, College of Pharmacy, Prince Sattam Bin Abdulaziz University, Al-Kharj 11942, Saudi Arabia
4 Department of Clinical Pharmacy, College of Pharmacy, Prince Sattam Bin Abdulaziz University, Al-Kharj 11942, Saudi Arabia
5 Pharmacology Department, Faculty of Medicine, Tanta University, Tanta 31527, Egypt
* Correspondence: a.alam@psau.edu.sa

Abstract: Essential oils (EO) are used as a natural remedy to treat various chronic diseases, although clinical evidence is lacking. In this context, we have endeavored to measure the percentage of chemical composition and biological efficacy of *Psidium guajava* (guava) leaf essential oil in treating oral infections and oral cancer. The essential oil obtained from hydrodistillation of *P. guajava* L. leaves was analyzed by gas chromatography–mass spectrometry (GC–MS). The activities of selected oral pathogens *Candida albicans* (*C. albicans*) and *Streptococcus mutans* (*S. mutants*) were studied in vitro and in silico. MTT assay was used to test for anticancer activity against human oral epidermal carcinoma (KB). GC–MS showed that the main components of PGLEO were limonene (38.01%) and β-caryophyllene (27.98%). Minimum inhibitory concentrations (MICs) of 0.05–0.1% were demonstrated against *C. albicans* and *S. mutans*. Antimicrobial activity against *C. albicans* and *S. mutans*, as shown by molecular linkage analysis, revealed that the main metabolites, limonene and β-caryophyllene, potentially inhibited the receptors of *C. albicans* and *S. mutans*. PGLEO showed significant ($p < 0.001$) anticancer activity (45.89%) at 200 μg/mL compared to doxorubicin (47.87%) with an IC_{50} value of 188.98 μg/mL. The outcomes of the present study suggest that PGLEO has promising antimicrobial and anticancer activities and could be a useful source for developing a natural therapeutic agent for oral infections and oral cancer.

Keywords: *Psidium guajava*; essential oil; leaves; oral infection; molecular coupling; cytotoxic effects

Citation: Alam, A.; Jawaid, T.; Alsanad, S.M.; Kamal, M.; Balaha, M.F. Composition, Antibacterial Efficacy, and Anticancer Activity of Essential Oil Extracted from *Psidium guajava* (L.) Leaves. *Plants* **2023**, *12*, 246. https://doi.org/10.3390/plants12020246

Academic Editors: Hazem Salaheldin Elshafie, Ippolito Camele and Adriano Sofo

Received: 5 November 2022
Revised: 12 December 2022
Accepted: 23 December 2022
Published: 5 January 2023

1. Introduction

Psidium guajava L. (*P. guajava*), a tropical and subtropical plant of the Myrtaceae family, is commonly called guava and is used for nutritional and medicinal purposes. In traditional medicine, the fruits, leaves, and juice of guava are commonly used to treat oral and dental infections, pain, and other diseases [1]. The leaves of this plant are traditionally used in the African and Asian continents to treat various infectious diseases [2]. The aqueous leaf extract is effective against various Gram-positive/Gram-negative bacteria, fungi, viruses, and protozoa [3].

The use of *P. guajava* is found in various traditional systems of medicine, including the Ayurvedic system of medicine of India, the traditional Chinese system of medicine, and the folk systems of medicine of the West Indies and Latin America. Guava leaves have been studied for various activities, including antioxidant, antidiabetic, antihyperlipidemic, antidiarrheal, cardioprotective, analgesic, nephrotoxic, and antimicrobial effects [4]. Toxicity studies have shown that *P. guajava* is safe to use [5]. The leaf extract of guava has been reported for its anticancer and antibacterial activities [6]. Recently, several terpenoid compounds have been isolated from the leaf extract of guava and evaluated. Among the isolated compounds, several meroterpenoids showed anticancer and antifungal activities [7,8]. The essential oil constituents depend on the chemical composition, geographical conditions, growing conditions, and collection time of the mother plant and its parts [9,10].

Guava leaves from different regions have different volatile compositions; for example, Chinese guava contains β-caryophyllene, copaen, azulene, eucalyptol, etc. [11]; Brazilian guava contains β-caryophyllene, β-elemene, β-selinol, and α-humine [12]; and Egyptian guava contains β-caryophyllene, trans-nerolidol, globulol, and D-limonene [13]. Recently, Indian guava was shown to contain limonene, caryophyllene, caryophyllene oxide, etc. [14]. In general, terpenoids, such as limonene, β-caryophyllene, 1,8-cineole, β-elemene, etc., may be responsible for antimicrobial and anticancer activities [15,16].

The extensive use of conventional antimicrobials in oral infections has led to antimicrobial resistance. The rate of antimicrobial resistance is higher than that of new drug development, which poses a medical challenge [17,18]. Guava leaves are commonly used to treat mouth ulcers in southern India, while in northern India, raw leaves and tender shoots of guava are used to treat toothache and mouth ulcers [18]. Traditional Cameroonians use guava leaves to treat dental infections, and guava twigs are chewed to relieve dental problems [19]. Extracts of guava leaves with various solvents have shown antibacterial activity against oral pathogens and dental plaque [20]. Various microorganisms are present in the oral cavity; *Streptococcus mutans* and *Candida albicans* play a significant role in developing oral cavity, caries, and periodontal infection [21]. Evaluating molecular linkage studies is an effective tool for drug design and development [20–22]. Human epithelial carcinoma cells (KB cell lines) are squamous cell carcinoma (SCC). They are the most common oral cavity cancer responsible for morbidity and mortality [23]. In a previous study, essential oil from 17 Thai medicinal plants was investigated against KB cell lines, including oil from guava leaves. It was reported that oil from guava leaves had the highest antiproliferative values, more than sweet basil and vincristine [24].

Considering the enormous amount of research on the antimicrobial and anticancer effects of guava essential oil, this study aimed to investigate the role of PGLEO in protecting against oral infections and oral cancer. In addition, molecular linkage studies against *S. mutans* and *C. albicans* receptors were performed to validate the mechanism of antimicrobial activity of the main compounds analyzed in PGLEO. In addition, the anticancer effect on human oral epidermal carcinoma cell lines (KB) was investigated using an MTT assay to determine the role of PGLEO in protecting against oral cancer and bacterial infections.

2. Results

2.1. Characterization of Essential Oil Composition through Gas Chromatography–Mass Spectrometry (GC–MS)

The *Psidium guajava* leaves produced 0.58% *w*/*w* of essential oil. Figure 1 shows the gas chromatography–mass spectrometry chromatogram of PGLEO. A total of 34 volatile metabolites were identified and are listed in Table 1. Among the volatile essential oils, the total concentration of monoterpenes was 40.37%, of which monoterpene hydrocarbons and oxygenated monoterpenes accounted for about 39.87% and 0.5%, respectively.

However, the total concentration of sesquiterpenes was 57.65%, of which about 49.54% were sesquiterpene hydrocarbons and 8.11% were oxygenated sesquiterpenes. Six main volatile constituents (<3%) were detected in the present sample, including five sesquiterpenes (β-caryophyllene (27.98%), copaene (6.25%), cubenene (4.80%), nerolidol (4.49%) and α-Caryophyllene (4.21%)) and one monoterpene (D-limonene (38.01%)). These compounds accounted for about 85.81% of the total percentage of volatile compounds.

Table 1. Volatile composition of the PGLEO by GC–MS analysis.

Volatile Composition	Types	RI (Lit)	RI (Obs)	Area %
α-Pinene	MH	949	948	0.71
Myrcene	MH	963	958	0.57
β-Ocimene	MH	986	976	0.58
D-Limonene	MH	1022	1018	38.01
Menthol	OM	1173	1164	0.20
Linalyl acetate	OM	1270	1272	0.30
α-Cubebene	SH	1347	1344	0.98
Copaene	SH	1377	1371	6.25
α-cedrene	SH	1415.1	1403	1.62
β-Caryophyllene	SH	1419	1414	27.98
α-Caryophyllene	SH	1446	1440	4.21
Cadina-3,5-diene	SH	1454	1448	0.23
γ-Muurolene	SH	1475	1465	1.19
β-Selinene	SH	1492	1490	0.19
α-Bisabolene	SH	1511	1500	1.29
Cubinene	SH	1515	1510	4.80
Cadina-1,3,5-triene	SH	1542	1537	0.80
Ledol	OS	1540	1530	0.96
Nerolidol	OS	1565	1564	4.56
Caryophyllene oxide	OS	1580	1570	2.17
α-Muurolol	OS	1611	1580	0.42
Cholestadiene	D	-	2390	0.22
2-Hexadecen-1-ol, 3,7,11,15-tetramethyl	NT	2114	2114	0.69
13-Hexyloxacyclotridec-10-en-2-one	NT	2325	2325	1.08
Monoterpene hydrocarbon (MH)			39.87	
Oxygenated monoterpenes (OM)			0.5	
Sesquiterpenes hydrocarbon (SH)			49.54	
Oxygenated sesquiterpenes (OS)			8.11	
Diterpenes (D)			0.22	
Nonterpene (NT)			1.77	
Total identification (34 components)			100.01%	

RI (Obs): calculated retention index; RI (Lit) = retention indices according to Adams (Adams, 2007).

2.2. Antibacterial and Antifungal Activities against Oral Bacteria

The results of PGLEO antimicrobial assays against *S. mutans* and *C. albicans* are summarized in Table 2. An inhibition zone of 6 mm in diameter was considered a positive result. The zone of inhibition study outcomes showed that PGLEO exhibited significant inhibitory activity against *S. mutans* (18.5 ± 0.7 mm) and *C. albicans* (21.5 ± 0.5 mm) at 10%. The outcomes of minimum inhibitory concentration (MIC: 0.05–0.1%) showed that PGLEO exhibited almost equal activity against both *S. mutans* and *C. albicans*.

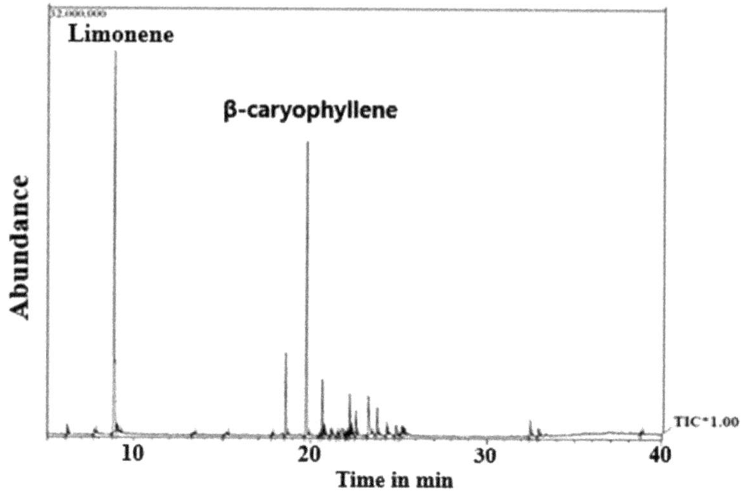

Figure 1. GC–MS chromatogram of PGLEO.

Table 2. Antimicrobial activity of *Psidium guajava*.

Concentration (%)	Zone of Inhibition (mm, Mean ± SD)	
	S. mutans	*C. albicans*
0.25	Less than 6mm	Less than 6mm
0.5	9.5 ± 0.7	10 ± 0.4
7.5	15.5 ± 0.7	16.5 ± 0.7
10	18.5 ± 0.7	21.5 ± 0.5
Minimum inhibitory concentration (MIC)	0.05–0.1%	0.05–0.1%

Values were expressed as mean ± SD (standard deviation) calculation (n = 3).

The cell viability (time-kill) assays were performed using PGLEO, and the results are shown in Figures 2 and 3. The outcomes of the control group (untreated *S. mutans*) exhibited approximately 6 to 9 (log10 CFU/mL) growth. In contrast, the test group (PGLEO treated) showed that the growth of *S. mutans* decreased dramatically in the first 4 to 8 h. It remained constant at approximately 2.5 to 3.5 (log10 CFU/mL) and 1.8 to 2.6 (log10 CFU/mL) with 1XMIC (0.1%) and 2XMIC (0.2%) treatment, respectively (Figure 2). Similarly, untreated *C. albicans* grew at 6 to 8.7 (log10 CFU/mL). After treatment with PGLEO, the growth of *C. albican* decreased dramatically in the first 4 to 8 h. It remained constant at approximately 3 to 3.5 (log10 CFU/mL) and 2 to 2.4 (log10 CFU/mL) for 1XMIC (0.1%) and 2XMIC (0.2%) treatment, respectively (Figure 3). The results indicate that PGLEO exhibits a lethal effect on *S. mutans* and *C. albicans* equally. The plot of both samples measured at the 2XMIC stage differed from that at the 1XMIC stage for both microorganisms. PGLEO showed a rapid killing effect with 2XMIC.

2.3. Molecular Coupling Studies

Two important volatile compounds, D-limonene and β-caryophyllene, were selected for molecular coupling studies. Molecular coupling was performed against antioxidant target enzymes such as *S. mutans* antigen I/ II (PDB: 1JMM) and antifungal target enzymes such as N-myristoyl transferase receptors (PDB: 1IYL) to identify critical ligand–protein interactions. Table 3 shows the results of binding energy, inhibition constant, and amino acid residues.

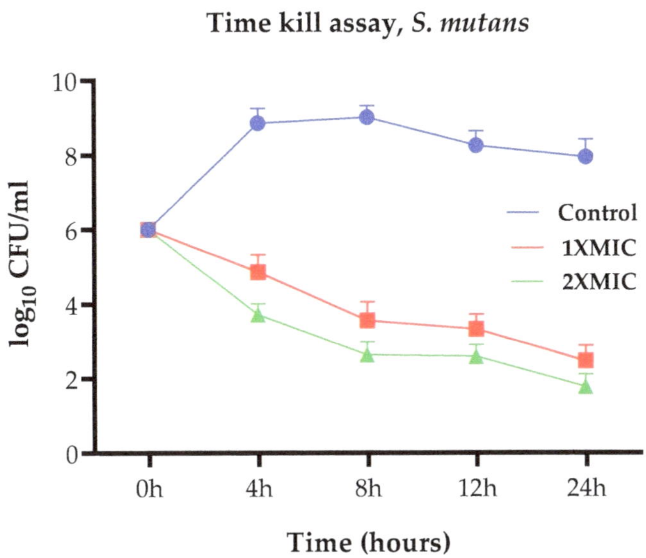

Figure 2. Time-kill assays of *S. mutans*.

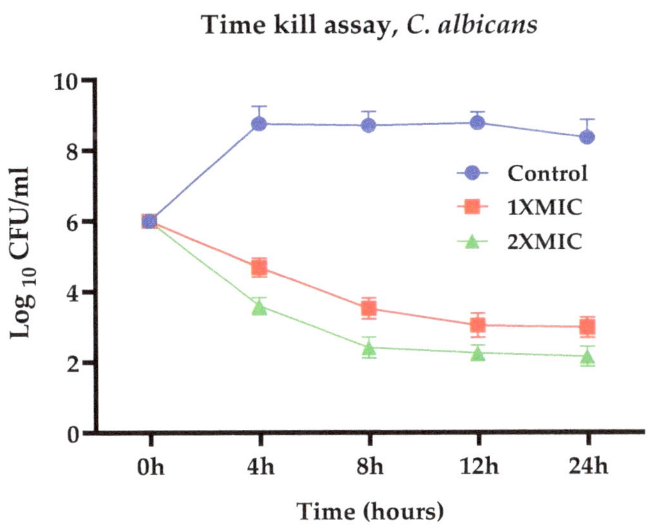

Figure 3. Time-kill assays of *C. albicans*.

Table 3. Binding energy (ΔG; kcal/mol) and inhibition constant (Ki; μM) for limonene (LIM) and caryophyllene (CRP) with target proteins.

Targets (PDB)	ΔG		Ki		Residues of Amino Acid	
	LIM	CRP	LIM	CRP	D-Limonene (LIM)	β-Caryophyllene (CRP)
1JMM	−7.42	−7.39	785.97	876.32	PHE A:192, ILE A:345, ALA A:319	ILE A:345, ALA A:319
1IYL	−8.03	−8.21	752.10	675.29	TYR A:287, TYR A:158, PHE A:50, PHE A:272, PHE A:48, PHE A:173	PHE A:173, HIS A:160, TRP A:165

Figure 4 presents the ligand interaction of D-limonene (1-2) and β-caryophyllene (3-4) with proteins. The primary metabolite, limonene, displayed a high binding affinity to *Candida* receptors 1IYL with docking scores of 8.03, while the second major metabolite, β-caryophyllene, demonstrated higher docking scores of 8.21.

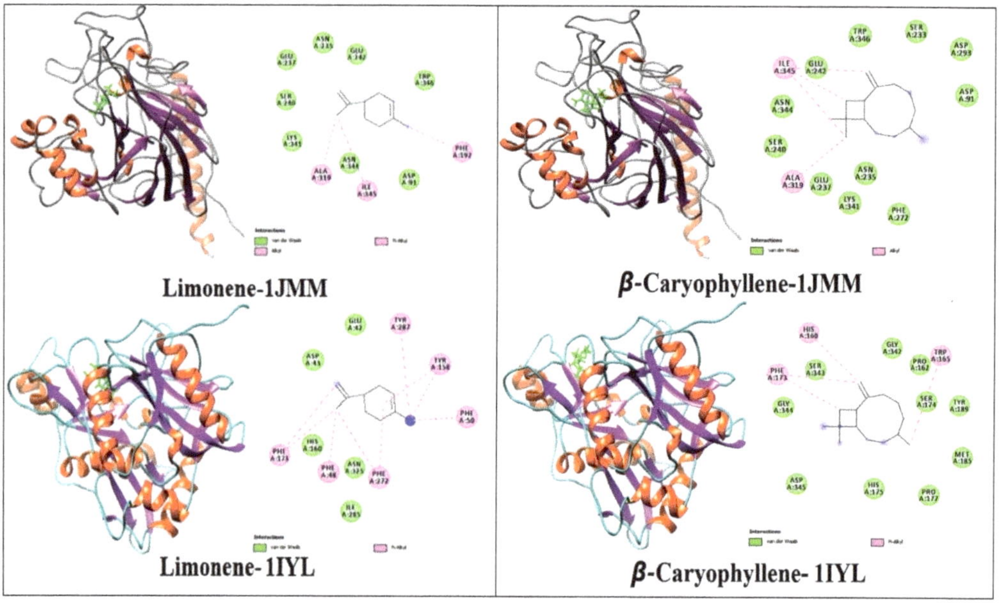

Figure 4. Interaction of the 1JMM and 1IYL protein with the ligand D-limonene and β-caryophyllene.

S. mutans receptor 1JMM and *Candida* receptor 1IYL showed a binding energy of 7.42 and −7.39 kcal/mol against limonene and β-caryophyllene, respectively. A similar trend in linkage values was observed for both limonene and β-caryophyllene. The Ki range of limonene and β-caryophyllene was highest against the Candida target 1IYL (640.45 μM) and moderate against the *S. mutans* target (752.10 μM). The Ki value against the *S. mutans* target (1JMM) was lower than against 1IYL.

2.4. Inhibition of KB Cells from the Growth of Oral Cancer by PGLEO

Figure 5 shows the effects of PGLEO treatment at 25 to 200 μg/mL on cell viability measured by MTT assay from KB. PGLEO treatment significantly suppressed production of KB cells after 24 h of treatment (Figure 5A). In this study, the IC_{50} value of PGLEO was measured using a linear regression equation, i.e., Y = Mx + C. Here, Y = 50 and M and C values were derived from the viability graph. The survival of KB cells was significantly ($p < 0.001$) reduced in a dose-dependent manner with an IC_{50} value of 188.43 ± 8.26 μg/mL (Figure 5a). PGLEO treatment resulted in irregular morphology of KB cells, which shrank and exhibited a round shape, and reduced the viability of KB cells (Figure 5b–f).

The MTT assay is a colorimetric assay for determining cell proliferation and cytotoxicity based on reducing the yellow-colored water-soluble tetrazolium dye MTT to formazan crystals. Mitochondrial lactate dehydrogenase produced by living cells reduces MTT to insoluble formazan crystals. After dissolution in a suitable solvent, they show a purple color; the intensity is proportional to the number of viable cells and can be measured spectrophotometrically at 570 nm. The observations in the statistical data of the cell cytotoxicity study by MTT indicated that PGLEO had significant cytotoxic potential against KB cell lines and test compounds.

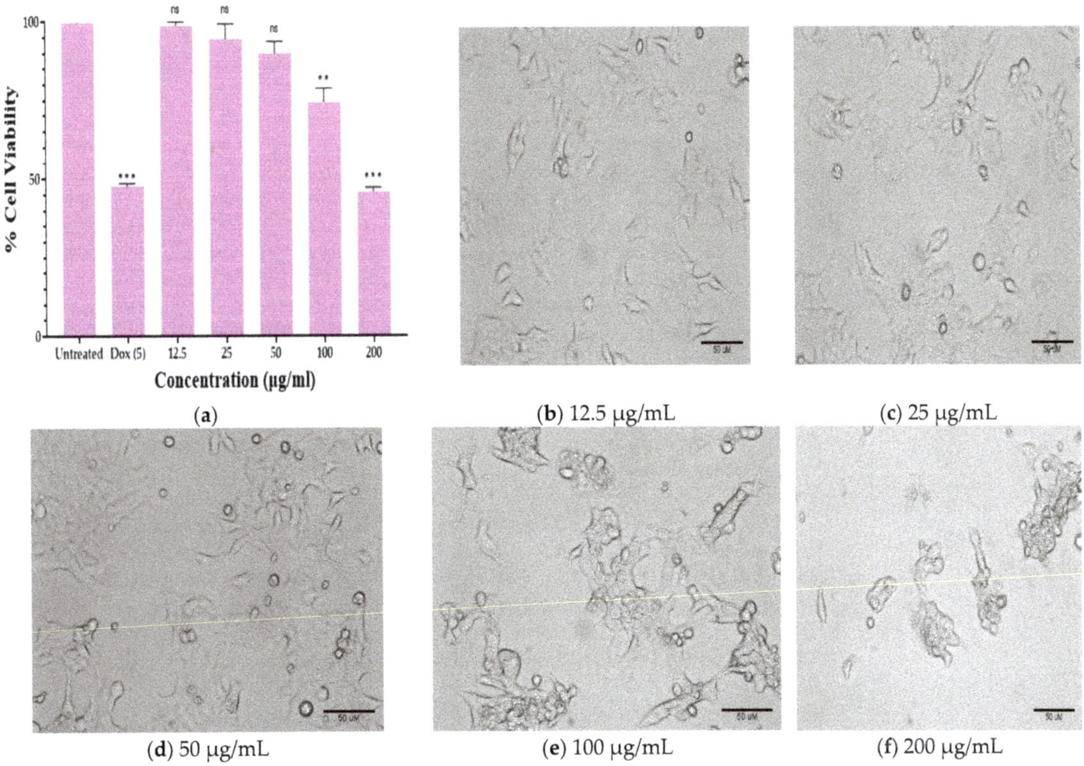

Figure 5. (**a**) Effect of PGLEO on cell viability and morphological characteristics; ** $p < 0.05$ and *** $p < 0.001$ compared to control. (**b–f**) KB cells treated with PGLEO at various concentrations (12.5 to 200 µg/mL) for 24 h.

3. Discussion

The percentage yield of PGLEO was similar to a previous study with qualitatively different volatile compositions [25,26]. The major constituents of essential oil vary depending on the geographical location, cultivar, and cultivar of the parent plant. Several studies have been conducted on the essential oil of North Indian guava leaves. The results showed similar proportions of sesquiterpenes and monoterpenes but differences in the quality and quantity of constituents [26]. Soliman et al. (2016) reported a higher proportion of monoterpenes than sesquiterpenes; in contrast to the present study, sesquiterpenes were identified as the main constituent [27]. Has-san et al. studied the PGLEO of six cultivars from the same farms in Egypt. They showed that the Red Malaysian and White Indian cultivars contained limonene as the significant volatile compound. In contrast, El-Qanater, Early, El-Sabahya El-Gedida, and Red Indian had β-caryophyllene as the significant volatile compound [13].

Several studies have reported that caryophyllene is an essential component of PGLEO. Weli et al. studied indigenous PGLEO from the Sultanate of Oman (October 2012) and found that caryophyllene, viridiflorene, and farnesene were the major volatile compounds [28,29]. Chen and Yen (2007) studied the PGLEO of cultivars from Taiwan and found that β-caryophyllene, α-pinene, and 1,8-cineole were the major volatile compounds [30]. Satyal et al. (2015) investigated the essential oil composition of guava leaves from Kathmandu, Nepal. They reported that (E)-nerolidol and (E)-caryophyllene were the major compounds, which is in agreement with the results of the present study [31]. Silva et al. (2019) investigated the essential oil composition of guava leaves from Rio Verde Go, Brazil. They reported antibacterial activity against *S. mutans* that was similar to the present study, but the major volatile compositions of β-caryophyllene, α-humulene, aromadendrene oxide, δ-selinene, and selin-11-en-4α-ol were different [32].

Chalannavar et al. (2014) studied the essential oil composition of guava leaves from South Africa. They reported that sesquiterpenes, caryophyllene oxide, and caryophyllene were the main constituents, in contrast to the present study [33]. In addition, some studies have reported that limonene is the main constituent of PGLEO [34]. Chaturvedi et al. recently analyzed the composition of Indian guava essential oil. They reported that the quality and quantity of limonene, (E)-caryophyllene, and caryophyllene oxide were relatively different from those observed in the present study [14]. Sacchetti et al. studied the essential oil composition of guavas from the Philippines and found that limonene, α-pinene, β-caryophyllene, and longicyclene were the most significant compounds [35]. The highest contents of the monoterpenes limonene and α-pinene were found in PGLEO from Ecuador. In contrast, Tunisian guava leaf oil was found to have the highest content of the sesquiterpenes vi-ridiflorol and trans-caryophyllene [36]. Most studies have reported that caryophyllene is one of the main components of guava essential oil; the fragrance of guava leaves could be due to the combined effect of β-caryophyllene and other compounds [14,16,37].

Diseases found to have associations with the oral cavity, especially dental diseases but also diseases such as cardiovascular disease, diabetes, rheumatoid arthritis, pneumonia, and Alzheimer's disease, are the main causes of high morbidity and mortality [38]. The oral cavity, especially saliva and dental plaque, of patients with periodontal disease or other oral infections seems to be a logical source of oral pathogens. Among the various microorganisms in the oral cavity, the caries bacteria *S. mutans* and the fungus *C. albicans* play an essential role in developing oral cavity and dental infections [39]. *S. mutans* is most commonly associated with dental caries [40]. Bhushan et al. (2014) reported the antidermatophytic activity of PGLEO against various fungal strains, suggesting that it has excellent antifungal potential [41]. Gonçalves et al. (2008) studied the antimicrobial activity of guava essential oils and reported that they have inhibitory activity against Gram-positive and Gram-negative bacteria [42]. Several studies have investigated the anticarcinogenic activity of PGLEO and reported the potential effects of Streptococcus sp. and Candida sp. against various oral pathogens and dental plaque, supporting the results of the present study [43,44]. The antimicrobial activities (*S. mutans* and *C. albicans*) of PGLEO may be attributed to a high percentage of significant compounds such as limonene, β-caryophyllene, or various other components [45–48].

Docking (molecular) studies have been widely used to evaluate the pharmacological properties of natural products; they explain the likely mechanisms of action and the binding pathways within protein pockets [49]. Previous studies have reported the role of *C. albicans* in dental plaque, caries lesions, and other types of oral infections [50–52]. Limonene and β-caryophyllene may have mechanisms against *C. albicans* N-myristoyl transferase) and *S. mutans* antigen III [53,54]. The molecular docking simulation results support the inhibitory activities of PGLEO against *S. mutans* and *C. albicans*. Previous in vitro antifungal and antibacterial studies have demonstrated the efficacy of PGLEO against oral infections [2,11,14,25]. The activity of PGLEO may be due to the presence of the significant compounds limonene and β-caryophyllene, which strongly inhibit the N-myristoyl transferase of *C. albicans* and the *S. mutans* antigen III and may thus contribute to the eradication of oral infections.

The cytotoxic effect and dose-dependent cell toxic effect of PGLEO on the cells of KB was confirmed by MTT assay. The cytotoxic effect of PGLEO was observed in the cells of KB, suggesting that volatile compounds have increased cytotoxicity in KB oral cancer cells. The IC_{50} value of doxorubicin (0.18 ± 0.13 μg/mL) has been reported for KB cells in a previous study [55]; the IC_{50} value of PGLEO is very high compared to standard doxorubicin. Our results are in agreement with previous reports [56]. Working with Thai *Psidium guajava*,

Manosroi et al. (2006) showed the significant cytotoxic effect of PGLEO at different doses and its antiproliferative effect against KB cells for 24 h. In contrast, the suppressive effect was greater than vincristine and essential basil oil [24]. In the present work, in vitro and in silico studies were carried out on oral microorganisms (*C. albicans* and *S. mutans*), and the anticancer effect of PGLEO on human mouth epidermal carcinoma KB cell line was investigated. The outcome of the study shows the role of PGLEO in overcoming oral problems such as dental caries and oral cancer. PGLEO may be a useful source in future drug development, especially for managing oral conditions.

4. Materials and Methods

4.1. Chemicals

The human oral adenocarcinoma cell line KB was obtained from the National Center for Cell Science (NCCS, Pune). Dulbecco's modified Eagle's high-glucose medium (DMEM-HG; Cat No: AL007), fetal bovine serum (FBS), MTT reagent, and Dulbecco's phosphate-buffered saline (D-PBS) were purchased from HiMedia Laboratories (Mumbai, India). Dimethyl sulfoxide (DMSO) and doxorubicin were purchased from Sigma-Aldrich (Mumbai, India). All media used for the antimicrobial studies were purchased from HiMedia Laboratories (Mumbai, India). Analytical-grade chemicals were purchased from HiMedia, Sigma Aldrich, and Merck (Mumbai, India), and 96-well plates for cell culture were purchased from Corning (Kennebunk, ME, USA).

4.2. Extraction of Essential Oils

Guava (*P. guajava*) leaves were collected from Integral College (Lucknow, Uttar Pradesh, India) in November 2021. The sample was authenticated (IU/PHAR/HRBD 22/03) and deposited at the College of Pharmacy, Integral College. The leaves were dried and powdered. Then, the essential oil was extracted using a Clevenger-type apparatus by the hydrodistillation method as described in a previous study [57]. The extracted essential oil was dried with anhydrous sodium carbonate; it was weighed and stored in an airtight amber vial at −20 °C (under refrigeration) until use.

4.3. Characterization of PGLEO Using Gas Chromatography–Mass Spectrometry

Chemical characterization was performed by gas chromatography and mass spectrometry. A gas chromatograph–mass spectrometer model QP2010 (Shimadzu, Tokyo, Japan) equipped with an autosampler and autoinjector (AOC-20si) was used for this study. The procedures for injection, separation, and identification of compounds, as well as the temperature of the column oven, were based on previous reports with some modifications [58,59]. The essential oil was dissolved in hexane (10 mg/mL), and the system was operated at 70 eV. This solution (1 µL) was injected in split mode (1:10 ratio) onto a Rtx-5MS column (Shimadzu, Japan) with the following dimensions: stationary phase film thickness of 0.25 µm, length of 30 m, and diameter of 0.25 mm. The injection temperature was set at 260 °C. The oven temperature was started at 50 °C for 2 min after injection and then increased at 5 °C/min to 180 °C for 1 min, followed by an increase at 15 °C/min to 280 °C, where the column was held for 15 min. The helium gas flow rate with a column head pressure of 69 kPa was set to 1.21 mL/min. Mass spectra were obtained in the range of 40 to 650 m/z. Essential oil components were identified based on a search (National Institute of Standards and Technology, NIST 14), the calculation of retention indices relative to homologous series of n-alkane (C7-C30), and a comparison of their mass spectra libraries with data from the mass spectra in the literature [60,61].

4.4. Effects of PGLEO on Oral Pathogen

The antimicrobial activity of PGLEO was tested against *Streptococcus mutans* (MTCC 389) and the fungus *C. albicans* (MTCC 9215). The antimicrobial activity of PGLEO was tested using the well diffusion method [62] with some modifications, and each experiment was evaluated in triplicate. Nutrient agar and Sabouraud dextrose agar medium were used for the antimicrobial assay. *S. mutans* cultures were separately inoculated into the nutrient broth and cultured at 37 °C for 18 h, while *C. albicans* subcultures were inoculated into Sabouraud dextrose broth and cultured at 37 °C for 48 h. The suspensions of all test organisms were diluted separately with a phosphate buffer (pH 7.4) to obtain 1×10^8 colony-forming units/mL of microbial suspension.

The inoculum (50 µL) from the broth was poured onto a fresh, sterile, solidified agar medium plate using a micropipette. Four 6 mm wells were drilled into the inoculated medium using a sterile cork borer. In addition, 2% dimethyl sulfoxide (DMSO) was used for sample preparation. Each well was filled with a 50 µL sample at different concentrations (0.25%, 5%, 7.5%, and 10%). Amoxicillin and fluconazole (0.001% in 2% DMSO) were used as positive controls for antibacterial and antifungal analyses, respectively, while 2% DMSO was used as a negative control. Plates were labelled and samples were placed in the wells of the plates and incubated at 37 °C for 24 and 48 h for bacteria and fungi, respectively.

The diameter of the inhibition (inhibition zone) around the wells was measured, and the mean ± standard deviation was calculated for each of the three assays. Different concentrations (0.0125, 0.025, 0.05, 0.1, and 0.2%) of essential oil were prepared in 2% DMSO and used to determine the minimum inhibitory concentration (MIC) using the broth dilution method [63]. Next, 100 µL of inoculum was added to each diluted tube, and the control tubes (no bacterial inoculation) were evaluated simultaneously. All tubes with antibacterial activity were incubated at 37 °C for 24 h, and the tubes with antifungal activity were incubated at 37 °C for 48 h; the lowest concentration resulted in no visible growth.

To investigate the bactericidal and fungicidal activity of PGLEO, the time-kill assay of Foudah et al. was performed [57]. *S. mutans* cultures were incubated on nutrient broth agar, while *C. albicans* was inoculated into Sabouraud dextrose broth and cultured at 37 °C for 8 h. The suspension was centrifuged and resuspended in saline at 10^6 CFU/mL. The suspensions of *S. mutans* and *C. albicans* were treated with broth medium containing different concentrations of PGLEO according to the MIC, mixed, and cultured at 37 °C. Samples were removed from the culture at selected time intervals (0, 4, 8, 16, and 24 h), diluted with saline, and recultured in a broth medium. After incubation of the plates at 37 °C for 24 h, the colony-forming unit/mL (CFU/mL) was calculated and a graph of CFU/mL versus time was plotted to calculate the time-kill assay.

4.5. Molecular Coupling Effects of Major Compounds on the Receptors for S. mutans *and* C. albicans

In this study, the two primary compounds, D-limonene and β-caryophyllene, were selected for molecular coupling to evaluate their effects on *S. mutans* and *C. albicans*. The AutoDock (ADT) tool created the file from the conventional RCSB PDB files [64]. For the molecular linkage study, 1JMM: region V of *S. mutans* antigen III was selected for the inhibitory effect of *S. mutans* and 1IYL: *C. albicans* N-myristoyl transferase with nonpeptide inhibitor was selected for the inhibitory effect of *C. albicans* [65]. To prepare the 3D input files, only the protein part of the PDB structures was selected by removing unwanted atoms, ions, and molecules [66]. All target proteins were energetically minimized by applying the CHARMm force field [67,68]. The 2D structures of selected ligands, including β-caryophyllene (CID: 5281515) and D-limonene (CID: 440917), in standard data format (SDF) were retrieved from the PubChem database [69]. SDF-2D was converted to PDB-3D using the BIOVIA discovery studio visualizer [70], and all ligands were energy minimized following the same protocol as the receptor molecules [68].

The molecular interactions of the ligands (β-caryophyllene and D-limonene) and the 3D structure of the protein molecules were performed using ADT to determine their potential binding affinities. The receptors and ligands, a grid parameter file, and docking parameter PDF files were created to perform the docking experiments. The grid frame around the protein molecule was drawn with variable grid points on the x, y, and z axes and a maximum distance (1.00) between two consecutive grids to provide sufficient space for ligand movement. Ten runs were performed for each ligand as standard. The minimum binding free energy and inhibition constant (Ki) was considered selective parameters to choose one of the best poses of the ligands bound to the binding cleft of the receptor molecules [71].

4.6. Cytotoxicity Assay

Human mouth epidermal carcinoma (KB) cells were maintained in high glucose medium supplemented with 10% FBS (fetal bovine serum) and 1% antibiotic–antifungal solution in an atmosphere of 5% CO_2 and 18–20% O_2 at 37 °C in a CO_2 incubator (Healforce, Shanghai, China) and subcultured every two days. The human oral adenocarcinoma cell line was established according to a previous experiment [69] with some modifications. First, 200 μL cell suspension was seeded into a 96-well plate (2×10^4 cells/well), and the cells were allowed to grow for approximately 24 h. Then, appropriate concentrations of the test samples were added and incubated for 24 h at 37 °C in a 5% CO_2 atmosphere. The plates were removed from the incubator, the spent medium was removed, and 3-(4,5-dimethylthiazol-2-yl)-2,5-diphenyl-2H-tetrazolium bromide (MTT reagent) was added at a final concentration of 0.05% of total volume. The plate was wrapped with aluminum foil and incubated for 3 h. The MTT reagent was then removed, 100 μL DMSO was carefully added, and the absorbance of the plates was read using an enzyme-linked immunosorbent assay (ELISA) plate reader (Biorad-PW41, Hercules, CA USA) at 570 nm. Percentage cell viability was calculated using the formula below, and IC_{50} was determined using a linear regression equation Y = Mx + C, where Y = 50 and M and C values were derived from the viability diagram.

$$\%\text{ cell viability} = [\text{Abs of treated cells}/\text{Abs of untreated cells}] \times 100 \quad (1)$$

Morphological changes and cell death of apoptotic cells were studied by fluorescence microscopy. At different concentrations (0.00125%–0.02%), cells were visualized after treatment of KB with PGLEO using fluorescence microscopy (Leica Microsystems GmbH) at a magnification of 100×. Cells with condensed and fragmented nuclei were considered apoptotic cells.

4.7. Statistical Analysis

Data were analyzed using GraphPad InStat (La Jolla, CA, USA) software and are presented as mean ± standard deviation. The essential oil concentration that provides 50% inhibition (IC_{50} value) was calculated using the graph by plotting the inhibition percentage versus the essential oil concentration.

5. Conclusions

This study investigated the effects of PGLEO on common oral pathogens and oral cancer. The oil showed solid antibacterial properties against *S. mutans* and *C. albicans*. Molecular docking studies of the primary compounds (limonene and β-caryophyllene) of PGLEO with the proteins 1JMM and 1IYL revealed theoretical inhibitory effects on the target proteins of *S. mutans* and *C. albicans*, consistent with the present antimicrobial activities, as indicated by their significant protein–ligand interaction energy. In addition to antibacterial activity, PGLEO also exhibited oral anticancer effects. Therefore, the present results support the potential use of PGLEO as an oral hygiene, anticariogenic, and anticarcinogenic agent.

Author Contributions: Conceptualization, A.A. and T.J.; methodology, A.A.; software, M.K.; validation, M.K., M.F.B. and T.J.; formal analysis, A.A.; investigation, T.J. and M.F.B.; resources, A.A.; data curation, S.M.A.; writing—original draft preparation, A.A.; writing—review and editing, M.K.; visualization, M.F.B. and S.M.A.; supervision, A.A.; project administration, S.M.A.; funding acquisition, M.F.B. and S.M.A. All authors have read and agreed to the published version of the manuscript.

Funding: This study is supported via funding from Prince Sattam bin Abdulaziz University Project Number (PSAU/2023/R/1444).

Institutional Review Board Statement: Not applicable.

Informed Consent Statement: Not applicable.

Data Availability Statement: Not applicable.

Conflicts of Interest: The authors declare no conflict of interest.

References

1. Daswani, P.G.; Gholkar, M.S.; Birdi, T.J. *Psidium guajava*: A single plant for multiple health problems of rural Indian population. *Pharmacogn. Rev.* **2017**, *11*, 167. [CrossRef] [PubMed]
2. Biswas, B.; Rogers, K.; McLaughlin, F.; Daniels, D.; Yadav, A. Antimicrobial activities of leaf extracts of guava (*Psidium guajava* L.) on two gram-negative and gram-positive bacteria. *Int. J. Microbiol.* **2013**, *2013*, 746165. [CrossRef] [PubMed]
3. Morais-Braga, M.F.B.; Carneiro, J.N.P.; Machado, A.J.T.; dos Santos, A.T.L.; Sales, D.L.; Lima, L.F.; Figueredo, F.G.; Coutinho, H.D.M. *Psidium guajava* L., from ethnobiology to scientific evaluation: Elucidating bioactivity against pathogenic microorganisms. *J. Ethnopharmacol.* **2016**, *194*, 1140–1152. [CrossRef] [PubMed]
4. Deguchi, Y.; Miyazaki, K. Anti-hyperglycemic and anti-hyperlipidemic effects of guava leaf extract. *Nutr. Metab.* **2010**, *7*, 9. [CrossRef] [PubMed]
5. Manekeng, H.T.; Mbaveng, A.T.; Ntyam Mendo, S.A.; Agokeng, A.J.D.; Kuete, V. Evaluation of Acute and Subacute Toxicities of *Psidium guajava* Methanolic Bark Extract: A Botanical with in Vitro Antiproliferative Potential. *Evid. Based Complement Altern. Med.* **2019**, *2019*, 13. [CrossRef] [PubMed]
6. Ashraf, A.; Sarfraz, R.A.; Rashid, M.A.; Mahmood, A.; Shahid, M.; Noor, N. Chemical composition, antioxidant, antitumor, anticancer, and cytotoxic effects of *Psidium guajava* leaf extracts. *Pharm. Biol.* **2016**, *54*, 1971–1981. [CrossRef]
7. Qin, X.J.; Yu, Q.; Yan, H.; Khan, A.; Feng, M.Y.; Li, P.P.; Hao, X.J.; An, L.K.; Liu, H.Y. Meroterpenoids with Antitumor Activities from Guava (*Psidium guajava*). *J. Agric. Food Chem.* **2017**, *65*, 4993–4999. [CrossRef]
8. Liu, X.C.; Lin, D.M.; Liu, M.; Zhang, M.; Li, Q.; Wang, J.; Xu, L.L.; Gao, Y.; Yang, J. Chemical constituents of *Psidium guajava* and their antitumor and antifungal activities. *China J. Chin. Mater. Med.* **2021**, *46*, 3877–3885. [CrossRef]
9. Figueiredo, A.C.; Barroso, J.G.; Pedro, L.G.; Scheffer, J.J.C. Factors affecting secondary metabolite production in plants: Volatile components and essential oils. *Flavour Fragr. J.* **2008**, *23*, 213–226. [CrossRef]
10. Hassan, F.U.; Arshad, M.A.; Ebeid, H.M.; Rehman, M.S.U.; Khan, M.S.; Shahid, S.; Yang, C. Phytogenic Additives Can Modulate Rumen Microbiome to Mediate Fermentation Kinetics and Methanogenesis Through Exploiting Diet–Microbe Interaction. *Front. Vet. Sci.* **2020**, *7*, 575801. [CrossRef]
11. Wang, L.; Wu, Y.; Huang, T.; Shi, K.; Wu, Z. Chemical Compositions, Antioxidant and Antimicrobial Activities of Essential Oils of *Psidium guajava* L. Leaves from Different Geographic Regions in China. *Chem. Biodivers.* **2017**, *14*, e1700114. [CrossRef] [PubMed]
12. de Souza, W.F.C.; de Lucena, F.A.; de Castro, R.J.S.; de Oliveira, C.P.; Quirino, M.R.; Martins, L.P. Exploiting the chemical composition of essential oils from *Psidium cattleianum* and *Psidium guajava* and its antimicrobial and antioxidant properties. *J. Food Sci.* **2021**, *86*, 4637–4649. [CrossRef] [PubMed]
13. Hassan, E.M.; El Gendy, A.E.N.G.; Abd-Elgawad, A.M.; Elshamy, A.I.; Farag, M.A.; Alamery, S.F.; Omer, E.A. Comparative Chemical Profiles of the Essential Oils from Different Varieties of *Psidium guajava* L. *Molecules* **2021**, *26*, 119. [CrossRef] [PubMed]
14. Chaturvedi, T.; Singh, S.; Nishad, I.; Kumar, A.; Tiwari, N.; Tandon, S.; Saikia, D.; Verma, R.S. Chemical composition and antimicrobial activity of the essential oil of senescent leaves of guava (*Psidium guajava* L.). *Nat. Prod. Res.* **2021**, *35*, 1393–1397. [CrossRef]
15. Fasola, T.R.; Oloyede, G.K.; Aponjolosun, B.S. Chemical composition, toxicity and antioxidant activities of essential oils of stem bark of Nigerian species of guava (*Psidium guajava* Linn.). *Excli J.* **2011**, *10*, 34–43.
16. Kumar, M.; Tomar, M.; Amarowicz, R.; Saurabh, V.; Sneha Nair, M.; Maheshwari, C.; Sasi, M.; Prajapati, U.; Hasan, M.; Singh, S.; et al. Guava (*Psidium guajava* L.) leaves: Nutritional composition, phytochemical profile, and health-promoting bioactivities. *Foods* **2021**, *10*, 752. [CrossRef]
17. Sierra, J.M.; Fusté, E.; Rabanal, F.; Vinuesa, T.; Viñas, M.; Sierra, J.M.; Fusté, E.; Rabanal, F.; Vinuesa, T. Expert Opinion on Biological Therapy An overview of antimicrobial peptides and the latest advances in their development. *Expert Opin. Biol. Ther.* **2017**, *17*, 663–676. [CrossRef]
18. Pradhan, B.K.; Badola, H.K. Ethnomedicinal plant use by Lepcha tribe of Dzongu valley, bordering Khangchendzonga Biosphere Reserve, in North Sikkim, India. *J. Ethnobiol. Ethnomed.* **2008**, *4*, 22. [CrossRef]
19. Okwu, D.E.; Ekeke, O. Phytochemical screening and mineral composition of chewing sticks in South Eastern Nigeria. *Glob. J. Pure Appl. Sci.* **2003**, *9*, 335–338. [CrossRef]
20. Varghese, J.; Ramenzoni, L.L.; Shenoy, P.; Nayak, U.Y.; Nayak, N.; Attin, T.; Schmidlin, P.R. In vitro evaluation of substantivity, staining potential, and biofilm reduction of guava leaf extract mouth rinse in combination with its anti-inflammatory effect on human gingival epithelial keratinocytes. *Materials* **2019**, *12*, 3903. [CrossRef]
21. Cavalcanti, I.M.G.; Del Bel Cury, A.A.; Jenkinson, H.F.; Nobbs, A.H. Interactions between *Streptococcus oralis*, *Actinomyces oris*, and *Candida albicans* in the development of multispecies oral microbial biofilms on salivary pellicle. *Mol. Oral Microbiol.* **2017**, *32*, 60–73. [CrossRef] [PubMed]

22. Alam, A.; Jawaid, T.; Alam, P. In vitro antioxidant and anti-inflammatory activities of green cardamom essential oil and in silico molecular docking of its major bioactives. *J. Taibah Univ. Sci.* **2021**, *15*, 757–768. [CrossRef]
23. Sireesha, D.; Reddy, B.S.; Reginald, B.A.; Samatha, M.; Kamal, F. Effect of amygdalin on oral cancer cell line: An in vitro study. *J. Oral Maxillofac. Pathol.* **2019**, *23*, 104–107. [CrossRef] [PubMed]
24. Manosroi, J.; Dhumtanom, P.; Manosroi, A. Anti-proliferative activity of essential oil extracted from Thai medicinal plants on KB and P388 cell lines. *Cancer Lett.* **2006**, *235*, 114–120. [CrossRef]
25. Arain, A.; Sherazi, S.T.H.; Mahesar, S.A.; Sirajuddin. Essential Oil from *Psidium guajava* Leaves: An excellent source of β-caryophyllene. *Nat. Prod. Commun.* **2019**, *14*, 1934578X19843007. [CrossRef]
26. Borah, A.; Pandey, S.K.; Haldar, S.; Lal, M. Chemical Composition of Leaf Essential Oil of *Psidium guajava* L. from North East India. *J. Essent. Oil Bear. Plants* **2019**, *22*, 248–253. [CrossRef]
27. Soliman, F.M.; Fathy, M.M.; Salama, M.M.; Saber, F.R. Comparative study of the volatile oil content and antimicrobial activity of *Psidium guajava* L. and *Psidium cattleianum* Sabine leaves. *Bull. Fac. Pharm. Cairo Univ.* **2016**, *54*, 219–225. [CrossRef]
28. Siani, A.C.; Souza, M.C.; Henriques, M.G.M.O.; Ramos, M.F.S. Anti-inflammatory activity of essential oils from *Syzygium cumini* and *Psidium guajava*. *Pharm. Biol.* **2013**, *51*, 881–887. [CrossRef]
29. Weli, A.; Al-Kaabi, A.; Al-Sabahi, J.; Said, S.; Hossain, M.A.; Al-Riyami, S. Chemical composition and biological activities of the essential oils of *Psidium guajava* leaf. *J. King Saud Univ. Sci.* **2019**, *31*, 993–998. [CrossRef]
30. Chen, H.Y.; Yen, G.C. Antioxidant activity and free radical-scavenging capacity of extracts from guava (*Psidium guajava* L.) leaves. *Food Chem.* **2007**, *101*, 686–694. [CrossRef]
31. Satyal, P.; Paudel, P.; Lamichhane, B.; Setzer, W.N. Leaf essential oil composition and bioactivity of *Psidium guajava* from Kathmandu, Nepal. *Am. J. Essent. Oils Nat. Prod.* **2015**, *3*, 11–14.
32. Silva, E.A.J.; Estevam, E.B.B.; Silva, T.S.; Nicolella, H.D.; Furtado, R.A.; Alves, C.C.F.; Souchie, E.L.; Martins, C.H.G.; Tavares, D.C.; Barbosa, L.C.A.; et al. Antibacterial and antiproliferative activities of the fresh leaf essential oil of *Psidium guajava* L. (Myrtaceae). *Braz. J. Biol.* **2019**, *79*, 697–702. [CrossRef] [PubMed]
33. Chalannavar, R.K.; Venugopala, K.N.; Baijnath, H.; Odhav, B. The Chemical Composition of Leaf Essential Oils of *Psidium guajava* L. (White and Pink fruit forms) from South Africa. *J. Essent. Oil Bear. Plants* **2014**, *17*, 1293–1302. [CrossRef]
34. Joseph, B.; Priya, R.M. Phytochemical and biopharmaceutical aspects of *Psidium guajava* (L.) essential oil: A review. *Res. J. Med. Plant* **2011**, *5*, 432–442. [CrossRef]
35. Sacchetti, G.; Maietti, S.; Muzzoli, M.; Scaglianti, M.; Manfredini, S.; Radice, M.; Bruni, R. Comparative evaluation of 11 essential oils of different origin as functional antioxidants, antiradicals and antimicrobials in foods. *Food Chem.* **2005**, *91*, 621–632. [CrossRef]
36. Khadhri, A.; El Mokni, R.; Almeida, C.; Nogueira, J.M.F.; Araújo, M.E.M. Chemical composition of essential oil of *Psidium guajava* L. growing in Tunisia. *Ind. Crop Prod.* **2014**, *52*, 29–31. [CrossRef]
37. Jassal, K.; Kaushal, S.; Rashmi; Rani, R. Antifungal potential of guava (*Psidium guajava*) leaves essential oil, major compounds: Beta-caryophyllene and caryophyllene oxide. *Arch. Phytopathol. Plant Prot.* **2021**, *54*, 2034–2050. [CrossRef]
38. Bui, F.Q.; Almeida-da-Silva, C.L.C.; Huynh, B.; Trinh, A.; Liu, J.; Woodward, J.; Asadi, H.; Ojcius, D.M. Association between periodontal pathogens and systemic disease. *Biomed. J.* **2019**, *42*, 27–35. [CrossRef]
39. Metwalli, K.H.; Khan, S.A.; Krom, B.P.; Jabra-Rizk, M.A. *Streptococcus mutans, Candida albicans*, and the Human Mouth: A Sticky Situation. *PLoS Pathog.* **2013**, *9*, e1003616. [CrossRef]
40. Gross, E.L.; Beall, C.J.; Kutsch, S.R.; Firestone, N.D.; Leys, E.J.; Griffen, A.L. Beyond *Streptococcus mutans*: Dental Caries Onset Linked to Multiple Species by 16S rRNA Community Analysis. *PLoS ONE* **2012**, *7*, e47722. [CrossRef]
41. Bhushan, G.; Sharma, S.K.; Kumar, S.; Tandon, R.; Singh, A.P. In-vitro antidermatophytic activity of essential oil of *Psidium guajava* (Linn.). *Indian J. Pharm. Biol. Res.* **2014**, *2*, 57–59. [CrossRef]
42. Gonçalves, F.A.; Andrade Neto, M.; Bezerra, J.N.S.; Macrae, A.; De Sousa, O.V.; Fonteles-Filho, A.A.; Vieira, R.H.S.D.F. Antibacterial activity of guava, *Psidium guajava* Linnaeus, leaf extracts on diarrhea-causing enteric bacteria isolated from seabob shrimp, *Xiphopenaeus kroyeri* (Heller). *Rev. Inst. Med. Trop. Sao Paulo* **2008**, *50*, 11–15. [CrossRef]
43. Ibrahim, S.A.; Yang, G.; Song, D.; Tse, T.S.F. Antimicrobial effect of guava on *Escherichia coli* O157:H7 and *Salmonella typhimurium* in liquid medium. *Int. J. Food Prop.* **2011**, *14*, 102–109. [CrossRef]
44. Vieira, D.R.P.; Amaral, F.M.M.; Maciel, M.C.G.; Nascimento, F.R.F.; Libério, S.A.; Rodrigues, V.P. Plant species used in dental diseases: Ethnopharmacology aspects and antimicrobial activity evaluation. *J. Ethnopharmacol.* **2014**, *155*, 1441–1449. [CrossRef] [PubMed]
45. Thakre, A.; Zore, G.; Kodgire, S.; Kazi, R.; Mulange, S.; Patil, R.; Shelar, A.; Santhakumari, B.; Kulkarni, M.; Kharat, K.; et al. Limonene inhibits *Candida albicans* growth by inducing apoptosis. *Med. Mycol.* **2018**, *56*, 565–578. [CrossRef] [PubMed]
46. Sabulal, B.; Dan, M.; Kurup, R.; Pradeep, N.S.; Valsamma, R.K.; George, V. Caryophyllene-rich rhizome oil of *Zingiber nimmonii* from South India: Chemical characterization and antimicrobial activity. *Phytochemistry* **2006**, *67*, 2469–2473. [CrossRef]
47. Yoo, H.J.; Jwa, S.K. Inhibitory effects of β-caryophyllene on *Streptococcus mutans* biofilm. *Arch. Oral Biol.* **2018**, *88*, 42–46. [CrossRef]
48. Subramenium, G.A.; Vijayakumar, K.; Pandian, S.K. Limonene inhibits streptococcal biofilm formation by targeting surface-associated virulence factors. *J. Med. Microbiol.* **2015**, *64*, 879–890. [CrossRef]

49. Mahmud, S.; Uddin, M.A.R.; Zaman, M.; Sujon, K.M.; Rahman, M.E.; Shehab, M.N.; Islam, A.; Alom, M.W.; Amin, A.; Akash, A.S.; et al. Molecular docking and dynamics study of natural compound for potential inhibition of main protease of SARS-CoV-2. *J. Biomol. Struct. Dyn.* **2021**, *39*, 6281–6289. [CrossRef]
50. Ghasempour, M.; Sefidgar, S.A.A.; Eyzadian, H.; Gharakhani, S. Prevalence of *Candida albicans* in dental plaque and caries lesion of early childhood caries (ECC) according to sampling site. *Casp. J. Intern. Med.* **2011**, *2*, 304–308.
51. Vila, T.; Sultan, A.S.; Montelongo-Jauregui, D.; Jabra-Rizk, M.A. Oral candidiasis: A disease of opportunity. *J. Fungi* **2020**, *6*, 15. [CrossRef] [PubMed]
52. Hasan, S.; Danishuddin, M.; Khan, A.U. Inhibitory effect of zingiber officinale towards *Streptococcus mutans* virulence and caries development: In vitro and in vivo studies. *BMC Microbiol.* **2015**, *15*, 1. [CrossRef] [PubMed]
53. Hannan, N.J.; Brownfoot, F.C.; Cannon, P.; Deo, M.; Beard, S.; Nguyen, T.V.; Palmer, K.R.; Tong, S.; Kaitu'U-Lino, T.J. Resveratrol inhibits release of soluble fms-like tyrosine kinase (sFlt-1) and soluble endoglin and improves vascular dysfunction—Implications as a preeclampsia treatment. *Sci. Rep.* **2017**, *7*, 1819. [CrossRef] [PubMed]
54. Ryu, N.H.; Park, K.R.; Kim, S.M.; Yun, H.M.; Nam, D.; Lee, S.G.; Jang, H.J.; Ahn, K.S.; Kim, S.H.; Shim, B.S.; et al. A hexane fraction of guava leaves (*Psidium guajava* L.) induces anticancer activity by suppressing AKT/mammalian target of rapamycin/ribosomal p70 S6 kinase in human prostate cancer cells. *J. Med. Food* **2012**, *15*, 231–241. [CrossRef]
55. Yang, T.; Xu, L.; Li, B.; Li, W.; Ma, X.; Fan, L.; Lee, R.J.; Xu, C.; Xiang, G. Antitumor activity of a folate receptor-targeted immunoglobulin G-doxorubicin conjugate. *Int. J. Nanomed.* **2017**, *2017*, 2505–2515. [CrossRef]
56. Foudah, A.I.; Alqarni, M.H.; Alam, A.; Salkini, M.A.; Alam, P.; Alkholifi, F.K.; Yusufoglu, H.S. Determination of chemical composition, in vitro and in silico evaluation of essential oil from leaves of *Apium graveolens* grown in Saudi Arabia. *Molecules* **2021**, *26*, 7372. [CrossRef]
57. Matulyte, I.; Marksa, M.; Ivanauskas, L.; Kalveniene, Z.; Lazauskas, R.; Bernatoniene, J. GC-MS analysis of the composition of the extracts and essential Oil from *Myristica fragrans* Seeds Using Magnesium Aluminometasilicate as Excipient. *Molecules* **2019**, *24*, 1062. [CrossRef]
58. Moura, E.d.S.; Faroni, L.R.D.; Heleno, F.F.; Rodrigues, A.A.Z. Toxicological stability of *Ocimum basilicum* essential oil and its major components in the control of *Sitophilus zeamais*. *Molecules* **2021**, *26*, 6483. [CrossRef]
59. Adams, R.P. *Identification of Essential Oil Components by Gas Chromatography/Mass Spectrometry*, 4th ed.; Allured Publishing Corporation: Carol Stream, IL, USA, 2007.
60. Waterman, P.G. Identification of essential oil components by gas chromatography/mass spectroscopy. *Biochem. Syst. Ecol.* **1996**, *6*, 594. [CrossRef]
61. Ashraf, S.A.; Al-Shammari, E.; Hussain, T.; Tajuddin, S.; Panda, B.P. In-vitro antimicrobial activity and identification of bioactive components using GC–MS of commercially available essential oils in Saudi Arabia. *J. Food Sci. Technol.* **2017**, *54*, 3948–3958. [CrossRef]
62. Jorgensen, J.H.; Hindler, J.F. New consensus guidelines from the Clinical and Laboratory Standards Institute for antimicrobial susceptibility testing of infrequently isolated or fastidious bacteria. *Clin. Infect. Dis.* **2007**, *44*, 280–286. [CrossRef] [PubMed]
63. Shakeel, E.; Akhtar, S.; Khan, M.K.A.; Lohani, M.; Arif, J.M.; Siddiqui, M.H. Molecular docking analysis of aplysin analogs targeting survivin protein. *Bioinformation* **2017**, *13*, 293–300. [CrossRef] [PubMed]
64. da Silva, D.R.; Deps, T.D.; Sakaguchi, O.A.S.; Costa, E.M.M.d.B.; dos Santos, C.A.O.; e Silva, J.P.R.; da Silva, B.D.; Ribeiro, F.F.; Mendonça-Júnior, F.J.B.; da Silva, A.C.B. *Molecular Docking of Phytochemicals against Streptococcus Mutans Virulence Targets: A Proteomic Insight into Drug Planning*; Intech Open: London, UK, 2022. [CrossRef]
65. Foudah, A.I.; Alqarni, M.H.; Alam, A.; Salkini, M.A.; Ross, S.A.; Yusufoglu, H.S. Phytochemical Screening, In Vitro and In Silico Studies of Volatile Compounds from *Petroselinum crispum* (Mill) Leaves Grown in Saudi Arabia. *Molecules* **2022**, *27*, 934. [CrossRef] [PubMed]
66. Brooks, B.R.; Brooks, C.L.; Mackerell, A.D.; Nilsson, L.; Petrella, R.J.; Roux, B.; Won, Y.; Archontis, G.; Bartels, C.; Boresch, S.; et al. CHARMM: The biomolecular simulation program. *J. Comput. Chem.* **2009**, *30*, 1545–1614. [CrossRef]
67. Steffen, C.; Thomas, K.; Huniar, U.; Hellweg, A.; Rubner, O.; Schroer, A. TmoleX—A graphical user interface for TURBOMOLE. *J. Comput. Chem.* **2010**, *31*, 2967–2970. [CrossRef]
68. Kim, S.; Thiessen, P.A.; Bolton, E.E.; Chen, J.; Fu, G.; Gindulyte, A.; Han, L.; He, J.; He, S.; Shoemaker, B.A.; et al. PubChem substance and compound databases. *Nucleic Acids Res.* **2016**, *44*, D1202–D1213. [CrossRef]
69. Qasaymeh, R.M.; Rotondo, D.; Oosthuizen, C.B.; Lall, N.; Seidel, V. Predictive binding affinity of plant-derived natural products towards the protein kinase g enzyme of *Mycobacterium tuberculosis* (Mtpkng). *Plants* **2019**, *8*, 477. [CrossRef]
70. Ajijur, R.; Salman, A.; Ahmad, K.M.K. Combinatorial design to decipher novel lead molecule against mycobacterium tuberculosis. *Biointerface Res. Appl. Chem.* **2021**, *11*, 12993–13004. [CrossRef]
71. Cha, J.D.; Kim, J.Y. Essential oil from *Cryptomeria japonica* induces apoptosis in human oral epidermoid carcinoma cells via mitochondrial stress and activation of caspases. *Molecules* **2012**, *17*, 3890–3901. [CrossRef]

plants

Article

Diversity of *Citrullus colocynthis* (L.) Schrad Seeds Extracts: Detailed Chemical Profiling and Evaluation of Their Medicinal Properties

Merajuddin Khan *, Mujeeb Khan, Khaleel Al-hamoud, Syed Farooq Adil, Mohammed Rafi Shaik and Hamad Z. Alkhathlan

Department of Chemistry, College of Science, King Saud University, P.O. Box 2455, Riyadh 11451, Saudi Arabia
* Correspondence: mkhan3@ksu.edu.sa; Tel.: +966-11-4675910

Abstract: Seeds and fruits of *Citrullus colocynthis* have been reported to possess huge potential for the development of phytopharmaceuticals with a wide range of biological activities. Thus, in the current study, we are reporting the potential antimicrobial and anticancer properties of *C. colocynthis* seeds extracted with solvents of different polarities, including methanol (M.E.), hexane (H.E.), and chloroform (C.E.). Antimicrobial properties of *C. colocynthis* seeds extracts were evaluated on Gram-positive and Gram-negative bacteria, whereas, anticancer properties were tested on four different cell lines, including HepG2, DU145, Hela, and A549. All the extracts have demonstrated noteworthy antimicrobial activities with a minimum inhibitory concentration (MIC) ranging from 0.9–62.5 μg/mL against *Klebsiella planticola* and *Staphylococcus aureus*; meanwhile, they were found to be moderately active (MIC 62.5–250 μg/mL) against *Escherichia coli* and *Micrococcus luteus* strains. Hexane extracts have demonstrated the highest antimicrobial activity against *K. planticola* with an MIC value of 0.9 μg/mL, equivalent to that of the standard drug ciprofloxacin used as positive control in this study. For anticancer activity, all the extracts of *C. colocynthis* seeds were found to be active against all the tested cell lines (IC_{50} 48.49–197.96 μg/mL) except for the chloroform extracts, which were found to be inactive against the HepG2 cell line. The hexane extract was found to possess the most prominent anticancer activity when compared to other extracts and has demonstrated the highest anticancer activity against the DU145 cell line with an IC_{50} value of 48.49 μg/mL. Furthermore, a detailed phytoconstituents analysis of all the extracts of *C. colocynthis* seeds were performed using GC–MS and GC–FID techniques. Altogether, 43 phytoconstituents were identified from the extracts of *C. colocynthis* seeds, among which 21, 12, and 16 components were identified from the H.E., C.E., and M.E. extracts, respectively. Monoterpenes (40.4%) and oxygenated monoterpenes (41.1%) were the most dominating chemical class of compounds from the hexane and chloroform extracts, respectively; whereas, in the methanolic extract, oxygenated aliphatic hydrocarbons (77.2%) were found to be the most dominating chemical class of compounds. To the best of our knowledge, all the phytoconstituents identified in this study are being reported for the first time from the *C. colocynthis*.

Keywords: plant extracts; *Citrullus colocynthis*; chemical profiling; biological activities; phytoconstituents

Citation: Khan, M.; Khan, M.; Al-hamoud, K.; Adil, S.F.; Shaik, M.R.; Alkhathlan, H.Z. Diversity of *Citrullus colocynthis* (L.) Schrad Seeds Extracts: Detailed Chemical Profiling and Evaluation of Their Medicinal Properties. *Plants* **2023**, *12*, 567. https://doi.org/10.3390/plants12030567

Academic Editors: Hazem Salaheldin Elshafie, Ippolito Camele and Adriano Sofo

Received: 15 December 2022
Revised: 15 January 2023
Accepted: 23 January 2023
Published: 26 January 2023

1. Introduction

Folk medicine has long been dependent on plants, which are considered as a crucial source of bioceuticals for the treatment and prevention of innumerable diseases for generations [1,2]. Albeit immense progress in modern medicine, a huge chunk of population across the globe, and particularly, people of a low- income category, are still dependent on natural product-based traditional methods of treatment for curing a variety of ailments [3,4]. The heavy use of these treatment methods is mainly derived by ancient knowledge, local credence, effectiveness, and low-cost [5]. Indeed, the recent emergence of pandemics has greatly renewed interests in the application of natural products, including plant-based materials and their compounds as nutraceuticals [6–8]. Traditionally, plant materials are dried, crushed, or extracted to generate products that are often referred as botanical medicines, herbal medicines, or phytotherapeuticals [9,10]. In such a way, a variety of commercial drugs have been fabricated from plant sources, and indeed, in the discovery of novel pharmaceuticals, plant materials offer several benefits due to their abundance in nature and wide geographic distribution [11,12]. Contrary to the use of extracts, modern medicines mostly rely on single substances, to ensure consistent efficacy and quality [13]. Nevertheless, modern drug discovery methods are still largely dependent on the process of extraction from natural products, modification of currently applied phytotherapeuticals, design and synthesis of molecules mimicking phytoconstituents, etc. [14].

Notably, plants generate therapeutic secondary metabolites to protect themselves from harmful pathogenic microorganisms, insects, and other creatures, which are referred as phytochemicals [15,16]. Most of the phytochemicals often possess antimicrobial properties, due to which they are also capable of protecting humans and animals against several infectious diseases caused by microorganisms or toxins [17,18]. These phytochemicals include several classes of compounds, such as phytosterols, terpenoids, flavonoids, alkaloids, phenolics, carotenoids, organic acids and proteases inhibitors, etc., which possess natural therapeutic properties and offer the best template for future pharmaceutical development [19,20]. Phytochemicals are typically extracted using a variety of techniques, including Soxhlet extraction, maceration, supercritical fluid extraction, subcritical water, and ultrasound mediated extractions, etc. [21].

Among these, solvent extraction is the most popular technique, which is efficacious and easy to use, and thus, widely applied for the extraction of therapeutic secondary metabolites from plants and other natural resources [22,23]. In this technique, the contents and yield of secondary metabolites varies with the type of solvents used for the extraction; this is due to the difference in the polarities and other physicochemical properties of the solvents [24]. For example, polar solvents facilitate the isolation of phenolic constituents and their glycosidic derivatives and saponins etc.; meanwhile, non-polar solvents are typically used to extract fatty acids and steroids, etc. Moreover, temperature, extraction time, amount of solvent with respect to plant material, part of the plant used, as well as the preparation method of plant materials, also have a significant effect on the quality and quantity of the resulting secondary metabolites [25]. Often, these parameters also have a strong influence on the biological properties of phytoconstituents, which is well documented in several studies [26]. Indeed, scientists have usually adopted diverse extraction techniques, solvents, and other parameters to obtain a variety of different and effective bioactive compounds [27]. This is typically achieved by the comparison of the biological properties, including the antimicrobial and anticancer potential of the extract of same part of the plant extracted in different solvents [28].

For example, Chiavaroli et al., have extracted and screened the leaf and bark extract of *Rhizophora racemosa* G. Mey. using different solvents and extraction methods [29]. Among these extracts, the methanolic leave and bark extracts, which were obtained by both the homogenizer-assisted extraction and maceration extraction method, have demonstrated an abundance of phenolics, flavonoids, and other phenolic acids, due to which they exhibited effective radical scavenging, total antioxidant and reducing potential. Similarly, in our previous study, we have investigated the effect of extraction solvents on the biological potential of therapeutic secondary metabolites [30]. To do that, the plant material of *Artemisia judaica* was extracted using three different solvents, including hexane, chloroform, and methanol [30]. In continuation of our previous research, we have selected *Citrullus colocynthis*, which is an important medicinal plant and has been used for centuries in traditional medicine for the treatment of various ailments [31].

C. colocynthis is an herbaceous desert plant consisting of perennial roots and vine-like stems [32]. It belongs to the family Cucurbitaceae, and it is native to the arid sandy areas of West Asia, Arabia, tropical Africa, and the Mediterranean [33]. This plant contains a battery of biologically active substances, including glycosides, flavonoids, alkaloids, fatty acids, and essential oils, etc. [34]. Different parts of *C. colocynthis* have long been used for treating various diseases, such as its fruit pulp (dried), which is effective in treating indigestion and gastroenteritis, while its fruit is known to possesses antioxidant, antimicrobial, and anti-inflammatory properties [35,36]. Moreover, the other pharmacological potential of the *C. colocynthis* include anti-diabetic, anthelmintic, analgesic, anti-allergic, and anti-cancerous properties, etc. [37]. Therefore, to investigate the effect of solvents on the biological potential of *C. colocynthis*, in this study, the plant material was extracted using different solvents, such as, methanol (M.E.), hexane (H.E.), and chloroform (C.E.). The phytoconstituents of each extract was determined separately using gas chromatographic methods. In addition, the antimicrobial and anticancer properties of each extract were assessed individually against several microorganisms and cell lines, respectively.

2. Results and Discussions

2.1. Chemo-Profiling of Different Extracts

Bioactive secondary metabolites are important for the physiology of both plants and humans, as they protect them by acting as anti-oxidants against oxidative stress [38]. In this regards, numerous studies have been reported in the literature describing the important biological properties of secondary metabolites including anti-microbial and anti-cancer properties [39,40]. Thus, the chemical characterization analysis of phytoconstituents of different extracts of the seeds of *C. colocynthis*, which are extracted with solvents of varied polarities, including methanol (M.E.), hexane (H.E.), and chloroform (C.E.), is undertaken. Moreover, the biological potentials including the antibacterial and anticancer properties of each individual extract were also evaluated. The solvent extractions of the seeds of *C. colocynthis* from Saudi Arabia was carried out at room temperature using conventional percolation technique, as described in earlier reports [30], and is shown in Figure 1.

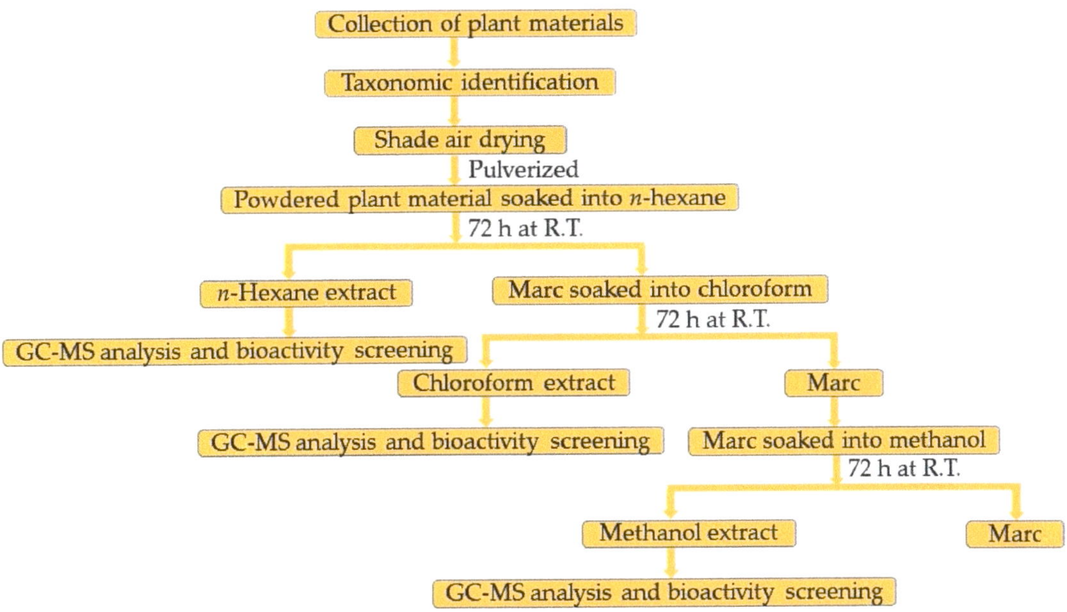

Figure 1. Flowchart for the preparation of *C. colocynthis* seeds extracts and their bioactivity screening.

The extraction was carried out individually by using an initial 500 g of seeds of *C. colocynthis* in each solvent, which yielded 11.2 g, 50.0 g, and 70.0 g of seed extract in M.E., H.E., and C.E., respectively. Notably, the extraction processes have yielded an almost similar color (dark brown) of extracts in all the solvents; however, their amount was slightly different due to the nature and quantity of the secondary metabolites extracted. For instance, C.E. extraction has resulted in higher yield, due to the high solubility of the long range of phytoconstituents including medium-polar and polar compounds in the chloroform solution. The chemo-profiling of all the extracts was carried out by GC–MS and GC–FID techniques, which has resulted in the recognition of a total of 21, 12, and 16 components from the H.E., C.E., and M.E. extracts, respectively. All the recognized phytochemicals generated from different extracts and their respective proportions are provided in the Table 1 based on the elution order of the compound from the column (HP-5MS).

Table 1. Identified chemical constituents from various extracts of *C. colocynthis* seeds of Saudi Arabia.

Peaks	Compounds *	M.F	CAS No.	R.T. (min)	LRI	H.E. %	C.E. %	M.E. %
1	*cis*-2-Pentenol	$C_5H_{10}O$	1576-95-0	5.187	772	-	-	0.7
2	Toluene	C_7H_8	108-88-3	5.394	778	3.6	-	-
3	Capronaldehyde	$C_6H_{12}O$	66-25-1	5.996	796	-	-	0.2
4	1-Octene	C_8H_{16}	111-66-0	6.034	797	-	-	0.4
5	**Isopropyl butanoate**	$\mathbf{C_7H_{14}O_2}$	**638-11-9**	**7.466**	**840**	**10.4**	**-**	**-**
6	*p*-Xylene	C_8H_{10}	106-42-3	8.422	869	2.8	-	-
7	*o*-Xylene	C_8H_{10}	95-47-6	9.251	894	1.6	-	-
8	Santolina triene	$C_{10}H_{16}$	2153-66-4	9.806	909	1.6	-	-
9	Isobutyl isobutyrate	$C_8H_{16}O_2$	97-85-8	9.958	913	1.6	-	-
10	*α*-Thujene	$C_{10}H_{16}$	2867-05-2	10.43	926	1.5	-	-
11	Benzaldehyde	C_7H_6O	100-52-7	11.662	958	-	1.6	-
12	***α*-Pinene**	$\mathbf{C_{10}H_{16}}$	**80-56-8**	**10.754**	**934**	**30.6**	**-**	**-**
13	Sabinene	$C_{10}H_{16}$	3387-41-5	12.29	974	1.6	-	-
14	Pseudocumene	C_9H_{12}	95-63-6	13.058	994	2.7	-	-
15	Undecane	$C_{11}H_{24}$	1120-21-4	17.098	1100	-	-	1.1
16	***δ*3-Carene**	$\mathbf{C_{10}H_{16}}$	**13466-78-9**	**13.696**	**1011**	**5.1**	**-**	**-**
17	***o*-Methylacetophenone**	$\mathbf{C_9H_{10}O}$	**577-16-2**	**18.527**	**1138**	**10.8**	**-**	**-**
18	Dodecane	$C_{12}H_{26}$	112-40-3	20.783	1200	1.4	-	-
19	Ethyl phenyl acetate	$C_{10}H_{12}O_2$	101-97-3	22.604	1253	-	3.1	-
20	2*E*-Decenal	$C_{10}H_{18}O$	3913-81-3	22.971	1263	-	3.4	-
21	**Thymol**	$\mathbf{C_{10}H_{14}O}$	**499-75-2**	**23.988**	**1293**	**-**	**37.2**	**3.0**
22	Filifolide-A	$C_{10}H_{14}O_2$	-	24.83	1318	-	3.9	-
23	Tetradecane	$C_{14}H_{30}$	629-59-4	27.483	1400	2.8	-	1.5
24	Coumarin	$C_9H_6O_2$	91-64-5	28.542	1434	-	-	0.3
25	2-Methyl butyl benzoate	$C_{12}H_{16}O_2$	52513-03-8	28.696	1439	-	-	0.5
26	*α*-Guaiene	$C_{15}H_{24}$	3691-12-1	28.795	1443	-	-	0.2
27	*β*-Ionol	$C_{13}H_{22}O$	22029-76-1	31.038	1517	-	4.8	-
28	Caryophyllene oxide	$C_{15}H_{24}O$	1139-30-6	33.252	1593	1.4	-	-
29	Hexadecane	$C_{16}H_{34}$	544-76-3	33.459	1600	3.1	-	-
30	8-Cedren-13-ol	$C_{15}H_{24}O$	18319-35-2	35.846	1686	-	2.8	-
31	Octadecane	$C_{18}H_{38}$	593-45-3	38.831	1800	1.7	-	-
32	7-Hydroxycoumarin	$C_9H_6O_3$	93-35-6	39.837	1840	3.9	-	-
33	**Methyl hexadecanoate**	$\mathbf{C_{17}H_{34}O_2}$	**112-39-0**	**41.982**	**1927**	**-**	**5.6**	**18.3**
34	*n*-Hexadecanoic acid	$C_{16}H_{32}O_2$	57-10-3	42.789	1960	-	-	1.6
35	Ethyl hexadecanoate	$C_{18}H_{36}O_2$	628-97-7	43.603	1994	3.3	-	-
36	**8,11-Octadecadienoic acid, methyl ester**	$C_{19}H_{34}O_2$	**56599-58-7**	**45.91**	**2091**	**-**	**13.0**	**28.6**
37	**(*Z*)-9-Octadecenoic acid methyl ester**	$C_{19}H_{36}O_2$	**112-62-9**	**46.029**	**2096**	**-**	**5.3**	**20.4**
38	Methyl oleate	$C_{19}H_{36}O_2$	112-62-9	46.319	2108	2.1	-	-
39	***n*-Octadecanoic acid, methyl ester**	$C_{19}H_{38}O_2$	**112-61-8**	**46.599**	**2120**	**-**	**-**	**7.4**
40	Ethyl linoleate	$C_{20}H_{36}O_2$	544-35-4	47.605	2162	1.9	-	-

Table 1. *Cont.*

Peaks	Compounds *	M.F	CAS No.	R.T. (min)	LRI	H.E. %	C.E. %	M.E. %
41	Tetracosane	$C_{24}H_{50}$	646-31-1	53.297	2400	-	2.1	1.4
42	***trans*-Ferruginyl acetate**	$C_{22}H_{32}O_2$	**15340-79-1**	**53.562**	**2411**	-	**8.1**	-
43	**6-Ketoferruginol**			**54.625**	**2456**	-	-	**9.6**
	Monoterpenes hydrocarbons					40.4	-	-
	Oxygenated monoterpenes					-	41.1	3.0
	Sesquiterpene hydrocarbons					-	-	0.2
	Oxygenated sesquiterpenes					1.4	7.6	-
	Aliphatic hydrocarbons					9	2.1	4.4
	Oxygenated aliphatic hydrocarbons					19.3	27.3	77.2
	Aromatics					21.5	4.7	0.5
	Diterpenoids					-	8.1	9.6
	Others					3.9	-	0.3
	Total identified					**95.5**	**90.9**	**95.2**

* Components are recorded as per their order of elution from HP-5MS column; compounds higher than 5.0% are highlighted in boldface; LRI = linear retention index computed with reference to the *n*-alkanes mixture (C7–C30) on HP-5MS column; H.E. = hexane extract of *C. colocynthis* seeds; C.E. = chloroform extract of *C. colocynthis* seeds; M.E. = methanol extract of *C. colocynthis* seeds.

According to the results in the Table 1, the H.E. extract was dominated by monoterpene hydrocarbons; whereas, the C.E. and M.E. extracts contained oxygenated monoterpenes and oxygenated aliphatic hydrocarbons as major chemical class of compounds, respectively. Particularly, the M.E. extract consisted of 77.2% of oxygenated aliphatic hydrocarbons; meanwhile, H.E. and C.E. extracts contained an almost similar amount of monoterpene hydrocarbons, i.e., 40.4% and oxygenated monoterpene hydrocarbons, i.e., 41.1%, respectively. Apart from the major chemical classes of phytoconstituents, each individual extract contains different chemical categories of compounds as subsequent chemical classes. For example, besides monoterpene hydrocarbons, the H.E. extract consisted of oxygenated aliphatic hydrocarbons and aromatics in an almost similar percentage, i.e., 19.3 and 21.5%, respectively. In the case of the C.E. extract, the oxygenated aliphatic hydrocarbons were present at a distant second position, which was recorded at 27.3%. In addition to these, the C.E. extract also contains diterpenoids, oxygenated sesquiterpenes, aliphatic hydrocarbons, and aromatics; however, chemical classes of these compounds were present in minor amount (<10% each). On the other hand, the M.E. extract does not contain other chemical classes of compounds in large quantities; after their major class of compounds, the other classes of phytoconstituents in M.E. extract are present in minor quantity just below 20% of total constituents. The categories of chemical compounds include oxygenated monoterpenes (3%), aliphatic hydrocarbons (4%), diterpenoids (9.6%), and others (<1%).

A comprehensive analysis of the phytoconstituents of all of the extract of seeds of *C. colocynthis* has revealed a total presence of 43 phytoconstituents; out of these components, 21, 12, and 16 components were identified from the H.E., C.E., and M.E. extracts, respectively. Among these, the H.E. extract clearly stands out with maximum number of phytoconstituents. The total ion chromatogram of each extracts of the *C. colocynthis* seed extracts are given in Figure 2. The H.E. extract was mostly dominated by α-pinene (30.6) followed by o-methylacetophenone (10.8%), isopropyl butanoate (10.4%), and δ3-carene (5.1%), while the remaining compounds, such as *p*-xylene, pseudocumene, tetradecane, hexadecane, methyl hexadecanoate, ethyl hexadecanoate, and methyl oleate are all present in a minor quantity of less than 5%. In the case of the C.E. extract, the major compound was identified as thymol (37.2%), which is followed by 8,11-octadecadienoic acid methyl ester (13.0%), *trans*-ferruginyl acetate (8.1%), and *β*-ionol (4.8%). The minor components of the same extract include filifolide-A (3.9%), ethyl phenyl acetate (3.1%), 2*E*-decenal (3.4%), 8-cedren-13-ol (2.8%), and tetracosane (2.1%). Whereas, the M.E. extract is mainly dominated by the 8,11-octadecadienoic acid methyl ester (28.6%) followed by the (Z)-9-

octadecenoic acid methyl ester (20.4%), methyl hexadecanoate (18.3%), 6-ketoferruginol (9.6%), *n*-octadecanoic acid, methyl ester (7.4%), thymol (3.0%), and others, are present in an insignificant amount.

It is noteworthy that all the major compounds found in all three different extracts of *C. colocynthis* seeds, such as α-pinene, thymol, 8,11-octadecadienoic acid methyl ester and others (Figure 3), have been found to be distinct to the plant species collected from Riyadh region in KSA. These phytoconstituents have not been found in *C. colocynthis* collected from other regions of the world, as shown in Table 2. For example, the plant collected from the city of Tangier in Morocco has demonstrated the presence of nonadienal (15.4%), linalool propanoate (14.3%), and 2,4-decadienal (7.8%) as major constituents [41]. Whereas, the Indian species of the *C. colocynthis* collected from different cities have shown the presence of 2-methyl,4-heptanone (48.0), 3-methyl,2-heptanone (12.9), *n*-hexadecanoic acid (12.4), and morphine (9.1) as dominant compounds [42,43]. Particularly, the three major compounds found in each separate extract, such as α-pinene, thymol, and 8,11-octadecadienoic acid methyl ester in H.E., C.E., and M.E. extracts, respectively, have been known to possess excellent biological properties. These compounds are distinct to the plant species investigated in this study, and thus the seeds extract of the *C. colocynthis* collected from Riyadh are expected to demonstrate decent biological properties when compared to the same species gathered from other regions of the world.

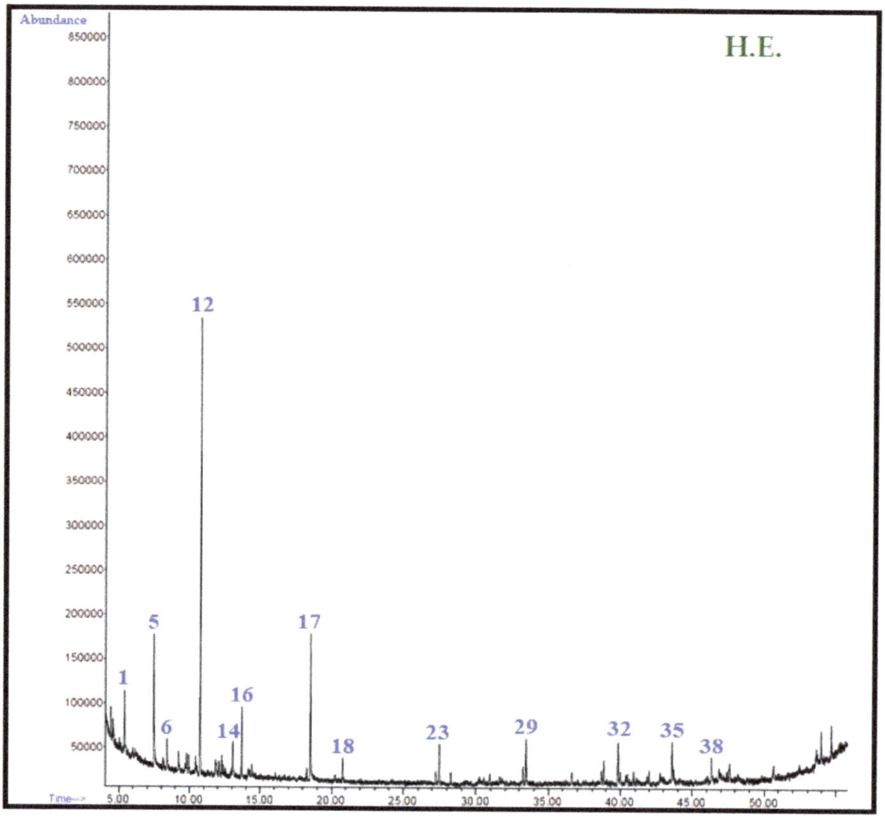

Figure 2. *Cont.*

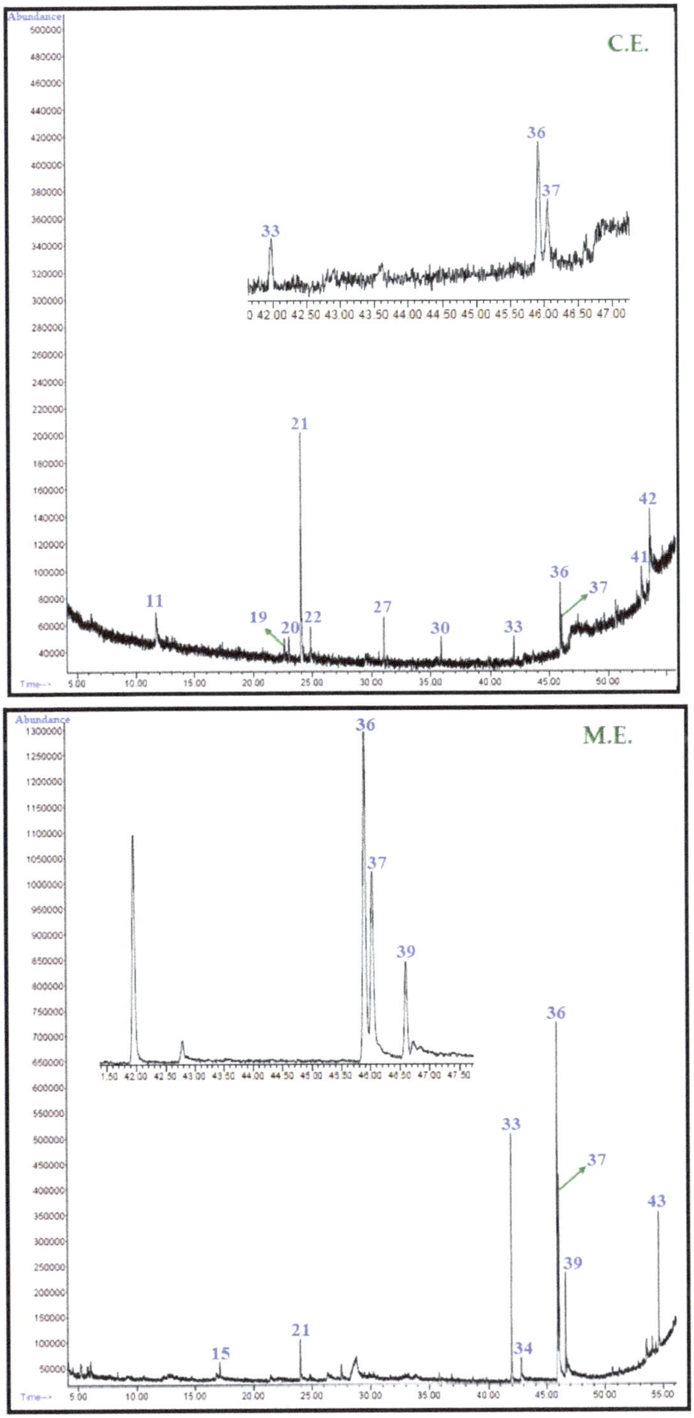

Figure 2. Total ion chromatogram (TIC) of *n*-hexane (H.E.), chloroform (C.E.), and methanol (M.E.) extracts of *C. colocynthis* seeds.

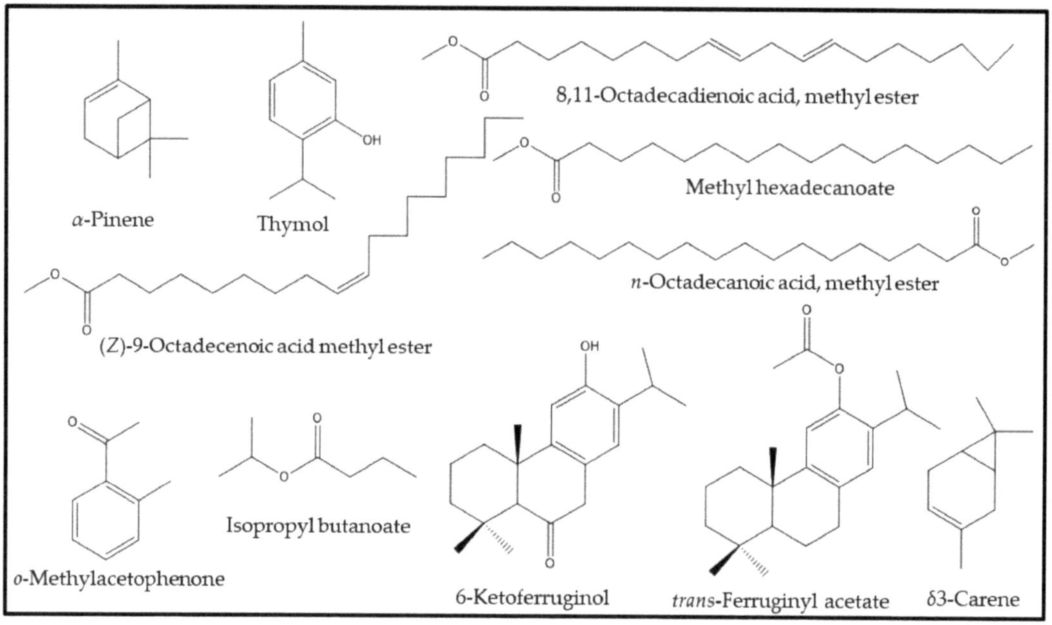

Figure 3. Chemical structure of most dominating identified compounds from *C. colocynthis* seeds extracts.

Table 2. Major components of *C. colocynthis* from different parts of the world.

No.	Country	City	Major Components (%)	Reference
1.	Morocco	Tangier	Nonadienal (15.4), linalool propanoate (14.3), 2,4-decadienal (7.8), pentadecane (7.2), hexanal (4.5), and butylated hydroxy anisol (4.3).	[44]
2.	India	Parangipettai	2-Methyl,4-heptanone (48.0), 3-Methyl,2-heptanone (12.9), trimethylsilylmethanol (9.1), pentane, 1-propoxy (6.5), and 2-pentanol, 5-methoxy-2-methyl- (5.3).	[45]
		Southern Haryana	*n*-Hexadecanoic acid (12.4), morphine (9.1), narceine (10.3), isoquinoline, 1-[(3,4-dimethoxyphenyl)methyl]-6,7-dimethoxy- (7.5), codein (6.6), and glycerol (5.5).	[46]
3.	Saudi Arabia	Riyadh	Thymol (3.0–37.2), α-pinene (0–30.0), 8,11-octadecadienoic acid, methyl ester (13.0–28.6), *(Z)*-9-octadecenoic acid methyl ester (0–20.4), methyl hexadecanoate (5.6–18.3), *o*-methylacetophenone (0–10.8), isopropyl butanoate (0–10.4), 6-ketoferruginol (0–9.6), *trans*-ferruginyl acetate (0.0–8.1), *n*-octadecanoic acid, methyl ester (0–7.4), *trans*-sabinyl acetate (0–6.0).	Present study

Typically, α-pinene is an important secondary metabolite, which is mainly found in essential oils from different plants, such as the *Piper nigrum* or *Juniperus* species [47]. α-pinene is a monoterpene, which consists of hydrophobic and volatile properties with fresh pine scent and woody flavour [48]. This compound has been known to possess excellent antimicrobial properties against various Gram-positive and Gram-negative bacterial strains, including the methicillin-resistant *Staphylococcus aureus* [49]. Additionally, α-pinene has been reported to demonstrate anticancer properties against the human ovarian cancer PA-1 cell line [50]. On the other hand, thymol, which is a monoterpene phenol mainly found in essential oils of the plants from Lamiaceae family (*Thymus*, *Ocimum*, *Origanum*, and *Monarda* genera) is also surprisingly found in the C.E. extract of the *C. colocynthis* seeds [51]. Mainly, the thymol-based plant species are used as flavouring and preservative agents and has also been recognized as "safe" (GRAS) or as approved food additives [52]. Thymol is known to possesses excellent anti-inflammatory, anti-microbial, anti-oxidant, and antifungal properties, besides being beneficial for the cardiovascular system [53]. The solvent-based variation in major constituents is not new. Indeed, plants demonstrate a huge difference in their phytochemical constituents, which are typically based on a variety of different factors, such as geographic location, genetic variations, ecological and environmental factors, etc. [39,54]. Similarly, different experimental conditions, such as the solvent, temperature, and pH of extraction processes also have serious effects on the quality and quantity of the phytomolecules. The difference in the major constituents may possibly have different synergistic interactions, which ultimately determine the biological activities of plant-based materials [55].

For example, in a recent study, the monoterpenes' thymol demonstrated direct antibacterial activity against the *S. aureus* IS-58 strain [56]. Additionally, thymol has also been recognized as anti-tumor agent, which is demonstrated in a recent study during the evaluation of the cell viability and apoptosis in U-87 cells treated with thymol at different concentrations. The half-maximal inhibitory concentration (IC_{50}) of thymol in the U-87 cells was 230 μM, while on a normal cell line it did not exhibit any cytotoxic effect at the same concentration [57]. Besides the two biologically active phytoconstituents, α-pinene and thymol, another major constituent, 8,11-octadecadienoic acid, as the methyl ester found in M.E. extract, has also demonstrated excellent biological properties in previously reported studies [58]. For example, the ethyl acetate extract of the seeds of *Acacia nilotica* Linn, which contained 11-Octadecenoic acid, methyl ester as major compound, has demonstrated excellent activities against the several tested microbes with zones of inhibition diameters ranging from 27–32 mm against *Salmonela* typhi, *E. coli*, *Streptococcus feacalis*, *S. aureus*, and so on [59].

2.2. Antibacterial Properties

In order to test the antimicrobial efficacy of seeds' extracts of *C. colocynthis*, all the different extracts, including H.E., C.E., and M.E. extracts, were employed against both Gram-positive bacterial strains, such as *S. aureus* and *M. luteus*, and Gram-negative bacterial strains, such as *K. planticola* and *E. coli*, respectively. Whereas, ciprofloxacin, which is commonly prescribe as antibiotic, has been used as a control during this study. The extracts have delivered mixed results, i.e., the extracts were active against both Gram-positive and Gram-negative bacterial strains; however, neither of them demonstrated good activities against *E. coli*, which is Gram-negative bacteria. For instance, both H.E. and C.E. extracts were active against *S. aureus* and *K. planticola*, which are Gram-positive and Gram-negative bacteria, respectively. Whereas, the M.E. extract demonstrated antibacterial activity only against a single bacterial strain, i.e., *K. planticola*. Among these extracts, the H.E. extract has demonstrated superb antibacterial activity against *K. planticola*, which is almost comparable to the commercially available antibiotic. Meanwhile, the other extracts demonstrated very mild activities towards the tested strains.

The results further revealed that the C.E. extract exhibited mild activity against *S. aureus* with 7.8 μg/mL, but demonstrated excellent potential towards *K. planticola*, with

1.9 μg/mL. Meanwhile, the H.E. extract exhibited decent activity against *S. aureus* with 3.9 μg/mL; whereas, it demonstrated the best of all antibacterial activities against the *K. planticola*, with 0.9 μg/mL, which was equal to the activity of the commercially available antibiotic, i.e., ciprofloxacin (cf. Table 3). On the other hand, the M.E. was the least active extract among the tested materials, and demonstrated decent activity towards a single strain, which is 7.8 μg/mL towards the *K. planticola*. Notably, all the extracts studied have demonstrated very mild activity against the *M. luteus* and *E. coli*, except the slightly decent activity of the M.E. extract against *M. luteus*, with 62.2 μg/mL.

Table 3. Antimicrobial activity of various extracts of *C. colocynthis* seeds against Gram-positive and Gram-negative bacteria.

Tested Extracts of *C. colocynthis* Seeds	Minimum Inhibitory Concentration (μg/mL)			
	Gram-Positive		Gram-Negative	
	S. aureus MTCC 96	*M. luteus* MTCC 2470	*K. planticola* MTCC 530	*E. coli* MTCC 739
M.E.	62.5	62.5	7.8	>250
H.E.	3.9	>250	0.9	>250
C.E.	7.8	>250	1.9	>250
Ciprofloxacin	0.9	0.9	0.9	0.9

Among all the extracts, the best antimicrobial activity was demonstrated by he H.E. extract against *K. planticola*, with 0.9 μg/mL, which is same as the commercially available antibiotic (ciprofloxacin). Similarly, the decent antibacterial activity of the H.E. extract was also observed against *S. aureus*, with 0.9 μg/mL. However, the same extract demonstrated very mild activity against the other two species of bacteria, namely *M. leuteus* and *E. coli*. These results are not surprising, as the H.E. extract contains high amount of α-pinene, which is already known to exhibit strong antibacterial activities against several bacterial strains, including *K. planticola* and *S. aureus* [60]. For example, the essential oil of *Baccharis reticulata*, which contains a high amount of α-pinene, demonstrated excellent antibacterial activities against *S. aureus* with a MIC value of 256 μg/mL; whereas, the same extract did not have a significant effect on *E. coli* and other bacteria [61]. α-pinene in a pure form demonstrates excellent antibacterial activities against a large number of bacterial strains; however, when present in the extract or essential oils together with other phytoconstituents, it selectively targets the bacterial species. This can be attributed to the presence of antagonist phytoconstituents, which may stop the action or effect of α-pinene. As in the H.E. extract, besides α-pinene, isopropyl butanoate, *o*-methylacetophenone, and many other phytochemicals are present, which may function as antagonist phytomolecules. Similar results were also obtained the in case of the C.E. extract, which consisted of thymol as the major constituent, and is already known to have demonstrated excellent antibacterial properties towards several strains [62]. For example, thymol-rich essential oils of *Oliveria decumbens* (Apiaceae) collected from different Iranian populations demonstrated high antibacterial properties against a variety of Gram-positive and Gram-negative bacteria. The essential oils obtained from *Oliveria decumbens* (Apiaceae) collected from the Behbahan region of Iran exhibited a MIC value of 1.0 μg/mL against *S. aureus* (Gram-positive), while the thymol-rich C.E. in this study demonstrated a near similar antibacterial property with a MIC value of 7.8 μg/mL [63]. The gas chromatographic-mass spectrometry analysis put in evidence four main volatile constituents, such as thymol (20.3–36.4%), However, very little has been published with regards to the biological properties of 8,11-Octadecadienoic acid, methyl ester, which is a major constituent of M.E. extract. This is also reflected in our study, where the M.E. extract has demonstrated the least antibacterial activities when compared to the other extracts.

2.3. Anticancer Properties

Besides antibacterial properties, the seed extracts of *C. colocynthis* were also evaluated for their potential anticancer properties, which is explored against a variety of cell lines, including HepG2 (hepatic cancer cells), DU145 (prostate cancer cells), Hela (cervical cancer cells), and A549 (human lung cancer cells). During this study, doxorubicin, a prescription anticancer drug, was employed as a control, which is a commercially available anticancer drug (cf. Table 4). All the studied extracts have demonstrated diverse anticancer activities against different cell lines, which are provided in Table 4. When compared to the controlled drug used in this study, which has shown IC_{50} (μg/mL) of less than one (<1) against all the studied cell lines, neither of the extracts has demonstrated the activity, which is close to the value of doxorubicin. The M.E. extract has exhibited IC_{50} values of 126.6, 91.9, 99.9, 70.1 μg/mL against HepG2, DU145, Hela, and A549, respectively. Whereas, the H.E. and C.E. extracts have demonstrated the IC_{50} values of 177.0 and no activity, and 48.4 and 53.3, 197.2 and 83.8, 82.9 and 154.0 μg/mL against the same cell lines, respectively. However, the IC_{50} values are insignificant when compared to the controlled drug, and, according to a reported study on the extensive screening of several extracts from a variety of plants, a plant extract is usually considered to possess in vitro cytotoxic activity when the IC_{50} (concentration that causes a 50% cell kill) value is less than 20 μg/mL for the extract [64].

Table 4. Anticancer activity of various extracts of *C. colocynthis* seeds against various cancer cell lines.

Tested Extracts of *C. colocynthis* Seeds	IC_{50} (μg/mL)			
	HepG2	**DU145**	**Hela**	**A549**
M.E.	126.65 ± 11.48	91.94 ± 7.88	99.96 ± 9.70	70.18 ± 1.17
H.E.	177.05 ± 4.84	48.49 ± 0.50	197.28 ± 9.45	82.99 ± 6.5
C.E.	NA	53.32 ± 1.59	83.87 ± 4.61	154.05 ± 14.25
Doxorubicin	0.72 ± 0.012 (μM)	0.36 ± 0.01 (μM)	0.8 ± 0.71 (μM)	0.55 ± 0.16 (μM)

Results are expressed as mean ± SD, NA = No activity.

Taking this into account, only the H.E. and C.E. extract has demonstrated very mild 48.4 and 53.3 μg/mL, respectively, against a single cell line, namely, DU145. Nevertheless, the major constituents of these extracts, including α-Pinene and thymol, have been reported to demonstrate considerable anticancer properties against a battery of cell lines. For example, pine needle oil from the crude extract of pine needles, which consists of large amount of α-Pinene, has exhibited considerable inhibitory effect on hepatoma carcinoma BEL-7402 cells, with an inhibitory rate of 79.3% in vitro and 69.1% in vivo [65]. Similarly, the crude extract of *Trachyspermum ammi* consisting of thymol in large amount has also shown potential cytotoxic activity in the breast cancer cell line MCF-7. The MTT assay demonstrated that the IC_{50} values of thymol on MCF-7 cells for 48 h and 72 h were 54 and 62 μg/mL, respectively [66]. These values are close to the IC_{50} values of the C.E. extract of *C. colocynthis* seeds (53.2 μg/mL), which has thymol as major constituents.

Despite the mild anticancer properties of all the studied extracts against different cancer cell lines, the data still demonstrate a clear trend for the selection of extracts for the activity guided isolation of phytomolecules, which is essential in the quest of finding biologically active phytoconstituents [67]. In this regard, no prior reports on the comprehensive analysis of the anticancer properties of the seed extracts of *C. colocynthis* in different solvents with varying polarities have been reported in the literature. However, few reports have appeared on the anticancer potential of the essential oils of the seed of *C. colocynthis*, which have demonstrated reasonable anticancer properties towards colorectal cancer cell lines, with IC_{50} values varying between 4 and 7 mg/mL [68]. Meanwhile, the other studies have reported that the seed and pulp extracts (extracted using a single solvent) of the fruit of *C. colocynthis* were effective against various cancer cell lines [69]. However, in this study, we have employed three different solvents with varying polarities to isolate the extracts of the *C. colocynthis* seeds, which have delivered notable results with different major constituents in a different solvent extract. Similar to the antibacterial results, the M.E. extract with 8,11-Octadecadienoic acid, methyl ester as major constituent has not demonstrated decent anticancer activities with the IC_{50} values of more 75 μg/mL against almost all the cell lines studied.

3. Materials and Methods

3.1. Plant Material

Entire aerial parts of the *C. colocynthis* grown in the region of Taif, a city in the Mecca Province of southwest Saudi Arabia, were procured in May 2020. Identifications of *C. colocynthis* were authenticated by Dr. Rajakrishnan Rajagopal from the herbarium division of King Saud University. A specimen sample (24,531) of *C. colocynthis* is retained in the herbarium division of the King Saud University.

3.2. Chemicals

All the chemicals including methanol, chloroform, and *n*-hexane were of analytical grade and purchased from Sigma–Aldrich, Hamburg, Germany. Pure volatile constituents or enriched fractions of volatile constituents, such as thymol, δ3-carene, α-pinene (Alfa Aesar, Lancashire, UK), *n*-hexadecanoic acid, caryophyllene oxide, (*Z*)-9-Octadecenoic acid methyl ester, and 8,11-Octadecadienoic acid, methyl ester (enriched fractions), were available and used for co-injection/comparative analysis.

3.3. Preparation of C. colocynthis Seeds Extracts

Procured aerial parts of *C. colocynthis* were air-dried at room temperature. The fruits, leaves, and stem of the plant were separated and subjected to drying separately until a constant weight was achieved. From the fruits of *C. colocynthis*, the seeds were carefully removed and then ground using a grinder. The resultant *C. colocynthis* seeds (500 g) were first percolated with *n*-hexane (550 mL) three times at room temperature. After *n*-hexane extraction, the marc was again subjected to extraction three times with $CHCl_3$ (550 mL). Finally, the same extraction process was repeated using the residual marc with methanol (550 mL) for three times at room temperature. Notably, each time, the extraction process was carried out for 3 days for all the solvents employed. The resultant *n*-hexane, chloroform, and methanol extracts of *C. colocynthis* seeds were separately dried under a vacuum at 40 °C until the solvents were completely removed using a Buchi rotary evaporating system (Rotavapor R-215, Buchi, Flawil, Switzerland) equipped with a vacuum controller (V-850) and vacuum pump (V-700). These separately dried *n*-hexane, $CHCl_3$, and methanol extracts were used for the screening of anticancer and antimicrobial activities, as well as for the GC analysis.

3.4. GC and GC–MS Analysis of C. colocynthis Seeds Extracts

In order to identify the chemical constituents of the extracts of *C. colocynthis* seeds, the dried extracts of *n*-hexane and $CHCl_3$ extracts were dissolved in diethyl ether, whereas the methanol extract was dissolved in methanol and subjected to GC–FID and GC–MS analyses. The GC–MS system was equipped with stationary phase columns (HP-5MS) employing the same method, as described previously [70]. Detailed methodology is given in Supplementary Materials (Scheme S1). The identified constituents from *n*-hexane, $CHCl_3$, and methanol extracts of *C. colocynthis* seeds and their relative percentages are provided in Table 1, and the constituents are listed according to their elution order on the HP-5MS column.

3.5. Calculation of Linear Retention Indices (LRIs)

LRI values of chemical constituents of *C. colocynthis* seeds extracts were determined following a previously reported method [70], and they are listed in Table 1. The detailed methodology is provided in Supplementary Materials (Scheme S2).

3.6. Identification of Volatile Chemical Components

Identification of the chemical constituents of *C. colocynthis* seeds extracts were carried out through an analysis on a HP-5MS column, as described previously [70]. Detailed methodology for the identification of chemical constituents is provided in the Supplementary Materials (Scheme S3) [71–73]. GC–MS chromatograms for the identified constituents of *n*-hexane, chloroform, and methanol extracts of *C. colocynthis* seeds on the HP-5MS column are given in Figure 2.

3.7. Evaluation of Antimicrobial and Anticancer Activity

3.7.1. Antimicrobial Activity

Antimicrobial activity of the *C. colocynthis* seeds extracts was examined using the well diffusion method [74] towards a panel of four pathogenic bacterial strains, including *Staphylococcus aureus* MTCC 96, *Micrococcus luteus* MTCC 2470, *Escherichia coli* MTCC 739, and *Klebsiella planticola* MTCC 530. The four pathogenic reference strains were spread on the surface of the Mueller–Hinton agar Petri plates with 0.1 mL of previously prepared microbial suspensions individually containing 1.0×10^7 CFU/mL (equal to 0.5 McFarland standard). Using a cork borer, the wells of the 6.0 mm diameter were prepared in the media plates, and the prepared test extracts at a dosage range of 250–0.48 µg/well were added in each well under sterile conditions in a laminar air flow chamber. The standard antibiotic solution of Ciprofloxacin at a dose range of 250–0.48 µg/well and the well containing dimethyl sulfoxide (DMSO) served as positive and negative controls, respectively. The plates were incubated for 24 h at 37 °C, and the well containing the least concentration showing the inhibition zone was considered as the minimum inhibitory concentration (MIC). All experiments were carried out in duplicates and mean values are represented.

3.7.2. Anticancer Activity

Cytotoxicity of *C. colocynthis* seeds extracts was assessed against the human lung adenocarcinoma cell line (A549), human hepatocarcinoma cell line (HepG2), human cervical cancer cell line (HeLa), and human prostate cancer cell line (DU145) using the MTT assay [75]. Briefly, 1×10^4 exponentially growing cells were seeded into each 96-well plate (counted by Trypan blue exclusion dye method) and allowed to grow until 60–70% confluence; then, different concentrations of test extracts were added to the culture medium along with negative (DMSO) and positive controls (Doxorubicin). The plates were incubated for 48 h in a CO_2 incubator at 37 °C with a 90% humidified atmosphere and 5% CO_2. Then, the media of the wells were replaced with 90 µL of fresh serum-free media and 10 µL of MTT (5 mg/mL of PBS), and the plates were incubated at 37 °C for 2 h. The media was discarded and allowed to dry for 30 min. Later, 100 µL of DMSO was added in each well to dissolve the purple formazan crystals, and the absorbance was recorded at 570 nm using Spectra Max plus 384 UV-Visible plate reader (Molecular Devices, Sunnyvale, CA, USA). Each test compound was examined at various concentrations in triplicate, and the results are expressed as a mean with standard deviation (mean ± SD), (n = 3). One-way ANOVA and Dunnett's post-comparison test were used to analyse the data for significant differences (test vs. control). The statistical significance for the experiment was set at $p < 0.05$.

4. Conclusions

In this study, the effect of the polarity of the extraction solvents on the phytochemical contents and biological potential of the extracts of the seeds of *C. colocynthis* were explored. To do this, three different solvents, including M.E., C.E., and H.E., were selected to isolate the phytoconstituents of the studied plant material. The contents of all the studied extracts were vastly different with respect to their major components, and the H.E. and C.E. extracts demonstrated α-pinene and thymol as their major constituents, respectively; whereas, the M.E. extract demonstrated the presence of 8,11-octadecadienoic acid, methyl ester in a large quantity. Out of all the extracts, the H.E. and C.E. extracts clearly stood out in terms of their major constituents and their biological potential. Particularly, the H.E. extract, consisting of

α-pinene (30.6%), demonstrated a superior antimicrobial activity against most of the strains studied and indeed, in the case of *K. planticola*, it demonstrated excellent antibacterial activity, which was almost equal to the commercially available antibiotic. Therefore, the *C. colocynthis* seed extracts may offer a variety of phytopharmaceutical, food products, and other commercial entities in the form of biologically active pure phytomolecules such as α-pinene and thymol.

Supplementary Materials: The following supporting information can be downloaded at: https://www.mdpi.com/article/10.3390/plants12030567/s1. Scheme S1. Gas Chromatography (GC) and Gas Chromatography−Mass Spectrometry (GC-MS) Analysis of *C. colocynthis* seeds Extracts; Scheme S2. Linear retention indices (LRIs); Scheme S3. Identification of volatile components.

Author Contributions: M.K. (Merajuddin Khan) designed the project; M.K. (Merajuddin Khan), M.R.S., S.F.A. and M.K. (Mujeeb Khan) helped to draft the manuscript; M.K. (Merajuddin Khan), K.A.-h. and H.Z.A. carried out the preparation of plant extract and characterization of the plant extract material; M.K. (Merajuddin Khan) and M.R.S. carried out the experimental part; H.Z.A. provided scientific guidance. All authors have read and agreed to the published version of the manuscript.

Funding: This work was funded by the National Plan for Science, Technology and Innovation (MAARIFAH), King Abdul-Aziz City for Science and Technology, Kingdom of Saudi Arabia, grant Number 14-MED1227-02.

Data Availability Statement: Data are contained within the article.

Acknowledgments: This work was funded by the National Plan for Science, Technology and Innovation (MAARIFAH), King Abdul-Aziz City for Science and Technology, Kingdom of Saudi Arabia, grant Number 14-MED1227-02.

Conflicts of Interest: The authors declare no conflict of interest.

References

1. Makhuvele, R.; Naidu, K.; Gbashi, S.; Thipe, V.C.; Adebo, O.A.; Njobeh, P.B. The use of plant extracts and their phytochemicals for control of toxigenic fungi and mycotoxins. *Heliyon* **2020**, *6*, e05291. [CrossRef] [PubMed]
2. Junsongduang, A.; Kasemwan, W.; Lumjoomjung, S.; Sabprachai, W.; Tanming, W.; Balslev, H. Ethnomedicinal knowledge of traditional healers in Roi Et, Thailand. *Plants* **2020**, *9*, 1177. [CrossRef] [PubMed]
3. Uttra, A.M.; Ahsan, H.; Hasan, U.H.; Chaudhary, M.A. Traditional medicines of plant origin used for the treatment of inflammatory disorders in Pakistan: A review. *J. Tradit. Chin. Med.* **2018**, *38*, 636–656.
4. Zhao, Z.; Li, Y.; Zhou, L.; Zhou, X.; Xie, B.; Zhang, W.; Sun, J. Prevention and treatment of COVID-19 using Traditional Chinese Medicine: A review. *Phytomedicine* **2021**, *85*, 153308. [CrossRef]
5. Tesfaye, S.; Asres, K.; Lulekal, E.; Alebachew, Y.; Tewelde, E.; Kumarihamy, M.; Muhammad, I. Ethiopian medicinal plants traditionally used for the treatment of cancer, part 2: A review on cytotoxic, antiproliferative, and antitumor phytochemicals, and future perspective. *Molecules* **2020**, *25*, 4032. [CrossRef]
6. Khan, M.; Adil, S.F.; Alkhathlan, H.Z.; Tahir, M.N.; Saif, S.; Khan, M.; Khan, S.T. COVID-19: A global challenge with old history, epidemiology and progress so far. *Molecules* **2020**, *26*, 39. [CrossRef]
7. Pastor, N.; Collado, M.C.; Manzoni, P. Phytonutrient and nutraceutical action against COVID-19: Current review of characteristics and benefits. *Nutrients* **2021**, *13*, 464. [CrossRef]
8. Alesci, A.; Aragona, M.; Cicero, N.; Lauriano, E.R. Can nutraceuticals assist treatment and improve COVID-19 symptoms? *Nat. Prod. Res.* **2022**, *36*, 2672–2691. [CrossRef]
9. Houghton, P.J. Old yet new—Pharmaceuticals from plants. *J. Chem. Educ.* **2001**, *78*, 175. [CrossRef]
10. Salehi, B.; Zucca, P.; Sharifi-Rad, M.; Pezzani, R.; Rajabi, S.; Setzer, W.N.; Varoni, E.M.; Iriti, M.; Kobarfard, F.; Sharifi-Rad, J. Phytotherapeutics in cancer invasion and metastasis. *Phytother. Res.* **2018**, *32*, 1425–1449. [CrossRef]
11. Musthaba, S.M.; Baboota, S.; Ahmed, S.; Ahuja, A.; Ali, J. Status of novel drug delivery technology for phytotherapeutics. *Expert Opin. Drug Deliv.* **2009**, *6*, 625–637. [CrossRef] [PubMed]
12. Cravotto, G.; Boffa, L.; Genzini, L.; Garella, D. Phytotherapeutics: An evaluation of the potential of 1000 plants. *J. Clin. Pharm. Ther.* **2010**, *35*, 11–48. [CrossRef]
13. Dave, V.; Yadav, R.B.; Ahuja, R.; Yadav, S. Formulation design and optimization of novel fast dissolving tablet of chlorpheniramine maleate by using lyophilization techniques. *Bull. Fac. Pharm. Cairo Univ.* **2017**, *55*, 31–39. [CrossRef]
14. Cragg, G.M.; Newman, D.J. Natural products: A continuing source of novel drug leads. *Biochim. Biophys. Acta BBA Gen. Subj.* **2013**, *1830*, 3670–3695. [CrossRef] [PubMed]

15. Zaynab, M.; Fatima, M.; Abbas, S.; Sharif, Y.; Umair, M.; Zafar, M.H.; Bahadar, K. Role of secondary metabolites in plant defense against pathogens. *Microb. Pathog.* **2018**, *124*, 198–202. [CrossRef] [PubMed]
16. Rajput, V.D.; Singh, R.K.; Verma, K.K.; Sharma, L.; Quiroz-Figueroa, F.R.; Meena, M.; Gour, V.S.; Minkina, T.; Sushkova, S.; Mandzhieva, S. Recent developments in enzymatic antioxidant defence mechanism in plants with special reference to abiotic stress. *Biology* **2021**, *10*, 267. [CrossRef] [PubMed]
17. Gorlenko, C.L.; Kiselev, H.Y.; Budanova, E.V.; Zamyatnin Jr, A.A.; Ikryannikova, L.N. Plant secondary metabolites in the battle of drugs and drug-resistant bacteria: New heroes or worse clones of antibiotics? *Antibiotics* **2020**, *9*, 170. [CrossRef] [PubMed]
18. Asimuddin, M.; Shaik, M.R.; Adil, S.F.; Siddiqui, M.R.H.; Alwarthan, A.; Jamil, K.; Khan, M. Azadirachta indica based biosynthesis of silver nanoparticles and evaluation of their antibacterial and cytotoxic effects. *J. King Saud Univ. Sci.* **2020**, *32*, 648–656. [CrossRef]
19. Guerriero, G.; Berni, R.; Muñoz-Sanchez, J.A.; Apone, F.; Abdel-Salam, E.M.; Qahtan, A.A.; Alatar, A.A.; Cantini, C.; Cai, G.; Hausman, J.-F. Production of plant secondary metabolites: Examples, tips and suggestions for biotechnologists. *Genes* **2018**, *9*, 309. [CrossRef] [PubMed]
20. Jain, C.; Khatana, S.; Vijayvergia, R. Bioactivity of secondary metabolites of various plants: A review. *Int. J. Pharm. Sci. Res* **2019**, *10*, 494–504.
21. Jones, W.P.; Kinghorn, A.D. Extraction of plant secondary metabolites. In *Natural Products Isolation*; Springer: Cham, Switzerland, 2006; pp. 323–351.
22. De Silva, G.O.; Abeysundara, A.T.; Aponso, M.M.W. Extraction methods, qualitative and quantitative techniques for screening of phytochemicals from plants. *Am. J. Essent. Oils Nat. Prod.* **2017**, *5*, 29–32.
23. Shaik, M.; Albalawi, G.; Khan, S.; Khan, M.; Adil, S.; Kuniyil, M.; Al-Warthan, A.; Siddiqui, M.; Alkhathlan, H.; Khan, M. "Miswak" based green synthesis of silver nanoparticles: Evaluation and comparison of their microbicidal activities with the chemical synthesis. *Molecules* **2016**, *21*, 1478. [CrossRef] [PubMed]
24. Matrose, N.A.; Obikeze, K.; Belay, Z.A.; Caleb, O.J. Impact of spatial variation and extraction solvents on bioactive compounds, secondary metabolites and antifungal efficacy of South African Impepho [*Helichrysum odoratissimum* (L.) Sweet]. *Food Biosci.* **2021**, *42*, 101139. [CrossRef]
25. Dai, J.; Mumper, R.J. Plant phenolics: Extraction, analysis and their antioxidant and anticancer properties. *Molecules* **2010**, *15*, 7313–7352. [CrossRef]
26. Ahmed, E.; Arshad, M.; Khan, M.Z.; Amjad, M.S.; Sadaf, H.M.; Riaz, I.; Sabir, S.; Ahmad, N. Secondary metabolites and their multidimensional prospective in plant life. *J. Pharmacogn. Phytochem.* **2017**, *6*, 205–214.
27. Žlabur, J.Š.; Žutić, I.; Radman, S.; Pleša, M.; Brnčić, M.; Barba, F.J.; Rocchetti, G.; Lucini, L.; Lorenzo, J.M.; Domínguez, R. Effect of different green extraction methods and solvents on bioactive components of chamomile (*Matricaria chamomilla* L.) flowers. *Molecules* **2020**, *25*, 810. [CrossRef] [PubMed]
28. Adnan, M.; Oh, K.K.; Azad, M.O.K.; Shin, M.H.; Wang, M.-H.; Cho, D.H. Kenaf (*Hibiscus cannabinus* L.) leaves and seed as a potential source of the bioactive compounds: Effects of various extraction solvents on biological properties. *Life* **2020**, *10*, 223. [CrossRef] [PubMed]
29. Chiavaroli, A.; Sinan, K.I.; Zengin, G.; Mahomoodally, M.F.; Bibi Sadeer, N.; Etienne, O.K.; Cziáky, Z.; Jekő, J.; Glamočlija, J.; Soković, M. Identification of chemical profiles and biological properties of Rhizophora racemosa G. Mey. extracts obtained by different methods and solvents. *Antioxidants* **2020**, *9*, 533. [CrossRef] [PubMed]
30. Khan, M.; Khan, M.; Al-Hamoud, K.; Adil, S.F.; Shaik, M.R.; Alkhathlan, H.Z. Comprehensive Phytochemical Analysis of Various Solvent Extracts of Artemisia judaica and Their Potential Anticancer and Antimicrobial Activities. *Life* **2022**, *12*, 1885. [CrossRef]
31. Hussain, A.I.; Rathore, H.A.; Sattar, M.Z.; Chatha, S.A.; Sarker, S.D.; Gilani, A.H. *Citrullus colocynthis* (L.) Schrad (bitter apple fruit): A review of its phytochemistry, pharmacology, traditional uses and nutritional potential. *J. Ethnopharmacol.* **2014**, *155*, 54–66. [CrossRef]
32. Li, Q.-Y.; Munawar, M.; Saeed, M.; Shen, J.-Q.; Khan, M.S.; Noreen, S.; Alagawany, M.; Naveed, M.; Madni, A.; Li, C.-X. *Citrullus colocynthis* (L.) Schrad (Bitter Apple Fruit): Promising traditional uses, pharmacological effects, aspects, and potential applications. *Front. Pharmacol.* **2021**, *12*, 791049. [CrossRef] [PubMed]
33. Pravin, B.; Tushar, D.; Vijay, P.; Kishanchnad, K. Review on *Citrullus colocynthis*. *Int. J. Res. Pharm. Chem* **2013**, *3*, 46–53.
34. Rahimi, R.; Amin, G.; Ardekani, M.R.S. A review on *Citrullus colocynthis* Schrad.: From traditional Iranian medicine to modern phytotherapy. *J. Altern. Complement. Med.* **2012**, *18*, 551–554. [CrossRef]
35. Marzouk, B.; Marzouk, Z.; Haloui, E.; Fenina, N.; Bouraoui, A.; Aouni, M. Screening of analgesic and anti-inflammatory activities of *Citrullus colocynthis* from southern Tunisia. *J. Ethnopharmacol.* **2010**, *128*, 15–19. [CrossRef] [PubMed]
36. Hameed, B.; Ali, Q.; Hafeez, M.; Malik, A. Antibacterial and antifungal activity of fruit, seed and root extracts of *Citrullus colocynthis* plant. *Biol. Clin. Sci. Res. J.* **2020**, *2020*, 1. [CrossRef]
37. Ostovar, M.; Akbari, A.; Anbardar, M.H.; Iraji, A.; Salmanpour, M.; Ghoran, S.H.; Heydari, M.; Shams, M. Effects of *Citrullus colocynthis* L. in a rat model of diabetic neuropathy. *J. Integr. Med.* **2020**, *18*, 59–67. [CrossRef]
38. Cavazos, P.; Gonzalez, D.; Lanorio, J.; Ynalvez, R. Secondary metabolites, antibacterial and antioxidant properties of the leaf extracts of *Acacia rigidula* benth. and Acacia berlandieri benth. *SN Appl. Sci.* **2021**, *3*, 1–14. [CrossRef]
39. Khan, M.; Khan, S.T.; Khan, M.; Mousa, A.A.; Mahmood, A.; Alkhathlan, H.Z. Chemical diversity in leaf and stem essential oils of *Origanum vulgare* L. and their effects on microbicidal activities. *AMB Express* **2019**, *9*, 176. [CrossRef]

40. Kurnia, D.; Ajiati, D.; Heliawati, L.; Sumiarsa, D. Antioxidant properties and structure-antioxidant activity relationship of Allium species leaves. *Molecules* **2021**, *26*, 7175. [CrossRef]
41. Bourhia, M.; Bouothmany, K.; Bakrim, H.; Hadrach, S.; Salamatullah, A.M.; Alzahrani, A.; Khalil Alyahya, H.; Albadr, N.A.; Gmouh, S.; Laglaoui, A. Chemical profiling, antioxidant, antiproliferative, and antibacterial potentials of chemically characterized extract of citrullus colocynthis L. seeds. *Separations* **2021**, *8*, 114. [CrossRef]
42. Gurudeeban, S.; Ramanathan, T.; Satyavani, K. Characterization of volatile compounds from bitter apple (*Citrullus colocynthis*) using GC-MS. *Int. J. Chem. Anal. Sci.* **2011**, *2*, 108–110.
43. Singh, S.; Devi, B. Estimation of phytoconstituents from *Citrullus colocynthis* (L.) schrad roots extract by GC-MS spectroscopy. *Int. J. Sci. Res.* **2016**, *7*, 648–652.
44. El-Shazly, A.; Hafez, S.; Wink, M. Comparative study of the essential oils and extracts of Achillea fragrantissima (Forssk.) Sch. Bip. and Achillea santolina L.(Asteraceae) from Egypt. *Die Pharm. Int. J. Pharm. Sci.* **2004**, *59*, 226–230.
45. Alsohaili, S.A.; Al-fawwaz, A.T. Composition and antimicrobial activity of Achillea fragrantissima essential oil using food model media. *Eur. Sci. J.* **2014**, *10*, 156–165.
46. Alsohaili, S. Seasonal variation in the chemical composition and antimicrobial activity of essential oil extracted from Achillea fragrantissima grown in Northern-Eastern Jordanian desert. *J. Essent. Oil-Bear. Plants* **2018**, *21*, 139–145. [CrossRef]
47. Allenspach, M.; Valder, C.; Flamm, D.; Grisoni, F.; Steuer, C. Verification of chromatographic profile of primary essential oil of Pinus sylvestris L. combined with chemometric analysis. *Molecules* **2020**, *25*, 2973. [CrossRef]
48. Allenspach, M.; Steuer, C. α-Pinene: A never-ending story. *Phytochemistry* **2021**, *190*, 112857. [CrossRef]
49. Utegenova, G.A.; Pallister, K.B.; Kushnarenko, S.V.; Özek, G.; Özek, T.; Abidkulova, K.T.; Kirpotina, L.N.; Schepetkin, I.A.; Quinn, M.T.; Voyich, J.M. Chemical composition and antibacterial activity of essential oils from *Ferula* L. species against methicillin-resistant *Staphylococcus aureus*. *Molecules* **2018**, *23*, 1679. [CrossRef]
50. Hou, J.; Zhang, Y.; Zhu, Y.; Zhou, B.; Ren, C.; Liang, S.; Guo, Y. α-Pinene induces apoptotic cell death via caspase activation in human ovarian cancer cells. *Med. Sci. Monit. Int. Med. J. Exp. Clin. Res.* **2019**, *25*, 6631. [CrossRef]
51. Marchese, A.; Orhan, I.E.; Daglia, M.; Barbieri, R.; Di Lorenzo, A.; Nabavi, S.F.; Gortzi, O.; Izadi, M.; Nabavi, S.M. Antibacterial and antifungal activities of thymol: A brief review of the literature. *Food Chem.* **2016**, *210*, 402–414. [CrossRef]
52. Alagawany, M.; Farag, M.R.; Abdelnour, S.A.; Elnesr, S.S. A review on the beneficial effect of thymol on health and production of fish. *Rev. Aquac.* **2021**, *13*, 632–641. [CrossRef]
53. Kachur, K.; Suntres, Z. The antibacterial properties of phenolic isomers, carvacrol and thymol. *Crit. Rev. Food Sci. Nutr.* **2020**, *60*, 3042–3053. [CrossRef] [PubMed]
54. Kokkini, S.; Karousou, R.; Vokou, D. Pattern of geographic variations of *Origanum vulgare* trichomes and essential oil content in Greece. *Biochem. Syst. Ecol.* **1994**, *22*, 517–528. [CrossRef]
55. Chorianopoulos, N.; Kalpoutzakis, E.; Aligiannis, N.; Mitaku, S.; Nychas, G.-J.; Haroutounian, S.A. Essential oils of Satureja, Origanum, and Thymus species: Chemical composition and antibacterial activities against foodborne pathogens. *J. Agric. Food. Chem.* **2004**, *52*, 8261–8267. [CrossRef] [PubMed]
56. Sousa Silveira, Z.d.; Macêdo, N.S.; Sampaio dos Santos, J.F.; Sampaio de Freitas, T.; Rodrigues dos Santos Barbosa, C.; Júnior, D.L.d.S.; Muniz, D.F.; Castro de Oliveira, L.C.; Júnior, J.P.S.; Cunha, F.A.B.d. Evaluation of the antibacterial activity and efflux pump reversal of thymol and carvacrol against *Staphylococcus aureus* and their toxicity in *Drosophila melanogaster*. *Molecules* **2020**, *25*, 2103. [CrossRef]
57. Qoorchi Moheb Seraj, F.; Heravi-Faz, N.; Soltani, A.; Ahmadi, S.S.; Talebpour, A.; Afshari, A.R.; Ferns, G.A.; Bahrami, A. Thymol has anticancer effects in U-87 human malignant glioblastoma cells. *Mol. Biol. Rep.* **2022**, *49*, 9623–9632. [CrossRef] [PubMed]
58. Mittermeier, V.K.; Dunkel, A.; Hofmann, T. Discovery of taste modulating octadecadien-12-ynoic acids in golden chanterelles (*Cantharellus cibarius*). *Food Chem.* **2018**, *269*, 53–62. [CrossRef]
59. Shoge, M.; Amusan, T. Phytochemical, antidiarrhoeal activity, isolation and characterisation of 11-octadecenoic acid, methyl ester isolated from the seeds of *Acacia nilotica* Linn. *J. Biotechnol. Immunol.* **2020**, *2*, 1–12.
60. Dhar, P.; Chan, P.; Cohen, D.T.; Khawam, F.; Gibbons, S.; Snyder-Leiby, T.; Dickstein, E.; Rai, P.K.; Watal, G. Synthesis, antimicrobial evaluation, and structure–activity relationship of α-pinene derivatives. *J. Agric. Food. Chem.* **2014**, *62*, 3548–3552. [CrossRef]
61. Freitas, P.R.; de Araújo, A.C.J.; dos Santos Barbosa, C.R.; Muniz, D.F.; da Silva, A.C.A.; Rocha, J.E.; de Morais Oliveira-Tintino, C.D.; Ribeiro-Filho, J.; da Silva, L.E.; Confortin, C. GC-MS-FID and potentiation of the antibiotic activity of the essential oil of *Baccharis reticulata* (ruiz & pav.) pers. and α-pinene. *Ind. Crops. Prod.* **2020**, *145*, 112106.
62. Zhu, Z.; Min, T.; Zhang, X.; Wen, Y. Microencapsulation of Thymol in Poly (lactide-co-glycolide)(PLGA): Physical and Antibacterial Properties. *Materials* **2019**, *12*, 1133. [CrossRef] [PubMed]
63. Khoshbakht, T.; Karami, A.; Tahmasebi, A.; Maggi, F. The variability of thymol and carvacrol contents reveals the level of antibacterial activity of the Essential Oils from different accessions of *Oliveria decumbens*. *Antibiotics* **2020**, *9*, 409. [CrossRef] [PubMed]
64. Almehdar, H.; Abdallah, H.M.; Osman, A.-M.M.; Abdel-Sattar, E.A. In vitro cytotoxic screening of selected Saudi medicinal plants. *J. Nat. Med.* **2012**, *66*, 406–412. [CrossRef]
65. Jo, H.; Cha, B.; Kim, H.; Brito, S.; Kwak, B.M.; Kim, S.T.; Bin, B.-H.; Lee, M.-G. α-pinene enhances the anticancer activity of natural killer cells via ERK/AKT pathway. *Int. J. Mol. Sci.* **2021**, *22*, 656. [CrossRef]

66. Seresht, H.R.; Albadry, B.J.; Al-mosawi, A.K.M.; Gholami, O.; Cheshomi, H. The cytotoxic effects of thymol as the major component of trachyspermum ammi on breast cancer (MCF-7) cells. *Pharm. Chem. J.* **2019**, *53*, 101–107. [CrossRef]
67. Alkhathlan, H.; Khan, M.; Abdullah, M.; AlMayouf, A.; Badjah-Hadj-Ahmed, A.Y.; AlOthman, Z.; Mousa, A. Anticorrosive assay-guided isolation of active phytoconstituents from *Anthemis pseudocotula* extracts and a detailed study of their effects on the corrosion of mild steel in acidic media. *RSC Adv.* **2015**, *5*, 54283–54292. [CrossRef]
68. Al-Hwaiti, M.S.; Alsbou, E.M.; Abu Sheikha, G.; Bakchiche, B.; Pham, T.H.; Thomas, R.H.; Bardaweel, S.K. Evaluation of the anticancer activity and fatty acids composition of "Handal"(*Citrullus colocynthis* L.) seed oil, a desert plant from south Jordan. *Food Sci. Nutr.* **2021**, *9*, 282–289. [CrossRef] [PubMed]
69. Joshi, G.; Kaur, J.; Sharma, P.; Kaur, G.; Bhandari, Y.; Kumar, R.; Singh, S. P53-mediated anticancer activity of *Citrullus colocynthis* extracts. *Nat. Prod. J.* **2019**, *9*, 303–311. [CrossRef]
70. Khan, M.; Al-Saleem, M.S.; Alkhathlan, H.Z. A detailed study on chemical characterization of essential oil components of two Plectranthus species grown in Saudi Arabia. *J. Saudi Chem. Soc.* **2016**, *20*, 711–721. [CrossRef]
71. Acree, T.; Arn, H. Gas Chromatography-Olfactometry (GCO) of Natural Products. Flavornet and Human Odor Space, Sponsored by DATU Inc. 2004. Available online: http://www.flavornet.org (accessed on 25 December 2022).
72. NIST Mass Spectrometry Data Center, W.E.W. Director "Retention Indices". In *NIST Chemistry WebBook; NIST Standard Reference Database Number, 69*; Linstrom, P.J., Mallard, W.G., Eds.; National Institute of Standards and Technology: Gaithersburg, MD, USA, 2020; p. 20899. [CrossRef]
73. Adams, R.P. *Identification of Essential Oil Components by Gas Chromatography/Mass Spectrometry*, 4th ed.; Allured Publishing Corporation: Carol Stream, IL, USA, 2007.
74. Swapnaja, K.J.M.; Yennam, S.; Chavali, M.; Poornachandra, Y.; Kumar, C.G.; Muthusamy, K.; Jayaraman, V.B.; Arumugam, P.; Balasubramanian, S.; Sriram, K.K. Design, synthesis and biological evaluation of diaziridinyl quinone isoxazole hybrids. *Eur. J. Med. Chem.* **2016**, *117*, 85–98. [CrossRef]
75. Hansen, M.B.; Nielsen, S.E.; Berg, K. Re-examination and further development of a precise and rapid dye method for measuring cell growth/cell kill. *J. Immunol. Methods* **1989**, *119*, 203–210. [CrossRef] [PubMed]

plants

Article

Biological Activity of *Cupressus sempervirens* Essential Oil

Lucia Galovičová [1,*], Natália Čmiková [1], Marianna Schwarzová [1], Milena D. Vukic [1,2], Nenad L. Vukovic [2], Przemysław Łukasz Kowalczewski [3], Ladislav Bakay [4], Maciej Ireneusz Kluz [5], Czeslaw Puchalski [5], Ana D. Obradovic [6], Miloš M. Matić [6] and Miroslava Kačániová [1,5,*]

1 Institute of Horticulture, Faculty of Horticulture and Landscape Engineering, Slovak University of Agriculture, Tr. A. Hlinku 2, 94976 Nitra, Slovakia
2 Department of Chemistry, Faculty of Science, University of Kragujevac, 34000 Kragujevac, Serbia
3 Department of Food Technology of Plant Origin, Poznań University of Life Sciences, 31 Wojska Polskiego St., 60-624 Poznań, Poland
4 Institute of Landscape Architecture, Faculty of Horticulture and Landscape Engineering, Slovak University of Agriculture, Tr. A. Hlinku 2, 94976 Nitra, Slovakia
5 Department of Bioenergetics and Food Analysis, Institute of Food Technology and Nutrition, University of Rzeszow, Zelwerowicza 4, 35-601 Rzeszow, Poland
6 Department of Biology and Ecology, Faculty of Science, University of Kragujevac, 34000 Kragujevac, Serbia
* Correspondence: l.galovicova95@gmail.com (L.G.); miroslava.kacaniova@gmail.com (M.K.)

Abstract: The aim of this study was to evaluate the antioxidant, antibiofilm, antimicrobial (in situ and in vitro), insecticidal, and antiproliferative activity of *Cupressus sempervirens* essential oil (CSEO) obtained from the plant leaf. The identification of the constituents contained in CSEO was also intended by using GC and GC/MS analysis. The chemical composition revealed that this sample was dominated by monoterpene hydrocarbons α-pinene, and δ-3-carene. Free radical scavenging ability, performed by using DPPH and ABTS assays, was evaluated as strong. Higher antibacterial efficacy was demonstrated for the agar diffusion method compared to the disk diffusion method. The antifungal activity of CSEO was moderate. When the minimum inhibitory concentrations of filamentous microscopic fungi were determined, we observed the efficacy depending on the concentration used, except for *B. cinerea* where the efficacy of lower concentration was more pronounced. The vapor phase effect was more pronounced at lower concentrations in most cases. Antibiofilm effect against *Salmonella enterica* was demonstrated. The relatively strong insecticidal activity was demonstrated with an LC50 value of 21.07% and an LC90 value of 78.21%, making CSEO potentially adequate in the control of agricultural insect pests. Results of cell viability testing showed no effects on the normal MRC-5 cell line, and antiproliferative effects towards MDA-MB-231, HCT-116, JEG-3, and K562 cells, whereas K562 cells were the most sensitive. Based on our results, CSEO could be a suitable alternative against different types of microorganisms as well as suitable for the control of biofilms. Due to its insecticidal properties, it could be used in the control of agricultural insect pests.

Keywords: *Cupressus sempervirens*; MALDI-TOF MS Biotyper; biofilm; insecticidal activity; vapor phase

Citation: Galovičová, L.; Čmiková, N.; Schwarzová, M.; Vukic, M.D.; Vukovic, N.L.; Kowalczewski, P.Ł.; Bakay, L.; Kluz, M.I.; Puchalski, C.; Obradovic, A.D.; et al. Biological Activity of *Cupressus sempervirens* Essential Oil. *Plants* **2023**, *12*, 1097. https://doi.org/10.3390/plants12051097

Academic Editors: Hazem Salaheldin Elshafie, Ippolito Camele and Adriano Sofo

Received: 4 February 2023
Revised: 24 February 2023
Accepted: 25 February 2023
Published: 1 March 2023

1. Introduction

Essential oils (EOs) constitute only a fraction of the total plant mass, but they are responsible for significant properties of aromatic plants. Composed of hundreds of biologically active compounds, EOs are effective due to their complex composition. Hydrocarbons such as sesquiterpenes and monoterpenes are dominant, while among oxygenated compounds, aldehydes, ketones, phenols, oxides, alcohols, and acetates are the most abundant. Esters and ethers can also be found in high amounts. Hydrocarbons and oxygenated compounds significantly influence the odor and taste characteristics of EOs [1]. EOs are considered safe substances and can be used as potential antimicrobials [2]. It has also recently been found that EOs can show comparable, and in some cases even higher, efficacy than currently used antimicrobials [3,4].

Cupressus sempervirens L. [Cupressaceae] is an evergreen tree with a distinctive aroma, rich in essential oils. It has been used for centuries in traditional medicine for its expectorant, antiseptic properties. It is most commonly used to treat coughs, bronchitis, diabetes, boils, and laryngitis, but also inflammation and toothache [5]. *C. sempervirens* also exhibit significant biological activities including antioxidant, antimicrobial, and insecticidal effects [6].

Today, food safety is emphasized by manufacturers, regulators, and consumers alike [7]. Despite significant modernization of production techniques, contamination of food with food spoilage-causing pathogens and microorganisms is a global problem [8]. Microbial burden poses a risk to food sustainability as well as alimentary diseases, increasing the burden on public health [9]. In this regard, EOs can be a useful alternative for applications in food preservation.

A biofilm is an organized cluster of microorganisms surrounded by an extracellular polymeric matrix that they produce. The problem with biofilms is that they reduce the effectiveness of antibiotic therapy [10]. Biofilm formation is a mechanism of survival in adverse conditions, and for this reason, it is almost ubiquitous [11]. Literature data show that biofilm formation is influenced, among others, by the characteristics of the contact surface [12]. Biofilms can cause many problems in different industries, including food, medicine, and agriculture.

There are numerous conventional techniques for identifying bacteria, and many of them can be time-consuming, complicated, and expensive [13–15]. Considering the urgency of antibiotic resistance and the struggle with bacterial biofilms, the development of quick and accurate techniques is of high importance [15,16]. In recent years, the MALDI-TOF MS technique has been widely used in microbiology to study different types of bacteria (Gram-positive, Gram-negative, mycobacteria, and anaerobic). The advantages of this method are mainly reflected in its precision, speed, simplicity, and reproducibility, as well as in its price [13–15].

This study aimed to evaluate the antioxidant, antibiofilm, antimicrobial, insecticidal, and antiproliferative activity of *Cupressus sempervirens* essential oil (CSEO). Preliminary testing regarding the antimicrobial activity of the CSEO was performed with disc diffusion and minimum inhibitory concentration assays. To further elucidate the antimicrobial effects of this EO, we have performed an *in situ* assay on kohlrabi as a food model. This assay is performed to assess its potential application as a natural preservative. To our knowledge, there are no previous reports on this species of EO that demonstrate its antimicrobial efficiency in food preservation with an *in situ* assay. Next, we examined the antibiofilm activity of CSEO against *Salmonella enterica* biofilms formed on stainless steel and plastic surfaces using MALDI-TOF MS Biotyper. Additionally, since literature data was lacking, using the MTT assay we determined the antiproliferative activity of this EO towards human breast cancer (MBA-MB-231), human colon cancer (HCT-116), human choriocarcinoma (JEG-3), and chronic myelogenous leukemia (K562). For purpose of examining the biocompatibility of the tested EO, we also evaluated the effects of CSEO on the healthy human lung fibroblast (MRC-5) using the same assay. Finally, by employing GC and GC/MS techniques we have determined the chemical composition of the constituents contained in CSEO.

2. Results

2.1. Chemical Composition

Results presented in Table 1 show the volatile composition of *C. sempervirens* essential oil (CSEO). Thirty-nine compounds were identified, which represents 98.9% of the total. Monoterpene hydrocarbons were present in high abundance (90.7%), with α-pinene (40.5%), and δ-3-carene (24.4%) as the major compounds. Other identified monoterpene hydrocarbons were α-terpinolene (8.6%), limonene (4.3%), β-myrcene (3.0%), sabinene (2.1%), and β-pinene (1.7%). From the class of oxygenated monoterpenes (2.4% of the total), monoterpene alcohol, 4-terpineol, was detected with an abundance of 1.9% of the total. Moreover, from the class of sesquiterpenes (5% of the total), sesquiterpene alcohol, cedrol, was dominant (2.4%). Other compounds present in this EO sample were detected in quantities below 1.5%.

2.2. Antioxidant Activity

The antioxidant potential of CSEO was determined by the means of the neutralization of stable DPPH radical and ABTS radical cation. The obtained results are presented in Table 2. Based on the IC_{50} value, the radical scavenging capacity of the tested EO was found to be stronger than that of the reference compound Trolox. 10 μL of this EO was able to neutralize 76.32 ± 0.43% of DPPH radical, which is equivalent to the 6.46 ± 0.18 TEAC, and 91.53 ± 0.16% of ABTS radical cation (4.92 ± 0.06 TEAC).

Table 1. Chemical composition of CSEO.

No	RT	RI (lit)	RI (calc.) [a]	Compound [b]	%
				Monoterpenes	
				monoterpene hydrocarbons	
1	7.22	926	925	Tricyclene	0.2
2	7.33	930	929	α-Thujene	0.9
3	7.69	939	940	α-Pinene	40.5
4	8.01	952	952	α-Fenchene	0.9
5	8.16	954	953	Camphene	0.5
6	8.28	967	957	Verbenene	tr [c]
7	8.98	975	975	Sabinene	2.1
8	9.17	979	980	β-Pinene	1.7
9	9.60	990	990	β-Myrcene	3
10	10.45	1011	1012	δ-3-Carene	24.4
11	10.70	1017	1018	α-Terpinene	0.8
12	10.80	1024	1021	*p*-Cymene	Tr
13	11.02	1026	1026	*o*-Cymene	0.8
14	11.23	1029	1032	Limonene	4.3
15	11.29	1029	1033	β-Phellandrene	0.4
16	11.48	1050	1048	(*E*)-β-ocimene	Tr
17	12.43	1059	1060	γ-Terpinene	1.2
18	13.42	1088	1086	α-Terpinolene	8.6
19	13.83	1091	1090	*p*-Cymenene	Tr
				Subtotal	90.3
				oxygenated monoterpenes	
				monoterpene ethers	
20	11.35	1031	1035	1,8- Cineole	Tr
21	23.30	1244	1242	Carvacrol methyl ether	Tr
				summ	Tr
				monoterpene alcohols	
22	14.31	1096	1099	Linalool	0.5
23	16.48	1139	1140	trans-Pinocarveol	Tr
24	19.02	1177	1182	4-Terpineol	1.9
25	19.39	1182	1188	*p*-Cymen-8-ol	Tr
26	19.86	1188	1191	α-Terpineol	0.5
				summ	2.9
				monoterpene ketones	
27	16.89	1146	1145	Camphor	tr
28	17.48	1159	1157	Karahanaenone	0.3
29	18.33	1171	1171	Umbellulone	tr
				summ	0.3
				monoterpene esters	
30	24.30	1285	1285	Isobornyl acetate	0.4
				summ	0.4
				subtotal	3.6

Table 1. *Cont.*

No	RT	RI (lit)	RI (calc.) [a]	Compound [b]	%
				sesquiterpenes	
				sesquiterpene hydrocarbons	
31	28.07	1375	1373	α-Ylangene	tr
32	29.69	1411	1412	α-Cedrene	0.6
33	29.85	1419	1417	(*E*)-Caryophyllene	0.5
34	30.01	1420	1421	β-Cedrene	0.1
35	31.17	1454	1453	α-Humulene	0.2
36	31.86	1484	1471	α-Amorphene	0.2
37	32.06	1481	1476	Germacrene D	0.6
38	33.26	1513	1511	γ-Cadinene	0.4
				subtotal	2.6
				oxygenated sesquiterpenes	
				sesquiterpene alcohols	
39	35.89	1600	1608	Cedrol	2.4
				subtotal	2.4
				total	98.9

[a] Values of retention indices on HP-5MS column; [b] Identified compounds; [c] tr-compounds identified in amounts less than 0.1%.

Table 2. In vitro antioxidant activity of CSEO.

	% of Inhibition	TEAC (mg/L)	Trolox (IC_{50}) (mg/L)
$DPPH^{\bullet}$	76.32 ± 0.43	6.46 ± 0.18	4.39 ± 0.13
$ABTS^{\bullet+}$	91.53 ± 0.16	4.92 ± 0.06	2.96 ± 0.01

2.3. Antimicrobial Activity In Vitro

2.3.1. Disc Diffusion Method

A weak antimicrobial activity was observed towards one gram-negative bacterium–*P. aeruginosa* (zone of inhibition of 4.33 ± 1.15 mm), and two gram-positive bacteria–*S. aureus* and *E. faecalis* (inhibition zones of 2.67 ± 0.58 mm and 1.00 ± 0.00 mm, respectively). A weak antimicrobial activity was also observed against the yeast *C. krusei* with a zone of inhibition of 3.67 ± 0.58 mm and the biofilm-producing bacterium *S. enterica* with a zone of inhibition of 4.00 ± 1.00 mm. Moderate antimicrobial activities were detected against two gram-negative bacteria, *Y. enterocolitica* (5.67 ± 0.58 mm) and *S. enterica* (6.67 ± 0.58 mm), two species of yeast, *C. glabrata* (8.33 ± 0.58 mm) and C. *tropicalis* (6.33 ± 0.58 mm), and against all the microscopic filamentous fungi tested, the zones of inhibition ranged from 6.33 ± 0.58 mm for *A. flavus* to 7.67 ± 0.58 mm for *P. citrinum*. A strong antimicrobial effect was observed for the gram-positive bacterium *B. subtilis* (10.33 ± 1.53 mm) and the yeast *C. albicans* (11.33 ± 1.15 mm). Further details on the antimicrobial activity are summarized in Table 3.

2.3.2. Minimal Inhibition Concentration

Low MIC 50 (40.92–93.80 μL/mL) and MIC 90 (82.87–131.49 μL/mL) values were detected for two gram-negative bacteria, *Y. enterocolitica* and *S. enterica*, and two yeasts, *C. krusei,* and *C. tropicalis*. Intermediate MIC 50 (154.31–187.31 μL/mL) and MIC 90 (199.21–289.10 μL/mL) values were observed for *C. glabrata* and the biofilm-producing bacterium *S. enterica*. High MIC 50 (202.11–374.02 μL/mL) and MIC 90 (397.64–401.67 μL/mL) values were observed for *P. aeruginosa, B. subtilis, S. aureus, E. faecalis* and *C. albicans*. Compared to the results obtained with the disk diffusion method, a higher efficiency against gram-positive bacteria and biofilm-producing bacteria was observed in the agar microdilution method. Further details of the minimum inhibitory concentrations are summarized in Table 4.

Table 3. Antimicrobial activity of CSEO with disc diffusion method.

Microorganism	Inhibition Zone	Activity of EO	Control
Gram-negative bacteria			
Pseudomonas aeruginosa	4.33 ± 1.15	*	25.00 ± 0.03
Yersinia enterocolitica	5.67 ± 0.58	**	24.00 ± 0.08
Salmonella enterica subsp. *enterica* ser. Enteritidis	6.67 ± 0.58	**	28.00 ± 0.06
Salmonella enterica biofilm	4.00 ± 1.00	*	25.00 ± 0.02
Gram-positive bacteria			
Bacillus subtilis	10.33 ± 1.53	***	26.00 ± 0.05
Staphylococcus aureus subsp. *aureus*	2.67 ± 0.58	*	24.00 ± 0.08
Enterococcus faecalis	1.00 ± 0.00	*	25.00 ± 0.08
Yeasts			
Candida krusei	3.67 ± 0.58	*	24.00 ± 0.09
Candida albicans	11.33 ± 1.15	***	26.00 ± 0.08
Candida tropicalis	6.33 ± 0.58	**	25.00 ± 0.02
Candida glabrata	8.33 ± 0.58	**	28.00 ± 0.04
Fungi			
Aspergillus flavus	6.33 ± 0.58	**	29.00 ± 1.00
Botrytis cinerae	6.67 ± 0.58	**	30.00 ± 1.00
Penicillium citrinum	7.67 ± 0.58	**	27.00 ± 1.50

* Weak activity (zone 1–5 mm); ** Moderate activity (zone 5–10 mm); *** Strong activity (over 10 mm); antibiotics used as a control: cefoxitin for G^- bacteria, gentamicin for G^+ bacteria, fluconazole for microscopic filamentous fungi.

Table 4. Antimicrobial activity of CSEO.

Microorganism	MIC50 (μL/mL)	MIC90 (μL/mL)
Gram-negative bacteria		
Pseudomonas aeruginosa	202.11	401.67
Yersinia enterocolitica	93.80	99.91
Salmonella enterica subsp. *enterica* ser. Enteritidis	40.92	82.87
Salmonella enterica biofilm	187.31	199.21
Gram-positive bacteria		
Bacillus subtilis	374.02	397.64
Staphylococcus aureus subsp. *aureus*	202.11	401.67
Enterococcus faecalis	248.24	413.93
Yeasts		
Candida albicans	93.80	99.91
Candida glabrata	374.02	397.64
Candida krusei	71.30	131.49
Candida tropicalis	154.31	289.10

The minimum inhibitory concentrations of CSEO against three species of microscopic filamentous fungi (*Penicillium citrinum*, *Botrytis cinerea*, and *Aspergillus flavus*) were evaluated by a different method because mycelial growth is difficult to observe in the agar microdilution method. A modification of the disk diffusion method was used where the inhibition zones formed around the disk were measured when different concentrations of

CSEO diluted in DMSO were applied. The largest inhibition zone was recorded against *P. citrinum* at the highest concentration tested, 500 μL/mL (7.67 ± 1.53 mm), and with decreasing CSEO concentration the inhibition zones decreased resulting in the inhibition zone of 6.33 ± 1.53 mm at a concentration of 62.5 μL/mL. A more pronounced effect was observed against *B. cinerea* when lower concentrations were used. The largest inhibition zone of 8.67 ± 1.15 mm was observed at an applied concentration of 62.5 μL/mL, and the highest concentration tested, 500 μL/mL, an inhibition zone of only 4.67 ± 0.58 mm was observed. Against *A. flavus*, the highest inhibition zone of 7.33 ± 0.58 mm was observed at the highest CSEO concentration of 500 μL/mL, and the inhibition zones decreased as the concentration decreased. A more detailed summary of the results is shown in Table 5.

Table 5. Minimum inhibitory concentrations of CSEO against microscopic fungi.

Fungi	Concentration (μL/mL)	Inhibition Zone (mm)
P. citrinum	500	7.67 ± 1.53
	250	7.33 ± 0.58
	125	7.00 ± 1.00
	62.5	6.33 ± 1.53
B. cinerea	500	4.67 ± 0.58
	250	7.33 ± 2.31
	125	7.67 ± 0.58
	62.5	8.67 ± 1.15
A. flavus	500	7.33 ± 0.58
	250	5.33 ± 0.58
	125	4.67 ± 0.58
	62.5	2.67 ± 0.58

2.4. *In Situ Antimicrobial Activity*

In situ antifungal activity analysis demonstrated the efficacy of all tested concentrations against microscopic filamentous fungi, while the efficacy of lower tested concentrations was higher. In situ antibacterial efficacy was demonstrated for all bacterial and yeast species except the biofilm-producing bacterium *S. enterica*, for which only low efficacy was observed at the lowest concentration of 62.5 μL/mL, and growth stimulation occurred at the other concentrations tested. For most of the bacteria and yeasts, as with the microscopic filamentous fungi, we observed higher efficacy with the application of the lower concentration of CSEO. More detailed results are shown in Table 6.

2.5. *Antibiofilm Activity*

The MALDI-TOF MS Biotyper mass spectrometer was used to analyze the differences in mass spectra of the biofilm-producing bacterium *S. enterica*. The anti-biofilm effect of CSEO was evaluated on days 3, 5, 7, 9, 12, and 14 in biofilms developing on stainless steel and plastic surfaces and compared with the control biofilms developing without EO treatment. The control planktonic spectra were used for comparison with the experimental groups from the different surfaces since the evolution of the planktonic control spectra was identical to the control spectra on the surfaces analyzed on each day.

The development of the primary biofilm of the experimental group during day 3 (Figure 1A) was almost identical to that of the control group; no significant changes were observed in the mass spectra of the two groups. As the development progressed to day 5 of the experiment (Figure 1B), there was a difference between the experimental and control groups suggesting an influence of CSEO on *S. enterica* biofilm homeostasis. From day 7 of the experiment to the end of the experiment (Figure 1C–F), the changes in the experimental groups were significant. Significant visual differences can be observed between the experimental and control spectra. Based on these findings, we conclude that CSEO has a significant disrupting effect on the homeostasis of *S. enterica* biofilm thereby leading to its inhibition. Since the biofilm stage (day 5) was already relatively early affected, CSEO has a significant potential to be applied as an anti-biofilm agent.

Table 6. In situ analysis of the antimicrobial activity of the vapor phase of CSEO in kohlrabi.

Bacteria		Bacterial Growth Inhibition (%)			
		The Concentration of CSEO in µL/mL			
		62.5	125	250	500
Gram-negative	*P. aeroginosa*	33.59 ± 2.01 [d]	22.86 ± 1.62 [c]	12.52 ± 1.06 [b]	4.60 ± 1.16 [a]
	Y. enterocolitica	44.74 ± 0.95 [d]	33.53 ± 1.97 [c]	23.34 ± 1.50 [b]	15.00 ± 2.26 [a]
	S. enterica	84.82 ± 3.00 [d]	64.78 ± 2.57 [c]	43.85 ± 1.84 [b]	23.63 ± 1.95 [a]
	S. enterica biofilm	7.05 ± 1.10 [d]	−25.00 ± 2.41 [c]	−44.04 ± 1.62 [b]	−76.03 ± 2.78 [a]
Gram-positive	*B. subtilis*	45.53 ± 2.24 [c]	15.22 ± 1.36 [b]	6.69 ± 1.11 [a]	86.93 ± 2.00 [d]
	S. aureus	35.74 ± 1.06 [c]	14.89 ± 2.25 [b]	43.59 ± 2.21 [d]	7.19 ± 1.33 [a]
	E. faecalis	64.22 ± 1.29 [d]	13.81 ± 2.70 [a]	22.44 ± 0.55 [b]	33.71 ± 1.70 [c]
Yeasts	*C. albicans*	44.33 ± 2.21 [c]	76.15 ± 2.25 [d]	36.30 ± 1.91 [b]	15.63 ± 2.11 [a]
	C. glabrata	34.03 ± 1.53 [c]	14.29 ± 1.45 [a]	43.77 ± 2.05 [d]	24.03 ± 1.28 [b]
	C. krusei	54.33 ± 2.15 [c]	24.51 ± 1.38 [a]	43.78 ± 1.95 [b]	74.96 ± 1.69 [d]
	C. tropicalis	16.78 ± 1.95 [a]	34.93 ± 2.79 [c]	25.74 ± 2.81 [b]	56.63 ± 2.68 [d]
Microscopic fungi	*A. flavus*	64.48 ± 2.81 [a]	54.93 ± 1.54 [d]	13.60 ± 1.78 [c]	23.67 ± 2.01 [b]
	B. cinerea	65.36 ± 2.01 [d]	46.77 ± 2.72 [c]	34.48 ± 2.06 [b]	24.96 ± 2.71 [a]
	P. citrinum	65.54 ± 2.31 [c]	54.34 ± 2.07 [d]	44.18 ± 1.47 [b]	24.03 ± 1.54 [a]

One-Way ANOVA, Individual letters [(a–d)] in the upper case indicate the statistical differences between the concentrations; $p \leq 0.05$; the negative values indicate a probacterial activity of the essential oil against the growth of microbial strains.

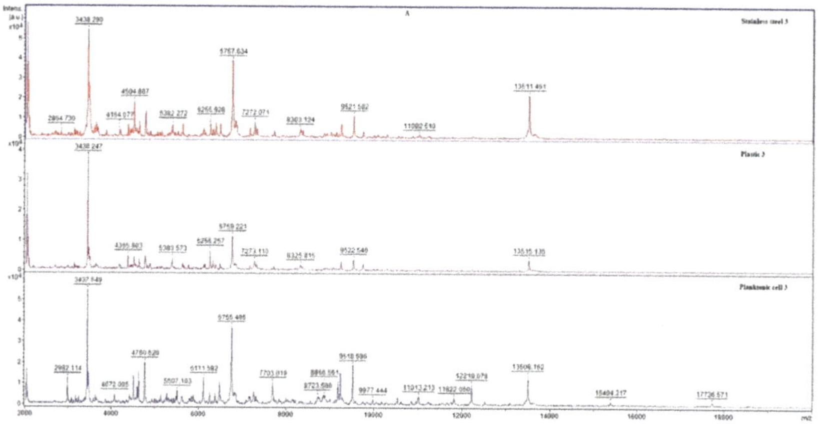

Figure 1. *Cont.*

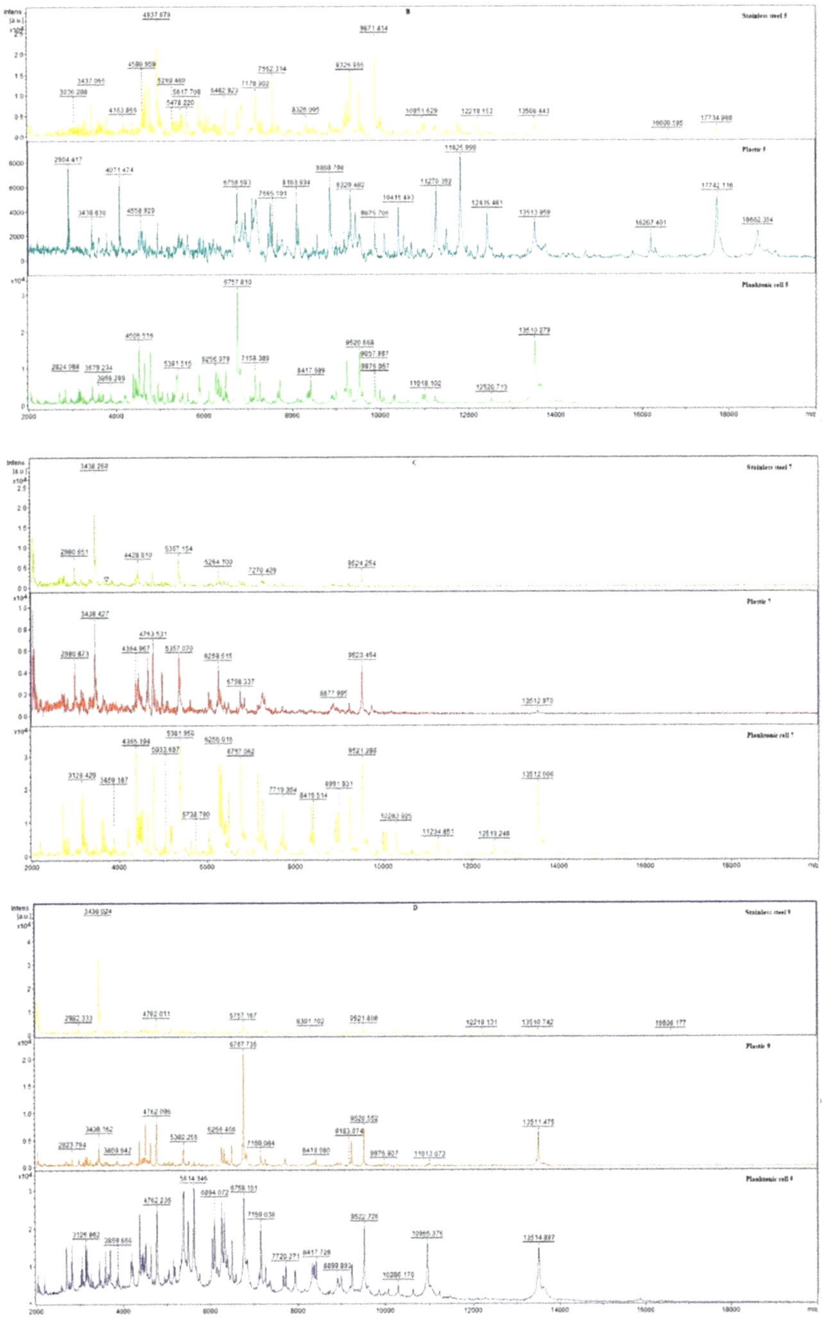

Figure 1. *Cont.*

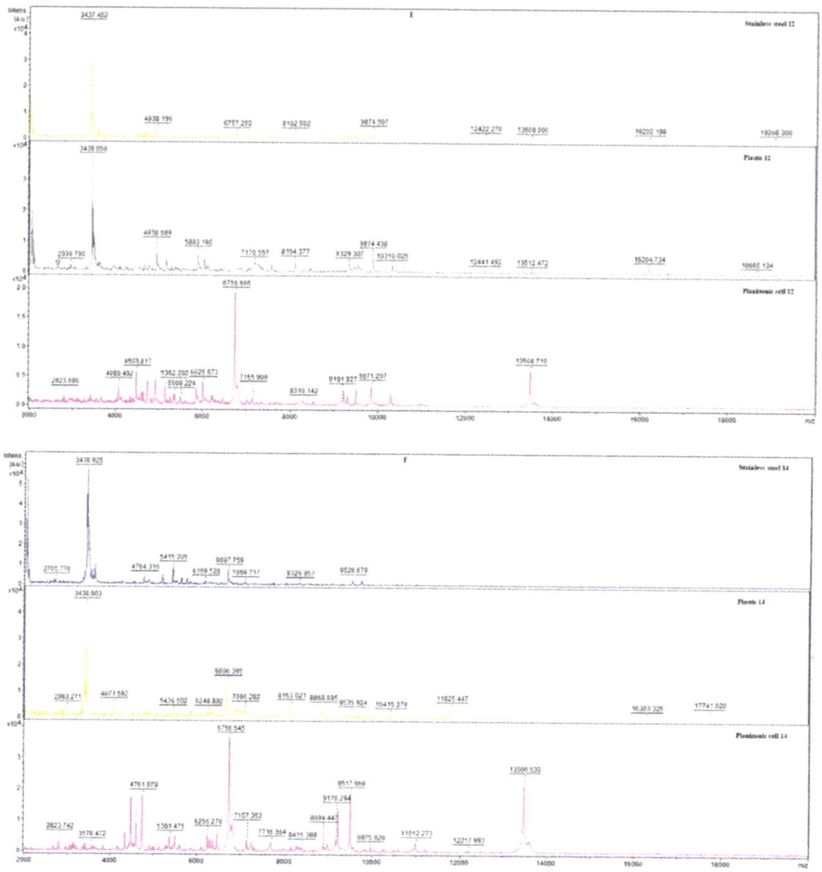

Figure 1. Representative MALDI-TOF mass spectra of *S. enterica*: (**A**) 3rd day, (**B**) 5th day, (**C**) 7th day, (**D**) 9th day, (**E**) 12th day, and (**F**) 14th day.

The dendrogram constructed from the individual MSP distances of the control and experimental mass spectra serves as a visualization of the interrelatedness of biofilms on different surfaces (Figure 2). The constructed dendrogram shows that the shortest MSP distances between the control and experimental groups were observed during days 3 and 5 of the experiment (SES 3;5 and SEP 3;5) along with the control groups throughout the length of the experiment indicating that very little change in the molecular structure of the biofilm was occurring at this time under the influence of CSEO. From day 7 of the experiment (SES 7 and SEP 7) onwards, there was a significant lengthening of the MSP distances of the experimental group from the controls, which is evidence of the inhibitory effect of CSEO against the *S. enterica* biofilm. This trend persisted for the rest of the duration of the experiment. The most significant increases in MSP distances for the experimental groups were observed during days 12 and 14 of the experiment (SES 12;14 and SEP 12;14) indicating significant biofilm inhibition. Based on this evidence, we conclude that CSEO has the potential to act as an antibiofilm agent against *S. enterica* biofilm which confirms the findings inferred from the mass spectra analysis.

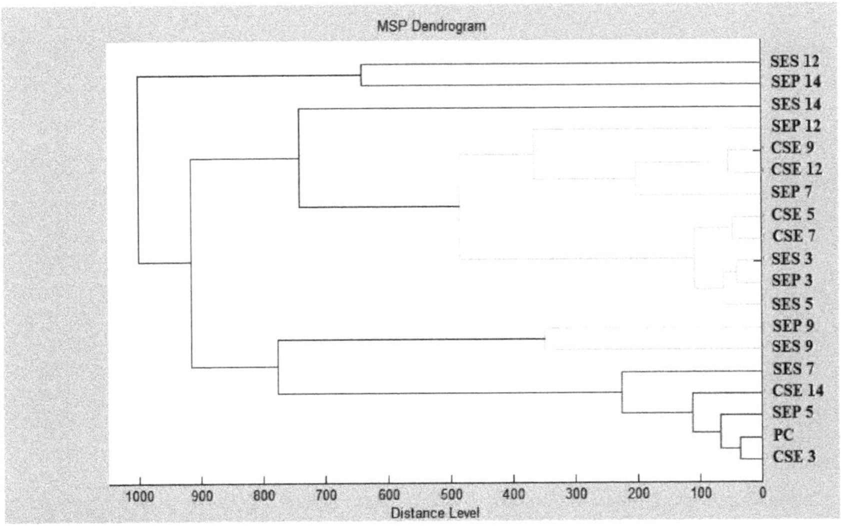

Figure 2. Dendrogram of *S. enterica* biofilm progress after CSEO exposition. SE–*S. enterica*; C-control; S–stainless-steel; P-plastic; PC-planktonic cells.

2.6. Insecticidal Activity of CSEO

CSEO showed relatively strong insecticidal activity (Table 7) towards *O. lavaterae* with the highest concentration tested showing 100% insecticidal activity, killing all individuals, a concentration of 50% (prepared by dilution in 0.1% polysorbate solution) had 80% insecticidal activity and even a concentration of 25% CSEO had more than 50% insecticidal activity killing half of the individuals. At the lower concentrations tested, insecticidal activities of 36.66% and 10% were observed. The lowest concentration tested 3.125% did not show insecticidal activity as well as the control group in which all individuals survived. Out of the obtained results we have calculated lethal concentrations that are found to be for LC50 21.07% and for LC90 78.21%.

Table 7. Insecticidal activity of CSEO.

Concentration (%)	Number of Living Individuals	Number of Dead Individuals	Insecticidal Activity (%)
100	0	30	100.00
50	6	24	80.00
25	14	16	53.33
12.5	19	11	36.66
6.25	27	3	10.00
3.125	30	0	0.00
Control group	30	0	0.00

2.7. Cell Viability Assay

Additionally, the MTT assay was used to evaluate the effects of CSEO on the viability of human breast cancer (MBA-MB-231), human colon cancer (HCT-116), human choriocarcinoma (JEG-3), chronic myelogenous leukemia (K562), and normal human lung fibroblast (MRC-5) cell lines. During the treatment that lasted 24 h and 72 h, the cells were treated with different concentrations of CSEO (1, 10, 20, 50, 100, and 200 μg/mL), and obtained results are presented in Figures 3 and 4.

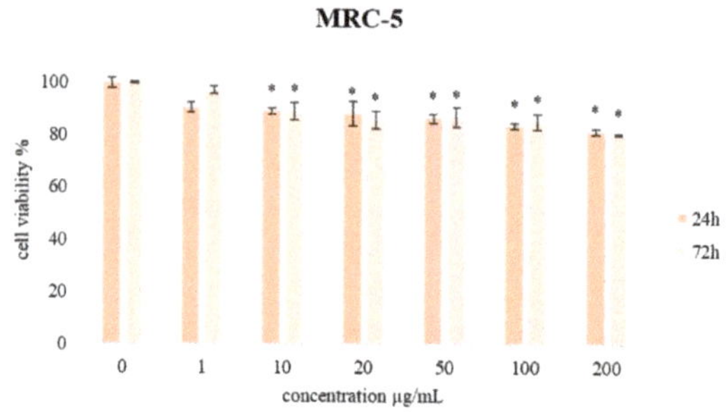

Figure 3. The effects of six concentrations of CSEO on MRC-5 cell viability after 24 h and 72 h of treatment. Results are presented as the mean of three independent experiments ± standard error; * $p < 0.05$ relative to control.

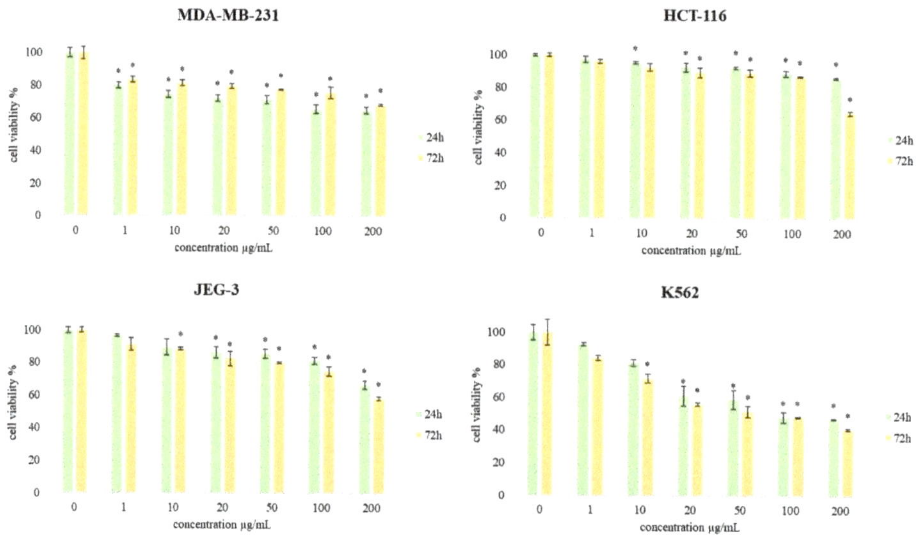

Figure 4. The effects of six concentrations of CSEO on MDA-MB-231, HCT-116, JEG-3, and K562 cell viability after 24 h and 72 h of treatment. Results are presented as the mean of three independent experiments ± standard error: * $p < 0.05$ relative to control.

The selectivity of the CSEO toward cancer cells was examined by treatment of normal human lung fibroblast cell line MRC-5. Obtained results are presented in Figure 3. The proliferation level of cultivated MRC-5 cells was higher than 80.25% compared to non-treated cells in a concentration of 200 µg/mL, demonstrating that the biocompatibility of CSEO is acceptable. Accordingly, this analysis demonstrated that the tested essential oil did not have a non-specific toxic effect. These results qualify this EO sample as suitable for further evaluation of its antiproliferative effects.

Figure 4 shows the results of cell viability testing of the human breast cancer (MBA-MB-231), human colon cancer (HCT-116), human choriocarcinoma (JEG-3), and chronic myelogenous leukemia (K562) cell lines. Overall, results indicate that all treatments exert time and dose-dependent decrease in cell viability in all cell lines, but the most sensitive was Kthe 562 cell line (47.07%). Considering the presented results, a general conclusion can be made that CSEO induces a significant antiproliferative effect in the tested cells.

Comparing the results, it can be concluded that K562 cells were the most sensitive to the treatment with CSEO, suggesting that the tested essential oil could be most effectively used for the treatment of chronic myelogenous leukemia. Towards K562 cells, IC_{50} values below the maximal tested concentration of 200 μg/mL were observed for both treatment times-24 h (105.06 μg/mL) and 72 h (66.56 μg/mL).

For all six tested concentrations of CSEO, the viability of exposed cancer cells was significantly decreased compared to the control cells, and multifold lower when compared to non-cancer human lung fibroblast cells MRC-5. In addition, the cell viability after 72 h was further decreased when compared to short-term treatment, suggesting a time-dependent effect.

3. Discussion

A large number of scientific findings dealing with the chemical composition of EO concluded that the amount of volatiles present in plant matter varies depending on the region in which the plant was growing, the extraction method used, the genetic background of the species, as well as environmental factors such as altitude, climate, soils, and precipitations [17]. Previous studies on *C. sempervirens* essential oil chemical composition reported 10–67 compounds identified depending on the plant organ used for EO extraction [17–31]. EOs obtained from the leaves of this species are characterized by α-pinene and δ-3-carene as major components [17,18,21,22,24,28,31]. These results are in agreement with the ones obtained in this study. Considering the volatiles present in minor amounts, our sample was characterized by a notable amount of α-terpinolene, limonene, β-myrcene, sabinene, and cedrol. Selim et al. reported limonene and α-terpinolene being the next most abundant compounds, while some other studies show a high abundance of cedrol, α-terpenylacetate, myrcene, and β-caryophyllene [17,21,22,24].

The tests commonly used to examine the antioxidant capacity of essential oils are DPPH and ABTS radical scavenging assays. The presence of compounds with different functional groups in complex mixtures such as EOs as well as the analytical method used can lead to varying results. Previous reports indicate moderate to the strong antioxidant potential of the *C. sempervirens* EOs. Ben Nouri et al. determined the strong potency of this EO toward both DPPH radical and ABTS radical cation [18]. However, some other authors showed moderate radical-scavenging activity toward the DPPH radical [29,32–34]. The results obtained in this study show that 10 μL of CSEO can neutralize 76.32 ± 0.43% of DPPH radical, and 91.53 ± 0.16% of ABTS radical cation (4.92 ± 0.06 TEAC). Compared to the IC_{50} value of the standard Trolox which is estimated at 4.39 ± 0.13 mL/L for DPPH radical and 2.96 ± 0.01 mL/L for ABTS radical cation we can conclude that the sample investigated in this study is a strong antioxidant against DPPH radical and ABTS radical cation. The antioxidant activity of essential oils is mainly described as a synergistic or antagonistic effect of two or more of its components [18].

Mazari et al. [22] determined the diameters of inhibition zones including the paper disc (6 mm) for *P. aeruginosa* (7 mm), *E. faecalis* (9 mm), *S. aureus* (10.3 mm), *B. cereus* (7.6 mm), and *E. coli* (9.3 mm) after application of CSEO. Considering that the authors measured the diameter of the inhibition zone using a disc (6 mm) and in our study, the radius of the inhibition zone was measured from the edge of the disc to the edge of the zone so the inhibition zones detected by us for *P. aeruginosa* are larger than those in the referenced study. We detected similar inhibition zones for *E. faecalis* and *S. aureus*. Elansary et al. [29] also evaluated the antimicrobial activity of CSEO by disc diffusion method by measuring the diameter of inhibition zones with the disc (5 mm) and detected inhibition zones for *P. aeruginosa* no inhibition zone was observed, *B. subtilis* (18 mm), *S. aureus* (13 mm). Again, if we consider that the authors measured the diameter, and in our work, the radius was measured, the zones of inhibition reported in the mentioned paper were larger compared to our study except for *P. aeruginosa* in the case of which we determined an inhibition zone of 4.33 mm. Layal et al. [35] determined the diameters of the zones of inhibition in the disc rim (5 mm) for *P. aeruginosa* (7 mm) and *S. aureus* (12 mm) after the application of CSEO. The inhibition zones detected by us were higher for *P. aeruginosa* and lower for *S. aureus*. The differences between our findings and the results of other authors may be due to the different origins of the microorganisms as well as the different origins of CSEO and its acquisition method.

Emami et al. [31] analyzed the antifungal activity of CSEO on 8 microscopic filamentous fungi (*Fusarium culmorum, Fusarium oxysporum, Fusarium equisiti, Fusarium verticillioides, Fusarium nygamai, Botrytis cinerea*, and *Microvar. Alternochium* var. *nivale*), and against all the tested species strong antifungal effects were observed at low concentrations. These findings are consistent with our results which reflect a significant antifungal activity of CSEO against microscopic filamentous fungi. Tekaya-Karoui et al. [36] analyzed the antifungal effect of CSEO against 10 species of microscopic filamentous and also reported strong antifungal effects. In agreement with our findings, the authors concluded that CSEO has a significant potential to act as an antifungal agent. Mazari et al. [22] determined the antifungal activity of CSEO against 3 species of filamentous microscopic fungi (*Aspergillus flavus, Fusarium oxysporum*, and *Rhizopus stolonifer*) and observed very large zones of inhibition (31–80 mm). In our work, we detected significantly smaller inhibition zones which nonetheless support the claim that CSEO has significant potential for the control of microscopic filamentous fungi. Badawy et al. [37] reported a significant antifungal effect of CSEO against *B. cinerea* at low concentrations, and they believe that CSEO could be suitable for controlling this plant pathogen. Our findings are in agreement with this, as the effect against *B. cinerea* observed by us was very strong.

Elansary et al. [29] determined minimal inhibitory concentrations of CSEO for *B. subtilis* and reported MIC values of less than 250 μg/mL. Their results for *S. aureus* and *P. aeruginosa* were 500 μg/mL and 2000 μg/mL, respectively. Since the authors do not distinguish between MIC 50 and MIC 90 as in our work, it is difficult to compare the given values, but for *B. subtilis* their MIC values were lower than MIC 50 and MIC 90 in our work, for *S. aureus* and *P. aeruginosa*, on the other hand, the MIC values are significantly higher than MIC 50 and MIC 90 in our work. Rguez et al. [38] determined the MIC values of CSEO for *S. aureus* (50 μg/mL) and *P. aeruginosa* (250 μg/mL). The findings of these authors are significantly lower for *S. aureus* than in our work but the MIC value of 250 μg/mL for *P. aeruginosa* is in the range of our MIC values of 50 and MIC 90. Taghreed [39] determined MIC values of 90 for *C. albicans* (0.42 μg/mL), *C. krusei* (64 μg/mL), and *C. glabrata* (64 μg/mL) after the application of CSEO, and these values are significantly lower than those determined in our work. The differences between the referenced and our results could have been caused by the different essential oil soils as well as the different microbial strains used for the analyses.

Ismail et al. [40] found that CSEO showed significant growth inhibition of all tested fungal species (*Fusarium culmorum, Fusarium oxysporum, Fusarium equisiti, Fusarium verticillioides, Fusarium nygamai, Botrytis cinerea, Microdochium nivale* var. *nivale*, and *Alternaria* sp.). Moreover, they determined CSEO to be effective against *B. cinerea* in a lower concentration. This finding is in agreement with our observations. Pansera et al. [41] observed the inhibitory effect of CSEO against conidial germination of several microscopic fungi, including *B. cinerea*, at very low concentrations, and concluded that this EO has the potential for use as an antifungal agent. The findings of the authors of this study support our findings.

Reyes-Jurado et al. [42] evaluated the effect of the vapor phase of many essential oils as potential antimicrobial agents and concluded that the results are encouraging and suggest possible applications in food preservation. Kloucek et al. [43] observed that filamentous microscopic fungi were more sensitive to vapor application than bacteria and yeasts. This statement is in agreement with our findings since higher efficiency was observed for filamentous microscopic fungi at lower EO concentrations. In a work by Kačániová et al. [44], a significantly higher efficacy of vapor phase application of EO compared to contact application was found. This is confirmed by our findings as the efficiency was higher at lower concentrations compared to contact application. In their work, Vimalaha et al. [45] evaluated the effect of vapor phase CSEO against viruses and bacteria, reporting promising results and suggesting that certain components contained in CSEO make its effect more pronounced in the vapor phase. These results are in accordance with our findings on the effectiveness of the CSEO vapor phase.

Jose et al. [46], who used microscopy and the crystal violet test, observed a significant eradication of *Klebsiella pneumoniae* biofilm after CSEO application. Despite the use of different species of biofilm-producing bacteria and other methods, these results are in agreement with our findings on the suitability of CSEO for the control of biofilm-producing bacteria. Rehman [47] investigated the antibiofilm effect of CSEO against *B. subtilis*, *P. multocida*, and *E. coli* and reported a very strong inhibitory effect determined using the crystal violet spectrophotometric assay. Despite methodological differences, our findings are in agreement with this observation. Pedroso et al. [48] reported that CSEO showed antibiofilm activity against all *Candida albicans* species tested which indicated that it could be an adjuvant in the treatment of *Candida albicans* infections associated with biofilm. Their results were obtained using a minimum inhibitory concentration test for biofilm and the MTT assay. Although they did not use MALDI TOF to evaluate the antibiofilm activity, as was the case in our study, they came to the same conclusion on the suitability of CSEO for combating biofilm. Jose et al. also confirmed the efficacy of CSEO against biofilms. Our findings are consistent with the previously published results on the efficacy of CSEO against biofilms. Silva [49] reported the antibiofilm effect of CSEO against biofilm produced by *Candia* species using an MTT assay. Despite the differences in the methods used, the author came to the same conclusion that CSEO has the potential to inhibit biofilm. Pereira et al. [16] reported that the use of MALDI TOF MS is sensitive enough to detect phenotypic changes in biofilm progression, and is also able to detect some distinct characteristics related to the surface on which the bacteria were grown. Kırmusaoğlu [50] states that the use of MALDI TOF MS is a very suitable method to study biofilm because extracellular polymeric substances (EPS) do not only contain polysaccharides, but also proteins such as extracellular enzymes. These expressed proteins localized in the EPS matrix can be detected and characterized by MALDI TOF MS.

Langsi et al. [51], evaluated the insecticidal activity of CSEO against insect pests of maize noting high efficacy. Our findings are in agreement with those of these authors. Pinto et al. [52] observed strong insecticidal effects of CSEO against *T. absoluta* in a dose- and exposure-duration-dependent manner. These findings are consistent with our observations, since in our experiment the effect decreased as a function of the concentration used. Almadii and Nenaah [24] reported that CSEO caused significant insecticidal bioactivity against *Culex quinquefasciatus*. Our findings on insecticidal activity are in agreement with the findings of other authors. Saad et al. [53] evaluated the efficacy of CSEO at different concentrations (50, 100, 200, 300, and 400 μg/mL) against *Tribolium castaneum* individuals, recording efficacies ranging from 42% to 20%. Compared to this author, the efficacy of the essential oil tested by us was higher. Ulukanli et al. [54] also reported good insecticidal efficacy of CSEO against *Ephestia kuehniella*. The authors' findings support our findings that EO has the potential to be used as an insecticidal agent. Borotova et al. [55] stated that essential oils are a suitable alternative to synthetic insecticides.

Studies on the antiproliferative activity of CSEO are scarce. However, some reports show that it exerts a cytotoxic effect against NB4, HL-60, and EACC cell lines with LC_{50} values in concentrations ranging from 333.79 μg/mL to 372.43 μg/mL [34]. The results obtained in our study show that CSEO has antiproliferative activity towards human breast cancer MDA-MB-231, colon cancer HCT-116 cell line, choriocarcinoma JEG-3, and chronic myelogenous leukemia K562 cell line, and at the same time does not affect the viability of the human lung fibroblast cell line MRC-5. Moreover, the obtained data imply that K562 cells are the most sensitive to the treatment, suggesting that the essential oil tested in our study could be used for the treatment of chronic myelogenous leukemia.

4. Materials and Methods

4.1. Essential Oil

Cupressus sempervirens essential oil (CSEO) was purchased for research purposes from the Slovak company Hanus s.r.o. The essential oil (obtained from plant leaf) was stored at 4 °C without access to light during the whole duration of the experiments.

4.2. Microorganisms

Microorganisms purchased from the Czech Collection of Microorganisms (Brno, Czech Republic) were used for the analysis of antimicrobial activity. Analyses were performed on three gram-negative bacteria (*Pseudomonas aeruginosa* CCM 3955, *Yersinia enterocolitica* CCM 7204, *Salmonella enterica* subsp. *enterica* ser. Enteritidis CCM 4420), and 3 gram-positive bacteria (*Bacillus subtilis* CCM 1999, *Staphylococcus aureus* subsp. *aureus* CCM 2461, *Enterococcus faecalis* CCM 4224,). Analyses were also performed on 4 species of yeasts of the genus Candia (*Candida krusei* CCM 8271, *Candida albicans* CCM 8261, *Candida tropicalis* CCM 8223, *Candida glabrata* CCM 8270). A biofilm-producing strain of *S. enterica* isolated from a chicken sample during previous studies was used for the analysis of antibiofilm activity. Antifungal activity was evaluated on 3 fungal species (*Penicillium citrinum*, *Botrytis cinerea*, and *Aspergillus flavus*) that were isolated from grape samples. The isolated biofilm-producing bacterial strain and the tested fungal species were subjected to 16S rRNA sequencing and were also identified by MALDI-TOF MS Biotyper.

4.3. Chemical Characterization of CSEO by Gas Chromatography/Mass Spectrometry (GC/MS) and Gas Chromatography (GC-FID)

The volatile detection of *C. sempervirens* essential oil has been performed using GC and GC/MS analysis. For this purpose, used was Agilent Technologies (Palo Alto, Santa Clara, CA, USA) 6890 N gas chromatograph. The chromatograph was equipped with corresponding quadrupole mass spectrometer 5975 B (Agilent Technologies, Santa Clara, CA, USA), and operated with an interfaced HP Enhanced ChemStation software (Agilent Technologies). Separation of volatile compounds was performed by employing the HP-5MS capillary column (30 m × 0.25 mm × 0.25 μm). A sample of essential oil has been prepared by dilution with hexane (10% solution), and the injection volume was 1 μL. As carrier gas helium 5.0 was used, with the flow rate of 1mL/min. The temperature of the split/splitless injector, the MS source, and the MS quadruple was set at 280 °C, 230 °C, and 150 °C respectively. The mass scan range was 35–550 amu at 70 eV. The solvent delay time was 3.2 min for essential oil sample analysis, while in the case of n-alkanes (C7–C35), the solvent delay time was sat at 2.1 min to obtain the retention index for n-heptane which is identified at 2.6 min. The temperature conditions of the performed analysis were set as follows: from 50 °C to 90 °C (with a rate of increase of 3 °C/min), held for 4 min at 90 °C, from 90 °C to 130 °C (with a rate of increase of 4 °C/min), held for 1 min at 130 °C, from 130 °C to 290 °C (with a rate of increase of 5 °C/min). The total run time was 60 min, and the split ratio was 40.8:1. Volatile constituents were identified by the means of their retention indices (RI) comparison, as well as the reference spectra reported in the literature and the ones stored in the MS library (Wiley7Nist) [56,57]. Using GC-FID with the same HP-5MS capillary column performed was semi-quantification of the components taking into consideration amounts higher than 0.1%.

4.4. Antioxidant Activity

4.4.1. DPPH Assay

The antioxidant activity of CSEO was determined using 2,2-diphenyl-1-picrylhydrazyl (DPPH, Sigma Aldrich, Germany). A stock solution of DPPH (0.025 g/L DPPH dissolved in methanol) was diluted by adding methanol to an absorbance of 0.7 at 515 nm. The analysis was carried out in 96-well microplates where 190 μL of diluted DPPH solution (absorbance 0.7) was injected into the well followed by the addition of 10 μL of CSEO. The prepared plate was incubated for 30 min at laboratory temperature without access to light on a shaker plate (MS 3 digital, IKA®, Deutschland, Germany) at 1000 rpm. The antioxidant activity of CSEO was expressed using the percentage of DPPH radical inhibition. $(A0 - AA)/A0 \times 100$ was used for the calculation, where A0 was the absorbance of DPPH and AA was the absorbance of the sample. The relationship between the antioxidant activity and the reference substance Trolox (Sigma Aldrich, Schnelldorf, Germany) dissolved in methanol (Uvasol® for spectroscopy, Merck, Darmstadt, Germany) was calculated over the concentration range 0–5 µg/mL which served as a standard. After constructing the calibration curve, the total antioxidant activity was expressed as the TEAC value. The results were presented as mean values ± standard deviation (SD) of three independent measurements.

4.4.2. ABTS Assay

ABTS [2,20-azinobis(3-ethylbenzothiazoline-6-sulfonic acid) diammonium] radical cation was generated according to the already described procedure [58]. The prepared radical cation was diluted prior to the analysis up to an absorbance value of 0.7 at 744 nm. The 190 µL of this solution was mixed with 10 µL of EO (in a 96-well microtiter plate) for 30 min with continuous shaking at 1000 rpm at room temperature in the dark. A decrease in absorbance at 744 nm was registered and the results are presented as a percentage of ABTS inhibition using the same equation as in the previous section. All measurements were performed in triplicate. Methanol was used as blank, and Trolox as the standard reference substance. Results were expressed as % of inhibition as well as according to the calibration curve of Trolox (TEAC). The results were presented as mean values ± standard deviation (SD) of three independent measurements.

4.5. Antibacterial Activity

4.5.1. Disc Diffusion Method

The disk diffusion method was used to determine the antimicrobial activity in the form of inhibition zones of CSEO. The inoculum of each microorganism was prepared 24 h in advance using Mueller Hinton broth (MHB, Oxoid, Basingstoke, UK) at 37 °C for bacteria and Sabouraud dextrose broth (SDB, Oxoid, Basingstoke, UK) at 25 °C for yeasts. The pre-cultured inoculum was adjusted to an optical density of 0.5 McFarland standard (1.5×10^8 CFU/mL) by dilution with distilled water using a densitometer (BIOSAN, Latvian Republic). 100 µL of treated inoculum was applied to Petri dishes (PD) with Mueller Hinton agar (MHA, Oxoid, Basingstoke, UK) and spread thoroughly with an L-stick. Sterile blank discs (Oxoid, Basingstoke, UK) with a diameter of 6 mm were then placed on the PD. 10 µL of CSEO was applied to each disc. Samples were placed in thermostats according to the respective conditions for bacteria (37 °C) and yeast (25 °C) for 24h. Cefoxitin (Oxoid, Basingstoke, UK) was used as a positive control for gram-positive bacteria, gentamicin (Oxoid, Basingstoke, UK) for gram-negative bacteria, and one antifungal fluconazole (Oxoid, Basingstoke, UK) was used as a positive control for microscopic filamentous fungi. A solution of 0.1% dimethyl sulfoxide (Centralchem, Bratislava, Slovakia) served as a negative control. The radii of the inhibition zones (from disc edge to zone edge) formed by CSEO were measured at three locations and the standard deviation was then calculated. To evaluate the inhibition zones formed by the antibiotic control, the diameter of the inhibition zone (from zone edge to zone edge, including the 6 mm disc) was measured.

To assess the antimicrobial activity of CSEO, the criteria for very strong activity were an inhibition zone of more than 10 mm, for moderate activity an inhibition zone of more than 5 mm, and for weak activity an inhibition zone of less than 5 mm. All measurements were performed in triplicate.

4.5.2. Minimum Inhibitory Concentration (MIC)

The inoculum of each microorganism was prepared 24 h in advance using Mueller Hinton broth (MHB, Oxoid, Basingstoke, UK) at 37 °C for bacteria and Sabouraud dextrose broth (SDB, Oxoid, Basingstoke, UK) at 25 °C for yeast. The precultured inoculum was adjusted to an optical density of 0.5 McFarland standard by dilution with the appropriate broth using a densitometer (BIOSAN, Latvian Republic). The microbial inoculum was cultured for 24 h in Mueller Hinton broth (MHB, Oxoid, Basingstoke, UK) at 37 °C for bacteria and Sabouraud dextrose broth (SDB, Oxoid, Basingstoke, UK) at 25 °C for yeast. 100 μL of modified inoculum with an optical density of 0.5 McFarland standard (McF) was injected into a 96-well microtiter plate. A concentration gradient of CSEO was generated by serial dilution with a concentration range of 500 μL/mL to 0.244 μL/mL in the wells. MHB/SDB with EO was used as a negative control, and inoculum without addition was used as a positive control for maximum growth. At time 0h, the plates were measured with a Glomax spectrophotometer (Promega Inc., Madison, WI, USA) at 570 nm. Subsequently, the plates were placed in thermostats for 24h at the respective temperatures.

After 24 h incubation, absorbance was again measured using the spectrophotometer at 570 nm. The growth of microorganisms after 24 h was calculated. Subsequently, MIC 50 and MIC 90 values were calculated using logit analysis. The test was performed in triplicate.

The minimum inhibitory concentrations of CSEO against three species of microscopic filamentous fungi (*Penicillium citrinum, Botrytis cinerea,* and *Aspergillus flavus*) were evaluated by a different method because mycelial growth is difficult to observe by the agar microdilution method. Four concentrations (500, 250, 125, and 62.5 μL /mL) of CSEO were prepared by dilution in 0.1 % DMSO solution. The inoculum from the 24-hour culture was adjusted to 0.5 McF (1.5×10^8 CFU/mL) by dilution with distilled water. 100 μL of inoculum was applied to the PD with SDA spread with an L-stick. Sterile blank discs were placed on the PDs and 10 μL of the appropriate concentration of CSEO was applied. The dishes were placed in a thermostat at 25 °C for 5 days. After cultivation, the inhibition zone radii were measured, and the mean inhibition zone and standard deviation were calculated for the respective concentration and the microscopic filamentous fungus. Antimicrobial activity was performed in triplicate.

4.6. *In Situ Antimicrobial Activity*

Seven species of bacteria of which 4 are gram-negative (*P. aeruginosa, Y. enterocolitica, S. enterica* subsp. *enterica* ser. Enteritidis, biofilm-forming *S. enterica*), 3 gram-positive bacteria (*B. subtilis, S. aureus* subsp. *aureus, E. faecalis*), 4 yeasts (*C. krusei, C. albicans, C. tropicalis, C. glabrata*) and 3 filamentous microscopic fungi (*P. citrinum, B. cinerea, A. flavus*) were used to analyze the antimicrobial effect of CSEO in situ in the vapor phase. A commercially available vegetable species (kohlrabi) was used as a model food. MHA and SDA were poured into a 60 mm diameter PD according to the species of microorganism (also into the lid). A slice of the model food with a thickness of about 0.5 cm was placed on the agar. The microbial inoculum (preparation described above) was applied to the slice using a bacteriological needle puncture. Dilutions of CSEO in ethyl acetate prepared concentrations of 62.5–500 μL/mL. The test concentrations were applied in a volume of 100 μL on sterile filter paper and placed in a PD cap. The PDs were hermetically sealed and placed in thermostats according to the respective culture conditions of the microorganisms used for 7 days.

After incubation, the experiment was evaluated using stereological methods in ImageJ software. The bulk density (vv) of bacterial colonies was estimated and the grid points that contained colonies (P) and those (p) that were in the reference space (growth medium used) were counted. The bulk density of bacterial colonies was therefore calculated as follows: vv (%) = P/p. The antibacterial activity of EO was defined as the percentage of bacterial growth inhibition (BGI):

$$BGI = [(C - T)/C] \times 100$$

where C and T are the bacterial growth (expressed as volume/volume) in the control and treatment groups, respectively. Negative results represent growth stimulation.

4.7. Antibiofilm Activity

MALDI-TOF MS Biotyper mass spectrometry was used to evaluate the effect of plant essential oil on the degradation of the *Salmonella enterica* biofilm protein profile. The experiment was conducted in 50 mL polypropylene centrifuge tubes. MHB culture medium in a volume of 20 mL was added to the tubes, and then the test surface species were placed in the form of strips about 1 cm thick and 5 cm long. Our analysis was performed on a stainless-steel surface and a plastic surface. For the experimental groups, the culture medium was enriched and supplemented with 0.1% (*w*/*v*) CSEO. A bacterial inoculum of 100 μL adjusted to an optical density of 0.5 McF was added to both groups. The samples were placed on an incubation shaker (GFL 3031, Germany) at 37 °C and 170 rpm.

Samples were analyzed by MALDI-TOF MS Biotyper mass spectrometry on days 3, 5, 7, 9, 12, and 14 of the experiment. Biofilm samples were collected with a sterile cotton swab from both surfaces and applied to a MALDI-TOF metal target plate. Planktonic cells obtained from the culture medium were also subjected to analysis. From the culture medium, 300 μL were collected and centrifuged at 12,000 rpm for 1 min. The supernatant was poured off and the pellet was washed 3 times with 30 μL of ultrapure water. Finally, the clean pellet was resuspended in 30 μL of ultrapure water and 1 μL was applied to a MALDI-TOF target plate. The samples were allowed to dry at room temperature.

After drying, the samples were overlaid with 1 μL of the a-cyano-4-hydroxycinnamic acid matrix (10 mg/mL). After crystallization of the matrix, the samples were analyzed using a MALDI-TOF MicroFlex (Bruker Daltonics, Billerica, MA, USA) with linear and positive mode settings with a range of m/z 200–2000. Using automated analysis, the same similarities were used to generate a standard global spectrum (MSP). Based on the Euclidean distance, 19 spectra were generated in MALDI Biotyper 3.0 and subsequently merged into a dendrogram [59].

4.8. Insecticidal Activity

Thirty individuals of *Oxycarenus lavaterae* were placed in the PD with vents. A circle of filter paper was placed in the lid of the PD on which the appropriate concentration (50, 25, 12.5, 6.25, and 3.125%) of CSEO was prepared by dilution in 0.1% polysorbate solution in a volume of 100 μL. The PDs were sealed around the perimeter using parafilm and left at room temperature for 24 h. A 0.1% polysorbate solution was used as a control. After 24 h, the number of dead and live individuals was evaluated, and insecticidal activity was calculated. The experiment was carried out in triplicate. The values for LC50 and LC90 were calculated using Finney's Probit Analysis [60].

4.9. Determination of Cell Viability (MTT Assay)

The following reagents and chemicals were used: Dulbecco's Modified Eagle medium (DMEM), 10% fetal bovine serum (FBS), 0.4% Trypan blue, 0.25%, trypsin-EDTA, dimethyl sulfoxide (DMSO), 3-(4,5-Dimethylthiazol-2-yl)-2,5-diphenyltetrazoliumbromide (MTT), phosphate-buffered saline (PBS), All the chemicals and reagents used in this study were of the highest commercially available purity.

The human lung normal fibroblast cell line (MRC-5), breast cancer cell line (MDA-MB-231), colon cancer cell line (HCT-116), choriocarcinoma cell line (JEG-3), and chronic myelogenous leukemia cell line (K562) were obtained from American Tissue Culture Collection. These cells were propagated and maintained in DMEM and supplemented with 10% FBS and a combination of antibiotics (100 IU/mL penicillin and 100 μg/mL streptomycin). The cells were grown in a 75 cm^2 culture flask and supplied with 15 mL DMEM at a confluence of 70% to 80%. The cells were seeded in a 96-well microplate (10,000 cells per well) and cultured in a humidified atmosphere with 5% CO_2 at 37 °C. After 24 h of cell incubation, 100 μL of medium containing various doses of treatment (1 μg/mLto 200 μg/)mL was added to each well of the microplate, and the cells were incubated for 24 h and 72 h, after which the evaluation of cell viability was performed. Non-treated cells were used as control. The stock solution was prepared in the concentration of 10 mg/mL, while during the experiment used were concentrations of 1, 10, 20, 50, 100, and 200 μg/mL. The essential oil of *Cupressus semperiverens* was used in experiments. Above mentioned concentrations were obtained from the first stock solution by adding a certain volume of DMEM. All concentrations were used in triplicate for all the methods.

The viability of the cells was determined using an MTT assay [61]. Briefly, the cells were plated at a density of 10,000 cells/ mL (100 / μLwell) in 96-well plates with DMEM. After a period of incubation (24 h), at a temperature of 37 °C and 5% CO_2, the 6 different concentrations of essential oil (from 1 to 200 μg/mL) dissolved in DMEM, were added to each well (100 μL per well). The untreated cells (cultured only in a medium) served as a control. After 24 and 72 h of incubation, the cell viability was determined with an MTT assay where 20 μL of MTT (concentration of 5 mg/mL) was added to each well. MTT is a yellow tetrazolium salt that is reduced to purple formazan in the presence of mitochondrial dehydrogenase. During this reaction, which started approximately after three hours, the formed crystals were dissolved in 20 μL of DMSO. The color formed in the reaction was measured on an ELISA reader at a wavelength of 550 nm. The percentage of viable cells was calculated as the ratio between the absorbance at each dose of the treatment and the absorbance of the non-treated control multiplied by 100 to get a percentage. We also calculated the half-maximal inhibitory concentration (IC_{50}), defined as the concentration of tasted essential oil that inhibited cell growth by 50% when compared to the control. The IC_{50} values were calculated from the dose curves by the software CalcuSyn, Version 2.0.

4.10. Statistical Data Evaluation

One-way analysis of variance (ANOVA) was performed using Prism 8.0.1 (GraphPad Software, San Diego, CA, USA) followed by Tukey's test at $p < 0.05$. SAS® version 8 software was used to process the data. MIC values (concentration that caused 50% and 90% inhibition of bacterial growth) were determined by logit analysis. All data regarding the determination of cell viability were evaluated using IBM-SPSS 23 software for Windows (SPSS Inc., Chicago, IL, USA). The data were presented as a mean ± standard error (S.E.M). The statistical significance was determined using a Paired Sample-T test. The level of statistical significance was set at * $p < 0.05$.

5. Conclusions

The results of our study show that CSEO obtained from the commercial company Hanus s.r.o. produced in Slovakia shows good biological activity. The chemical composition evaluation revealed that CSEO was rich in α-pinene and δ-3-carene. DPPH and ABTS assay showed the better radical scavenging potential of this sample compared to the IC_{50} value Trolox. Higher antibacterial efficacy was determined using the agar diffusion method compared to the disc diffusion method. The antifungal activity of CSEO was weak to moderate depending on the concentration used. Also, the antibiofilm effect of this essential oil was demonstrated indicating its suitability as an alternative substance for combating biofilm-producing pathogens. The vapor phase effect was in most cases more pronounced at lower concentrations, but an inhibitory effect was observed for all the tested microorganisms at all tested concentrations, except for *S. enterica*. Based on our findings, we believe that CSEO could find application in storage extension and the protection of agricultural products in vapor application. The relatively strong insecticidal activity offers the possibility for the future use of CSEO in the control of agricultural insect pests. Results of cell viability testing showed that the viability of exposed cancer cells was significantly decreased compared to control cells in a time- and dose-dependent manner, and multifold lower when compared to non-cancer human lung fibroblast cells MRC-5. K562 cells were found to be the most sensitive to the treatment with CSEO, which suggests that the tested essential oil could be effectively used for the treatment of chronic myelogenous leukemia. Nonetheless, further tests are needed.

Author Contributions: Conceptualization, L.G., N.Č., M.S., M.D.V., N.L.V., A.D.O. and M.K.; Data curation, L.G., N.Č., M.S., M.D.V., N.L.V., P.Ł.K., L.B., M.I.K., C.P. and A.D.O.; Methodology, L.G., N.Č., M.S., M.D.V., N.L.V., L.B., M.M.M. and M.K.; Supervision, M.D.V., N.L.V., M.M.M. and M.K.; Writing–original draft, L.G., N.Č., M.S., M.D.V., N.L.V., P.Ł.K., L.B., M.I.K., C.P. and M.K. All authors have read and agreed to the published version of the manuscript.

Funding: This research was funded by the grant APVV-20-0058 "The potential of the essential oils from aromatic plants for medical use and food preservation".

Data Availability Statement: All data generated or analyzed during this study are included in this published article.

Acknowledgments: This work has been supported by the grants of the VEGA no. 1/0180/20, and by the Serbian Ministry of Education, Science, and Technological Development (Agreement Nos. 451-03-68/2023-01/200122).

Conflicts of Interest: The authors declare no conflict of interest.

References

1. Yousefi, M.; Rahimi-Nasrabadi, M.; Pourmortazavi, S.M.; Wysokowski, M.; Jesionowski, T.; Ehrlich, H.; Mirsadeghi, S. Supercritical Fluid Extraction of Essential Oils. *TrAC Trends Anal. Chem.* **2019**, *118*, 182–193. [CrossRef]
2. Stefanakis, M.K.; Touloupakis, E.; Anastasopoulos, E.; Ghanotakis, D.; Katerinopoulos, H.E.; Makridis, P. Antibacterial Activity of Essential Oils from Plants of the Genus Origanum. *Food Control* **2013**, *34*, 539–546. [CrossRef]
3. Bajpai, V.K.; Baek, K.-H. Biological Efficacy and Application of Essential Oils in Foods—A Review. *J. Essent. Oil Bear. Plants* **2016**, *19*, 1–19. [CrossRef]
4. Cano-Lamadrid, M.; Viuda-Martos, M.; García-Garví, J.M.; Clemente-Villalba, J.; Carbonell-Barrachina, Á.A.; Sendra, E. Polyphenolic Profile and Antimicrobial Potential of Peel Extracts Obtained from Organic Pomegranate (*Punica granatum* L.) Variety "Mollar De Elche". *Acta Hortic. Regiotect.* **2020**, *23*, 1–4. [CrossRef]
5. Nehdi, I.A. Cupressus Sempervirens Var. Horizentalis Seed Oil: Chemical Composition, Physicochemical Characteristics, and Utilizations. *Ind. Crops Prod.* **2013**, *41*, 381–385. [CrossRef]
6. Hasaballah, A.; Shehata, A.; Fouda, M.; Hassan, M.; Gad, M. The Biological Activity of Cupressus Sempervirens Extracts against Musca Domestica. *Asian J. Biol.* **2018**, *5*, 1–12. [CrossRef]
7. Bachir, R.G.; Benali, M. Antibacterial Activity of the Essential Oils from the Leaves of Eucalyptus Globulus against Escherichia Coli and Staphylococcus Aureus. *Asian Pac. J. Trop. Biomed.* **2012**, *2*, 739–742. [CrossRef]
8. Dwivedy, A.K.; Kumar, M.; Upadhyay, N.; Prakash, B.; Dubey, N.K. Plant Essential Oils against Food Borne Fungi and Mycotoxins. *Curr. Opin. Food Sci.* **2016**, *11*, 16–21. [CrossRef]
9. Gonelimali, F.D.; Lin, J.; Miao, W.; Xuan, J.; Charles, F.; Chen, M.; Hatab, S.R. Antimicrobial Properties and Mechanism of Action of Some Plant Extracts Against Food Pathogens and Spoilage Microorganisms. *Front. Microbiol.* **2018**, *9*, 1639. [CrossRef]
10. Jamal, M.; Ahmad, W.; Andleeb, S.; Jalil, F.; Imran, M.; Nawaz, M.A.; Hussain, T.; Ali, M.; Rafiq, M.; Kamil, M.A. Bacterial Biofilm and Associated Infections. *J. Chin. Med. Assoc.* **2018**, *81*, 7–11. [CrossRef]
11. Vestby, L.K.; Grønseth, T.; Simm, R.; Nesse, L.L. Bacterial Biofilm and Its Role in the Pathogenesis of Disease. *Antibiotics* **2020**, *9*, 59. [CrossRef]
12. Lee, K.-G.; Shibamoto, T. Antioxidant Property of Aroma Extract Isolated from Clove Buds [*Syzygium aromaticum* (L.) Merr. et Perry]. *Food Chem.* **2001**, *74*, 443–448. [CrossRef]
13. Alizadeh, M.; Yousefi, L.; Pakdel, F.; Ghotaslou, R.; Rezaee, M.A.; Khodadadi, E.; Oskouei, M.A.; Soroush Barhaghi, M.H.; Kafil, H.S. MALDI-TOF Mass Spectroscopy Applications in Clinical Microbiology. *Adv. Pharmacol. Pharm. Sci.* **2021**, *2021*, 9928238. [CrossRef]
14. Akimowicz, M.; Bucka-Kolendo, J. MALDI-TOF MS—Application in Food Microbiology. *Acta Biochim. Pol.* **2020**, *67*, 327–332. [CrossRef]
15. Elbehiry, A.; Aldubaib, M.; Abalkhail, A.; Marzouk, E.; ALbeloushi, A.; Moussa, I.; Ibrahem, M.; Albazie, H.; Alqarni, A.; Anagreyyah, S.; et al. How MALDI-TOF Mass Spectrometry Technology Contributes to Microbial Infection Control in Healthcare Settings. *Vaccines* **2022**, *10*, 1881. [CrossRef]
16. Pereira, F.D.E.S.; Bonatto, C.C.; Lopes, C.A.P.; Pereira, A.L.; Silva, L.P. Use of MALDI-TOF Mass Spectrometry to Analyze the Molecular Profile of Pseudomonas Aeruginosa Biofilms Grown on Glass and Plastic Surfaces. *Microb. Pathog.* **2015**, *86*, 32–37. [CrossRef]
17. Fadel, H.; Benayache, F.; Chalchat, J.-C.; Figueredo, G.; Chalard, P.; Hazmoune, H.; Benayache, S. Essential Oil Constituents of *Juniperus oxycedrus* L. and *Cupressus sempervirens* L. (Cupressaceae) Growing in Aures Region of Algeria. *Nat. Prod. Res.* **2021**, *35*, 2616–2620. [CrossRef]
18. Ben Nouri, A.; Dhifi, W.; Bellili, S.; Ghazghazi, H.; Aouadhi, C.; Chérif, A.; Hammami, M.; Mnif, W. Chemical Composition, Antioxidant Potential, and Antibacterial Activity of Essential Oil Cones of Tunisian *Cupressus sempervirens*. *J. Chem.* **2015**, *2015*, 538929. [CrossRef]

19. Argui, H.; Youchret-Zalleza, O.B.; Suner, S.C.; Periz, Ç.D.; Türker, G.; Ulusoy, S.; Ben-Attia, M.; Büyükkaya, F.; Oral, A.; Coskun, Y.; et al. Isolation, Chemical Composition, Physicochemical Properties, and Antibacterial Activity of *Cupressus sempervirens* L. Essential Oil. *J. Essent. Oil Bear. Plants* **2021**, *24*, 439–452. [CrossRef]
20. Tapondjou, A.L.; Adler, C.; Fontem, D.A.; Bouda, H.; Reichmuth, C. Bioactivities of Cymol and Essential Oils of Cupressus Sempervirens and Eucalyptus Saligna against Sitophilus Zeamais Motschulsky and Tribolium Confusum Du Val. *J. Stored Prod. Res.* **2005**, *41*, 91–102. [CrossRef]
21. Selim, S.A.; Adam, M.E.; Hassan, S.M.; Albalawi, A.R. Chemical Composition, Antimicrobial and Antibiofilm Activity of the Essential Oil and Methanol Extract of the Mediterranean Cypress (*Cupressus sempervirens* L.). *BMC Complement. Altern. Med.* **2014**, *14*, 179. [CrossRef] [PubMed]
22. Mazari, K.; Bendimerad, N.; Bekhechi, C.; Fernandez, X. Chemical Composition and Antimicrobial Activity of Essential Oils Isolated from Algerian *Juniperus phoenicea* L. and *Cupressus sempervirens* L. *J. Med. Plants Res.* **2010**, *4*, 959–964.
23. Jahani, M.; Akaberi, M.; Khayyat, M.H.; Emami, S.A. Chemical Composition and Antioxidant Activity of Essential Oils from *Cupressus sempervirens*. Var. *sempervirens*, *C. sempervirens*. Cv. Cereiformis and *C. sempervirens* Var. *Horizentalis*. *J. Essent. Oil Bear. Plants* **2019**, *22*, 917–931. [CrossRef]
24. Almadiy, A.A.; Nenaah, G.E. Bioactivity and Safety Evaluations of Cupressus Sempervirens Essential Oil, Its Nanoemulsion and Main Terpenes against Culex Quinquefasciatus Say. *Environ. Sci. Pollut. Res.* **2022**, *29*, 13417–13430. [CrossRef] [PubMed]
25. Asgary, S.; Naderi, G.A.; Shams Ardekani, M.R.; Sahebkar, A.; Airin, A.; Aslani, S.; Kasher, T.; Emami, S.A. Chemical Analysis and Biological Activities of *Cupressus sempervirens* Var. *Horiz. Essent. Oils Pharm. Biol.* **2013**, *51*, 137–144. [CrossRef]
26. Milos, M.; Radonic, A.; Mastelic, J. Seasonal Variation in Essential Oil Compositions of *Cupressus sempervirens* L. *J. Essent. Oil Res.* **2002**, *14*, 222–223. [CrossRef]
27. Kassem, F.F.; Harraz, F.M.; El-Sebakhy, N.A.; De Pooter, H.L.; Schamp, N.M.; Abou-Shleib, H. Composition of the Essential Oil of Egyptian *Cupressus sempervirens* L. Cones. *Flavour Fragr. J.* **1991**, *6*, 205–207. [CrossRef]
28. Boukhris, M.; Regane, G.; Yangui, T.; Sayadi, S.; Bouaziz, M. Chemical Composition and Biological Potential of Essential Oil from Tunisian *Cupressus sempervirens* L. *J. Arid. Land Stud.* **2012**, *22*, 329–332.
29. Elansary, H.O.; Salem, M.Z.M.; Ashmawy, N.A.; Yacout, M.M. Chemical Composition, Antibacterial and Antioxidant Activities of Leaves Essential Oils from *Syzygium cumini* L., *Cupressus sempervirens* L. and *Lantana camara* L. from Egypt. *J. Agric. Sci.* **2012**, *4*, 144–152. [CrossRef]
30. Rguez, S.; Djébali, N.; Ben Slimene, I.; Abid, G.; Hammemi, M.; Chenenaoui, S.; Bachkouel, S.; Daami-Remadi, M.; Ksouri, R.; Hamrouni-Sellami, I. Cupressus Sempervirens Essential Oils and Their Major Compounds Successfully Control Postharvest Grey Mould Disease of Tomato. *Ind. Crops Prod.* **2018**, *123*, 135–141. [CrossRef]
31. Emami, S.A.; Asili, J.; Rahimizadeh, M.; Fazly-Bazzaz, B.S.; Khayyat, M.H. Chemical and Antimicrobial Studies of *Cupressus sempervirens* L. and *C. horizentalis* Mill. Essential Oils. *Iran. J. Pharm. Sci.* **2006**, *2*, 103–108.
32. Orhan, I.E.; Tumen, I. Potential of Cupressus Sempervirens (*Mediterranean cypress*) in Health. In *The Mediterranean Diet*; Elsevier: Amsterdam, The Netherlands, 2015; pp. 639–647. ISBN 978-0-12-407849-9.
33. Sacchetti, R.; Luca, G.D.; Zanetti, F. Influence of Material and Tube Size on DUWLs Contamination in a Pilot Plant. *New Microbiol.* **2007**, *30*, 29–34.
34. Fayed, S.A. Chemical Composition, Antioxidant, Anticancer Properties and Toxicity Evaluation of Leaf Essential Oil of *Cupressus sempervirens*. *Not. Bot. Horti Agrobot. Cluj-Napoca* **2015**, *43*, 320–326. [CrossRef]
35. Layal, A.; Hassan, R.; Ahmad, K.; Hamid, B.S. Chemical Composition and Biological Potentials of Lebanese *Cupressus sempervirens* L. Leaves Extracts. *J. Med. Plants Res.* **2020**, *14*, 292–299. [CrossRef]
36. Tekaya-Karoui, A.; BOUGHALLEB, N.; Hammami, S.; Ben Jannet, H.; Gannoun, S. Chemical Composition and Antifungal Activity of Volatile Components from Woody Terminal Branches and Roots of *Tetraclinis articulate* (Vahl). Masters Growing in TUNISIA. *Afr. J. Plant Sci.* **2011**, *5*, 115–122.
37. Badawy, M.E.I.; Kherallah, I.E.A.; Mohareb, A.S.O.; Salem, M.Z.M.; Yousef, H.A. Chemical Composition and Antifungal Activity of Essential Oils Isolated from *Cupressus sempervirens* L. and *Juniperus phoenicea* L. Grown in Al-Jabel Al-Akhdar Region, Libya against Botrytis Cinerea. *Nat. Prod. J.* **2017**, *7*, 298–308. [CrossRef]
38. Rguez, S.; Essid, R.; Adele, P.; Msaada, K.; Hammami, M.; Mkadmini, K.; Fares, N.; Tabbene, O.; Elkahoui, S.; Portelli, D.; et al. Towards the Use of *Cupressus sempervirens* L. Organic Extracts as a Source of Antioxidant, Antibacterial and Antileishmanial Biomolecules. *Ind. Crops Prod.* **2019**, *131*, 194–202. [CrossRef]
39. Taghreed, I.A.; El-Hela, A.A.; El-Hefnawy, H.M.; Al-Taweel, A.M.; Perveen, S. Chemical Composition and Antimicrobial Activities of Essential Oils of Some Coniferous Plants Cultivated in Egypt. *Iran. J. Pharm. Res. IJPR* **2017**, *16*, 328–337.
40. Ismail, A.; Lamia, H.; Mohsen, H.; Samia, G.; Bassem, J. Chemical Composition, Bio-Herbicidal and Antifungal Activities of Essential Oils Isolated from Tunisian Common Cypress (*Cupressus sempervirens* L.). *J. Med. Plants Res.* **2013**, *7*, 1070–1080.
41. Pansera, M.R.; Silvestre, W.P.; Sartori, V.C. Bioactivity of *Cupressus sempervirens* and *Cupressus lusitanica* Leaf Essential Oils on *Colletotrichum fructicola*. *J. Essent. Oil Res.* **2023**, *35*, 51–59. [CrossRef]
42. Reyes-Jurado, F.; Navarro-Cruz, A.R.; Ochoa-Velasco, C.E.; Palou, E.; López-Malo, A.; Ávila-Sosa, R. Essential Oils in Vapor Phase as Alternative Antimicrobials: A Review. *Crit. Rev. Food Sci. Nutr.* **2020**, *60*, 1641–1650. [CrossRef] [PubMed]
43. Kloucek, P.; Smid, J.; Frankova, A.; Kokoska, L.; Valterova, I.; Pavela, R. Fast Screening Method for Assessment of Antimicrobial Activity of Essential Oils in Vapor Phase. *Food Res. Int.* **2012**, *47*, 161–165. [CrossRef]

44. Kačániová, M.; Galovičová, L.; Valková, V.; Ďuranová, H.; Borotová, P.; Štefániková, J.; Vukovic, N.L.; Vukic, M.; Kunová, S.; Felsöciová, S.; et al. Chemical Composition and Biological Activity of *Salvia officinalis* Essential Oil. *Acta Hortic. Regiotect.* **2021**, *24*, 81–88. [CrossRef]
45. Vimalanathan, S.; Hudson, J. Anti-Influenza Virus Activity of Essential Oils and Vapors. *Am. J. Essent. Oils Nat. Prod.* **2014**, *2*, 47–53.
46. Jose, W.A.N.; Renata, A.C.; Magda, T.D.C.; Theodora, T.A.C. Antibiofilm Activity of Natural Substances Derived from Plants. *Afr. J. Microbiol. Res.* **2017**, *11*, 1051–1060. [CrossRef]
47. Rehman, R. Fast and Effective Identification of the Bioactive Compounds from Thuja and Cypress via Bioactivity Guided Isolation Approach. Ph.D. Thesis, University of Agriculture Faisalabad, Faisalabad, Pakistan, 2019.
48. Pedroso, R.D.S.; Balbino, B.L.; Andrade, G.; Dias, M.C.P.S.; Alvarenga, T.A.; Pedroso, R.C.N.; Pimenta, L.P.; Lucarini, R.; Pauletti, P.M.; Januário, A.H.; et al. In Vitro and In Vivo *Anti-Candida* spp. Activity of Plant-Derived Products. *Plants* **2019**, *8*, 494. [CrossRef]
49. Silva, R.A.D. Combinação de óleos Essenciais Entre si e com Clotrimazol na Inibição e Erradicação de Biofilmes Formados por Espécies de Candida [Combining Essential Oils with Each Other and with Clotrimazole Prevents the Formation of Candida Biofilms and Eradicates Mature Biofilms]. Master's Thesis, Universidade Federal de Uberlândia, Uberlândia, Brazil, 2022. [CrossRef]
50. Kırmusaoğlu, S. (Ed.) *Antimicrobials, Antibiotic Resistance, Antibiofilm Strategies and Activity Methods*; IntechOpen: Rijeka, Croatia, 2019; ISBN 978-1-78985-789-4.
51. Langsi, D.J.; Tofel, H.K.; Fokunang, C.N.; Suh, C.; Eloh, K.; Caboni, P.; Nukenine, E.N. Insecticidal Activity of Essential Oils of *Chenopodium ambrosioides* and *Cupressus sempervirens* and Their Binary Combinations on *Sitophilus zeamais*. *GSC Biol. Pharm. Sci.* **2018**, *3*, 024–034. [CrossRef]
52. Pinto, E.; Vale-Silva, L.; Cavaleiro, C.; Salgueiro, L. Antifungal Activity of the Clove Essential Oil from Syzygium Aromaticum on Candida, Aspergillus and Dermatophyte Species. *J. Med. Microbiol.* **2009**, *58*, 1454–1462. [CrossRef]
53. Saad, N.Y.; Muller, C.D.; Lobstein, A. Major Bioactivities and Mechanism of Action of Essential Oils and Their Components: Essential Oils and Their Bioactive Components. *Flavour Fragr. J.* **2013**, *28*, 269–279. [CrossRef]
54. Ulukanli, Z.; Karabörklü, S.; Ates, B.; Erdogan, S.; Cenet, M.; Karaaslan, M.G. Chemical Composition of the Essential Oil from *Cupressus sempervirens* L. *Horizontalis* Resin in Conjunction with Its Biological Assessment. *J. Essent. Oil Bear. Plants* **2014**, *17*, 277–287. [CrossRef]
55. Borotová, P.; Galovičová, L.; Valková, V.; Ďúranová, H.; Vuković, N.; Vukić, M.; Babošová, M.; Kačániová, M. Biological Activity of Essential Oil from *Foeniculum vulgare*. *Acta Hortic. Regiotect.* **2021**, *24*, 148–152. [CrossRef]
56. Adams, R.P. *Identification of Essential Oil Components by Gas Chromatography/Mass Spectrometry*; Allured Publishing Corporation: Carol Stream, IL, USA, 2007.
57. Van Den Dool, H.; Kratz, P.D. A Generalization of the Retention Index System Including Linear Temperature Programmed Gas—Liquid Partition Chromatography. *J. Chromatogr. A* **1963**, *11*, 463–471. [CrossRef]
58. Proestos, C.; Lytoudi, K.; Mavromelanidou, O.; Zoumpoulakis, P.; Sinanoglou, V. Antioxidant Capacity of Selected Plant Extracts and Their Essential Oils. *Antioxidants* **2013**, *2*, 11–22. [CrossRef]
59. Kačániová, M.; Terentjeva, M.; Galovičová, L.; Ivanišová, E.; Štefániková, J.; Valková, V.; Borotová, P.; Kowalczewski, P.Ł.; Kunová, S.; Felšöciová, S.; et al. Biological Activity and Antibiofilm Molecular Profile of Citrus Aurantium Essential Oil and Its Application in a Food Model. *Molecules* **2020**, *25*, 3956. [CrossRef]
60. Finney, D.J. Probit Analysis. 2nd ed. By D. J. Finney. Cambridge University Press, New York, 1952. 22.5 × 14 cm. xiv + 318 pp. *J. Am. Pharm. Assoc. Sci. Ed.* **1952**, *41*, 627. [CrossRef]
61. Mosmann, T. Rapid Colorimetric Assay for Cellular Growth and Survival: Application to Proliferation and Cytotoxicity Assays. *J. Immunol. Methods* **1983**, *65*, 55–63. [CrossRef]

plants

MDPI

Article

Chemical Diversity of *Artemisia rutifolia* Essential Oil, Antimicrobial and Antiradical Activity

Elena P. Dylenova [1], Svetlana V. Zhigzhitzhapova [1], Elena A. Emelyanova [1], Zhargal A. Tykheev [1,*], Daba G. Chimitov [2], Danaya B. Goncharova [1] and Vasiliy V. Taraskin [1]

1 Baikal Institute of Nature Management, Siberian Branch, Russian Academy of Sciences, 670047 Ulan-Ude, Russia; edylenova@mail.ru (E.P.D.); zhig2@yandex.ru (S.V.Z.); emelianowa.elena2312@mail.ru (E.A.E.); danaydomi5@gmail.com (D.B.G.); vvtaraskin@binm.ru (V.V.T.)

2 Institute of General and Experimental Biology, Siberian Branch, Russian Academy of Sciences, 670047 Ulan-Ude, Russia; dabac@mail.ru

* Correspondence: gagarin199313@gmail.com; Tel.: +7-3012-43-36-76

Abstract: This paper presents the results of the study of the composition of the essential oil (EO) of *Artemisia rutifolia* by the GC/MS method as well as its antimicrobial and antiradical activities. According to the PCA-analysis, these EOs can be conditionally divided into "Tajik" and "Buryat-Mongol" chemotypes. The first chemotype is characterized by the prevalence of α- and β-thujone, and the second chemotype by the prevalence of 4-phenyl-2-butanone, camphor. The greatest antimicrobial activity of *A. rutifolia* EO was observed against Gram-positive bacteria and fungi. The EO showed high antiradical activity with an IC_{50} value of 17.55 μL/mL. The presented first data on the composition and activity of the EO of *A. rutifolia* of the Russian flora indicate the prospects of the species as a raw material for the pharmaceutical and cosmetic industry.

Keywords: *Artemisia rutifolia*; essential oil; chemical composition; monoterpenes; sesquiterpenes; PCA-analysis; chemotypes; antibacterial activity; antiradical activity

Citation: Dylenova, E.P.; Zhigzhitzhapova, S.V.; Emelyanova, E.A.; Tykheev, Z.A.; Chimitov, D.G.; Goncharova, D.B.; Taraskin, V.V. Chemical Diversity of *Artemisia rutifolia* Essential Oil, Antimicrobial and Antiradical Activity. *Plants* **2023**, *12*, 1289. https://doi.org/10.3390/plants12061289

Academic Editors: Hazem Salaheldin Elshafie, Ippolito Camele and Adriano Sofo

Received: 20 February 2023
Revised: 4 March 2023
Accepted: 10 March 2023
Published: 13 March 2023

1. Introduction

Essential oils are a mixture of volatile flavor substances belonging to different classes of organic compounds (terpenes, their oxygenated derivatives, aromatic and aliphatic compounds). These compounds can pass through biological membranes to exert antioxidant, antimicrobial, antifungal, anti-inflammatory, antiviral, and other effects [1], making EOs widely used in the pharmaceutical and cosmetic industries and increasing the demand for new natural sources of EOs.

Plants of the *Artemisia* L. genus, which grow abundantly in arid and semi-arid regions of Asia, can serve as a reliable natural source of EOs. A promising species is *Artemisia rutifolia* Steph. ex Spreng. (family Asteraceae Bercht. Et J. Presl., section *Absinthium* (Mill.) D.-C.), which is a semi-shrub, up to 80 cm tall with strongly branched, woody perennial stems covered with brownish grey, cracked bark [2]. It grows in Afghanistan, Kazakhstan, Kyrgyzstan, Mongolia, Nepal, Pakistan, Russia (Western and Eastern Siberia), Tajikistan, and Western Asia [3], in mountain steppes, rocky slopes, and screes [4]. On the territory of Baikal Siberia, *A. rutifolia* is a relict species [5], the life expectancy of which can reach 80–90 years [6]. In Kyrgyzstan folk medicine, fresh leaves have been used for toothache, and a decoction for sore throat, heart, and stomach diseases [7]. The therapeutic value of the species exhibited is due to the variety of biologically active substances it contains.

The isolation of sesquiterpene lactones (guyanolides, germacranolides, and costic acid derivatives) from the aerial part of *A. rutifolia* has been reported [8–10]. Another study reported that methanol, chloroform, and hexane extracts of *A. rutifolia* leaves contained polyphenolic compounds (organic acids, myricetin, and quercetin) and also exhibited antimicrobial and antioxidant activities [11]. The following terpenes were isolated by gas chromatography and identified by their IR spectra from the EO of *Artemisia rutifolia*: 1,8-cineole, α-, β-thujones, (+)-camphor, (−)-α-terpineol, and (−)-terpinen-4-ol [12]. However, studies of the essential oil composition of *A. rutifolia* were generally incomplete and related to plants growing in scattered populations from Tajikistan [13] and Mongolia [12,14–16].

This article is the first to investigate the chemical composition of the EO of *A. rutifolia*, growing in Buryatia (Russia), its antimicrobial and antiradical activities, and to conduct comparative chemometric analysis.

2. Results and Discussion

2.1. EOs Component Composition

The yield of EOs from the aerial part of *A. rutifolia* growing in Buryatia (Russia) was 1.82% (v/w) of dry weight. The chemical composition of the obtained EOs was investigated using the GC-MS technique (Figure 1). Forty components have been identified in the EO of *A. rutifolia*, most of which are represented by mono- and sesquiterpenoids, and are listed in Table 1. The dominant components were: 4-phenyl-2-butanone (34.98%), 1,8-cineol (16.53%), camphor (16.67%), also in significant quantities were found: terpinen-4-ol (3.71%), 4-phenyl-2-butanol (3.58%), α-terpineol (3.51%), α-methyl-benzenepropanol acetate (3.43%), bicyclogermacrene (2.06%), and germacrene D (1.02%). Monoterpenes (51.26%), especially the oxygenated ones (45.28%), made up the largest proportion of all components.

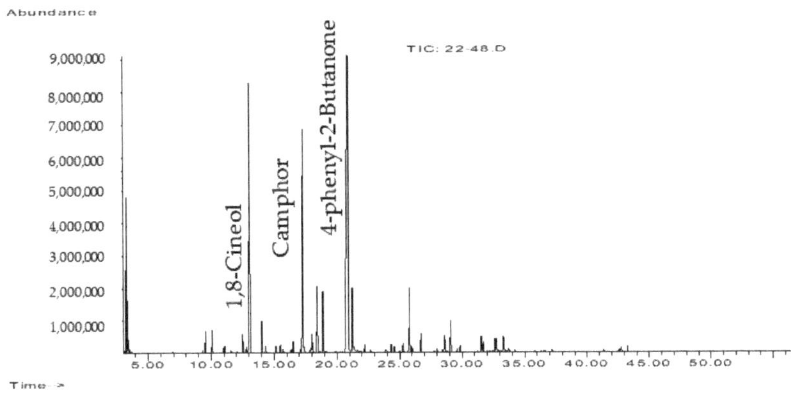

Figure 1. GC-MS total ion chromatogram of *A. rutifolia* EOs.

Table 1. Chemical composition of EOs extracted from the aerial parts of *A. rutifolia* from different countries.

No.	RI *	Rt	Component	Peak Area (%)	Molecular Formula
1	921	9.17	Tricyclene	0.05	$C_{10}H_{16}$
2	926	9.32	α-Thujene	0.12	$C_{10}H_{16}$
3	932	9.56	α-Pinene	0.98	$C_{10}H_{16}$
4	947	10.08	Camphene	1.06	$C_{10}H_{16}$
5	973	10.95	Sabinene	0.22	$C_{10}H_{16}$
6	975	11.06	β-Pinene	0.37	$C_{10}H_{16}$
7	990	11.55	2,3-dehydro-1,8-Cineol	0.15	$C_{10}H_{16}O$
8	1004	12.04	α-Phellandrene	0.05	$C_{10}H_{16}$
9	1017	12.49	α-Terpinene	0.89	$C_{10}H_{16}$
10	1024	12.78	*p*-Cymol	0.35	$C_{10}H_{14}$
11	1031	13.07	1,8-Cineol	16.53	$C_{10}H_{18}O$
12	1058	14.04	γ-Terpinene	1.52	$C_{10}H_{16}$
13	1066	14.33	*trans*-Sabinene hydrate	0.37	$C_{10}H_{18}O$

Table 1. *Cont.*

No.	RI *	Rt	Component	Peak Area (%)	Molecular Formula
14	1088	15.13	Terpinolene	0.37	$C_{10}H_{16}$
15	1098	15.48	*cis*-Sabinene hydrate	0.44	$C_{10}H_{18}O$
16	1103	15.70	Filifolone	0.19	$C_{10}H_{14}O$
17	1121	16.35	*cis-p*-Menth-2-en-1-ol	0.26	$C_{10}H_{18}O$
18	1126	16.49	Chrysanthenone	0.64	$C_{10}H_{14}O$
19	1141	17.06	*trans-p*-Menth-2-en-1-ol	0.23	$C_{10}H_{18}O$
20	1144	17.29	Camphor	16.67	$C_{10}H_{16}O$
21	1162	17.88	Pinocarvone	0.31	$C_{10}H_{14}O$
22	1166	17.99	Borneol	1.17	$C_{10}H_{18}O$
23	1177	18.41	Terpinen-4-ol	3.71	$C_{10}H_{18}O$
24	1191	18.88	α-Terpineol	3.51	$C_{10}H_{18}O$
25	1241	20.91	4-phenyl-2-Butanol	3.58	$C_{10}H_{14}O$
26	1247	21.25	4-phenyl-2-Butanone	34.95	$C_{10}H_{12}O$
27	1287	22.21	Bornyl acetate	0.38	$C_{12}H_{20}O_2$
28	1306	24.30	α-Terpineol formate	0.40	$C_{11}H_{18}O_2$
29	1359	24.56	Eugenol	0.32	$C_{10}H_{12}O_2$
30	1378	25.26	α-Copaene	0.48	$C_{15}H_{24}$
31	1418	25.79	α-methyl-Benzenepropanol acetate	3.43	$C_{12}H_{16}O_2$
32	1422	26.70	Caryophyllene	0.97	$C_{15}H_{24}$
33	1456	27.77	Humulene	0.07	$C_{15}H_{24}$
34	1464	28.00	*allo*-Aromadendrene	0.19	$C_{15}H_{24}$
35	1477	28.41	Selina-4,11-diene	0.13	$C_{15}H_{24}$
36	1484	28.61	Germacrene D	1.02	$C_{15}H_{24}$
37	1500	29.09	Bicyclogermacrene	2.06	$C_{15}H_{24}$
38	1517	29.59	γ-Cadinene	0.16	$C_{15}H_{24}$
39	1580	31.52	Spathulenol	1.10	$C_{15}H_{24}O$
40	1586	31.70	Caryophyllene oxide	0.60	$C_{15}H_{24}O$
		Total oxygenated hydrocarbons		41.96	
		Total monoterpenes		51.26	
		Total sesquiterpenes		6.78	
		Total hydrocarbons		41.96	

* RI, retention indices: experimental, for our data (RI, retention index as determined on a HP-5MS column using the homologous series of n-hydrocarbons).

2.2. Chemical Diversity of EOs

Comparative analysis of the obtained data and the literature review [13–16] (Appendix A) showed that the EO of plants growing in Buryatia was similar to the EOs of Mongolian plant populations in the content of the major components, but quite different from the EOs of plants from Tajikistan.

Thus, the dominant components in the EOs of *A. rutifolia* from the Muminobod and Yovon regions of Tajikistan were α-thujone (20.9–36.6%), β-thujone (36.1–47.3%), 1,8-cineol (3.2–11.7%), myrcene (0.3–2.8%), *p*-cymol (0.9–1.8%), *cis*-piperitone epoxide (0.9–2.0%), and germacrene D (1.8–2.8%). More than 90% of all components were monoterpenoids, mainly oxygenated (85.5–92.4%).

In contrast, the dominating components of the EOs of plants from the Mongolian populations were: 4-phenyl-2-butanone (33.1%), carvacrol methyl ether (29.58%), camphor (2.13–22.4%), 1,8-cineol (4.63–25.13%), 4-phenyl-2-butanol (3.4%), geraniol (2.91%), *p*-cymol (1.1–1.41%), α-terpineol (1–1.64%), α-thujone (0.7–3.38%), β-thujone (1.10–3.2%), and terpinen-4-ol (0.54–1.1%). Monoterpenoids (58.26–93.37%) also dominated among all of the other components.

Samples from Mongolia and Buryatia (compared to those from Tajikistan) were characterized by a sufficiently high content of camphor, which is used in creams, ointments, and lotions to relieve pain, irritation, itching, and has antifungal and antibacterial properties [17].

Using PCA to compare our own and the literature data on the content of the major components of *A. rutifolia* EO, it was shown that these EOs can currently be conditionally divided into the "Tajik" and "Buryat–Mongol" chemotypes (Figure 2).

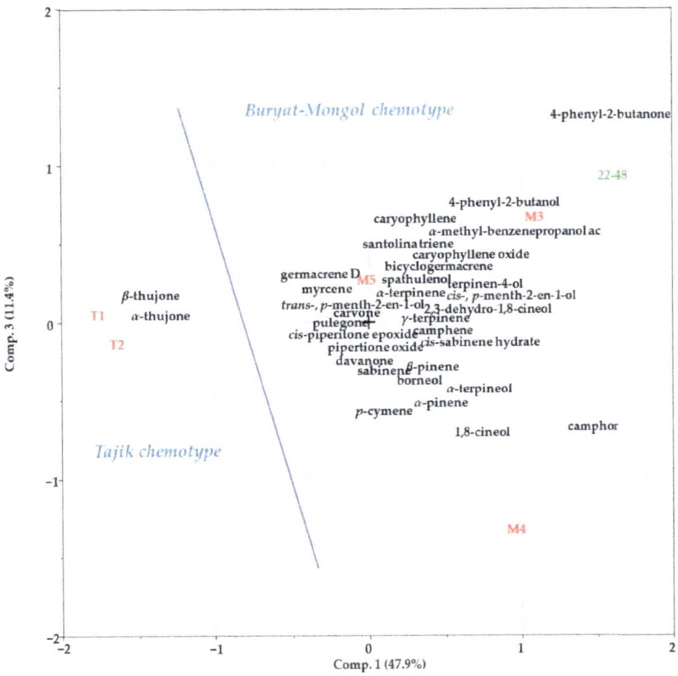

Figure 2. PCA biplot (principal component 1–principal component 3) for the data on the composition of *A. rutifolia* EOs.

The "Tajik" EOs were characterized by the prevalence of α- and β-thujone, while the "Buryat-Mongolian" chemotype was characterized by a high content of 4-phenyl-2-butanone, camphor (Figure 3).

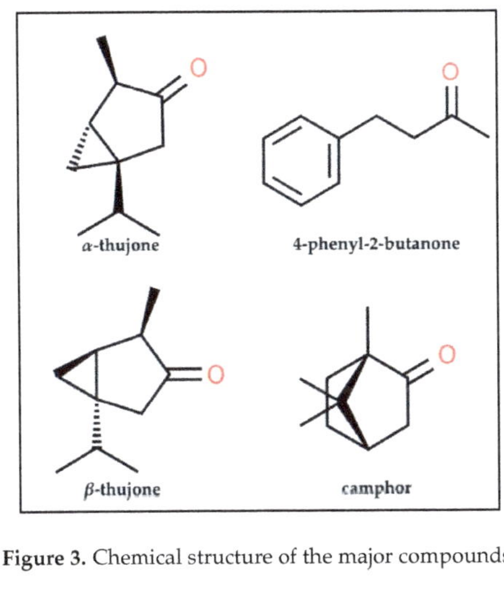

Figure 3. Chemical structure of the major compounds forming different chemotypes of *A. rutifolia*.

For example, in the EOs of *A. rutifolia* from the flora of Tajikistan [13] the content of α- and β-thujone was rather high: α-thujone (20.9–36.6%) and β-thujone (36.1–47.3%), whereas in the EOs of the plants from Mongolian populations [14–16] they were found in smaller amounts: α-thujone (0.70–3.38%), β-thujone (1.10–3.20%). However, they were not found in the plants of the Buryat flora.

It should be noted that studies of isomeric thujones (α- and β-) were previously initiated because wormwood is widely used to flavor alcoholic beverages. The most famous alcoholic beverage, absinthe, is made from *Artemisia absinthium*.

Thujones are known to be the main constituents of the EOs of *A. absinthium* [18]. It has neurotoxicity manifested by hyperactivity, tremors, and tonic convulsions [19]. The effects of thujone on the human body are related to the inhibition of GABAA receptors, leading to dose-dependent excitation and convulsions, with (−)-α-thujone having a greater ability to induce convulsions than the (+)-β-isomer; it is more likely that the convulsive effect of thujone acts on a specific receptor system [20]. For this reason, isomeric thujones were long thought to be responsible for the manifestation of the so-called "wormwood epilepsy".

Modern studies show that it is the additional components (apart from the main one—ethyl alcohol) of industrially produced absinthe that do not seem to have any harmful effects on health, leaving aside the effects of ethanol on the body. Absinthe has an exceptionally high alcohol content (>50% vol.). This can lead to serious health and social problems, but it is not unique to this drink. So-called "absinthism" cannot be clearly distinguished from chronic alcoholism [21].

In general, thujones (monoterpene ketones) are natural constituents of the EOs of plants of the genus *Artemisia* (*A. absinthium, A. campestris, A. alba, A. incana, A. pontica, A. santolinifolia, A. santonicum, A. spicigera, A. vulgaris*), Salvia (*S. fruticosa, S. lavandulifolia, S. officinalis, S. sclarea, S. triloba*), Thuja (*T. occidentalis, T. orientalis*), etc. [19]. However, the assessment of thujone toxicity remains poorly studied, the most important aspects of which are the relationships between dose, concentration, and effect in humans.

The content of thujones in the EO of *A. absinthium* can vary within a wide range. On this basis, thujone and sabinyl acetate EOs of *A. absinthium* were distinguished [22]. Thujone-containing and thujone-free forms are also characteristic of other wormwood species (e.g., *A. campestris* [23], *A. molinieri* [24]).

On the other hand, the discovery of thujone-free forms of *A. rutifolia* growing in Buryatia is important for the creation of safer medicines, cosmetics, food supplements, and therapeutic foods based on them. In addition, it allows us to understand the influence of environmental conditions on thujone biosynthesis. The currently available amount of information on the composition of EOs of *A. rutifolia* does not allow us to draw detailed conclusions, but we note that the formation of chemotypes occurs under the influence of a long-term and relatively uniform action of certain climatic conditions. In the course of evolution, changes in the composition of enzymes occur by replacing one or more amino acids. If the modified enzyme produces a useful product for the plant, these changes are fixed in the genes [25].

At the biochemical level, mechanisms are formed to synthesize a specific set of enzymes that contribute to the production of EO components of one or another chemotype. The biosynthesis of thujones has been studied in detail for only a few species. It is known that the first monoterpene in this transformation chain is sabinene, whose formation is catalyzed by the enzyme sabinene synthase. Furthermore, isomeric thujones are formed from isomeric sabinols, probably also from (+)-sabinone [26].

The territories of Tajikistan, Mongolia, and Buryatia (Russia), where *A. rutifolia* grows, belong to the arid zone of Asia. The territories of Buryatia and Mongolia belong to the eastern (and Tajikistan—to the western) longitudinal sector of the arid continental zone of Asia, where the most arid territory is Mongolia. The *eastern boundary* of the extremely arid deserts of southern Mongolia and northern China, which have no analogues in Eurasia, passes here at about 105 degrees east latitude. The harsh natural conditions are particularly pronounced in areas of high aridity in the continental winter climate zone. At the same time, the area where the plants were collected in Tajikistan is on the border of the *western sector*: the interaction of various circulation processes leads to a strong variability in the moisture regimes (there is almost no precipitation in summer). However, the climate of a particular area was influenced by meso- and microclimatic factors in addition to the macroclimatic factors.

The area of plant collection in Mongolia is located in the Great Lakes basin, the mesoclimate of which is close to the semi-arid climate of Buryatia [27]. Thus, these places where the raw materials were collected can be ranked as follows (in the order of increasing the aridity of growing conditions of plants in summer) Buryatia → Mongolia → Tajikistan.

The increasing aridity of climatic conditions will likely lead to the biosynthesis of thujones. In addition, other sabinene derivatives, *trans*- and *cis*-sabinene hydrates, have been found in small amounts in the EOs of *A. rutifolia* growing in the territories of Buryatia and Mongolia; in Mongolian plants, sabinyl acetate was found. These compounds probably block thujone biosynthesis.

2.3. Antimicrobial Activity

The antimicrobial activity of *A. rutifolia* EO was experimentally determined using the disc diffusion method against Gram-positive bacteria (*Streptococcus pyogenes, Staphylococcus aureus, Bacillus cereus*), Gram-negative bacteria (*Escherichia coli, Pseudomonas aeruginosa, Salmonella enterica*), and fungi (*Aspergillus niger, Candida albicans*).

The antimicrobial activity of the samples was evaluated by the diameter of the growth inhibition zones of the test strains (mm). Each sample was tested in three replicates. The test results of the antimicrobial activity of the samples are shown in Table 2.

Table 2. Antimicrobial activity of essential oil from the aerial part of *Artemisia rutifolia* against Gram-positive, Gram-negative bacteria, and fungi.

Tested Substance	Zone of Inhibition, mm							
	Gram-Positive			Gram-Negative			Fungi	
	Streptococcus pyogenes	*Staphylococcus aureus*	*Bacillus cereus*	*Pseudomonas aeruginosa*	*Salmonella enterica*	*Escherichia coli*	*Candida albicans*	*Aspergillus niger*
Essential oil	14	14	14	0	12	13	11	21
Positive control *	25	28	24	26	27	27	46	37

* Positive control: norfloxacin was for the Gram-positive bacteria; ceftazidime for the Gram-negative bacteria; fluconazole for the fungi.

The results indicate the greatest antimicrobial activity of *A. rutifolia* EO against Gram-positive bacteria (*Streptococcus pyogenes, Staphylococcus aureus, Bacillus cereus*) and fungi (*Aspergillus niger, Candida albicans*), with pronounced activity against Aspergillus niger.

To a lesser extent, the growth inhibition of Gram-negative bacteria (*Salmonella enterica subsp. enterica, Escherichia coli*) was observed. *Pseudomonas aeruginosa* proved to be the most resistant to the EO: no growth inhibition was observed.

The greatest antimicrobial activity of *A. rutifolia* EO from Buryatia was observed against the Gram-positive bacteria and fungi, which is consistent with the literature data. For example, the minimum inhibitory activity (MIC) and minimum bactericidal concentration (MBC) of the EOs of *A. rutifolia* from Tajikistan were previously determined to be 10 mg/mL against *E. coli* ATCC 25922, and 5 mg/mL against MRSA NCTC 10442 [28].

EOs of *A. rutifolia* from Mongolia at a concentration of 150 mg/mL (or 3 μg/disc) inhibited the growth of *S. enterica* by 9.3 ± 0.76 mm, *B. subtillus* by 10.3 ± 0.58 mm, and *S. aureus* by 9.6 ± 1.5 mm, thus showed moderate antimicrobial activity [16]. The target for the antimicrobial action of the EO is probably the bacterial cell wall, which is known to be fundamentally different in structure in Gram-positive and Gram-negative bacteria. The cell wall of Gram-negative bacteria contains a strong lipid layer on its surface, with which the EOs lose their antimicrobial activity [29]. Therefore, the *A. rutifolia* EO is recommended for use as an antimicrobial agent against Gram-positive bacteria and fungi.

2.4. Antiradical Activity

In order to evaluate the possible antiradical potential of the EO of *A. rutifolia,* the DPPH test (2,2-diphenyl-1-picrylhydrazyl radical inhibition) was applied. To determine the antiradical properties of the EO, a kinetic curve was constructed using the IC_{50} value (Figure 4).

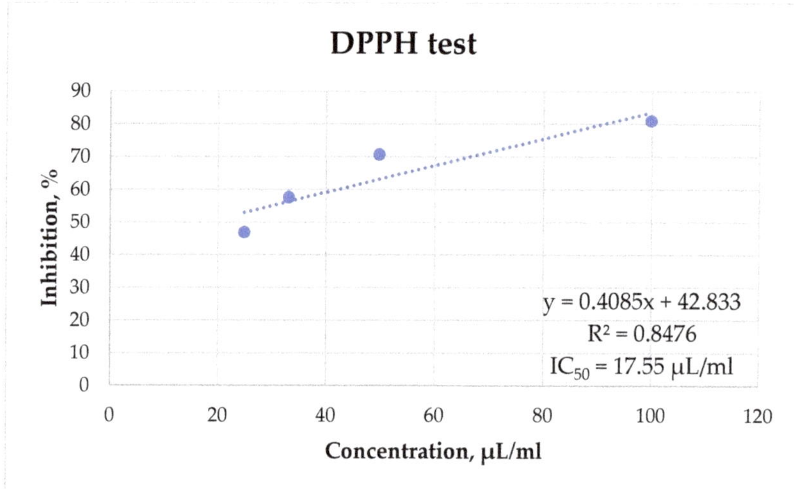

Figure 4. DPPH test of the antiradical activity of *A. rutifolia* EO.

According to the results of the test, it was found that the EO has high antiradical activity as the IC_{50} value was 17.55 μL/mL.

It is considered that the antioxidant potential of EOs is exhibited mainly due to the presence of oxygenated monoterpenes (especially of phenolic structure), while sesquiterpene hydrocarbons and their oxygenated derivatives have very low antioxidant activity [30]. The EOs from Tajikistan had a better antiradical potential (IC_{50} = 7.91 mg/mL) [28] compared to those from Buryatia (IC_{50} = 17.55 μL/mL) and the content of oxygenated monoterpenes was higher in the EOs of *A. rutifolia* from Tajikistan.

Previously, it has been shown that EOs exhibit much greater activity than their individual components, which may be due to the high percentage of major components, and synergism between the various components of the EO including minor ones [31]. For example, wormwood EOs, whose main components are camphor and 1,8-cineol, always show antiradical activity, while camphor and 1,8-cineol individually do not [32].

It has also been shown that cineol enrichment of the secondary oil fractions of bay laurel and the cube residue of rockrose enhances their antioxidant properties [1]. In the case of *A. rutifolia*, we believe that the EO of the plants from Buryatia has a higher antiradical activity due to the synergistic effect.

3. Materials and Methods

3.1. Plant Material Collection and EO Production

The aerial part of *A. rutifolia*, collected in 2022 in the Selenginsky District (Buryatia, Russia) during the vegetation period, was used as the object of study. The voucher specimens were identified by Dr. Oleg A. Anenkhonov and deposited at the Herbarium of Institute of General and Experimental Biology SB RAS (UUH 019695, 019696). Data on the sampling locations and EO yield are presented in Table 3 (compared to data from other studies).

Table 3. Origin of the plant material of *Artemisia rutifolia* and the yield of the essential oils from the aerial part.

Sample Code	Country	Locality	Collection Period	Latitude Longitude	Attitude (m)	Yield of the Essential Oil, *v*/*w* (%)	Source of Data
22–48	Russia	Surroundings of the Novoselenginsk Village, Selenginsky District, Buryatia	14.06.2022	N 51.25556 E 106.431389	549	1.82	Present study
T1	Tajikistan	Khonaobod Village, Muminobod region	02.05.2010	N 38.107547 E 69.966431	1200	0.50	[13]
T2	Tajikistan	Chormaghzak Village, Yovon region	25.07.2010	N 38.417502 E 69.172175	1300	0.80	[13]
M3	Mongolia	Middle Gobi Province	08.09.2007	– *	–	0.20	[15]
M4	Mongolia	–	–	–	–	1.20	[14]
M5	Mongolia	Khrakhiraa Mountain, Uvs aimag	09.2019	–	–	0.96	[16]

* Not specified.

EOs were obtained by hydrodistillation from air-dry raw materials (aboveground part of plants, for 3 h) in the year of raw material collection, according to OFS.1.5.3.0010.15 "Determination of essential oil content in medicinal plant raw materials and herbal drugs" with a modified Clevenger apparatus.

3.2. Gas Chromatography-Mass Spectrometry (GC-MS) Analysis and Principal Component Analysis (PCA)

The component composition of the EOs was determined by gas chromatography-mass spectrometry (GC-MS) using an Agilent 6890 gas chromatograph (Agilent Technologies, USA) with an HP 5973N mass-selective detector (Hewlett-Packard, Palo Alto, CA, USA) and an HP-5MS capillary column (30 m × 0.25 mm × 0.2 μm; Hewlett-Packard), as previously described in [33].

The principal component analysis (PCA) method was applied to the contents of the EO components (Sirius software package ver. 6.0, Pattern Recognition Systems, a/s, Norway).

3.3. Antiradical Activity

The antiradical activity of the EOs was determined by the DPPH test (using a stable radical, 2,2-diphenyl-1-picrylhydrazyl). Briefly, a DPPH solution (0.006% in 95% ethanol) was added to the EO of *A. rutifolia* (25–1000 μL/mL in ethyl alcohol) and incubated for 30 min in the dark at room temperature. The antiradical activity was then determined spectrophotometrically on a ClarioStar Plus multimode plate reader at 517 nm.

The antiradical activity (in % inhibition) was calculated using the formula:

$$\%\ \text{inhibition of DPPH radicals} = [(A_0 - A_1)/A_0] \times 100,$$

where A_0 is the absorbance of the control sample, A_1 is the absorbance of the test sample. The IC_{50} index was determined using regression analysis.

3.4. Antimicrobial Activity

The antimicrobial activity of the test samples was determined by the technique of diffusion in dense nutrient media. The inoculum was prepared by the direct suspension of the daily culture colonies of each test strain in a sterile isotonic solution to a density of 0.5 according to the McFarland turbidity standard, which approximately corresponds to a load of 1–2 × 10^8 CFU/mL. The resulting microbial suspension was applied evenly to the entire surface of the nutrient medium (agar) in three directions using a sterile cotton swab.

Mueller–Hinton agar was used as a nutrient medium for microorganisms with normal nutrient requirements, and Mueller–Hinton agar with the addition of 5% defibrinated blood was used for bacteria with complex nutrient requirements (*Streptococcus pyogenes*). After applying the microbial suspension, sterile paper discs were placed on the agar surface and 10 μL of the test samples was applied (one sample per disc). Factory paper discs with antimicrobial additives (norfloxacin for Gram-positive bacteria, ceftazidime for Gram-negative bacteria, fluconazole for fungi) were used as the positive controls.

Cultures were incubated at 37 °C (22 °C for molds and yeasts). The results were recorded after 24 h of incubation for bacteria and 48 h for mold and yeast. To determine the antimicrobial activity of the samples tested, the diameters of the microbial growth suppression zones around the disks were evaluated. Growth inhibition zones were measured to the nearest millimeter.

4. Conclusions

Thus, for the first time, the chemical composition and primary biological activities of the EOs of *A. rutifolia* collected in Buryatia were studied. The greatest antimicrobial activity of the EOs was noted with the Gram-positive bacteria (*Streptococcus pyogenes*, *Staphylococcus aureus*, *Bacillus cereus*) and fungi (*Aspergillus niger*, *Candida albicans*). In addition, it showed antiradical activity, and the IC_{50} index was 17.55 μL/mL. The obtained preliminary results of the antimicrobial and antiradical activities allow us to consider that *A. rutifolia* is a promising raw material for the pharmaceutical and cosmetic industries; however, it is necessary to carry out further studies.

The variability of plants growing within the natural habitat greatly affects the composition of essential oils. Despite the variability in the composition, the volatile substances of plants that form essential oils are the most important chemical markers that are used to solve the issues of chemosystematic or the taxonomic assignments of plants. The analysis of our own and the literature data showed that the EOs of *A. rutifolia* can be conditionally divided into "Tajik" and "Buryat-Mongol" chemotypes. The first chemotype is characterized by the prevalence of α- and β-thujone, and the second by the high content of 4-phenyl-2-butanone and camphor. The composition is highly variable and greatly depends on the geographical confinement.

Author Contributions: Conceptualization, E.P.D., S.V.Z., Z.A.T. and V.V.T.; Methodology, E.A.E., D.G.C. and D.B.G.; Formal analysis, E.P.D. and S.V.Z.; Investigation, E.A.E.. and D.B.G.; Writing—original draft preparation, E.P.D. and S.V.Z.; Writing—review and editing, Z.A.T., D.G.C. and V.V.T.; Supervision, S.V.Z., Z.A.T. and V.V.T. All authors have read and agreed to the published version of the manuscript.

Funding: This research was funded by the framework of the State Assignment of the Baikal Institute of Nature Management Siberian Branch of the Russian Academy of Sciences (BINM SB RAS, project no. FWSU-2021-0010, FWSU-2021-0004), Institute of General and Experimental Biology Siberian Branch of the Russian Academy of Sciences (IGEB SB RAS, project no. 121030900138-8), in the aspects of work of Interregional Scientific and Educational Center "Baikal". The resources of the Research Equipment Sharing Center of BINM SB RAS were used.

Data Availability Statement: All data generated or analyzed during this study are included in this published article.

Conflicts of Interest: The authors declare no conflict of interest.

Appendix A

Table A1. The chemical composition of EOs extracted from the aerial parts of *A. rutifolia* from different countries.

No.	Component	Peak Area (%)					
		Our Data	Literature Data				
			T1 [13]	T2 [13]	M3 [15]	M4 [14]	M5 [16]
	Monoterpene hydrocarbons						
1	Tricyclene	0.05				0.13	
2	α-Thujene	0.12	0.1			0.12	
3	α-Pinene	0.98	0.2	tr *	0.4	1.25	0.30
4	Camphene	1.06	0.1		0.8	0.34	0.21
5	Sabinene	0.22	0.3	0.4	0.1	0.27	0.82
6	β-Pinene	0.37	0.1	0.1	0.2	0.19	
7	α-Phellandrene	0.05	0.5	0.1			
8	α-Terpinene	0.89	0.2	0.2			
9	*p*-Cymol	0.35	1.8	0.9	1.1	1.41	
10	γ-Terpinene	1.52	0.5	0.4		0.19	
11	Terpinolene	0.37		0.1		0.1	
12	Santolina triene		0.1				22.38
13	Myrcene		2.8	0.3		0.15	21.84
14	Pseudo-limonene						5.27
15	Limonene					0.18	
16	*E*-β-Ocimene		tr				
	Total monoterpene hydrocarbons	5.98	2.1	2.5	2.6	4.33	50.82
	Oxygenated monoterpenes						
17	*cis*-Sabinene hydrate	0.44			0.4	0.11	
18	2,3-dehydro-1,8-Cineol	0.15			1.2		
19	1,8-Cineol	16.53	3.2	11.7	19.1	25.13	4.63
20	*trans*-Sabinene hydrate	0.37			0.3	0.20	
21	Filifolone	0.19					
22	*cis*-*p*-Menth-2-en-1-ol	0.26			1.5		
23	Chrysanthenone	0.64	0.1	0.8			
24	*trans*-*p*-Menth-2-en-1-ol	0.23	0.9	0.5			
25	Camphor	16.67	0.9	0.2	22.4	21.74	2.13
26	Pinocarvone	0.31	0.1	0.2		0.46	
27	Borneol	1.17	0.2	0.4	0.4	0.65	
28	Terpinen-4-ol	3.71	0.6	1.2	1.1	0.54	0.68
29	α-Terpineol	3.51	0.1	0.3	1	1.64	
30	Bornyl acetate	0.38		tr	0.1		
31	α-Terpineol formate	0.40					
32	Eugenol	0.32					
33	Santolina alcohol		0.4				
34	*trans*-2,3-epoxy Pinane					0.12	
35	Linalool					0.25	
36	α-Thujone		20.9	36.6	0.7	3.38	
37	β-Thujone		47.3	36.1	3.2	1.10	
38	Chryzanthenone					0.38	
39	*iso*-3-Thujanol		0.3	0.1			
40	*trans*-2-Pinanol					0.55	
41	*trans*-Verbenol				0.3		
42	*p*-Menth-3-en-1-ol		0.1	0.2			
43	Menthone		0.9				
44	Sabina ketone		0.2	0.3			
45	*cis*-Pinocamphone			0.1			
46	Thuj-3-en-10-al		0.2				
47	*p*-Cymen-8-ol			0.1	0.2		
48	*cis*-Piperitol		0.4			0.1	

Table A1. *Cont.*

No.	Component	Peak Area (%)					
		Our Data	Literature Data				
			T1 [13]	T2 [13]	M3 [15]	M4 [14]	M5 [16]
49	Myrtenol			0.3			
50	γ-Terpineol		tr				
51	*trans*-Piperitol		0.5	0.2			
52	*trans*-Carveol				0.2	0.1	
53	*m*-Cumenol		0.1	0.1			
54	*exo*-2-Pydroxycineol				2.3		
55	*nor*-Davanone			0.1			
56	Pulegone		1	0.3			
57	Carvone		0.9	0.1			
58	Carvacrol methyl ether					29.58	
59	Carvotanacetone		0.1	0.1			
60	Geraniol					2.91	
61	*cis*-Piperitone epoxide		2.0	0.9			
62	*cis*-Chrysanthenyl acetate		0.2	tr			
63	*iso*-3-Thujanol acetone		0.1	0.1			
64	*neoiso*-3-Thujanol acetone			0.1			
65	Sabinylacetate				0.9		
66	*p*-Cymen-7-ol		0.1	tr			
67	Thymol		0.7	0.2			
68	Carvacrol		0.9	0.4	0.1		
69	*Z*-Patchenol			0.2			
70	*cis*-Piperitol acetate		0.1	0.1			
71	Piperitone		0.1	0.1			
72	Pipertione oxide		1.4	tr			
73	*trans*-Carvylacetate				0.2		
74	α-Terpenylacetate				0.3	0.1	
75	*Z*-Jasmone		0.1	0.3	tr		
76	Methyleugenol				tr		
77	*E*-Ionone		0.1	tr			
	Total oxygenated monoterpenes	45.28	85.2	92.4	55.9	89.04	7.44
	Sesquiterpene hydrocarbons						
78	α-Copaene	0.48	0.1	tr			
79	Caryophyllene	0.97	0.4	0.1			7.19
80	Humulene	0.07					0.59
81	*allo*-Aromadendrene	0.19					
82	Selina-4,11-diene	0.13					0.97
83	Germacrene D	1.02	2.8	1.8			0.99
84	Bicyclogermacrene	2.06	0.5	0.8			
85	γ-Cadinene	0.16					0.48
86	α-Cedrene						0.22
87	β-Farnesene		0.2	0.1			
88	β-Chamigrene		0.1				
89	Valencene						1.20
90	Ledene					0.12	2.17
91	Aciphyllene						1.34
92	Bulnesene						0.84
93	β-Bisabolene		0.2				
94	δ-Cadinene		0.1	tr			0.42
95	β-Elemene					0.71	0.84
	Total sesquiterpene hydrocarbons	5.08	4.4	2.8	0	0.83	17.25
	Oxygenated sesquiterpenes						
96	Spathulenol	1.10	0.7	0.2	0.2	0.17	1.97
97	Caryophyllene oxide	0.60	0.2	0.1	0.1		5.82
98	dehydro-Sesquicineol						0.9

Table A1. *Cont.*

No.	Component	Peak Area (%)					
		Our Data	Literature Data				
			T1 [13]	T2 [13]	M3 [15]	M4 [14]	M5 [16]
99	Davana ether			0.1			
100	Davanone			1.3			
101	Viridiflorol		0.4				
102	Ledol		0.1				
103	Cedrol						2.27
104	Eremoligenol						0.69
105	Germacrene-D-1,10-epoxide		0.3				
106	α-Cadinol		0.1		0.1		
107	Germacra-4(15),5,10(14)-trien-1α-ol		0.1	0.1	0		
108	α-Bisabolol						0.42
109	4-Cuprenen-1-ol			tr			
110	Aciphilyc acid						0.91
	Total oxygenated sesquiterpenes	1.70	1.9	1.8	0.4	0.17	12.98
	Non-oxygenated hydrocarbons						
111	1-phenyl-2,4-Pentadiyne			0.1			
	Total non-oxygenated hydrocarbons	0	0	0.1	0	0	0
	Oxygenated hydrocarbons						
112	4-phenyl-2-Butanol	3.58			3.4		
113	4-phenyl-2-Butanone	34.95			33.1		
114	α-methyl-Benzenepropanol acetate	3.43					
115	(2*E*)-Hexenal			0.1			
116	Benzaldehyde				0.1		
117	1-Octen-3-ol		0.1				
118	(2*E*)-Dodecenal		0.2				
119	Phloacetophenone 2,4-dimethylether		0.3				
	Total oxygenated hydrocarbons	41.96	0.6	0.1	36.6	0	0
	Total monoterpenes	51.26	87.3	94.9	58.5	93.37	58.26
	Total sesquiterpenes	6.78	6.3	4.6	0.4	1.00	30.23
	Total hydrocarbons	41.96	0.6	0.2	36.6	0	0

* tr, trace amounts (less than 0.10%).

References

1. Atazhanova, G.A. *Terpenoids of Plant Essential Oils*; ICSPF: Moscow, Russia, 2008; pp. 90–127.
2. Krasnoborov, I.M. *Flora of Siberia, Vol. 13, Asteraceae (Compositae)*; Nauka: Novosibirsk, Russia, 1997; p. 123.
3. Flora of China. Available online: http://www.efloras.org/florataxon.aspx?flora_id=2&taxon_id=200023320 (accessed on 1 February 2023).
4. Budantsev, A.L. *Plant Resources of RUSSIA: Wild Flowering Plants, Their Composition and Biological Activity. Vol. 5. Asteraceae (Compositae) Family. Part 1. Genera Achillea—Doronicum*; Association of Scientific Publications KMK; SPb: Moscow, Russia, 2012; 317p.
5. Namzalov, B.B. Endemism and relict phenomena in the flora and vegetation of the steppe ecosystems of Baikalian Siberia. In *Biological Diversity of Baikalian Siberia*; Korsunov, V.M., Ed.; Nauka: Novosibirsk, Russia, 1999; pp. 184–192.
6. Voronin, V.I.; Razmakhina, T.B. *Artemisia rutifolia* (Asteraceae), a new indicator of long-term dynamics of atmospheric humidifying. *Botanicheskii Zhurnal* **2005**, *90*, 1565–1572.
7. Rastitel'nye Resursy SSSR. *Cvetkovye Rastenija, ih Himicheskij Sostav, Ispol'zovanie: Semejstvo Asteraceae (Plant Resources of the USSR. Flowering Plants, Their Chemical Composition, Use: Family Asteraceae)*; Sokolov, P.D., Ed.; Nauka: Saint-Petersburg, Russia, 1993; pp. 59–60.
8. Tan, R.X.; Jia, J.; Jakupovic, J.; Bohlmann, F.; Huntck, S. Sesquiterpene lactones from Artemisia rutifolia. *Phytochemistry* **1991**, *30*, 3033–3035. [CrossRef]
9. Jakupovic, J.; Tan, R.X.; Bohlmann, F.; Jia, Z.J.; Huneck, S. Sesquiterpene lactones from Artemisia rutifolia. *Phytochemistry* **1991**, *30*, 1714–1716. [CrossRef]
10. Tan, R.X.; Jia, Z.J. Sesquiterpenes from Artemisia rutifolia. *Phytochemistry* **1992**, *31*, 2534–2536. [CrossRef]

11. Ashraf, A.; Sarfraz, R.A.; Mahmood, A. Phenolic compounds' characterization of Artemisia rutifolia Spreng. from Pakistan flora and their relationship with antioxidant and antimicrobial attributes. *Int. J. Food Prop.* **2017**, *20*, 2538–2549. [CrossRef]
12. Shavarda, A.L. Essential oils of Mongolian plants: A study of the essential oils of Artemisia rutifolia. *Chem. Nat. Compd.* **1977**, *12*, 42–45. [CrossRef]
13. Sharopov, F.S.; Setzer, W.N. Thujone-rich essential oils of Artemisia rutifolia Stephan ex Spreng growing wild in Tajikistan. *J. Essent. Oil-Bear. Plants* **2011**, *14*, 136–139. [CrossRef]
14. Shatar, S.; Bodoev, N.V.; Zhigzhitzhapova, S.V.; Altantsetseg, S.; Namzalov, B.B. *Ether-Bearing Plants of the Selenga River Basin*; BSU Press: Ulan-Ude, Russia, 2006; 134p.
15. Trendafilova, A.; Shatar, S.; Todorova, M.; Altantsetseg, S. Essential oil composition of four Mongolian Artemisia species. *Compt. Rend. Acad. Bulg. Sci.* **2010**, *63*, 503–510.
16. Ayurzana, A.; Jambal, I.; Boldkhuu, N.; Batbayar, B.; Romanenk, E.P.; Shatar, A. Study on chemical composition of essential oil, ethanol extract, and anti-cancer, anti- bacterial activity of Artemisia rutifolia Steph.ex Spreng grown in Mongolia. *J. Pharm. Res. Int.* **2022**, *34*, 1–8. [CrossRef]
17. Kaltschmidt, B.P.; Ennen, I.; Greiner, J.F.W.; Dietsch, R.; Patel, A.; Kaltschmidt, B.; Kaltschmidt, C.; Hütten, A. Preparation of Terpenoid-Invasomes with Selective Activity against S. aureus and Characterization by Cryo Transmission Electron Microscopy. *Biomedicines* **2020**, *8*, 105. [CrossRef]
18. Nguyen, H.T.; Radacsi, P.; Gosztola, B.; Rajhart, P.; Nemeth, E.Z. Accumulation and composition of essential oil due to plant development and organs in wormwood (*Artemisia absinthium* L.). *Ind. Crops Prod.* **2018**, *123*, 232–237. [CrossRef]
19. Pelkonen, O.; Abass, K.; Wiesner, J. Thujone and thujone-containing herbal medicinal and botanical products: Toxicological assessment. *RTR* **2013**, *65*, 100–107. [CrossRef] [PubMed]
20. Dybowski, M.P.; Dawidowicz, A.L. The determination of α- and β-thujone in human serum—Simple analysis of absinthe congener substance. *Forensic Sci. Int.* **2016**, *259*, 188–192. [CrossRef] [PubMed]
21. Padosch, S.A.; Lachenmeier, D.W.; Kröner, L.U. Absinthism: A fictitious 19th century syndrome with present impact. *Subst. Abuse Treat. Prev. Policy* **2006**, *1*, 14. [CrossRef] [PubMed]
22. Nguyen, H.T.; Radácsi, P.; Rajhárt, P.; Németh, É. Variability of thujone content in essential oil due to plant development and organs from *Artemisia absinthium* L. and *Salvia officinalis* L. *J. Appl. Bot. Food Qual.* **2019**, *92*, 100–105. [CrossRef]
23. Judzentiene, A.; Budiene, J. Variability of *Artemisia campestris* L. essential oils from Lithuania. *J. Essent. Oil Res.* **2014**, *26*, 328–333. [CrossRef]
24. Masotti, V.; Juteau, F.; Bessière, J.M.; Viano, J. Seasonal and phenological variations of the essential oil from the narrow endemic species Artemisia molinieri and its biological activities. *J. Agric. Food Chem.* **2003**, *51*, 7115–7121. [CrossRef]
25. Verma, N.; Shukla, S. Impact of various factors responsible for fluctuation in plant secondary metabolites. *J. App. Res. Med. Aromat. Plants* **2015**, *2*, 105–113. [CrossRef]
26. Németh, É.Z.; Nguyen, H.T. Thujone, a widely debated volatile compound: What do we know about it? *Phytochem. Rev.* **2020**, *19*, 405–423. [CrossRef]
27. Beresneva, I.A. *Climates of the Arid Zone of Asia*; Nauka: Moscow, Russia, 2006; 287p.
28. Sharopov, F.; Braun, M.S.; Gulmurodov, I.; Khalifaev, D.; Isupov, S.; Wink, M. Antimicrobial, antioxidant, and anti-inflammatory activities of essential oils of selected aromatic plants from Tajikistan. *Foods* **2015**, *4*, 645–653. [CrossRef]
29. Raikova, S.V.; Golikov, A.G.; Shub, G.M.; Durnova, N.A.; Shapoval, O.G.; Rakhmetova, A.Y. Antimicrobial activity of essential oil of peppermint (*Mentha piperita* L.). *Saratov Sci. Med. J.* **2011**, *7*, 787–790.
30. Tepe, B.; Sokmen, M.; Akpulat, H.A.; Daferera, D.; Polissiou, M.; Sokmen, A. Antioxidative activity of the essential oils of *Thymus sipyleus* subsp. sipyleus var. sipyleus and *Thymus sipyleus* subsp. sipyleus var. rosulans. *J. Food Eng.* **2005**, *66*, 447–454. [CrossRef]
31. Zengin, H.; Baysal, A.H. Antibacterial and Antioxidant Activity of Essential Oil Terpenes against Pathogenic and Spoilage-Forming Bacteria and Cell Structure-Activity Relationships Evaluated by SEM Microscopy. *Molecules* **2014**, *19*, 17773–17798. [CrossRef] [PubMed]
32. Kordali, S.; Cakir, A.; Mavi, A.; Kilic, H.; Yildirim, A. Screening of chemical composition and antifungal and antioxidant activities of the essential oils from three Turkish Artemisia Species. *J. Agric. Food Chem.* **2005**, *53*, 1408–1416. [CrossRef] [PubMed]
33. Tykheev, Z.A.; Zhigzhitzhapova, S.V.; Zhang, F.; Taraskin, V.V.; Anenkhonov, O.A.; Radnaeva, L.D.; Chen, S. Constituents of Essential Oil and Lipid Fraction from the Aerial Part of *Bupleurum scorzonerifolium* Willd. (Apiaceae) from Different Habitats. *Molecules* **2018**, *23*, 1496. [CrossRef] [PubMed]

plants

MDPI

Article

Chemical Composition and Biological Activities of Essential Oils from *Origanum vulgare* Genotypes Belonging to the Carvacrol and Thymol Chemotypes

Paola Zinno [1,2,†], Barbara Guantario [1,†], Gabriele Lombardi [3], Giulia Ranaldi [1], Alberto Finamore [1], Sofia Allegra [1], Michele Massimo Mammano [4], Giancarlo Fascella [4], Antonio Raffo [1,*] and Marianna Roselli [1,*]

1 CREA-Research Centre for Food and Nutrition, Via Ardeatina, 546, 00178 Rome, Italy
2 Institute for the Animal Production System in the Mediterranean Environment, National Research Council, P.le E. Fermi 1, 80055 Portici, Italy
3 Department of Environmental Biology, Sapienza University, P.le Aldo Moro 5, 00185 Rome, Italy
4 CREA-Research Centre for Plant Protection and Certification, S.S. 113-Km 245.500, 90011 Bagheria, Italy
* Correspondence: antonio.raffo@crea.gov.it (A.R.); marianna.roselli@crea.gov.it (M.R.); Tel.: +39-06-51494573 (A.R.); +39-06-51494580 (M.R.)
† These authors contributed equally to the work.

Citation: Zinno, P.; Guantario, B.; Lombardi, G.; Ranaldi, G.; Finamore, A.; Allegra, S.; Mammano, M.M.; Fascella, G.; Raffo, A.; Roselli, M. Chemical Composition and Biological Activities of Essential Oils from *Origanum vulgare* Genotypes Belonging to the Carvacrol and Thymol Chemotypes. *Plants* **2023**, *12*, 1344. https://doi.org/10.3390/plants12061344

Academic Editors: Hazem Salaheldin Elshafie, Ippolito Camele and Adriano Sofo

Received: 23 February 2023
Revised: 10 March 2023
Accepted: 13 March 2023
Published: 16 March 2023

Abstract: The remarkable biological activities of oregano essential oils (EOs) have recently prompted a host of studies aimed at exploring their potential innovative applications in the food and pharmaceutical industries. The chemical composition and biological activities of EOs from two *Origanum vulgare* genotypes, widely cultivated in Sicily and not previously studied for their biological properties, were characterized. Plants of the two genotypes, belonging to the carvacrol (CAR) and thymol (THY) chemotypes and grown in different cultivation environments, were considered for this study. The chemical profiles, including the determination of enantiomeric distribution, of the EOs, obtained by hydrodistillation from dried leaves and flowers, were investigated by GC–MS. Biological activity was evaluated as antimicrobial properties against different pathogen indicator strains, while intestinal barrier integrity, reduction in pathogen adhesion and anti-inflammatory actions were assayed in the intestinal Caco-2 cell line. The chemical profile of the CAR genotype was less complex and characterized by higher levels of the most active compound, i.e., carvacrol, when compared to the THY genotype. The enantiomeric distribution of chiral constituents did not vary across genotypes, while being markedly different from that observed in *Origanum vulgare* genotypes from other geographical origins. In general, all EOs showed high antimicrobial activity, both in vitro and in a food matrix challenge test. Representative EOs from the two genotypes resulted not altering epithelial monolayer sealing only for concentrations lower than 0.02%, were able to reduce the adhesion of selected pathogens, but did not exert relevant anti-inflammatory effects. These results suggest their potential use as control agents against a wide spectrum of foodborne pathogens.

Keywords: oregano; *Origanum heracleoticum*; *Origanum vulgare ssp. viridulum* × *Origanum vulgare ssp. hirtum*; essential oil; enantiomers; antimicrobial activity; Caco-2 cells; intestinal permeability; reduction in pathogen adhesion; geographic origin

1. Introduction

Oregano is one of the most commercially valued aromatic plants worldwide, with more than 60 species used under this name having similar flavor profiles characterized mainly by cymyl compounds, such as carvacrol and thymol [1]. Beyond the widespread use of oregano as a culinary herb, its essential oil (EO) is currently used as an ingredient in flavoring formulations, dietary supplements and cosmetic and aromatherapeutic products [2]. A great deal of research has been recently dedicated to exploring other potential applications of oregano EOs in the food industry as food preservatives, based on their remarkable antimicrobial and antioxidant properties [3]. When considering such potential applications, it is important to take into account that foodborne diseases represent a serious and widespread threat to public health worldwide and require the adoption of strategies that ensure food safety and quality. An analysis of the European Rapid Alert System for Food and Feed (RASFF) Annual Report in 2020 [4] showed that a high number of notifications concerned microbiological contamination from pathogenic microorganisms in food of mostly animal origin, with a 37% increase in notifications in 2020 compared to 2019, when the largest number concerned *Salmonella* spp., *Listeria monocytogenes* and *Escherichia coli*. Concerning other potential applications in food production, a more recent field of research has focused on the use of oregano EOs as environmentally friendly alternatives to synthetic antibiotics in the aquaculture industry [5] or to synthetic pesticides as innovative products for crop protection [6]. Moreover, the potential of oregano EO as a protective agent in human chronic degenerative and infectious diseases, suggested by its several biological activities such as antimicrobial, antifungal, antiparasitic, anti-inflammatory, anticancer, antiproliferative, cytotoxic, has prompted a host of studies related to its potential pharmaceutical applications [7–9]. In particular, microbial resistance to antibiotics and chemotherapeutics is rapidly growing [10,11], requiring the search for alternative sources of antimicrobial compounds, both for prevention in human and veterinary medicine and for food preservation. Over the centuries, plants have been used for a wide variety of purposes, ranging from the treatment of infectious diseases to the extension of food shelf life, and been considered important sources of compounds with great chemical diversity, reflecting different biological properties [12]. In this context, oregano is considered among the plants whose extracts have the greatest antimicrobial effects [13]. Thus, a lot of efforts are currently underway to develop micro- and nano-encapsulation systems that enable the use of oregano EO in biotechnological and biomedical applications, by increasing its stability in aqueous media, thus improving its bioavailability, reducing its toxic effects, providing a controlled release and masking its strong aroma [14,15].

EOs obtained from *Origanum vulgare*, which is the main representative species of the genus *Origanum* L., are formed by a complex mixture of terpenes. Based on their proportions of cymyl, acyclic linalool/linalyl acetate and sabinyl compounds, three major chemotypes have been identified [16]. Within this complex chemical mixture, the two structurally similar cymyl compounds, carvacrol and thymol, are recognized as the main ones responsible for most of the biological activities of oregano EOs, and in particular, most of their antimicrobial activities [17]. It is generally assumed that their antibacterial action is exerted by inducing structural and functional damages to the cytoplasmic membrane of the target organism. However, a lot of evidence indicates that the biological activities of an EO may depend not only on the ratio in which the main active compounds are present but also on interactions between these and minor constituents in the oil [17]. From this perspective, it may be important to explore the chemical variability of oregano EO composition and to understand the influence of this variability on its biological activities.

One of the main sources of chemical variability in oregano EO is related to the genetic background of the plant. From this point of view, the species *Origanum vulgare*, native to and widespread in the Mediterranean region [18], encompasses many subspecies and hybrids, among which is the subspecies *Origanum heracleoticum* L. (Boissier) Hayek sensu Ietswaart [19], which is one of the most common growing in Sicily [20]. Information is available on the phytochemistry of this subspecies [20,21], but the biological activities of its EOs have not yet been investigated. On the contrary, no studies to the best of our knowledge have been conducted both on the chemical and biological features of the hybrid *Origanum vulgare ssp. viridulum* × *Origanum vulgare ssp. hirtum*, introduced into cultivation in Sicily in the last decade.

A lot of investigations have explored the intraspecific variability of the chemical composition of EOs from *Origanum vulgare* species, while a remarkable variation has also been observed within each subspecies, bearing to noticeable differences in the chemical profile and, possibly, in the associated biological properties [20,22]. Beyond genetic factors, this variability may also be due to environmental factors, such as, among others, geographical position [23] and altitude [24]. Regarding the chemical characterization of oregano EOs, very few data are available on the enantiomeric distribution of their main chiral compounds [25], even though this information may be useful not only for chemotaxonomical characterization, but also for authenticity issues. Changes in enantiomeric distribution could also influence the biological activities of oregano EO [26].

In the present study, a detailed characterization of the chemical composition and biological activities of EOs obtained from plants of the subspecies *Origanum heracleoticum* L. and the hybrid *Origanum vulgare ssp. viridulum* × *Origanum vulgare ssp. hirtum* is provided. The chemical characterization, besides their basic chemical profile, also included the determination of the enantiomeric distribution of chiral compounds present in their EOs and in their dried leaves and flowers. The study of their biological activities focused on antimicrobial, intestinal permeability and anti-inflammatory properties. To evaluate antimicrobial properties, different pathogen indicator strains were used, while the intestinal barrier integrity, reduction in pathogen adhesion and anti-inflammatory activity were evaluated in a widely used in vitro model of intestinal cells, the human Caco-2 cell line, differentiated on permeable filters. These conditions allowed for the reproduction of the environment of the intestinal mucosa, in which the epithelium separates the lumen from the inner body. A preliminary characterization carried out in our laboratory highlighted that the hybrid *Origanum vulgare ssp. viridulum* × *Origanum vulgare ssp. hirtum* was distinguished by a high content of carvacrol as the main EO constituent, thus allowing for a description of its EO as the carvacrol chemotype, whereas EOs from the subspecies *Origanum heracleoticum* L. are known to belong to the thymol chemotype [20]. Thus, the aim of the present study was to provide detailed information on the chemical and biological properties of the EOs from these two genotypes to promote alternative and innovative applications in food and pharmaceutical products suited to their specific properties. In addition, the information collected may contribute to a deeper understanding of the influence of varying chemical profiles, such as those with carvacrol or thymol as predominant constituents, on the biological activities of oregano EOs.

2. Results

2.1. Composition of EOs

GC–MS analyses enabled the detection of 32 volatile compounds at a level higher than 0.05% of total EO content (Table 1) in three EOs obtained from the same genotype of the hybrid *Origanum vulgare ssp. viridulum* × *Origanum vulgare ssp. hirtum* (CAR1, CAR2, CAR3) and in five EOs distilled from plants from the same genotype of the subspecies *Origanum heracleoticum* L. (THY1, THY2, THY3, THY4, THY5). Within each group of EOs, the plants obtained from the same genotype were grown on different farms located in a restricted area of southern Sicily. Thirty-one of the detected compounds were identified as monoterpene hydrocarbons (thirteen), oxygenated monoterpenes (eight), sesquiterpene hydrocarbons (eight) and two other compounds. The main feature of both groups of EO samples was the occurrence of carvacrol or thymol as the predominant constituent: in the EOs from the hybrid *Origanum vulgare ssp. viridulum* × *Origanum vulgare ssp. hirtum*, named the CAR group, carvacrol was the prominent compound, varying in amount from 81 to 85%, whereas in the EOs from the subspecies *Origanum heracleoticum* L., named the THY group, the main constituent was thymol, ranging in amount from 47 to 65%. Globally, the group of monoterpene hydrocarbons accounted from a higher proportion of EO content in the THY group (24–38%) than in the CAR group (12–14%): in both groups, γ-terpinene and p-cymene were the main constituents of the monoterpene hydrocarbon chemical class, being present at a higher level in the THY group (13–22% and 4–5% in THY, and 5–7% and 3% in CAR, respectively). Similarly, the sesquiterpene hydrocarbon content was higher in the THY group (3.9–6.9%) than in the CAR group (2.2–2.6%), with β-caryophyllene occurring as the main constituent (0.9–1.8% in THY and 1.9–2.4% in CAR). Conversely, the global level of oxygenated monoterpenes showed a higher content in the

CAR group (84–85%) than in the THY group (56–71%), this difference being dictated by the level of carvacrol/thymol. Interestingly, when 0.05% was established as the cut-off level for quantification, EOs of the THY group were characterized by a more complex chemical profile, formed by 30 compounds, whereas only 19 compounds were represented in the CAR group, in which a larger proportion of total content was formed by a single compound. Moreover, the chemical profiles of the EOs of the CAR group were quite similar to each other, whereas a higher level of variability was observed in the profile of the EOs of the THY group, suggesting a higher influence of the farm location on the chemical profile for this group (Figure 1). Diversification of chemical profiles in the THY group was mainly due to variability in the level of monoterpene hydrocarbons, such as α- and β-pinene and α-thujene (Figure 1).

Table 1. Chemical compositions of the EOs of the hybrid *Origanum vulgare ssp. viridulum* × *Origanum vulgare ssp. hirtum* (CAR1, CAR2, CAR3) and of the subspecies *Origanum heracleoticum* L. (THY1, THY2, THY3, THY4, THY5). Data are expressed as the % content of individual constituents.

No.	Compound Name	Class	Oregano Subspecies/Hybrid							
			Origanum vulgare ssp. viridulum × *Origanum vulgare ssp. hirtum*			*Origanum heracleoticum* L.				
			CAR1	CAR2	CAR3	THY1	THY2	THY3	THY4	THY5
1	α-thujene	MH	0.74	0.75	0.78	1.04	0.72	1.19	0.83	0.35
2	α-pinene	MH	0.32	0.32	0.33	0.39	0.29	0.46	0.32	0.16
3	1-octen-3-ol	Ot	0.15	0.15	0.14	-	-	-	-	-
4	3-octanone	Ot	0.07	0.08	0.06	-	-	-	-	-
5	sabinene	MH	-	-	-	-	-	0.07	-	-
6	β-pinene	MH	0.05	0.06	0.06	0.07	-	0.08	-	-
7	β-myrcene	MH	1.15	1.18	1.19	1.63	1.22	1.84	1.41	1.01
8	α-phellandrene	MH	0.14	0.17	0.17	0.27	0.19	0.30	0.21	0.14
9	α-terpinene	MH	1.09	1.24	1.37	3.45	2.45	3.92	2.92	2.21
10	*p*-cymene	MH	2.64	2.73	2.59	4.79	4.11	5.25	4.63	3.60
11	limonene	MH	0.24	0.28	0.28	0.42	0.32	0.48	0.36	0.28
12	*cis*-β-ocimene	MH	-	-	-	1.44	1.30	1.87	1.36	1.64
13	*tr*-β-ocimene	MH	-	-	-	0.20	0.17	0.26	0.18	0.22
14	γ-terpinene	MH	5.33	6.30	7.11	17.96	13.16	22.13	15.90	16.95
15	*cis*-sabinene hydrate	OM	0.29	0.29	0.28	0.18	0.15	0.23	0.13	0.15
16	terpinolene	MH	-	-	-	0.07	-	0.08	-	-
17	linalool	OM	-	-	-	0.30	0.32	0.41	0.32	0.39
18	borneol	OM	-	0.08	-	-	0.44	-	-	-
19	terpinen-4-ol	OM	0.29	0.35	0.32	0.57	-	0.63	0.49	0.43
20	thymol methyl ether	OM	-	-	-	0.70	0.86	1.30	1.06	2.14
21	carvacrol methyl ether	OM	-	-	-	4.24	3.52	4.63	3.82	3.48
22	thymol	OM	0.14	3.19	1.68	56.22	65.47	47.32	61.35	59.36
23	carvacrol	OM	84.70	80.59	81.32	0.93	0.68	1.28	0.86	0.50
24	β-caryophyllene	SH	2.45	1.93	2.08	1.20	1.11	1.54	0.95	1.80
25	α-humulene	SH	0.06	0.08	0.09	0.08	0.08	0.11	-	0.13
26	γ-muurolene	SH	-	-	-	0.09	0.08	-	-	0.08
27	germacrene D	SH	-	-	-	1.52	1.36	2.25	1.23	2.22
28	bicyclogermacrene	SH	-	-	-	0.15	0.13	0.20	0.12	0.24
29	β-bisabolene	SH	0.15	0.24	0.16	1.28	1.17	1.52	1.02	1.56

Table 1. *Cont.*

No.	Compound Name	Class	Oregano Subspecies/Hybrid							
			Origanum vulgare ssp. viridulum × *Origanum vulgare ssp. hirtum*			*Origanum heracleoticum* L.				
			CAR1	CAR2	CAR3	THY1	THY2	THY3	THY4	THY5
30	γ-cadinene	SH	-	-	-	0.15	0.12	0.10	0.08	0.12
31	δ-cadinene	SH	-	-	-	0.47	0.39	0.39	0.33	0.47
32	unknown	SH	-	-	-	0.19	0.18	0.17	0.14	0.24
Total content of compounds grouped in chemical classes										
Monoterpene Hydrocarbons (MH)			11.71	13.03	13.88	31.74	23.94	37.93	28.11	26.58
Oxygenated Monoterpenes (OM)			85.42	84.50	83.59	63.13	71.44	55.80	68.02	66.46
Sesquiterpene Hydrocarbons (SH)			2.66	2.25	2.33	5.13	4.61	6.27	3.87	6.85
Others			0.22	0.23	0.20	-	-	-	-	-

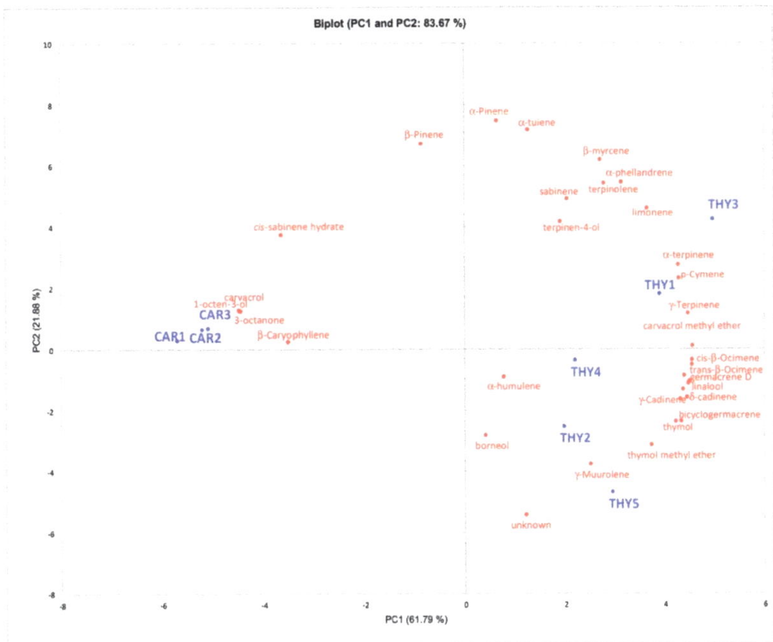

Figure 1. Principal component analysis (PCA) biplot of the first two PCs obtained in the dataset of the chemical compositions of EOs of the hybrid *Origanum vulgare ssp. viridulum* × *Origanum vulgare ssp. hirtum* (CAR1, CAR2, CAR3) and of the subspecies *Origanum heracleoticum* L. (THY1, THY2, THY3, THY4, THY5).

2.2. Enantiomeric Distribution of Chiral Compounds as Determined after Isolation from Oregano Dried Leaves and Flowers and in Distilled Oregano EOs

The enantiomeric distribution analysis of chiral volatile compounds of oregano leaves and flowers was performed both on the volatile isolates directly obtained by headspace solid-phase microextraction (HS-SPME) from the dried plant material (Table 2a) and on the corresponding EOs obtained from the plant material by distillation (Table 2b). GC analysis performed by an enantioselective GC column allowed for the detection of seven pairs of enantiomers in four monoterpene hydrocarbons, α-thujene, α-pinene, β-pinene, α-phellandrene and three oxygenated monoterpenes, linalool, terpinen-4-ol and α-terpineol. This last compound was not quantified in the GC analysis with an achiral column (Table 1) because its level was lower than 0.05%, but it was included in the enantioselective analyses because it was possible to accurately determine its two enantiomers. The analysis of the isolates obtained from the dried leaves and flowers by HS-SPME allowed for the determination of a higher number of enantiomeric pairs than in the analysis of the EOs, because the former isolates were more concentrated. In the case of α-thujene, it was not possible to identify the two enantiomers, due to unavailability of commercial standard pure compounds and lack of appropriate information from the literature of chromatographic data. The results obtained for both the dried aerial parts and EOs showed that the enantiomeric distribution was the same in plants belonging to the subspecies *Origanum heracleoticum* L. and the hybrid *Origanum vulgare ssp. viridulum* × *Origanum vulgare ssp. hirtum*, and it was not affected by the different environments of cultivation. Only in the case of the distribution of linalool enantiomers, as determined in the dried plant materials in one sample (CAR3), was there a slight difference in the distribution of enantiomers observed, with a 17% level of the S enantiomer instead of the 8–11% level observed in the other samples. Moreover, results obtained from the dried plant material showed less variability when compared to those obtained from the EOs. This could partly be due to the more concentrated isolate analyzed in the former case. Enantiomeric distribution as determined in the dried plant material was generally similar to that observed in the EOs, with some minor differences. For instance, the distribution of α-pinene enantiomers was, on average, 5:95 and 10:90 in the EOs and plant material, respectively; the distribution of β-pinene was 81:19 and 73:27; the distribution of terpinene-4-ol was 61:39 and 51:49; and the distribution of α-terpineol was 0:100 and 9:91. In all cases, a slightly higher enantiomeric excess was observed in the EOs than in the isolates from dried plant material.

Table 2. (**a**) Enantiomeric distributions (%) of chiral compounds determined after isolation by HS-SPME from oregano dried leaves and flowers of the hybrid *Origanum vulgare ssp. viridulum* × *Origanum vulgare ssp. hirtum* (CAR1, CAR2, CAR3) and of the subspecies *Origanum heracleoticum* L. (THY1, THY2, THY3, THY4, THY5). (**b**) Enantiomeric distributions (%) of chiral compounds determined from EOs of the hybrid *Origanum vulgare ssp. viridulum* × *Origanum vulgare ssp. hirtum* (CAR1, CAR2, CAR3) and the subspecies *Origanum heracleoticum* L. (THY1, THY2, THY3, THY4, THY5).

Compound/ Enantiomer		LRI [1]	Oregano Subspecies/Hybrid							
			Origanum vulgare ssp. viridulum × *Origanum vulgare ssp. hirtum*			*Origanum heracleoticum* L.				
			CAR1	CAR2	CAR3	THY1	THY2	THY3	THY4	THY5
(a)										
α-thujene	n.i. [2]	950	65	64	64	65	65	65	65	64
	n.i.	952	35	36	36	35	35	35	35	36
α-pinene	S/−	984	12	12	12	9	9	9	9	10
	R/+	988	88	88	88	91	91	91	91	90
β-pinene	R/+	1026	73	73	72	73	74	74	74	73
	S/−	1032	27	27	28	27	26	26	26	27
α-phellandrene	R/−	1036	5	5	5	4	4	3	3	4
	S/+	1039	95	95	95	96	96	97	97	96
linalool	R/−	1217	89	90	83	90	92	91	92	90
	S/+	1224	11	10	17	10	8	9	8	10

Table 2. *Cont.*

Compound/ Enantiomer		LRI [1]	Oregano Subspecies/Hybrid							
			Origanum vulgare ssp. viridulum × *Origanum vulgare ssp. hirtum*			*Origanum heracleoticum* L.				
			CAR1	CAR2	CAR3	THY1	THY2	THY3	THY4	THY5
terpinene-4-ol	S/+	1300	54	54	54	52	50	49	48	49
	R/−	1304	46	46	46	48	50	51	52	51
α-terpineol	R/+	1350	11	9	10	8	7	8	8	10
	S/−	1360	89	91	90	92	93	92	92	90
						(b)				
α-thujene	n.i. [2]	950	n.d.	n.d.	n.d.	n.d.	n.d.	n.d.	n.d.	n.d.
	n.i.	952	n.d.	n.d.	n.d.	n.d.	n.d.	n.d.	n.d.	n.d.
α-pinene	S/−	984	6	7	6	3	5	4	4	3
	R/+	988	94	93	94	97	95	96	96	97
β-pinene	R/+	1026	79	79	81	79	79	84	84	n.d.
	S/−	1032	21	21	19	21	21	16	16	n.d.
α-phellandrene	R/−	1036	n.d.	n.d.	n.d.	n.d.	n.d.	n.d.	n.d.	n.d.
	S/+	1039	n.d.	n.d.	n.d.	n.d.	n.d.	n.d.	n.d.	n.d.
linalool	R/−	1217	n.d.	n.d.	n.d.	n.d.	n.d.	n.d.	n.d.	n.d.
	S/+	1224	n.d.	n.d.	n.d.	n.d.	n.d.	n.d.	n.d.	n.d.
terpinene-4-ol	S/+	1300	58	59	58	63	62	61	63	62
	R/−	1304	42	41	42	37	38	39	37	38
α-terpineol	R/+	1350	0	0	0	0	0	0	0	n.d.
	S/−	1360	100	100	100	100	100	100	100	n.d.

[1] Linear retention index determined from the Cyclosil-B column stationary phase (30% Heptakis (2,3-di-O-methyl-6-O-t-butyl dimethylsilyl)-β-cyclodextrin in DB-1701). [2] It was not possible to identify the two enantiomers due to the unavailability of standard pure compounds and lack of information from the literature.

2.3. Antimicrobial Activity of EOs by Agar Spot Test against Pathogen Indicator Strains

EOs extracted from the hybrid *Origanum vulgare ssp. viridulum* × *Origanum vulgare ssp. hirtum* showed a relevant antimicrobial activity assayed by an agar spot test, with halos of various sizes (Table 3). In particular, all the pathogenic microorganisms considered were susceptible to the inhibitory activity of CAR1, CAR2 and CAR3, with inhibition halo radii exceeding 7.5 mm, with the exception of *L. monocytogenes* SA and *Pseudomonas fluorescens* B13, against which these oils had low activity.

Concerning the EOs extracted from *Origanum heracleoticum*, THY3 and THY5 were tested, as they resulted as more divergent within the THY cluster (Figure 1). These two EOs showed a medium activity against *P. putida* KT2240 and *P. fluorescens* B13 (radii between 5 and 7.5 mm) and high inhibitory activity against all other pathogenic and alterative microorganisms (Table 3).

On the basis of the results of the chemical characterization (Table 1) and antimicrobial activity by spot test (Table 3), CAR1 and THY5 were selected for further experiments, as they had higher carvacrol and thymol content among CAR1, CAR2 and CAR3 and among THY3 and THY5, respectively. Moreover, CAR1 and THY5 showed high antimicrobial activity against *L. monocytogenes* OH and *S.* Typhimurium LT2 and were thus chosen as representative of the most frequently reported pathogen related to foodborne diseases.

Table 3. Antimicrobial activity of oregano EOs against indicator pathogens (spot test).

Pathogen Indicator	Inhibition Halo Radius (mm)				
	CAR1	CAR2	CAR3	THY3	THY5
Listeria monocytogenes OH	+++	+++	+++	+++	+++
Listeria monocytogenes SA	+	+	+	+++	+++
Listeria monocytogenes CAL	+++	+++	+++	+++	+++
Listeria innocua 1770	++	++	++	+++	+++
Salmonella enterica Typhimurium LT2	+++	++	+++	+++	+++
Salmonella enterica Give	+++	+++	+++	+++	+++
Salmonella enterica Derby	+++	+++	+++	+++	+++
Enterotoxigenic *E. coli* (ETEC) K88	+++	+++	+++	+++	++
Pseudomonas putida WSC358	+++	++	+++	+++	++
Pseudomonas putida KT2240	+++	++	++	++	++
Pseudomonas fluorescens B13	+	+	+	++	++

Antimicrobial activity was considered high (+++) when the value of the inhibition halo radius (hr) was greater than 7.5 mm (hr > 7.5 mm), medium (++) when between 5 and 7.5 mm (5 mm $\leq$ hr $\leq$ 7.5 mm) and low (+) when less than 5 mm (hr < 5 mm).

2.4. *In Vitro Antimicrobial Activity of CAR1 and THY5 against L. monocytogenes OH and S. Typhimurium LT2 by Direct Contact Test*

The two EOs CAR1 and THY5 showed similar antimicrobial capacities by a direct contact test against *L. monocytogenes* OH (Figure 2A). In particular, at a lower EO concentration (0.12%), an approximately 3-log reduction in microbial concentration occurred as compared to samples exposed to 0% EO (C) and to 0% EO plus 0.125% Tween 80 (CTw). Proportionally to increased EO concentration (0.25 and 0.5%), a reduction of approximately 4 and 4.5 logs, respectively, was observed in the microbial titer, as compared to C. In addition, the treatment with CAR1 and THY5 at 0.25% induced a reduction in microbial concentration similar to that obtained with ampicillin, while for both oils, treatment with 0.5% induced a greater effect than ampicillin in reducing microbial load (Figure 2A). Similarly, the oils at the same extent showed antimicrobial activity by a direct contact test against *S.* Typhimurium LT2 (Figure 2B), which resulted in a reduction in the microbial titer of about 0.5 log CFU/mL when applied at a concentration of 0.12%. A dose–response effect was observed, as the 0.5% concentration induced a reduction of about 4 log CFU/mL. For *S.* Typhimurium LT2, the antimicrobial effects of the oils at a 0.5% concentration was similar to that of ampicillin (Figure 2B).

2.5. *Antimicrobial Effects of CAR1 and THY5 during a Challenge Test of Beef Minced Meat*

The antimicrobial effects of CAR1 and THY5 EOs against *L. monocytogenes* OH and *S.* Typhimurium LT2 was also evaluated in a complex food matrix of beef minced meat during storage at 4 °C for 7 days. The 0.5% concentration was used in order to reproduce the conditions of maximum antimicrobial concentration as in the direct contact test. The presence of CAR1 EOs caused a reduction of the bacterial population of both *L. monocytogenes* OH and *S.* Typhimurium LT2 by about 1 log CFU/g one day after contamination, remaining essentially unchanged until the third day (Figure 3A,B). In both sets of experiments, there was an increase in microbial load after seven days of storage both in the presence and absence of the EOs, suggesting an onset of recontamination of the minced meat. In parallel, the absence of *L. monocytogenes* OH and *S.* Typhimurium LT2 in the food matrix in the artificially uncontaminated sample was assessed.

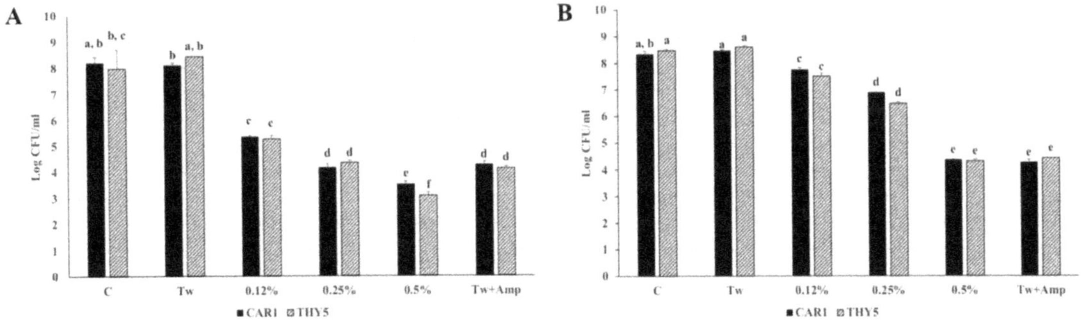

Figure 2. Direct contact test of CAR1 and THY5 antimicrobial activity (**A**) against *L. monocytogenes* OH and (**B**) *S.* Typhimurium LT2. Each EO was tested at 0.12, 0.25 and 0.5% concentrations. C: 0% EO; CTw: 0% EO + 0.125% Tween 80; CTw + Amp: 50 µg/mL ampicillin + 0.125% Tween 80. Bacterial cell viability is expressed as the geometric mean of CFU/mL ± SD of one experiment carried out in triplicate. Means without a common letter significantly differ, $p < 0.05$.

Regarding the antimicrobial activity of 0.5% THY5, a different effect was observed against the two pathogens tested. The microbial load of *L. monocytogenes* OH remained essentially unchanged after 1 and 2 days in the contaminated samples in the presence of EO, whereas a reduction of approximately 1 and 2 logs in the microbial load was observed at 3 and 7 days, respectively (Figure 3C). The minced meat in the absence of THY5 had a similar microbial titer of about 1×10^5 CFU/g during all storage timepoints.

The results of the challenge test set up with *S.* Typhimurium LT2 indicated that the antimicrobial effect of THY5 EO was already appreciable at the first day with a reduction of approximately 1.5 log CFU/g and a further decrease of 1 log CFU/g was observed at the following timepoints (2, 3 days). On day 7, an increase of about 0.6 log CFU/g suggested the start of recontamination of the minced meat (Figure 3D). Surprisingly, from the second day of storage at 4 °C, a reduction in the microbial titer was observed in the absence of EO.

2.6. Effects of CAR1 and THY5 on Caco-2 Cell Permeability

In order to evaluate if EOs exposure could perturb intestinal epithelial permeability, transepithelial electrical resistance (TEER) and phenol red apparent permeability (Papp) were measured in differentiated Caco-2 cells after treatment with several concentrations (ranging from 0.01 to 0.1%) of the two representative EOs CAR1 or THY5, dissolved in a complete culture medium for up to 24 h. The results show that only the 0.01 and 0.02% CAR1 treatments did not affect cell permeability up to 24 h, while the higher concentrations (0.025–0.03–0.05–0.1%) already induced a TEER drop after 1 h (less than 20% of the control), which was maintained until the end of the experiment (Figure 4A). This TEER decrease was associated with a biologically relevant phenol red Papp increase, as the corresponding values were in the order of magnitude of 6×10^{-6} cm s^{-1}, indicating that the tight junctions were open (Figure 4B). Indeed, permeability coefficients higher than 5×10^{-7} cm s^{-1} are considered to be indicative of compromised cell monolayers, while values above 1×10^{-6} cm s^{-1} indicate destroyed cell monolayer integrity [27]. Concerning THY5, a drop in TEER values was already observed after 1 h treatment with 0.03, 0.05 and 0.1% concentrations (Figure 4C). Differently from CAR1, the 0.025% concentration of THY5 induced a TEER decrease less rapidly, but regardless, was indicative of important and irreversible damage to the monolayer, since from 4 h treatment, the TEER values were lower than 50% of the control and they continued to decrease. At the lower concentrations tested (0.02 and 0.01%), the TEER values remained above 100% of the control up to 24 h, indicating the integrity maintenance of the monolayer (Figure 4C). The TEER data were confirmed by the results of the paracellular passage of phenol red, shown in Figure 4D. The Papp values were indeed higher than the threshold value (1×10^{-6} cm s^{-1}) for THY5 concentrations between 0.1 and 0.025%, while those related to the treatment with 0.02 and 0.01% were in the order of 1×10^{-7} cm s^{-1}, indicating that the tight junctions between cells were functionally sealed. From these data, it appears that among the concentrations tested, the highest concentration not damaging the Caco-2 cell monolayer was 0.02% for both EOs. This concentration was therefore used for the experiments of reduction in pathogen adhesion and for the evaluation of anti-inflammatory activity. The presence of 2% ethanol, corresponding to the concentration contained in the higher EO dilution tested, did not affect cell permeability for up to 24 h.

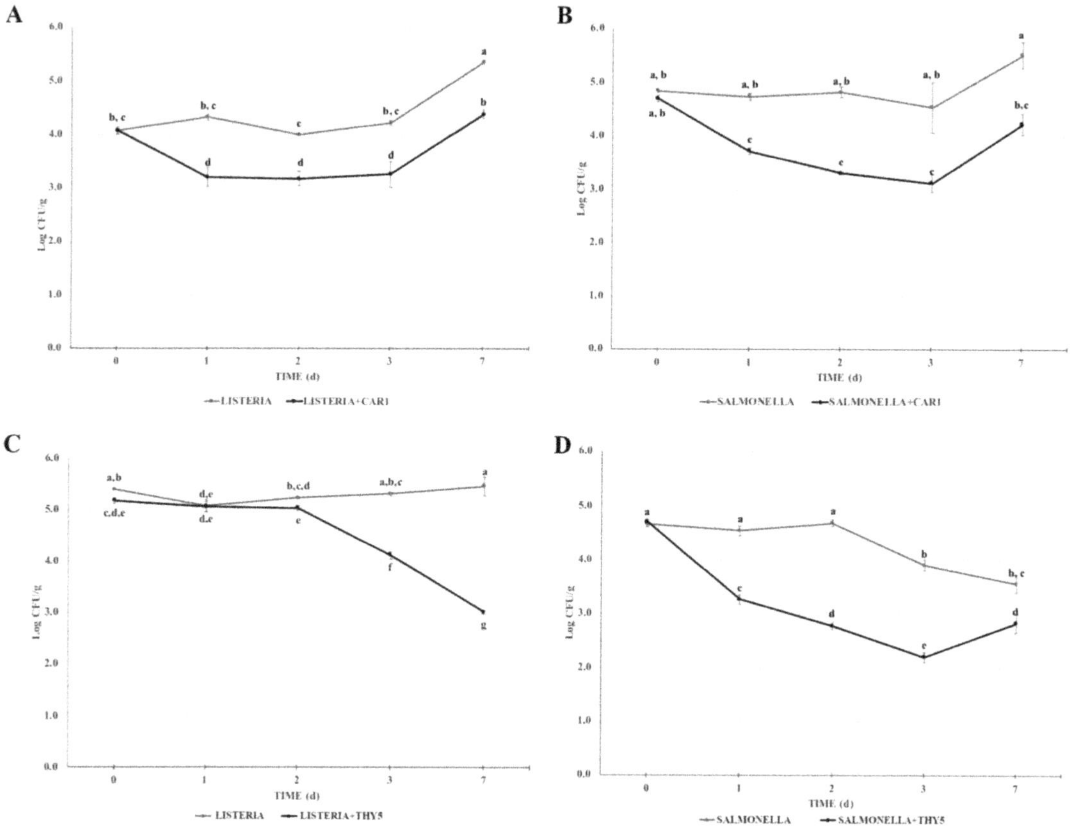

Figure 3. Challenge test of CAR1 against (**A**) *L. monocytogenes* OH and (**B**) *S.* Typhimurium LT2 and of THY5 against (**C**) *L. monocytogenes* OH and (**D**) *S.* Typhimurium LT2. The assay was performed in minced cow meat artificially contaminated with *L. monocytogenes* OH or *S.* Typhimurium LT2, with or without the addition of 0.5% CAR1 or THY5. The meat was stored at 4 °C for up to 7 days, homogenized at the indicated timepoints and appropriate dilutions were plated. Bacterial cell viability is expressed as the geometric mean of CFU/g ± SD of one experiment carried out in triplicate. Means without a common letter significantly differ, $p < 0.05$.

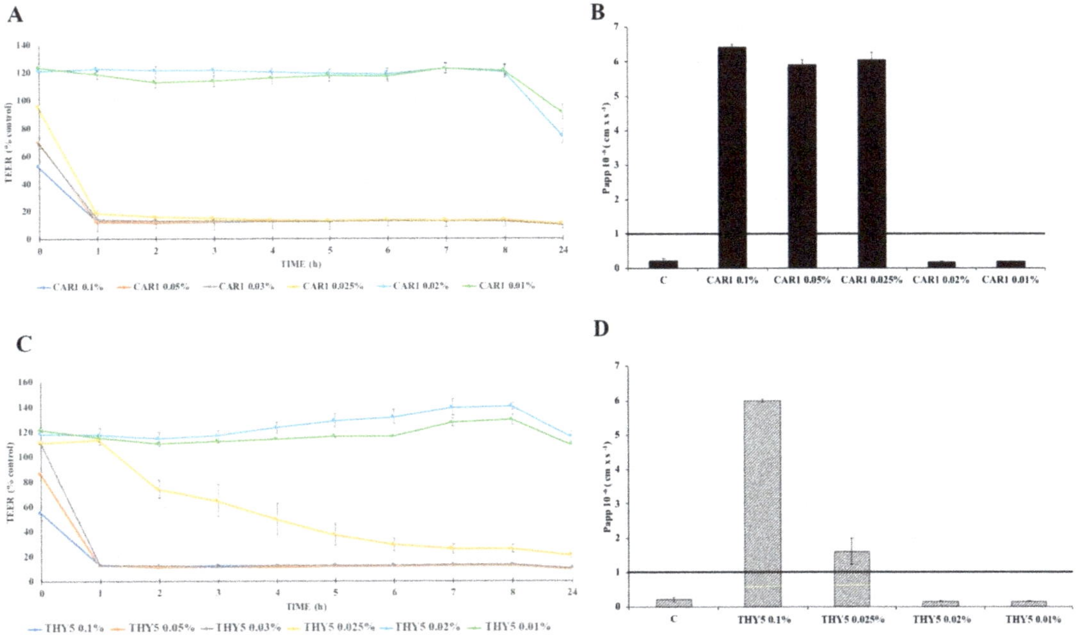

Figure 4. CAR1 and THY5 effects on Caco-2 cells monolayer integrity: Transepithelial electrical resistance (TEER) and phenol red apparent permeability (Papp). Cells were untreated (Control, C) or treated with different CAR1 or THY5 concentrations (0.01–0.1%). TEER values ((**A**) for CAR1 and (**C**) for THY5), recorded for up to 24 h, were calculated as the % of C for each timepoint. Phenol red Papp ((**B**) for CAR1 and (**D**) for THY5) was measured at 24 h and values are reported as cm s^{-1}. A black line set at 1×10^{-6} cm s^{-1} represents the Papp threshold, indicating destroyed cell monolayer integrity for values above. Values represent the mean ± SD of the two independent experiments, carried out in triplicate.

2.7. Reduction in Pathogen Adhesion to Caco-2 Cells by CAR1 and THY5

To evaluate the potential reduction ability exerted by 0.02% CAR1 and THY5 against pathogen adhesion on Caco-2 cells, the indicator strains *L. monocytogenes* OH and *S.* Typhimurium LT2 were selected. Both strains had a similar ability to adhere to Caco-2 cells, quantified in about 10% of the initial inoculum, which consisted of approximately 1×10^8 CFU/mL for each strain (Figure 5A,B). The presence of CAR1 significantly reduced the adhesion of *L. monocytogenes* OH as compared to the oil-free controls, with a decrease of approximately 1 log CFU/mL, while THY5 was not effective (approximately 0.5 log CFU/mL, Figure 5A).

Both CAR1 and THY5 were able to significantly reduce the number of adhered viable bacterial cells of *S.* Typhimurium LT2 to Caco-2 cells (Figure 5B). Thus, the two pathogens tested showed a different susceptibility to CAR1 and THY5.

Such adhesion reduction was not ascribable to the presence of 0.4% ethanol, as preliminary experiments did not show any effect of this ethanol concentration (contained in 0.02% EOs) on the number of viable bacterial cells adhering to the intestinal cell monolayer.

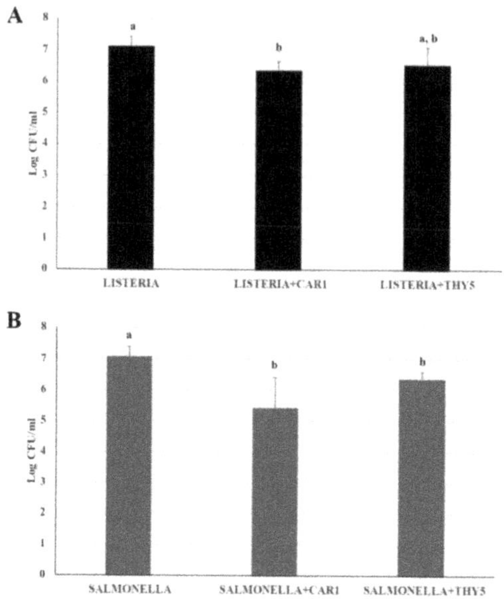

Figure 5. Adhesion reduction test for (**A**) *L. monocytogenes* OH and (**B**) *S.* Typhimurium LT2 in the presence of 0.02% CAR1 or THY5 in Caco-2 cells. Bacterial cell viability is expressed as the geometric mean of CFU/mL ± SD of the two independent experiments, carried out in triplicate. Means without a common letter significantly differ, $p < 0.05$.

2.8. Effects of CAR1 and THY5 on the NF-kB Pathway in Caco-2 Cells

To evaluate if CAR1 and THY5 could have a protective effect on a pro-inflammatory stimulus (TNF-α), the gene expression of some key genes of the NF-kB pathway was analyzed in Caco-2 cells. Epithelial integrity was not affected by experimental treatments with TNF-α or EOs, as indicated by the TEER values that remained comparable to the untreated control cells throughout all the experimental intervals.

As compared to the control, treatment with CAR1 induced a mild gene expression downregulation of the IkBα and IL-6 genes of the NF-kB pathway; however, this downregulation did not reach statistical significance, probably due to great data variability (Figure 6A). THY5 treatment did not affect gene expression (Figure 6B).

As expected, TNF-α treatment was able to induce a significant increase in the expression (up to 2 fold change) of all analyzed genes of the NF-kB pathway, except for IL-1α (Figure 6A,B). CAR1 pre-treatment before TNF-α addition induced a slight decrease in IkBα, IL-6 and IL-8 gene expression as compared to treatment with TNF-α alone. However, this remained as a trend ($p > 0.05$), as statistical significance could not be reached because of great data variability (Figure 6A). When the cells were pre-treated with THY5 before TNF-α addition, the gene expression levels of all analyzed genes were similar to TNF-α, even if an increasing trend could be observed for inflammatory IL-1α, IL-6 and IL-8 cytokines (Figure 6B).

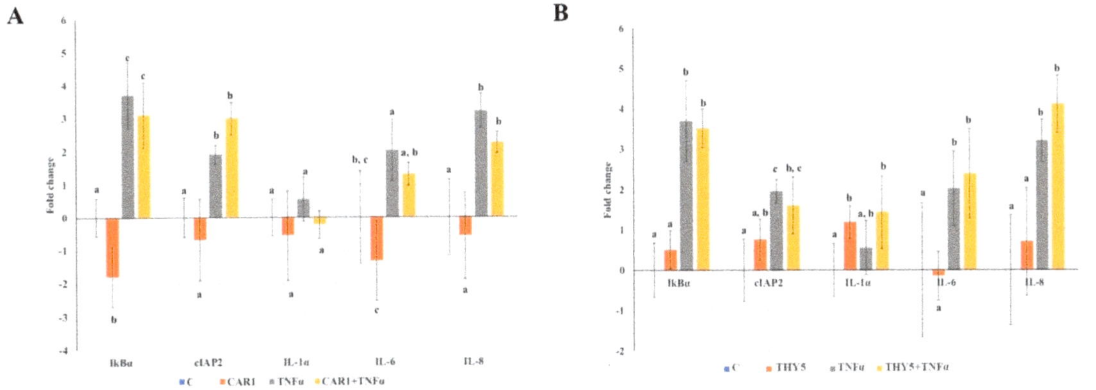

Figure 6. Study of NF-kB pathway gene expression in Caco-2 cells treated with (**A**) CAR1 and/or TNF-α and (**B**) THY5 and/or TNF-α. Caco-2 cells were pre-treated for 1 h with 0.02% CAR1 or THY5 and then treated for 1 h with 15 ng/mL TNF-α. IkBα, cIAP2, IL-1β, IL-6 and IL-8 gene expression was analyzed by real-time PCR. Data are expressed as the fold change (log2) ± SD of two independent experiments, carried out in triplicate. Means without a common letter significantly differ, $p < 0.05$.

3. Discussion

In the present study, eight different EOs distilled from plants obtained of two genotypes belonging to the subspecies *Origanum heracleoticum* L. and to the hybrid *Origanum vulgare ssp. viridulum* × *Origanum vulgare ssp. hirtum* were characterized according to their chemical composition and several biological activities. First, chemical analyses allowed for highlighting the differences in the profiles of the two genotypes and evaluating the variability within each genotype depending on the cultivation location. EOs from the two genotypes showed differences not only in their predominant compound, carvacrol or thymol, but also in their overall profile complexity, which was more pronounced in the thymol chemotype. The profile of the EOs from the *Origanum heracleoticum* L. subspecies, described as belonging to the thymol chemotype, was in accordance with the data reported in the literature on the same subspecies of wild and cultivated oregano plants grown in the same geographic area of Sicily [20,21,28]. No previous information was available in the literature, to the best of our knowledge, on the profile of the hybrid *Origanum vulgare ssp. viridulum* × *Origanum vulgare ssp. hirtum*, described as belonging to the carvacrol chemotype. In both groups, the occurrence of relatively high levels of γ-terpinene and p-cymene was consistent with the biosynthetic pathway proposed for the main compounds carvacrol/thymol, by which γ-terpinene is first converted to p-cymene and, in turn, further converted to carvacrol or thymol by two distinct enzymes [29]. Interestingly, while most monoterpene and sesquiterpene hydrocarbons appeared to be positively correlated among themselves and with thymol, β-caryophyllene was positively correlated with carvacrol and negatively correlated with the other sesquiterpenes. From the point of view of relationships with biological activities, it is interesting to note that the level of the most active EO compound was significantly higher in the hybrid *Origanum vulgare ssp. viridulum* × *Origanum vulgare ssp. hirtum*. than in the subspecies *Origanum heracleoticum* L., which might suggest a higher potential in the former. On the contrary, the profile of EOs from the subspecies *Origanum heracleoticum* L. was more complex than the other in terms of the number of minor constituents detected. Moreover, chemical characterization provided detailed information on the enantiomeric distribution of chiral compounds in the aerial parts of the considered oregano genotypes and in the corresponding EOs. This piece of information may be important not only because optical isomers can have different odors and, in general, different biological properties, but also because it provides useful information for uncovering the biosynthetic and geographical origins of the considered EOs as well as their authenticity and also for investigating technological treatments to which EOs could be subjected [30]. In the oregano EOs analyzed in this study, chiral compounds were all present at a relatively low level (<1%), whereas all major compounds were achiral. This would suggest a minor relevance of the enantiomeric characterization for these oregano genotypes. However, very little information is available in the literature on the enantiomeric distribution of chiral compounds in oregano EOs, and the only paper reporting experimental data showed a quite different picture as compared to our results, both in terms of chemical profiles and enantiomeric distributions [25]. In that study, EOs from native populations of *Origanum vulgare* collected in two different geographical locations in northwestern Himalaya and belonging to the linalool and cymyl chemotypes were investigated [25]. In that study, pairs of enantiomers were determined for four of the seven chiral compounds investigated in the present study, i.e., α-pinene, linalool, terpinene-4-ol and α-terpineol. In all cases, the enantiomeric distributions observed were markedly different between the two populations from Himalaya. Thus, the determination of enantiomeric distribution was confirmed to represent a valuable analytical tool for a detailed chemical characterization of oregano EOs, which may be useful for the recognition of geographical origin. Moreover, in that study, a marked effect of the distillation process on the observed enantiomeric distribution in the obtained EOs was postulated to explain some inversions of the ratios for some compounds between the two EOs investigated [25]. Data from the present study seemed to rule out the possibility of a marked effect of the distillation step on the enantiomeric distributions, while the minor differences observed between the EOs and dried aerial parts could be plausibly attributed mainly to differences in the analytical process, and, in particular, to the different concentration levels of individual VOCs in the isolates analyzed in the two cases. Moreover, an expected effect of the distillation process could be a higher extent of racemization, whereas on the contrary in this study, a slightly lower extent of racemization was observed in the EOs instead of the dried plant material.

On the other hand, the analysis performed on several biological activities of the EOs from the two genotypes provided a characterization useful for highlighting their possible distinguishing properties among different oregano genetic resources and for evaluating their potential for food and pharmaceutical applications.

All the EOs showed a marked dose-dependent antimicrobial activity against some pathogenic and spoilage bacteria, belonging to both Gram-negative and Gram-positive groups. In the direct contact in vitro test performed with CAR1 and THY5, a greater effect was observed against *L. monocytogenes* OH, a Gram-positive species, as compared to the Gram-negative *S.* Typhimurium LT2. These data are coherent with the antibacterial mechanism elucidated for thymol and carvacrol: these oxygenated monoterpenes, bearing a phenolic functional group, are able to interfere with the phospholipid bilayers, damaging the bacterial cell membrane and causing a decrease in ATP levels until the cell dies [3,31,32], while the hydrophilic nature of the outer membrane of Gram-negative bacteria exerts a shielding action towards EOs [33,34].

The antimicrobial activity of CAR1 and THY5 was also observed during a challenge test, in which a food matrix of minced beef was used to study the behaviour of a pathogenic or spoilage microorganism under specific storage conditions [35]. Although both CAR1 and THY5 showed antimicrobial activity, THY5′s effect appeared to be slower and more persistent than CAR1′s in the food contaminated by *L. monocytogenes* OH. The different dynamics of the two pathogens observed in this assay in the presence and absence of oil treatment could be partly ascribable to an inherent variability and complexity in the food matrix. The increase in microbial load observed after seven days of storage both in the presence and absence of the EOs could be due to multiple concomitant causes: e.g., recontamination of the ground meat or a decrease in the EO activity due to too long of a storage time.

Another Interesting aspect to consider in terms of antimicrobial activity is the effect of the minor components present in EOs [3]. Sivropoulou et al. [36] investigated the antimicrobial effect of EOs extracted from *Origanum vulgare ssp. hirtum*, *Origanum dictamnus* and a commercial oregano EO, concluding that these oils, although different in chemical composition, exhibited similar antibacterial activities. The authors attributed this behaviour to the action of minor compounds that could have their own activity or act in synergy or antagonism with the main ones. Several studies have reported synergistic activity among EO components: for example, the presence of carvacrol and *p*-cymene causes a major destabilization of the bacterial membrane and a reduction in membrane potential [37,38]. The results of the present study confirmed that the diversity in the chemical profile between the two genotypes did not produce marked differences in antibacterial activity.

The potential applications of EOs as food preservatives as well as food ingredients cannot be separated from the evaluation of their impact on intestinal health following their intake. Thus, in this study, we investigated possible adverse and beneficial actions of EOs on the intestinal mucosa since it represents the first site of impact between ingested food and the organism. To such purposes, human intestinal epithelial Caco-2 cells were used from a well characterized enterocyte-like cell line capable of expressing the morphological and functional differentiation features typical of mature enterocytes, including cell polarity and a functional brush border [39]. The Caco-2 cell line has been extensively used as a reliable in vitro system to study cell toxicity, inflammation and intestinal injury induced by pathogens [40].

The effect of oregano EOs on intestinal cells is quite unexplored, except for a couple of papers evaluating antioxidant activities [41,42]; yet, the few available studies are focused on isolated components, such as thymol and/or carvacrol, with a particular emphasis on cell toxicity [43]. Another study conducted on undifferentiated Caco-2 cells treated up to 48 h with relatively high concentrations (2.5 mM and below) of carvacrol and thymol demonstrated heavy perturbation of the cellular ultrastructure and apoptosis [44]. However, it should be emphasized that the activity of pure compounds (singly or in mixtures) can be very different from that of complete EOs, which also contain a variety of secondary compounds.

The incubation of Caco-2 cells in the presence of both CAR1 and THY5 caused a significative rapid and strong reduction in TEER values accompanied by a high passage of the paracellular marker phenol red at the higher concentrations tested (0.1%), approximately corresponding to 5.2 mM of carvacrol and 3.6 mM of thymol for CAR1 and THY5, respectively. Although this trend was relatively less sharp for the THY treatment at a 0.025% concentration, this sudden and strong toxic response likely indicated cell death occurrence, suggesting unspecific cytotoxicity effects, probably due to EO's ability to permeabilize mitochondrial membranes leading to cell death [45]. This was also confirmed by Fabian and co-workers [46], who observed a high necrosis rate in Caco-2 cells treated for 1 h with oregano EO at concentrations comparable to ours. However, at concentrations lower than 0.02%, both CAR1 and THY5 did not affect Caco-2 monolayer integrity for up to 24 h of incubation, indicating a complete absence of adverse effects under these experimental conditions and likely suggesting a level of safe concentrations at which EOs do not alter intestinal integrity. To the best of our knowledge, this is the first study evaluating the intestinal permeability effects of oregano EOs in human intestinal cells. These in vitro tests provide early and specific intestinal toxicity assays for identifying dietary compounds that may affect intestinal barrier function and represent a valuable tool for further characterizing oregano EO safety in humans, as already shown with oral toxicity tests performed on rats [47].

Data reported in the literature relating to oregano EO anti-inflammatory activity suggest their ability to reduce inflammation in several in vitro and in vivo models [48–50]; however, in our intestinal Caco-2 cell system, pre-incubation with EOs did not induce evident protection from TNF-α inflammatory challenge. Nevertheless, CAR1 preincubation followed by TNF-α exposure determined a weak downregulation of genes involved in the NF-kB pathway. This trend might confirm potential CAR1 anti-inflammatory activity in this experimental system, which could not be further assayed at a higher CAR1 concentration due to its toxicity. Finally, neither CAR1 nor THY5 displayed inflammatory action on their own, as incubation with these EOs did not induce any significant increase in the expression of the inflammatory genes tested, as compared to the control cells.

As reported by Di Vito et al. [31], the initial stages of pathogen adhesion to intestinal cells are crucial for the establishment of colonization, with the risk of being a reservoir of acute events; thus, the search for new inhibitors of bacterial adhesion to host tissues turns out to be very important. The results of the pathogen adhesion assay, described in the present work, showed that the addition of CAR1 and THY5 oils, despite being at low concentrations, was indeed able to reduce bacterial adhesion to the intestinal Caco-2 cell line, although THY5 was effective only against *S.* Typhimurium LT2. Our data are comparable to those of other research carried out on thyme EOs, in which both thymol (about 12%) and carvacrol (about 68%) were present; in particular, it was observed that both *Thymus capitatus* EO as well as thymol and carvacrol alone were able to reduce the adhesion of *E. coli* (ATCC35210) and *L. monocytogenes* (NCTC 7973) to the intestinal HT-29 cell line [51]. Previous work has demonstrated that carvacrol was able to reduce *Campylobacter jejuni* infection in epithelial cells [52], while thymol decreased the adhesion of various *E. coli* and *S. aureus* strains on vaginal epithelial cells [53]. Finally, both compounds were able to reduce the adhesion of three *L. monocytogenes* strains to Caco-2 cells [54]. Furthermore, in a recent study carried out on *Origanum vulgare* EO, a reduction in the adhesion of several *Salmonella* serotypes to Caco-2 cells was observed by a commercial mixture of EO containing both thymol and carvacrol; this effect was suggested to occur through the inhibition or disaggregation of bacterial biofilm [31]. The mechanisms underlying this effect can be multiple and ascribable to the antimicrobial activity of EOs against Gram-positive and Gram-negative pathogens, to their ability to alter the chemical–physical properties of the bacterial surface or microbial motility, as well as to the reduction in gene expression of some virulence factors [52–54]. Further experiments will be needed to fully understand the mechanisms underlying the protective effects exerted by oregano EOs.

4. Materials and Methods

4.1. Plant Materials

The plant material used for this study was obtained from eight farms located in south-western Sicily, in different areas of the province of Agrigento, at the following altitudes: farm 1 at 250 m a.s.l., 2 at 300 m, 3 at 200 m, 4 at 250 m, 5 at 300 m, 6 at 150 m, 7 at 250 m and 8 at 250 m. Plants from the hybrid *Origanum vulgare ssp. viridulum* × *Origanum vulgare ssp. hirtum* were provided by farms 1–3 (indicated in the text as CAR1, CAR2, CAR3), whereas plants from the subspecies *Origanum heracleoticum* L. were provided by farms 4–8 (THY1, THY2, THY3, THY4, THY5). It is not clear whether the subspecies *Origanum heracleoticum* L. should have been identified as *Origanum vulgare ssp. viridulum* or as *Origanum vulgare ssp. hirtum*, with contrasting identifications having been reported in the literature [20,21]. For both groups of oregano types, all plants were obtained by the propagation of the same genotype and then grown in different cultivation environments, i.e., the eight farms. On all the farms, oregano was grown according to the cultivation practice and crop management traditionally applied in these areas of Sicily [20]. Sampled plants were harvested in mid-June by cutting up to 5 cm above the soil level and then air-dried in the shade at an air temperature ranging between 25 and 30 °C, for about 10–15 days. The dried material was shipped to the CREA-Research Centre for Food and Nutrition laboratory, where all the analytical determinations were carried out.

4.2. Isolation of the EOs

The samples of air dried leaves and flowers (20 g) were subjected to hydrodistillation by a Clevenger-type apparatus for 3 h and then, the collected EO was dried over anhydrous sodium sulphate and stored under N_2 in a sealed vial at −24 °C until analysis.

4.3. GC–MS Analyses of EO Compositions

The EOs were diluted in methanol (1:20 v/v) before the GC analysis. GC analyses were performed on an Agilent 6890 5973 N, GC–MS system (Agilent Technologies, Palo Alto, CA, USA), provided with a quadrupole mass filter for mass spectrometric detection. GC separation was accomplished by a DB1-MS column (0.25 mm × 60 m, 0.5 μm film thickness, J&W; Agilent Technologies). The following chromatographic conditions were applied: injection of 1 μL volume, with a split ratio of 50:1 and injector temperature of 250 °C; the oven temperature program selected from 60 °C to 200 °C at 4 °C min^{-1} and then to 280 °C (5 min) at 50 °C min^{-1}; and the He carrier gas constant flow set at 1.5 mL min^{-1} corresponding to a linear velocity of 32 cm s^{-1}. The MS detector setting parameters were an electronic impact ionization mode set at 70 eV and the transfer line, source and quadrupole temperatures set at 300, 230, and 150 °C, respectively. For MS detection, the full scan mode was applied by selecting the mass range of 33–300 amu. The identification of EO constituents was performed by a comparison of the linear retention indices (LRI) and mass spectra of the chromatographic peaks with those obtained from a standard solution of pure reference compounds when commercially available (purchased from Merck, Sigma-Aldrich, Milan, Italy) (Table S1). Linear retention indices were determined by analyzing, under the same conditions used for the EOs, a standard solution of C7–C30 saturated alkanes and according to the equation proposed by van den Dool and Kratz. When a pure compound was not available, tentative identification was based on the comparison of determined linear retention indices with those reported in the literature [55], in the NIST Chemistry WebBook database [56] and by a comparison of mass spectra with those reported in the NIST/EPA/NIH Mass Spectral Library (Version 2.4, 2020). The percent content of the compounds was determined from their peak areas in the GC Total Ion Current profile. Only compounds at a level higher than 0.05% were quantified. Each oil sample was analyzed in duplicate.

4.4. Enantioselective GC–MS Analysis of Volatiles in Dried Leaves and Flowers and EO Constituents

The enantioselective analysis of chiral compounds was performed on the EO constituents and also on volatiles isolated from the dried leaves and flowers to investigate the effects on enantiomeric distribution due to the distillation procedure. The isolation of volatiles from the dried leaves and flowers was carried out by the headspace solid-phase microextraction (HS-SPME) technique. Briefly, a weighted amount of died plant material (1.5 g) was placed in a 15 mL vial capped with a PTFE/silicone septum (Supelco; Sigma-Aldrich) for HS-SPME. The extraction was carried out by exposing a 2 cm, 50/30 μm DVB/CAR/PDMS fiber (Supelco; Sigma-Aldrich) to the headspace of the plant material for 30 min, while keeping the extraction temperature at 30 °C by a water bath. At the end of the extraction, the fiber was immediately inserted into the GC split–splitless injection port for the desorption step and the GC run was started. A duplicate extraction of each cheese sample was carried out. GC analyses were performed using the same GC–MS system reported above, equipped with a chiral column of Cyclosil-B (0.25 mm × 30 m, 0.25 μm of film thickness, stationary phase: 30% Heptakis (2,3-di-*O*-methyl-6-*O*-t-butyl dimethylsilyl)-β-cyclodextrin in DB-1701, J&W; Agilent Technologies). The chromatographic conditions were as follows: desorption for 5 min in the GC injector provided with a 0.75 mm glass liner suitable for SPME and operating at 240 °C in the spitless mode; oven temperature program from 55 °C (2 min) to 190 °C at 4 °C/min, then to 220 °C (5 min) at 20 °C/min; helium used as carrier gas at a constant flow of 1.5 mL min^{-1} corresponding to a linear velocity of 45 cm s^{-1}; and MS detector parameters as stated above in Section 4.3. In the case of the analysis of the EO, 1 μL EO diluted in methanol (1:20 v/v) was injected and a slit ratio of 50:1 was set. All the other GC–MS conditions were the same as in the analysis of the HS-SPME isolates.

The identification of enantiomers was accomplished by a comparison with the mass spectra and linear retention indices (LRI) of authentic standards (Sigma-Aldrich). When authentic standards were not available, tentative identification was accomplished by a comparison with information reported in the NIST/EPA/NIH Mass Spectral Library (Version 2.4, 2020), in the literature [57] or kindly provided by Prof. Patrizia Rubiolo (personal communication). A standard solution of linear alkanes (C7–C30) was run under the same chromatographic conditions as the samples to determine the LRI of the detected compounds. Enantiomeric ratios were expressed as percent peak areas of the sum of the areas of the peaks of both enantiomers.

4.5. Indicator Microorganism Strains

The microorganisms used were *Salmonella enterica* serovar Typhimurium LT2 (DSMZ 18522; Braunschweig, Germany) and two isolates of *S. enterica* from chicken samples belonging to two different serovars (Derby and Give), provided by the Istituto Zooprofilattico Sperimentale del Mezzogiorno (Portici, Naples, Italy). The enterotoxigenic *Escherichia coli* strain K88 (ETEC, O149:K88ac) was provided by the Lombardia and Emilia Romagna Experimental Zootechnic Institute (Reggio Emilia, Italy). *Listeria monocytogenes* OH, *L. monocytogenes* CAL, *L. monocytogenes* SA and *L. innocua* 1770 were provided by the CREA-Research Centre for Animal Production and Aquaculture (Lodi, Italy), while *Pseudomonas putida* WSC358, *P. putida* KT2240 and *P. fluorescens* B13 were provided by Prof. Livia Leoni of Roma Tre University, Rome. With the exception of the ETEC growing in Luria-Bertani (LB) broth, Miller (DIFCO; Rodano (MI), Italy), all bacteria were routinely grown in tryptone soy broth (TSB; Oxoid, Basingstoke, UK) at their optimal growth temperature, which was 30 °C for the *Listeria* and *Pseudomonas* strains and 37 °C for the *Salmonella* and *Escherichia* strains.

4.6. Agar Spot Test

The spot-on-agar test was performed by spotting 2 μL EOs (CAR1, CAR2, CAR3, THY3, THY5) onto tryptone soy agar (TSA, 1.2%; Oxoid) plates previously seeded with 1×10^6 CFU/mL indicator strains at the exponential growth phase. Spotted plates were then incubated for 18 h and inhibition zones (radii of the microbial growth inhibition, halos) around the wells were measured in millimeters (mm), accordingly with Balouiri et al. [58]. In the same plates, the test efficacy was confirmed by adding 2 μL of 50 μg/mL ampicillin (Sigma-Aldrich). Antimicrobial capacity was considered high (+++) when the radius value of the inhibition halo was greater than 7.5 mm, medium (++) when it was between 5 and 7.5 mm and low (+) when it was less than 5 mm.

4.7. Direct Contact Test

The direct contact test was performed as described by Serio et al. [32] and modified as follows. Stock emulsions containing 4% (*v*/*v*) EOs (CAR1 or THY5) were prepared in tryptone salt solution (8.5 g/L NaCl, 1.0 g/L tryptone; Oxoid) with 1% Tween 80 (Sigma-Aldrich) added as an emulsifying agent and vortexed for 10 min. The emulsions were filter-sterilized (0.22 μm filter pore size). To study the effect of different EO concentrations, appropriate aliquots of stock emulsions were diluted in 10 mL PBS, to obtain final concentrations of 0–0.12–0.25–0.50% EOs.

EO dilutions were inoculated with a standard bacterial suspension (1×10^8 CFU/mL) of *L. monocytogenes* OH or *S.* Typhimurium LT2 indicator strains and incubated at 37 °C for 30 min. Bacterial cells were also exposed to 0% EO (control, C) and to 0% EO plus 0.125% Tween 80 with or without 50 μg/mL ampicillin, to confirm test efficacy and to exclude any antimicrobial effect of Tween 80, respectively (CTw + Amp or CTw). Bacteria were harvested by centrifugation for 10 min at 1100× *g* at 4 °C and appropriate dilutions were plated on tryptone soy agar to evaluate bacterial cell viability by colony counting, after the incubation of the plates for 18 h.

4.8. Challenge Test

Minced cow meat purchased at the supermarket was used to perform the challenge test. Meat portions (100 g each) were contaminated with *S.* Typhimurium LT2 or *L. monocytogenes* OH inocula (approximately 1×10^4 CFU/g), according to Zinno et al. [59]. To ensure the proper distribution of the pathogens, the inoculated samples were homogenized in Stomacher bags (Bag Mixer-400; Interscience, France) for 2 min at room temperature. After homogenization, the inoculated meat portions were split in two equal quantities (50 g each), one of which was treated with 0.5% CAR1 or THY5. In addition, one meat portion was not inoculated, nor oil-treated (control). The Stomacher bags were wrapped and stored under aerobic conditions at 4 °C for up to seven days.

Microbiological analysis of the populations of *L. monocytogenes* OH and *S.* Typhimurium LT2 was carried out after 0–1–2–3 and 7 days of refrigerated storage. At each sampling time, the aliquots of minced meat were aseptically 10-fold diluted in 0.9% NaCl and homogenized for 2 min at room temperature in Stomacher bags. The resulting slurries were serially diluted and plated in duplicate on selective OXFORD and xylose lysine deoxycholate (XLD) agar (Merck, Darmstadt, Germany) for *L. monocytogenes* OH and *S.* Typhimurium LT2, respectively. Populations of both bacteria were determined and quantified by colony counting, after incubation for 24 h.

Before inoculation with the two pathogens, the minced meat was also examined for the presence of any bacteria by an estimation of the total viable counts (TVCs) on plate count agar (PCA; Oxoid) after 72 h of incubation at 25 °C.

4.9. Intestinal Caco-2 Cell Culture

Caco-2 cells, obtained from INSERM (Paris, France), were subcultured at a low density [60] and used between passages 90–105. Cells were routinely maintained at 37 °C in a 95% 5% air/CO_2 atmosphere at 90% relative humidity in DMEM containing 25 mM glucose, 3.7 g/L $NaHCO_3$, 4 mM stable L-glutamine, 1% non-essential amino acids, 1×10^5 U/L penicillin and 100 mg/L streptomycin (all from Corning, Milan, Italy), supplemented with 10% heat-inactivated fetal bovine serum (Euroclone, Milan, Italy). Cells were seeded on polyethylene terephthalate permeable Transwell filters (Falcon™ 10.5 mm diameter, 0.4 μm pore size; Corning) at a density of 1×10^6 cells/filter and cultured for 17–21 days to allow complete differentiation. The medium was changed 3 times a week.

4.10. Cell Monolayer Permeability Assessments

Cell monolayer permeability was assayed by measuring the transepithelial electrical resistance (TEER), through a Millicell electrical resistance voltmeter (Merck Millipore, Darmstadt, Germany) on cells differentiated on Transwell filters, according to Sambuy et al. [39]. Only cell monolayers with TEER values higher than 1300 Ohm × cm^2 were used, as this value is indicative of correct cell differentiation, as identified in preliminary experiments. For the experiments, TEER was recorded every hour for up to 8 h, and then at 24 h, at the end of treatments. Cell permeability was also assessed at 24 h by measuring the paracellular passage of the phenol red marker, as reported by Monastra et al. [61]. Briefly, 0.5 mL of 1 mM phenol red was placed in the apical (AP) compartment of cell monolayers while 1 mL PBS containing Ca^{++} and Mg^{++} was placed in the basolateral (BL) compartment. After 1 h incubation at 37 °C, the BL medium was collected, injected with 0.1 N NaOH and read at 560 nm (Tecan Infinite M200 Microplate Reader; Tecan Italia, Milan, Italy) to determine the phenol red concentration. Phenol red apparent permeability (Papp) was calculated from the following formula: $Ct \times V_{BL}/\Delta t \times \cdot C_0 \times A$, where V_{BL} is the volume of the BL compartment (cm^3), A is the filter area (cm^2), Δt is the time interval (s), Ct is the phenol red concentration in the BL compartment at the end of the time interval and C_0 is the phenol red concentration in the AP compartment at the beginning. Phenol red Papp values below 1×10^{-6} cm s^{-1} were considered indicative of intact monolayers [27]. Thus, 1×10^{-6} cm s^{-1} was set as the threshold value, irrespectively of the statistical significance among samples.

4.11. Impact of Oregano EOs (CAR1 and THY5) on Intestinal Barrier Integrity

CAR1 and THY 5 were tested on Caco-2 cells to assay their potential effect on monolayer integrity through TEER and phenol red Papp measurements.

Several EO concentrations were added for up to 24 h to the AP compartment of Caco-2 cells differentiated on Transwell filters. To avoid possible interference with the fetal bovine serum proteins, cell monolayers were kept in a serum-free medium 16 h before the assay. The concentrations tested were 0.1–0.05–0.03–0.025–0.02–0.01% (obtained from a stock solution, diluted by 1:20 in ethanol from the initial preparation). As the control, the cells were also apically treated with 2% ethanol, corresponding to the concentration contained in the higher EO dilution tested.

4.12. *Pathogen Adhesion Assay to Caco-2 Cells*

For the adhesion assay, Caco-2 cells were seeded and differentiated in 24-well plates (Becton Dickinson, Milan, Italy) at 1×10^6 cells/well. Cells were placed in an antibiotic- and serum-free cell culture medium 16 h before the assay. On the day of the assay, overnight bacterial cultures of the pathogen indicator strains *L. monocytogenes* OH and *S.* Typhimurium LT2 were diluted 1:10 in appropriate media and grown for approximately 2 h up to the exponential growth phase. After monitoring the OD 600, bacterial cells were harvested by centrifugation at 5000× *g* for 10 min, resuspended in antibiotic- and serum-free cell culture medium and added to the cell monolayers at a concentration of 1×10^8 CFU/well (approximately 100:1 bacteria-to-cell ratio). CAR1 or THY5 were apically added together with the different indicator strains at a 0.02% concentration, identified as the highest nontoxic one from the Caco-2 permeability tests. Co-cultures of bacteria and Caco-2 cells were incubated at 37 °C for 1.5 h. Non-adhering bacteria were then removed by 5 washes with Hanks' Balanced Salt Solution (Corning) and cell monolayers were lysed with 1% Triton-X-100, according to Guantario et al. [62]. Adhering, viable bacterial cells were quantified by plating appropriate serial dilutions of Caco-2 lysates on different media. For *S.* Typhimurium LT2, plating was performed by inclusion in violet red bile glucose agar (VRBGA) medium, while for *L. monocytogenes*, OH cell lysates were plated on Oxford medium. Plates were incubated for 18 h.

4.13. *Inflammation Induction by TNF-α and EO Treatment for Gene Expression Analysis of NF-kB Pathway*

To assess the potential anti-inflammatory ability of CAR1 and THY5 EOs assayed in the present study, Caco-2 cells differentiated on Transwell filters were pre-incubated for 1 h with CAR1 or THY5 at the 0.02% concentration in the AP compartment. After media withdrawal, cells were challenged for an additional hour with the pro-inflammatory cytokine TNF-α (Invitrogen, Rodano, Milan, Italy), added to both the AP and BL compartments at 15 ng/mL. Other samples were treated only with EOs for 1 h or only with TNF-α for 1 h.

At the end of treatments, RNA was extracted from the Caco-2 cells by using miRNeasy Plus Mini Kit (Qiagen, Hilden, Germany). Nucleic acid concentration was determined by a NanoDrop® ND-2000 UV–VIS spectrophotometer (Thermo Fisher Scientific, Waltham, MA, USA) at an OD of 260 nm. All samples had an OD 260/OD 280 ratio higher than 1.8, corresponding to 90–100% of pure RNA. A total of 1 μg of RNA was reverse-transcribed into cDNA using ReadyScript™ cDNA Synthesis Mix (Sigma-Aldrich). The quantification of gene expression was determined by real-time PCR with a 7500 Fast Real-Time PCR System (Applied Biosystems, Waltham, MA, USA), using RT^2 SYBR® Green ROX qPCR Mastermix (Qiagen). Data were collected using 7500 software v2.0.5 and given as the threshold cycle (Ct). Ct values for each target and housekeeping gene were obtained and their difference was calculated (ΔCt). Primer efficiencies for all tested genes were similar. The comparative calculation, ΔΔCt, was used to find the difference in the expression levels between the control and treated samples. Data are expressed as the mean of log2 of fold change (FC) with respect to the control.

The following target genes were analyzed: inhibitor of nuclear factor kappa B (IkB)α, cellular inhibitor of apoptosis (cIAP)2, interleukin (IL)-1 α, IL-6, IL-8, and glyceraldehyde-3-phosphate dehydrogenase (GAPDH) was used as the housekeeping gene. Primer sequences are reported in Table S2.

4.14. Statistical Analysis

One-way ANOVA followed by a post-hoc Tukey's HSD test were used to evaluate the statistical significance, after performing Shapiro–Wilk's and Levene's tests, to verify normality and homogeneity of variance, respectively. Any p values < 0.05 were considered statistically significant. In the figures, mean values with different superscript letters significantly differ. Principal component analysis (PCA) was performed on mean values of the EO constituents levels, expressed as percentages, after normalization by arcsine transformation. The statistical analyses were executed with Microsoft Office Excel 2011 upgraded with XLSTAT (ver. 4 March 2014).

5. Conclusions

The characterization of the chemical and biological properties of native plant species with highly functional, nutraceutical and health protection properties may be important to promote innovative products with many possible applications (food functional, phytocosmetics, phytotherapy, plant defense), while contributing to the preservation of agricultural biodiversity in the Mediterranean area. The two *Origanum vulgare* genotypes investigated in the present study were selected as they are widely cultivated in Sicily, but little information was available up to now on the biological properties of their EOs.

The investigations carried out in the present study demonstrated a strong antimicrobial effect of EOs from the two genotypes on Gram-positive and -negative bacteria, both in vitro and in a challenge test on a food matrix, suggesting their potential use as control agents against a wide spectrum of foodborne pathogens, with the view to looking for efficient alternative strategies for foodborne disease containment. The experiments performed on Caco-2 cells, a suitable and validated human intestinal in vitro model, indicated that the EOs displayed essentially comparable biological effects, as they were both able to reduce pathogen adhesion to Caco-2 cells at a 0.02% concentration without inducing an inflammatory state. To the best of our knowledge, this is the first report focusing on the effect of oregano EOs on intestinal permeability, an aspect poorly explored in the literature, that lays the groundwork for further investigations aimed at a more comprehensive understanding of safety of use in humans. As a whole, the marked diversity in the chemical profile of EOs from the two genotypes did not produce relevant differences in the considered biological activities. This suggests the need for additional studies to clarify the contribution of individual oregano EO constituents to the investigated biological activities, along with the elucidation of the role of minor compounds and their possible synergistic interactions with the major active constituents, to develop a model for predicting biological activity based on chemical composition. Finally, from a chemical perspective, the analysis of enantiomeric distribution was also confirmed to be a valuable tool for the recognition of geographical origin in the case of oregano EOs.

Supplementary Materials: The following supporting information can be downloaded at: https://www.mdpi.com/article/10.3390/plants12061344/s1, Table S1: List of quantified constituents of oregano essential oils. Retention indices and method of identification used; Table S2: Target genes and primer sequences.

Author Contributions: Conceptualization, methodology and validation, M.R. and A.R.; investigation, formal analysis and visualization, P.Z., B.G., G.R., A.F., A.R. and M.R.; microbiology experiments, P.Z., B.G. and S.A.; cell culture experiments, G.R., A.F., S.A. and M.R.; statistical analysis, B.G.; chemical characterization analysis, A.R. and G.L.; project administration and funding acquisition, M.M.M. and G.F.; writing—original draft preparation, supervision and writing—review and editing, M.R., P.Z., B.G. and A.R. All authors have read and agreed to the published version of the manuscript.

Funding: This research was funded by the project "ESPAS—Valorisation des espèces végétales autochtones siciliennes et tunisiennes avec un intérêt nutritif et bon pour la santé", as part of the Program "IEV de Coopération Transfrontalière Italie—Tunisie 2014–2020".

Institutional Review Board Statement: Not applicable.

Informed Consent Statement: Not applicable.

Data Availability Statement: The data presented in this study are available upon request from the corresponding authors.

Conflicts of Interest: The authors declare no conflict of interest. The funders had no role in the design of the study; in the collection, analyses or interpretation of data; in the writing of the manuscript; or in the decision to publish the results.

References

1. Franz, C.; Novak, J. Sources of Essential Oils. In *Handbook of Essential Oils: Science, Technology, and Applications*; Baser, K.H.C., Buchbauer, G., Eds.; CRC Press, Taylor & Francis Group: Boca Raton, FL, USA, 2010; pp. 39–82.
2. Kintzios, S.E. Oregano. In *Handbook of Herbs and Spices*, 2nd ed.; Peter, K.V., Ed.; Woodhead Publishing Limited: Cambridge, UK, 2012; Volume 2, pp. 417–436.
3. Rodriguez-Garcia, I.; Silva-Espinoza, B.A.; Ortega-Ramirez, L.A.; Leyva, J.M.; Siddiqui, M.W.; Cruz-Valenzuela, M.R.; Gonzalez-Aguilar, G.A.; Ayala-Zavala, J.F. Oregano Essential Oil as an Antimicrobial and Antioxidant Additive in Food Products. *Crit. Rev. Food Sci. Nutr.* **2016**, *56*, 1717–1727. [CrossRef]
4. *The Rapid Alert System for Food and Feed—Annual Report 2020*; Publications Office of the European Union: Luxembourg, 2021; Available online: https://food.ec.europa.eu/system/files/2021-08/rasff_pub_annual-report_2020.pdf (accessed on 16 February 2023).
5. Dawood, M.A.O.; El Basuini, M.F.; Zaineldin, A.I.; Yilmaz, S.; Hasan, M.T.; Ahmadifar, E.; El Asely, A.M.; Abdel-Latif, H.M.R.; Alagawany, M.; Abu-Elala, N.M.; et al. Antiparasitic and Antibacterial Functionality of Essential Oils: An Alternative Approach for Sustainable Aquaculture. *Pathogens* **2021**, *10*, 185. [CrossRef]
6. Burggraf, A.; Rienth, M. *Origanum vulgare* Essential Oil Vapour Impedes *Botrytis Cinerea* Development on Grapevine (*Vitis vinifera*) Fruit. *Phytopathol. Mediterr.* **2020**, *59*, 331–344. [CrossRef]
7. Leyva-López, N.; Gutiérrez-Grijalva, E.; Vazquez-Olivo, G.; Heredia, J. Essential Oils of Oregano: Biological Activity beyond Their Antimicrobial Properties. *Molecules* **2017**, *22*, 989. [CrossRef]
8. Lombrea, A.; Antal, D.; Ardelean, F.; Avram, S.; Pavel, I.Z.; Vlaia, L.; Mut, A.-M.; Diaconeasa, Z.; Dehelean, C.A.; Soica, C.; et al. A Recent Insight Regarding the Phytochemistry and Bioactivity of *Origanum vulgare* L. Essential Oil. *Int. J. Mol. Sci.* **2020**, *21*, 9653. [CrossRef] [PubMed]
9. Pezzani, R.; Vitalini, S.; Iriti, M. Bioactivities of *Origanum vulgare* L.: An Update. *Phytochem. Rev.* **2017**, *16*, 1253–1268. [CrossRef]
10. Aljaafari, M.N.; AlAli, A.O.; Baqais, L.; Alqubaisy, M.; AlAli, M.; Molouki, A.; Ong-Abdullah, J.; Abushelaibi, A.; Lai, K.-S.; Lim, S.-H.E. An Overview of the Potential Therapeutic Applications of Essential Oils. *Molecules* **2021**, *26*, 628. [CrossRef] [PubMed]
11. Liu, Q.; Meng, X.; Li, Y.; Zhao, C.-N.; Tang, G.-Y.; Li, H.-B. Antibacterial and Antifungal Activities of Spices. *Int. J. Mol. Sci.* **2017**, *18*, 1283. [CrossRef] [PubMed]
12. Do Nascimento, L.D.; de Moraes, A.A.B.; da Costa, K.S.; Pereira Galúcio, J.M.; Taube, P.S.; Costa, C.M.L.; Neves Cruz, J.; de Aguiar Andrade, E.H.; de Faria, L.J.G. Bioactive Natural Compounds and Antioxidant Activity of Essential Oils from Spice Plants: New Findings and Potential Applications. *Biomolecules* **2020**, *10*, 988. [CrossRef]
13. Sakkas, H.; Papadopoulou, C. Antimicrobial Activity of Basil, Oregano, and Thyme Essential Oils. *J. Microbiol. Biotechnol.* **2017**, *27*, 429–438. [CrossRef]
14. Plati, F.; Paraskevopoulou, A. Micro- and Nano-Encapsulation as Tools for Essential Oils Advantages' Exploitation in Food Applications: The Case of Oregano Essential Oil. *Food Bioprocess Technol.* **2022**, *15*, 949–977. [CrossRef]
15. Pontes-Quero, G.M.; Esteban-Rubio, S.; Pérez Cano, J.; Aguilar, M.R.; Vázquez-Lasa, B. Oregano Essential Oil Micro- and Nanoencapsulation With Bioactive Properties for Biotechnological and Biomedical Applications. *Front. Bioeng. Biotechnol.* **2021**, *9*, 703684. [CrossRef] [PubMed]
16. Lukas, B.; Schmiderer, C.; Novak, J. Essential Oil Diversity of European *Origanum vulgare* L. (Lamiaceae). *Phytochemistry* **2015**, *119*, 32–40. [CrossRef]
17. Hyldgaard, M.; Mygind, T.; Meyer, R.L. Essential Oils in Food Preservation: Mode of Action, Synergies, and Interactions with Food Matrix Components. *Front. Microbiol.* **2012**, *3*, 12. [CrossRef]
18. De Falco, E.; Mancini, E.; Roscigno, G.; Mignola, E.; Taglialatela-Scafati, O.; Senatore, F. Chemical Composition and Biological Activity of Essential Oils of *Origanum vulgare* L. Subsp. *Vulgare* L. under Different Growth Conditions. *Molecules* **2013**, *18*, 14948–14960. [CrossRef]
19. Ietswaart, J.H. *A Taxonomic Revision of the Genus Origanum (Labiatae)*; Leiden University Press: The Hague, The Netherlands, 1980; 153p.
20. Napoli, E.; Giovino, A.; Carrubba, A.; How Yuen Siong, V.; Rinoldo, C.; Nina, O.; Ruberto, G. Variations of Essential Oil Constituents in Oregano (*Origanum vulgare* Subsp. *Viridulum* (=*O. Heracleoticum*) over Cultivation Cycles. *Plants* **2020**, *9*, 1174. [CrossRef]
21. Bonfanti, C.; Iannì, R.; Mazzaglia, A.; Lanza, C.M.; Napoli, E.M.; Ruberto, G. Emerging Cultivation of Oregano in Sicily: Sensory Evaluation of Plants and Chemical Composition of Essential Oils. *Ind. Crop. Prod.* **2012**, *35*, 160–165. [CrossRef]

22. Tuttolomondo, T.; Leto, C.; Leone, R.; Licata, M.; Virga, G.; Ruberto, G.; Napoli, E.M.; La Bella, S. Essential Oil Characteristics of Wild Sicilian Oregano Populations in Relation to Environmental Conditions. *J. Essent. Oil Res.* **2014**, *26*, 210–220. [CrossRef]
23. Kokkini, S.; Karousou, R.; Dardioti, A.; Krigas, N.; Lanaras, T. Autumn Essential Oils of Greek Oregano. *Phytochemistry* **1997**, *44*, 883–886. [CrossRef]
24. Giuliani, C.; Maggi, F.; Papa, F.; Maleci Bini, L. Congruence of Phytochemical and Morphological Profiles along an Altitudinal Gradient in *Origanum vulgare* ssp. *Vulgare* from Venetian Region (NE Italy). *Chem. Biodivers.* **2013**, *10*, 569–583. [CrossRef]
25. Bisht, D.; Chanotiya, C.S.; Rana, M.; Semwal, M. Variability in Essential Oil and Bioactive Chiral Monoterpenoid Compositions of Indian Oregano (*Origanum vulgare* L.) Populations from Northwestern Himalaya and Their Chemotaxonomy. *Ind. Crop. Prod.* **2009**, *30*, 422–426. [CrossRef]
26. Kasprzyk-Hordern, B. Pharmacologically Active Compounds in the Environment and Their Chirality. *Chem. Soc. Rev.* **2010**, *39*, 4466. [CrossRef]
27. Hubatsch, I.; Ragnarsson, E.G.E.; Artursson, P. Determination of Drug Permeability and Prediction of Drug Absorption in Caco-2 Monolayers. *Nat. Protoc.* **2007**, *2*, 2111–2119. [CrossRef]
28. Napoli, E.M.; Curcuruto, G.; Ruberto, G. Screening the Essential Oil Composition of Wild Sicilian Oregano. *Biochem. Syst. Ecol.* **2009**, *37*, 484–493. [CrossRef]
29. Novak, J.; Lukas, B.; Franz, C. Temperature Influences Thymol and Carvacrol Differentially in *Origanum* spp. (Lamiaceae). *J. Essent. Oil Res.* **2010**, *22*, 412–415. [CrossRef]
30. Rubiolo, P.; Sgorbini, B.; Liberto, E.; Cordero, C.; Bicchi, C. Essential Oils and Volatiles: Sample Preparation and Analysis. A Review. *Flavour Fragr. J.* **2010**, *25*, 282–290. [CrossRef]
31. Di Vito, M.; Cacaci, M.; Barbanti, L.; Martini, C.; Sanguinetti, M.; Benvenuti, S.; Tosi, G.; Fiorentini, L.; Scozzoli, M.; Bugli, F.; et al. *Origanum vulgare* Essential Oil vs. a Commercial Mixture of Essential Oils: In Vitro Effectiveness on *Salmonella* spp. from Poultry and Swine Intensive Livestock. *Antibiotics* **2020**, *9*, 763. [CrossRef]
32. Serio, A.; Chiarini, M.; Tettamanti, E.; Paparella, A. Electronic Paramagnetic Resonance Investigation of the Activity of *Origanum vulgare* L. Essential Oil on the *Listeria Monocytogenes* Membrane. *Lett. Appl. Microbiol.* **2010**, *51*, 149–157. [CrossRef]
33. Burt, S. Essential Oils: Their Antibacterial Properties and Potential Applications in Foods—A Review. *Int. J. Food Microbiol.* **2004**, *94*, 223–253. [CrossRef] [PubMed]
34. Holley, R.A.; Patel, D. Improvement in Shelf-Life and Safety of Perishable Foods by Plant Essential Oils and Smoke Antimicrobials. *Food Microbiol.* **2005**, *22*, 273–292. [CrossRef]
35. Govaris, A.; Solomakos, N.; Pexara, A.; Chatzopoulou, P.S. The Antimicrobial Effect of Oregano Essential Oil, Nisin and Their Combination against *Salmonella Enteritidis* in Minced Sheep Meat during Refrigerated Storage. *Int. J. Food Microbiol.* **2010**, *137*, 175–180. [CrossRef]
36. Sivropoulou, A.; Papanikolaou, E.; Nikolaou, C.; Kokkini, S.; Lanaras, T.; Arsenakis, M. Antimicrobial and Cytotoxic Activities of *Origanum* Essential Oils. *J. Agric. Food Chem.* **1996**, *44*, 1202–1205. [CrossRef]
37. Cristani, M.; D'Arrigo, M.; Mandalari, G.; Castelli, F.; Sarpietro, M.G.; Micieli, D.; Venuti, V.; Bisignano, G.; Saija, A.; Trombetta, D. Interaction of Four Monoterpenes Contained in Essential Oils with Model Membranes: Implications for Their Antibacterial Activity. *J. Agric. Food Chem.* **2007**, *55*, 6300–6308. [CrossRef]
38. Ultee, A.; Bennik, M.H.J.; Moezelaar, R. The Phenolic Hydroxyl Group of Carvacrol Is Essential for Action against the Food-Borne Pathogen *Bacillus cereus*. *Appl. Environ. Microbiol.* **2002**, *68*, 1561–1568. [CrossRef] [PubMed]
39. Sambuy, Y.; De Angelis, I.; Ranaldi, G.; Scarino, M.L.; Stammati, A.; Zucco, F. The Caco-2 Cell Line as a Model of the Intestinal Barrier: Infuence of Cell and Culture-Related Factors on Caco-2 Cell Functional Characteristics. *Cell Biol. Toxicol.* **2005**, *21*, 1–26. [CrossRef] [PubMed]
40. Lea, T. Caco-2 Cell Line. In *The Impact of Food Bioactives on Health*; Verhoeckx, K., Cotter, P., López-Expósito, I., Kleiveland, C., Lea, T., Mackie, A., Requena, T., Swiatecka, D., Wichers, H., Eds.; Springer International Publishing: Cham, Switzerland, 2015; pp. 103–111. ISBN 978-3-319-15791-7.
41. Kaschubek, T.; Mayer, E.; Rzesnik, S.; Grenier, B.; Bachinger, D.; Schieder, C.; König, J.; Teichmann, K. Effects of Phytogenic Feed Additives on Cellular Oxidative Stress and Inflammatory Reactions in Intestinal Porcine Epithelial Cells. *J. Anim. Sci.* **2018**, *96*, 3657–3669. [CrossRef]
42. Zou, Y.; Wang, J.; Peng, J.; Wei, H. Oregano Essential Oil Induces SOD1 and GSH Expression through Nrf2 Activation and Alleviates Hydrogen Peroxide-Induced Oxidative Damage in IPEC-J2 Cells. *Oxidative Med. Cell. Longev.* **2016**, *2016*, 5987183. [CrossRef] [PubMed]
43. Bimczok, D.; Rau, H.; Sewekow, E.; Janczyk, P.; Souffrant, W.B.; Rothkötter, H.-J. Influence of Carvacrol on Proliferation and Survival of Porcine Lymphocytes and Intestinal Epithelial Cells in Vitro. *Toxicol. In Vitro* **2008**, *22*, 652–658. [CrossRef]
44. Llana-Ruiz-Cabello, M.; Gutiérrez-Praena, D.; Pichardo, S.; Moreno, F.J.; Bermúdez, J.M.; Aucejo, S.; Cameán, A.M. Cytotoxicity and Morphological Effects Induced by Carvacrol and Thymol on the Human Cell Line Caco-2. *Food Chem. Toxicol.* **2014**, *64*, 281–290. [CrossRef]
45. Bakkali, F.; Averbeck, D.; Idaomar, M. Biological effects of essential oils—A review. *Food Chem. Toxicol.* **2008**, *46*, 446–475. [CrossRef]
46. Dušan, F.; Marián, S.; Katarína, D.; Dobroslava, B. Essential Oils—Their Antimicrobial Activity against Escherichia coli and Effect on Intestinal Cell Viability. *Toxicol. In Vitro* **2006**, *20*, 1435–1445. [CrossRef]

47. Llana-Ruiz-Cabello, M.; Maisanaba, S.; Puerto, M.; Pichardo, S.; Jos, A.; Moyano, R.; Cameán, A.M. A Subchronic 90-Day Oral Toxicity Study of *Origanum vulgare* Essential Oil in Rats. *Food Chem. Toxicol.* **2017**, *101*, 36–47. [CrossRef]
48. Bukovská, A.; Cikoš, Š.; Juhás, Š.; Il'ková, G.; Rehák, P.; Koppel, J. Effects of a Combination of Thyme and Oregano Essential Oils on TNBS-Induced Colitis in Mice. *Mediat. Inflamm.* **2007**, *2007*, 23296. [CrossRef]
49. Cappelli, K.; Sabino, M.; Trabalza-Marinucci, M.; Acuti, G.; Capomaccio, S.; Menghini, L.; Verini-Supplizi, A. Differential Effects of Dietary Oregano Essential Oil on the Inflammation Related Gene Expression in Peripheral Blood Mononuclear Cells From Outdoor and Indoor Reared Pigs. *Front. Vet. Sci.* **2021**, *8*, 602811. [CrossRef]
50. Cheng, C.; Zou, Y.; Peng, J. Oregano Essential Oil Attenuates RAW264.7 Cells from Lipopolysaccharide-Induced Inflammatory Response through Regulating NADPH Oxidase Activation-Driven Oxidative Stress. *Molecules* **2018**, *23*, 1857. [CrossRef]
51. Džamić, A.M.; Nikolić, B.J.; Giweli, A.A.; Mitić-Ćulafić, D.S.; Soković, M.D.; Ristić, M.S.; Knežević-Vukčević, J.B.; Marin, P.D. Libyan *Thymus capitatus* essential oil: Antioxidant, antimicrobial, cytotoxic and colon pathogen adhesion-inhibition properties. *J. Appl. Microbiol.* **2015**, *119*, 389–399. [CrossRef] [PubMed]
52. Van Alphen, L.B.; Burt, S.A.; Veenendaal, A.K.J.; Bleumink-Pluym, N.M.C.; van Putten, J.P.M. The Natural Antimicrobial Carvacrol Inhibits *Campylobacter jejuni* Motility and Infection of Epithelial Cells. *PLoS ONE* **2012**, *7*, e45343. [CrossRef] [PubMed]
53. Dal Sasso, M.; Culici, M.; Braga, P.C.; Guffanti, E.E.; Mucci, M. Thymol: Inhibitory Activity on Escherichia coli and *Staphylococcus aureus* Adhesion to Human Vaginal Cells. *J. Essent. Oil Res.* **2006**, *18*, 455–461. [CrossRef]
54. Upadhyay, A.; Johny, A.K.; Amalaradjou, M.A.R.; Ananda Baskaran, S.; Kim, K.S.; Venkitanarayanan, K. Plant-Derived Antimicrobials Reduce *Listeria monocytogenes* Virulence Factors In Vitro, and down-Regulate Expression of Virulence Genes. *Int. J. Food Microbiol.* **2012**, *157*, 88–94. [CrossRef] [PubMed]
55. Babushok, V.I.; Linstrom, P.J.; Zenkevich, I.G. Retention Indices for Frequently Reported Compounds of Plant Essential Oils. *J. Phys. Chem. Ref. Data* **2011**, *40*, 043101. [CrossRef]
56. NIST Chemistry WebBook. Available online: https://webbook.nist.gov/chemistry/ (accessed on 16 February 2023).
57. Liberto, E.; Cagliero, C.; Sgorbini, B.; Bicchi, C.; Sciarrone, D.; Zellner, B.D.; Mondello, L.; Rubiolo, P. Enantiomer Identification in the Flavour and Fragrance Fields by "Interactive" Combination of Linear Retention Indices from Enantioselective Gas Chromatography and Mass Spectrometry. *J. Chromatogr. A* **2008**, *1195*, 117–126. [CrossRef] [PubMed]
58. Balouiri, M.; Sadiki, M.; Ibnsouda, S.K. Methods for in Vitro Evaluating Antimicrobial Activity: A Review. *J. Pharm. Anal.* **2016**, *6*, 71–79. [CrossRef]
59. Zinno, P.; Devirgiliis, C.; Ercolini, D.; Ongeng, D.; Mauriello, G. Bacteriophage P22 to Challenge *Salmonella* in Foods. *Int. J. Food Microbiol.* **2014**, *191*, 69–74. [CrossRef] [PubMed]
60. Natoli, M.; Leoni, B.D.; D'Agnano, I.; D'Onofrio, M.; Brandi, R.; Arisi, I.; Zucco, F.; Felsani, A. Cell Growing Density Affects the Structural and Functional Properties of Caco-2 Differentiated Monolayer. *J. Cell. Physiol.* **2011**, *226*, 1531–1543. [CrossRef]
61. Monastra, G.; Sambuy, Y.; Ferruzza, S.; Ferrari, D.; Ranaldi, G. Alpha-Lactalbumin Effect on Myo-Inositol Intestinal Absorption: In Vivo and In Vitro. *CDD* **2018**, *15*, 1305–1311. [CrossRef]
62. Guantario, B.; Zinno, P.; Schifano, E.; Roselli, M.; Perozzi, G.; Palleschi, C.; Uccelletti, D.; Devirgiliis, C. In Vitro and in Vivo Selection of Potentially Probiotic Lactobacilli From Nocellara Del Belice Table Olives. *Front. Microbiol.* **2018**, *9*, 595. [CrossRef] [PubMed]

plants

MDPI

Article

Chemical Profiling and Biological Properties of Essential Oils of *Lavandula stoechas* L. Collected from Three Moroccan Sites: In Vitro and In Silico Investigations

Taoufiq Benali [1,2], Ahmed Lemhadri [1], Kaoutar Harboul [2], Houda Chtibi [2], Abdelmajid Khabbach [3], Si Mohamed Jadouali [4], Luisa Quesada-Romero [5], Said Louahlia [2], Khalil Hammani [2], Adib Ghaleb [6], Learn-Han Lee [7,*], Abdelhakim Bouyahya [8], Marius Emil Rusu [9,*] and Mohamed Akhazzane [10]

1 Environment and Health Team, Polydisciplinary Faculty of Safi, Cadi Ayyad University, Marrakech 46030, Morocco
2 Laboratory of Natural Resources and Environment, Polydisciplinary Faculty of Taza, Sidi Mohamed Ben Abdellah University of Fez, B.P. 1223 Taza-Gare, Taza 30050, Morocco
3 Laboratory of Biotechnology, Conservation and Valorisation of Natural Resources (BCVRN), Faculty of Sciences Dhar El Mahraz, Sidi Mohamed Ben Abdellah University, B.P. 1796, Fez 30003, Morocco
4 Department of Biotechnology and Analysis EST Khenifra, Sultan Moulay Sliman University, Khenifra 23000, Morocco
5 Escuela de Nutrición y Dietética, Facultad de Ciencias Para el Cuidado de la Salud, Universidad San Sebastián, General Lagos 1163, Valdivia 5090000, Chile
6 Laboratory of Analytical and Molecular Chemistry, Multidisciplinary Faculty of Safi, Cadi Ayyad University, Safi 46030, Morocco
7 Novel Bacteria and Drug Discovery Research Group (NBDD), Microbiome and Bioresource Research Strength (MBRS), Jeffrey Cheah School of Medicine and Health Sciences, Monash University Malaysia, Bandar Sunway, Subang Jaya 47500, Malaysia
8 Laboratory of Human Pathologies Biology, Department of Biology, Faculty of Sciences, Mohammed V University, Rabat 10100, Morocco
9 Department of Pharmaceutical Technology and Biopharmaceutics, Faculty of Pharmacy, Iuliu Hatieganu University of Medicine and Pharmacy, 400012 Cluj-Napoca, Romania
10 Engineering Laboratory of Organometallic and Molecular Materials and Environment, Faculty of Sciences Dhar El Mahraz, Sidi Mohamed Ben Abdellah University, Fez 30000, Morocco
* Correspondence: lee.learn.han@monash.edu (L.-H.L.); rusu.marius@umfcluj.ro (M.E.R.)

Citation: Benali, T.; Lemhadri, A.; Harboul, K.; Chtibi, H.; Khabbach, A.; Jadouali, S.M.; Quesada-Romero, L.; Louahlia, S.; Hammani, K.; Ghaleb, A.; et al. Chemical Profiling and Biological Properties of Essential Oils of *Lavandula stoechas* L. Collected from Three Moroccan Sites: In Vitro and In Silico Investigations. *Plants* **2023**, *12*, 1413. https://doi.org/10.3390/plants12061413

Academic Editor: Ain Raal

Received: 16 February 2023
Revised: 14 March 2023
Accepted: 20 March 2023
Published: 22 March 2023

Abstract: The aim of this study was the determination of the chemical compounds of *Lavandula stoechas* essential oil from Aknol ($LSEO_A$), Khenifra ($LSEO_K$), and Beni Mellal ($LSEO_B$), and *the* in vitro investigation of their antibacterial, anticandidal, and antioxidant effects, and in silico anti-SARS-CoV-2 activity. The chemical profile of LSEO was determined using GC-MS-MS analysis, the results of which showed a qualitative and quantitative variation in the chemical composition of volatile compounds including L-fenchone, cubebol, camphor, bornyl acetate, and τ-muurolol; indicating that the biosynthesis of essential oils of *Lavandula stoechas* (LSEO) varied depending on the site of growth. The antioxidant activity was evaluated using the ABTS and FRAP methods, our results showed that this tested oil is endowed with an ABTS inhibitory effect and an important reducing power which varies between 4.82 ± 1.52 and 15.73 ± 3.26 mg EAA/g extract. The results of antibacterial activity of $LSEO_A$, $LSEO_K$ and $LSEO_B$, tested against Gram-positive and Gram-negative bacteria, revealed that *B. subtilis* (20.66 ± 1.15–25 ± 4.35 mm), *P. mirabilis* (18.66 ± 1.15–18.66 ± 1.15 mm), and *P. aeruginosa* (13.33 ± 1.15–19 ± 1.00 mm) are the most susceptible strains to $LSEO_A$, $LSEO_K$ and $LSEO_B$ of which $LSEO_B$ exhibits bactericidal effect against *P. mirabilis*. furthermore The LSEO exhibited varying degrees of anticandidal activity with an inhibition zones of 25.33 ± 0.5, 22.66 ± 2.51, and 19 ± 1 mm for $LSEO_K$, $LSEO_B$, and $LSEO_A$, respectively. Additionally, the in silico molecular docking process, performed using Chimera Vina and Surflex-Dock programs, indicated that LSEO could inhibit SARS-CoV-2. These important biological properties of LSEO qualify this plant as an interesting source of natural bioactive compounds with medicinal actions.

Keywords: *Lavandula stoechas*; antioxidant; antibacterial; SARS-CoV-2; in silico; GC-MS-MS analysis; medicinal plants; docking

1. Introduction

Medicinal plants are of great interest as a source of bioactive molecules used to treat different human diseases [1]. Among the secondary metabolites, essential oils (EOs) have applications in the pharmaceutical and aromatic industries. It is for this reason that their use has increased during the last decade directly or indirectly in daily life [2]. Several studies have suggested the employment of EOs instead of synthetic chemicals in the treatment of human pathologies. The biosynthesis of bioactive products, the contents in EOs, and their biological effectiveness may vary based on many parameters, such as geographical variation, matrix use, phenological stages, seasonal variation, light availability, interaction, and anthropogenic activity [3–9]. Considering these factors, research is focused on identifying the optimal conditions to obtain EOs with rich content of bioactive molecules.

Lavandula stoechas L., Lamiaceae family is one of the 39 species in the *Lavandula* genus, and it is widely used throughout the Mediterranean region for its medicinal interests attributed to its bioactive compounds, including camphor, myrtenol, erythrodiol, lupeol, terpineol, eucalyptol, fenchone, luteolin, oleanolic acid, pinocarvyl acetate, and ursolic acid [10]. Lavender, due to its phytochemical composition, is a popular medicinal and aromatic plant commonly used in traditional medicine and food and cosmetic industries thanks to its key antioxidant, anti-inflammatory, and antimicrobial properties. According to ethnobotanical and ethnopharmacological investigations, *L. stoechas* is used in Morocco to treat inflammatory problems, nephrotic syndromes, and rheumatic diseases, as an antispasmodic agent, and to reduce pain. In Portugal, the aerial part is used to treat heartburn and sea-sickness and to enhance blood circulation [11,12]. In Turkey and Spain, it is used by women to regulate menstrual cycles as a carminative and antispasmodic [13,14]. The plants can also be used as an antidiabetic, to relieve kidney stones, and in the treatment of hypertension, epilepsy, migraine, and otitis [15,16]. From the point of view of the pharmacological activities of LSEO, several research works have evaluated their antimicrobial, antioxidant, antileishmanial, insecticidal, and anticancer activities [17–23]. These biological properties may be attributed to the high content of the LSEO fenchone/camphor chemotype. However, the results of the biological activities differ from one study to another, whose differences could be due to the quantitative and qualitative variation of the essential oil chemical composition [10], which might be influenced by the parameters mentioned above.

To the best of our knowledge, no study has been completed to report on the chemical profiles or biological activities of LSEO extracted from plants from different regions in Morocco. Therefore, the aim of this work was to evaluate, in vitro, the antioxidant, antibacterial, and anticandidal effects, as well as the anti-severe acute respiratory syndrome coronavirus 2 (SARS-CoV-2) action in silico of LSEO collected from three different Moroccan sites, Taza city (Northern Morocco), Khenifra, and Beni Mellal (Central Morocco), and thus find the optimal site to collect this species for use in alternative medicine or as a potential therapy in conventional medicine.

2. Materials and Methods

2.1. Collection of Plants and Isolation of Essential Oils

Plant samples were collected in April 2021 from three sites Sebt Malal Aknol, Aguelmous, and Moujd located in the provinces of Taza, Khenifra, and Beni Mellal, respectively. The identification of plants was achieved by Pr. Abdelmajid Khabbach in the Natural Resources and Environment Laboratory of the Polydisciplinary Faculty of Taza, Sidi Mohamed Ben Abdellah University of Fez. The dried leaves (100 g) were subjected to hydrodistillation using a Clevenger type apparatus for 3 h. The essential oil was stored at 4 °C until use.

2.2. GC-MS-MS Analysis of LSEO

The chemical composition of LSEO collected from the mentioned three sites was analyzed using GC-MS-MS analysis [24]. The investigation was performed on gas chromatography TQ8040 NX (Shimadzu, Tokyo, Japan) attached to a triple quadrupole, tandem mass spectrometer (GC-MS). Chromatography was conducted on an apolar, equipped with capillary column RTxi-5 Sil MS column (30 m × 0.25 mm ID × 0.25 μm). Purified helium was used as carrier gas, and the injection volume was 1 μL. Temperature of the source was 200 °C. The chromatographic system was programmed with splitless injection (split opening at 4 min), an injector temperature of 250 °C, and pressure of 37.1 kPa. Temperature was programmed with an initial temperature of 50 °C for 2 min, ramp 1 and ramp 2 were 5 °C/min to 160 °C for 2 min, and 5 °C/min to 280 °C for 2 min, respectively. The identification of each compound was based on its mass spectra (MS) and by computer matching with standard reference databases.

2.3. Antioxidant Activities

2.3.1. Free Radical Scavenging Activity by $ABTS^+$

The radical scavenging activity of LSEO against the radical $ABTS^+$ was evaluated according to the Brahmi et al. [25], with some modifications. First, the $ABTS^+$ solution was prepared at 7 mM concentration with potassium persulfate (2.45 mM); this solution was allowed in obscurity at room temperature for 12 h. Before tests, the $ABTS^+$ stock solution was diluted with methanol to an absorbance of 0.700 ± 0.020 at 734 nm. Then, 75 μL of test samples at different concentrations (31.12–500 μg/mL, prepared in methanol) were added to 925 μL of ABTS solution. The absorbance was measured at 734 nm using a spectrophotometer (SPECUVIS1, UV-Visible). Ascorbic acid was used as standard antioxidant.

The antioxidant activity (AA) was calculated using the following Formula (1):

$$AA\ (\%) = (Abs_{control} - Abs_{sample})/Abs_{control} \times 100 \quad (1)$$

where $Abs_{control}$ is the absorbance of the negative control, and Abs_{sample} is absorbance of the test sample.

2.3.2. Reducing Power Assay

The reducing power activity (FRAP) of LSEO was evaluated according to our previous study [26]. Indeed, the solution made up of the phosphate buffer (2.5 mL, 0.2 M, pH 6.6), potassium ferricyanide (2.5 mL), and the test samples (1 mL at 1 mg/mL dissolved in methanol) was prepared. To stop the reaction, trichloroacetic acid (10%) was added at a volume of 2.5 mL after incubation for 20 min at 50 °C (water bath). Then, the mixture was centrifuged at 3000 rpm/min for 10 min. Afterward, 2.5 mL supernatant was mixed with 0.5 mL of 0.1% ferric chloride and 2.5 mL of distilled water. Finally, absorbance was measured at 700 nm using a spectrophotometer (SPECUVIS1, UV-Visible). The reducing power is expressed in milligram equivalence of ascorbic acid per gram of extract (mg EAA/g).

2.4. Antibacterial Activity

2.4.1. Pathogen Bacteria and Growth Conditions

Antibacterial activity was performed against pathogen bacteria, including Gram-positive bacteria (*Bacillus subtilis* DSM 6633 and *Staphylococcus aureus* CECT 976) and Gram-negative bacteria (*Proteus mirabilis* INH, *Escherichia coli* K12, and *Pseudomonas aeruginosa* CECT 118), using the disc diffusion method as described in our previous work [1]. First, sterile disks (6 mm diameter) were applied onto the surface of the MHA, which was previously spread with the test inoculum concentrations, and were loaded with a volume of 12.5 μL of pure essential oil. Gentamicin (15 μg) served as a positive control and 10% dimethylsulfoxide (DMSO) as negative control. After incubation, the antibacterial effect was determined by calculating the diameter of inhibition zones.

2.4.2. Minimum Inhibitory Concentration and Minimum Bactericidal Concentration

The MIC values were evaluated in sterile 96-well microplate according to [27], with some modifications. First, 100 µL of Mueller–Hinton Broth (MHB) was distributed in all test wells except the first well in which a volume of 200 µL was added containing the LSEO with a concentration of 25 mg/mL in 10% DMSO. A series of doses varying from 25 to 0.097 mg/mL were prepared from the first to the ninth well. Then, 10 µL of the suspension from each well was removed and replaced by the inoculum test concentration, except the 10th well, which was used as sterility control. The last two wells (eleventh and twelfth) were considered as positive growth negative controls, which contained only MHB broth and 10% DMSO (*v*/*v*) without LSEO, respectively. Then, the plates were incubated at 37 °C for 24 h. After the incubation, a volume of 25 µL of an indicator of microorganism's growth was added to each well; 2,3,5-triphenyltetrazolium chloride (TTC) was prepared at a concentration of 5 mg/mL in sterile distilled water. The microplate was re-incubated at 37 °C for 30 min. The minimum bactericidal concentration (MBC) was determined by the inoculation in MHA of 10 µL of broth from the uncolored wells and incubated at 37 °C for 24 h.

2.5. *Anticandidal Effect*

The anticandidal activity of pure LSEO was evaluated against *Candida albicans*, which was cultured in YPGA medium (5 g yeast extract, 5 g peptone, 10 g glucose, and 15–18 g agar in 1 L) and incubated at 37 °C for 48 h. The effect was evaluated using disc diffusion method [28].

2.6. *Anti-SARS-CoV-2 In Silico*

2.6.1. Molecular Docking

Molecular modeling is an interesting in silico tool used to determine the stability of compounds and the interaction types responsible for antiviral biological activity. Different EO studies revealed antiviral activity against SARS-CoV-2 [29–31]. Two different software were used: Surflex-Dock and UCSF chimera in UCSF Chimera 1.13.1 [32,33]. The crystal structures were edited to remove water molecules, and all hydrogen atoms were added to the structure. For Surflex-Dock, protomol-specified residues in the protein were applied to determine the docked cavity of the receptors. All ligands were docked using automatic docking method, and total scores were expressed in -log10 (Kd) units to show binding affinities [32]. For UCSF Chimera, the 3D structure of both receptors (PDB:6lu7 and PDB:6vsb) were loaded to chimera window and prepared using Dock Prep mode. Polar hydrogens were added, and Gasteiger charges were calculated. The docking analyses of studied proteins were executed using the plug-in of Chimera Vina. The binding sites were identified using native ligand with a grid box of size 20 × 20 × 20 centered at x = 247.84, y = 255.31, z = 272.31 Å and x = −12.17, y = 13.96, z = 69.74 Å for both receptors PDB:6vsb and PDB:6lu7, respectively [34,35]. The native ligand was deleted before docking, and the conformations were searched with binding parameters of 3 kcal/mol as the maximum energy difference, 8 as exhaustiveness of search, and 9 as the number of binding modes. Root mean square deviation (RMSD) values were used to compare the ligand between the predicted and its corresponding crystal structure [36]. The lowest energy-minimized pose was used for further analysis. Discovery Studio 2016 software was utilized to visualize the different interactions of molecular docking results [37].

2.6.2. ADMET Properties

Pharmacokinetics is an important process that studies drug absorption, distribution, metabolism, excretion, and toxicity (ADMET). It is a fundamental concept to eliminate low drug candidates, which may present problems during in vivo studies, and it also determines the availability of a drug candidate [38]. ADME/T property predictions allow drug developers to understand the safety and efficacy of a drug candidate, as it is necessary for a drug developer to make a go/no-go decision in the late stages of preclinical and clinical programs. In this study, ADMET properties were determined using pkCSM online server [39].

2.6.3. Molecular Prediction

With the aim of determining the potential bioactive compounds that exist in *L. stoechas* plants and finding drug candidates against viral infections, molecular docking has been performed. Molecular docking is used to predict how receptors interact with bioactive compounds (ligands). Several studies investigated the bioactive compounds in plants that have potential to inhibit the proliferation of viruses [31,40,41]. Moreover, a new study reported that an inhibitor of HIV protease (nelfinavir) was predicted to be COVID-19 drug candidate using molecular docking [42].

The compounds docked were molecules found in high percentages in *L. stoechas* plants gathered from the interested regions in Morocco. These compounds were L-fenchone, camphor, bornyl acetate, cubebol, viridiflorol, and tau-muurolol.

Three-dimensional (3D) structures, main protease Mpro and spike glycoprotein targets of SARS-CoV-2, were retrieved from Protein Data Bank [34,35] in pdb formats. These proteins were chosen as receptors in molecular docking process. Water molecules and ligands that were still attached to the receptor were removed. The receptor was stored in the pdb, and polar hydrogen atoms were added. Docking preparations, analyses, and determination of hydrogen bonds (H-bonds) were conducted using two different software, Chimera 1.15 (vina) and sybyl-x 2.0 (Surflex-Doc). The visualization of receptor–ligand interactions was obtained using BIOVIA Discovery Studio Visualizer 2016 [37].

2.7. Statistical Analysis

All assays were done in triplicates. Values of each test were expressed as mean ± standard deviation (SD) and were subjected to analysis of variance (one-way ANOVA). The statistical analysis was performed using GraphPad Prism version 6.00 (GraphPad Inc., San Diego, CA, USA). Differences (between groups) were considered as statistically significant at $p < 0.05$.

3. Results and Discussion

3.1. Chemical Composition

The essential oil yields (*w*/*w*) were 1.84, 0.79, and 0.65% for $LSEO_K$, $LSEO_A$, and $LSEO_B$, respectively. The results of the GC-MS-MS analysis showed the richness of the plants collected from the three regions in volatile compounds with variability between the three essential oils analyzed. Indeed, $LSEO_A$ contains L-fenchone (14.39%), γ-1-cadinene aldehyde (10.61%), viridiflorol (8.54 %), bornyl acetate (8.39 %), and myrtenyl acetate (3.77%) as the main compounds or chemotypes (Table 1).

Table 1. Chemical composition of $LSEO_A$.

Peak Number	Compound	Retention Time	Area
1	L-fenchone	12.6	14.39
2	2-norbornanol	13.50	0.98
3	Camphor	14.42	23.80
4	Borneol	15.05	1.13
5	3-adamantan-1-yl-butan-2-one	15.280	1.72
6	Benzenemethanol, 4-(1-methylethyl)	15.47	1.18
7	2-pinen-10-ol	15.75	1.40
8	2-pinen-4-one	16.07	1.02
9	2-cylohexen-1-ol	16.38	0.94
10	*D*-carvone	17.08	0.64
11	Bornyl acetate	18.27	8.39
12	Myrtenyl acetate	19.30	3.77
13	α-cadino	23.72	0.64
14	Cubebol	23.81	1.63
15	Δ-cadinene	24.36	0.78
16	Cyclohexene, 1,3-diisopropenyl-6-methyl	25.12	1.45

Table 1. *Cont.*

Peak Number	Compound	Retention Time	Area
17	cis-.α.-copaene-8-ol	25.60	1.56
18	Caryophyllene oxide	26.21	0.89
19	Menthol	26.39	0.92
20	Viridiflorol	26.64	8.54
21	Acorenone B	26.75	1.03
22	Ledol	26.90	1.80
23	Humulane-1,6-dien-3-ol	27.04	4.56
24	Cedr-9-ene	27.52	1.35
25	τ-muurolol	27.95	2.7
26	Longiverbenone	28.78	1.32
27	β-copaen-4-ol	29.47	0.72
28	γ-1-cadinene aldehyde	32.99	10.61

$LSEO_B$ showed the presence of cubebol (22.68%), camphor (22.29%), borneol (5.15%), muurol-5-en-4-one <cis-14-nor-> (4.21%), L-fenchone (4.03%), and silphiperfol-5-ene (3.27%) as the main compounds (Table 2).

Table 2. Chemical composition of $LSEO_B$.

Peak Number	Compound	Retention Time	Area
1	*L*-fenchone	12.56	4.03
2	Linalool	12.95	2.48
3	Camphor	14.50	22.29
4	Pinocarvone	14.79	0.27
5	Borneol	15.14	5.15
6	*p*-menth-1-en-4-ol	15.33	1.93
7	Benzenemethanol, 4-(1-methylethyl)	15.54	2.21
8	Myrtenal	15.76	0.96
9	2-pinen-10-ol	15.82	0.54
10	Verbenone	16.14	2.66
11	2-cyclohexen-1-ol	16.42	0.96
12	D-carvone	17.10	0.41
13	Bornyl acetate	18.25	1.71
14	β-selinene	23.68	0.67
15	Myrtenyl acetate	24.04	0.46
16	cis-calamenene	24.43	2.79
17	Selina-3,7(11)-diene	25.08	2.00
18	Myrtenyl 2-methyl butyrate	25.28	0.41
19	Germacrene D-4-ol	25.48	0.5
20	1,3,3-trimethyl-2-(2-methylcyclopropyl)-1-cyclohexene	26.42	0.55
21	Eremophila ketone	26.60	1.13
22	2-octenoic acid	27.05	0.43
23	Cubebol	27.43	22.68
24	Aromadendrane-4,10-diol	27.54	0.55
25	τ-cadinol	27.97	2.63
26	Trans-valerenyl acetate	28.14	0.29
27	τ-muurolol	28.33	0.06
28	Silphiperfol-5-ene	28.66	3.27
29	Naphthalene, 1,6-dimethyl-4-(1-methylethyl)	28.78	1.11
30	Muurol-5-en-4-one <cis-14-nor->	29.20	4.21
31	δ-tridecalactone	29.54	0.44
32	1-naphthalenepropanol	29.80	2.94
33	Androstane-17,19-diol	30.53	0.29
34	Caryophyllene oxide	31.48	0.34
35	Neoisolongifolene	31.99	1.41
36	5-(7a-isopropenyl-4,5-dimethyl-octahydroinden-4-yl)-3-methyl-pent-2-en-1-ol	32.09	0.42
37	Longifolenaldehyde	32.53	0.58
38	Corymbolone	32.66	0.6
39	Myrtenyl acetate	35.81	0.83
40	Widdrol hydroxyether	36.06	0.31

However, τ-muurolol (18.44%), cubebol (16.07%), camphor (13.39), muurol-5-en-4-one (cis-14-nor-) (6.84), selina-3,7(11)-diene (4.5%), 3-adamantan-1-yl-butan-2-one (4.39%), borneol (3.26%), linalool (3.02%), and benzenemethanol, 4-(1-methylethyl) (3%) were the main compounds in $LSEO_K$ (Table 3).

Table 3. Chemical composition of $LSEO_K$.

Peak Number	Compound	Retention Time	Area
1	*L*-fenchone	12.55	1.88
2	Linalool	12.91	3.02
3	Camphor	14.43	13.39
4	Borneol	15.12	3.26
5	3-adamantan-1-yl-butan-2-one	15.33	4.39
6	Benzenemethanol, 4-(1-methylethyl)	15.60	3.00
7	2-pinen-10-ol	16.17	2.47
8	2-pinen-4-one	16.23	0.6
9	2-cyclohexen-1-ol	16.45	1.10
10	Verbenone	18.92	1.76
11	β-selinene	23.69	1.24
12	cis-calamenene	24.44	2.65
13	Selina-3,7(11)-diene	25.09	4.5
14	1,3,3-trimethyl-2-(2-methyl-cyclopropyl)-cyclohexene	26.43	0.65
15	Arctiol	27.07	1.20
16	Cubebol	27.40	16.07
17	τ-cadinol	27.96	2.08
18	τ-muurolol	28.54	18.44
19	Cedr-8(15)-en-9-ol	28.71	1.96
20	Naphthalene, 1,6-dimethyl-4-(1-methylethyl)	28.84	1.61
21	Muurol-5-en-4-one (cis-14-nor-)	29.29	6.84
22	δ-tridecalactone	29.58	1.74
23	1-naphthalenepropanol	29.81	2.10
24	2(3H)-naphthalenone	30.68	0.61
25	Caryophyllene oxide	31.51	0.57
26	Neoisolongifolene	32.00	0.81
27	Myrtenyl acetate	35.82	1.29
28	Methyl 5,9-docosadienoate	36.09	0.86

The literature reports supported these findings concerning other medicinal plants. Indeed, several studies have reported the chemical composition of LSEO, with some indicating that, in addition to the fenchone/camphor chemotypes, the chemical compositions of LSEO collected in Morocco and Greece contained 1,8-cineole and camphene, and α-cardinol, respectively, while others disclosed the presence of myrtenyl acetate, bornyl acetate, linalyl acetate, camphene, linalool, borneol, γ-terpinene, lavandulyl acetate, and caryophyllene as major compounds [10,18,43–46]. Besides the presence of some main compounds, our study clearly revealed the chemical composition quantitative and qualitative variability of *L. stoechas* plants collected from three different regions. This confirms the idea postulating that the environmental, climatic, and nutritional conditions of the same plant impact, quantitatively and qualitatively, the synthesis of secondary metabolites. Several previous works have revealed this causal link between the variation of external factors, such as temperature, humidity, soil, or climate type, metabolic pathways, and the chemical composition of EOs. Therefore, the nature of soil may induce different elicitor production, a group of molecules secreted by microorganisms in soil (at the rhizosphere), which stimulate and regulate the synthesis and accumulation of secondary metabolites in medicinal plants [47]. Moreover, it has been previously shown that environmental factors could change the synthesis of EOs via different epigenetic modifications or the alteration of gene expression involved in secondary metabolite anabolism [48,49].

It was also exposed that LSEO chemical compounds might vary between seasonal stages and plant parts (stems, leaves, and flowers) [50]. Indeed, the findings disclosed that

LSEO expressed volatile substances according to phenological stages and plant parts with remarkable variability.

3.2. Antioxidant Activity

The antioxidant activities of LSEO were examined using ABTS and FRAP tests. An ABTS radical scavenging assay, based on the transfer of both a hydrogen atom and an electron, measures the capacity of antioxidants to neutralize ABTS, a blue-green stable radical cation, enabling the quantification of the antioxidant capability of both hydrophilic and lipophilic compounds. The results show that $LSEO_K$ and $LSEO_B$ have a greater capacity to reduce ABTS compared to $LSEO_A$ (Figure 1). For a dose of 220 µg/mL, the percentage of inhibition exceeded 90%.

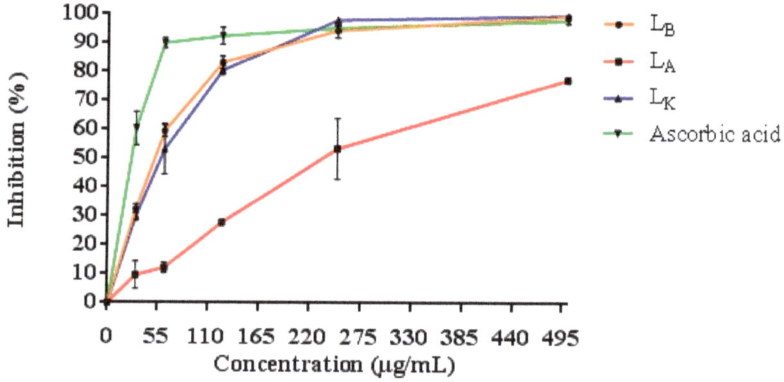

Figure 1. Scavenging activity of $LSEO_A$, $LSEO_B$, $LSEO_K$, and ascorbic acid.

The FRAP test is based on the transfer of one electron and measures the reduction of the ferric ion ($Fe3^+$)–ligand to the blue ferrous ($Fe2^+$) complex in acidic pH conditions to maintain iron solubility. For this test (Figure 2), $LSEO_B$ presented a significant value of 15.73 ± 3.26 mg EAA/g extract, while that of $LSEO_A$ and $LSEO_K$ were 6.91 ± 0.47, and 4.82 ± 1.52 mg EAA/g extract, respectively. The antioxidant potency of LSEO was previously evaluated, and the results demonstrated that they exert important antioxidant activities [18,21,51,52].

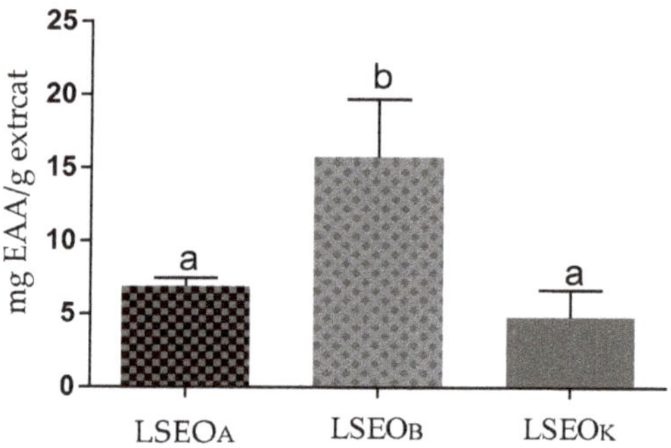

Figure 2. Ferric reducing antioxidant power of $LSEO_A$, $LSEO_B$, and $LSEO_K$ in mg of equivalent ascorbic acid/g of extract (values not sharing a common letter differ significantly at $p < 0.05$).

3.3. Antibacterial Activity

In vitro tests of the antibacterial effect of LSEO, using the filter paper disc diffusion and the microplate methods against microorganism tests, are summarized in Table 4. The findings revealed a variation in sensitivity between the bacteria tested. Concerning the Gram-positive bacteria, *B. subtilis* was the most sensitive strain to $LSEO_A$, $LSEO_K$, and $LSEO_B$ with a diameter of inhibition zone of 25 ± 4.35, 21.66 ± 2.08, and 20.66 ± 1.15 mm, respectively. Among the Gram-negative bacteria, *P. mirabilis* was significantly inhibited by $LSEO_K$ (22.66 ± 0.57 mm) compared to $LSEO_B$ and $LSEO_A$ with 20 ± 1.00 and 18.66 ± 1.15 mm, respectively. In addition, significant inhibition was exerted by $LSEO_K$ against *P. aeruginosa* (19 ± 1.00 mm) in comparison with $LSEO_B$ and $LSEO_A$ with 15.66 ± 0.57 and 13.33 ± 1.15 mm, respectively. The MBC/MIC values inform that $LSEO_A$, $LSEO_K$, and $LSEO_B$ exert a bacteriostatic effect versus all bacteria tested except $LSEO_B$, which exhibits a bactericide effect against *P. mirabilis*. From the point of view of the difference in the antibacterial potential of $LSEO_A$, $LSEO_K$, and $LSEO_B$, our results could be attributed to the qualitative and quantitative variation in their chemical composition with the active compounds.

Table 4. Antibacterial activity of $LSEO_A$, $LSEO_B$, and $LSEO_K$ determined by disc diffusion method and their minimum inhibitory concentration (MIC) and minimum bactericidal concentration (MBC) (mg/mL).

Strains	$LSEO_A$			$LSEO_K$			$LSEO_B$		
	DIZ *	MIC	MBC	DIZ	MIC	MBC	DIZ	MIC	MBC
S. aureus	6 ± 0.00 [a]	NT	NT	6 ± 0.00 [a]	NT	NT	7.66 ± 0.57 [a]	NT	NT
B. subtilis	25 ± 4.35 [a]	25	>50	21.66 ± 2.08 [a]	6.25	>50	20.66 ± 1.15 [a]	6.25	>50
P. aeruginosa	13.33 ± 1.15 [a]	NT	NT	19 ± 1.00 [b]	NT	NT	15.66 ± 0.57 [a]	NT	NT
P. mirabilis	18.66 ± 1.15 [a]	>50	>50	22.66 ± 0.57 [b]	12.25	>50	20 ± 1.00 [a]	12.5	25
E. coli	6 ± 0.00 [a]	12.5	>50	10.66 ± 0.57 [b]	3.12	>50	10 ± 0.00 [b]	3.12	>50

* The diameter of the inhibition zones (mm), including diameter of disc 6 mm, are given as mean ± SD of triplicate experiments; DIZ: Diameter Inhibition Zones; NT: not tested; within each line, different letters ([a,b]) indicate significant differences ($p < 0.05$).

Concerning the susceptibility of Gram-positive and Gram-negative bacteria, it has been revealed that the Gram-negative bacteria are less sensitive to plant extracts compared to Gram-positive bacteria because Gram-negative bacteria possess double membranes, which protect them versus the antibacterial products [1,53–55]. Our findings showed that $LSEO_A$, $LSEO_K$, and $LSEO_B$ were active against both Gram-negative (*P. aeruginosa* and *P. mirabilis*) and Gram-positive (*B. subtilis*) bacteria. These results may be related to the presence of a high content of active compounds with antibacterial potential. Many studies already confirmed that minor components in the EOs could have synergistic antimicrobial activity [56,57].

3.4. Anticandidal Effect

The in vitro anticandidal activity of the LSEO was qualitatively confirmed using the diameter of inhibition zones. The LSEO exhibited varying degrees of antifungal activity. The inhibition zones were 25.33 ± 0.5, 22.66 ± 2.51, and 19 ± 1 mm for $LSEO_K$, $LSEO_B$, and $LSEO_A$, respectively.

3.5. Anti-SARS-CoV-2 In Silico

Molecular docking study

Essential oils have shown promise as antiviral agents against several pathogenic viruses [58,59]. To gain structural insights and understand the binding mode of molecular structures and protein targets, we applied molecular docking processes that were previously described as an efficient in silico approach [60]. Various experiments revealed that EOs could contribute to preventing the entry of SARS-CoV-2 into the human body and investigated the efficacy of EO compounds in the prevention and treatment of COVID-19 [31,59,61–63]. Da Silva et al. used molecular docking analysis to determine the interaction of 171 essential oil components with SARS-CoV-2, showing that the compound with the best-normalized docking score to SARS-CoV-2 Mpro was the sesquiterpene hydrocarbon (E)-β-farnesene [64].

Two of the very well-characterized and promising drug targets are the main protease (Mpro; 3CLpro) and the papain-like protease (PLpro), which play key roles in viral replication and transcription [65]. They have been the main target of many vaccines as antibodies against this protein block the entry of the virus and inhibit viral replication [66]. There have been several molecular docking studies on these targets as well as EOs molecular docking with SARS-CoV-2 proteins [67–70]. Moreover, commercially available drugs have been confirmed using in silico methods [71,72].

As the chemical compositions of the researched *L. stoechas* plants gathered from the three regions were different and in order to determine the promising antiviral compounds against SARS-CoV-2, the molecular docking process was performed using Chimera Vina and Surflex-Dock programs. The redocking process of co-crystal ligands for both receptors showed low RMSD values (<1.5), which indicated the reliability of the applied docking process.

In silico molecular docking of the studied compounds, L-fenchone, camphor, bornyl acetate, cubebol, viridiflorol, and tau-muurolol, with the main protease Mpro and S-protein targets was applied. The results presented in Table 5 show that bornyl acetate and cubebol compounds have good binding affinities and an interesting scoring compared to chloroquine, a compound that has been known for quite a long time to inhibit the invasion of different viruses in cultured cells in vitro, including SARS-CoV and MERS-CoV [73–75].

Table 5. Molecular docking energy affinities of both receptors (pdb:6lu7 and pdb:6vsb) using Surflex-Dock and UCSF Chimera software.

Compounds	Surflex-Dock		UCSF Chimera	
	6lu7	6vsb	6lu7	6vsb
Cubebol	3.12	3.37	−5.5	−5.6
Bornyl acetate	3.92	3.55	−5.4	−5.3
L-fenchone	2.56	2.54	−4.2	−4.8
τ-muurolol	2.94	4.07	−5.3	−4.6
Viridiflorol	2.46	2.53	−5.5	−4.7
Camphor	2.60	3.26	−4.4	−4.4
Chloroquine	3.6	3.2	−5.7	−5.3

The molecular docking of each compound showed 10 different poses; the stable one presented in Figure 3 is the structure used for further studies.

The stable pose of bornyl acetate in the Mpro receptor pocket shown in Figure 4 presents the hydrogen bond with SER A:144 residue and pi-alkyl interactions with CYS A:145, and MET A:49 and MET A:165 residues, showing a score of 3.92 (−5.54 with Chimera Vina). Cubebol shows two hydrogen bonds with SER A:144, CYS A:145 residues, pi-alkyl interactions with MET A:49, and HIS A:41 and LEU A:27 residues, with a score of 3.12 (−5.5 using Chimera Vina). The compounds tau-muurolol and camphor are stabilized by the hydrogen bond, with LEU A:141 and GLU A:166 residues, respectively, but the presence of an unfavorable interaction with the SER A:144 residue for the tau-muurolol compound destabilized its inhibition compared to the rest of compounds. L-fenchone and viridiflorol are stabilized with different pi-alkyl interactions.

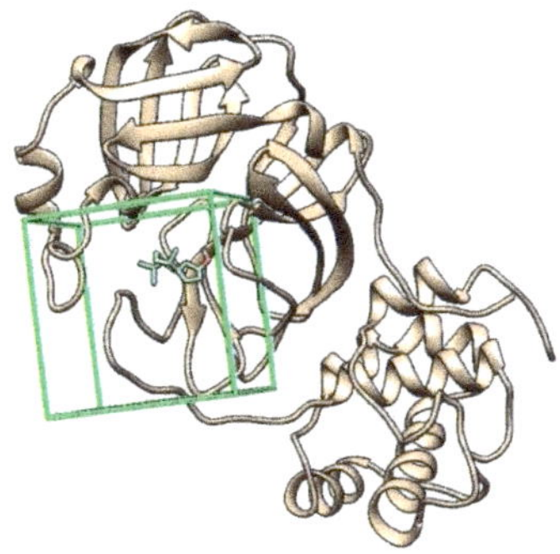

(**a**)

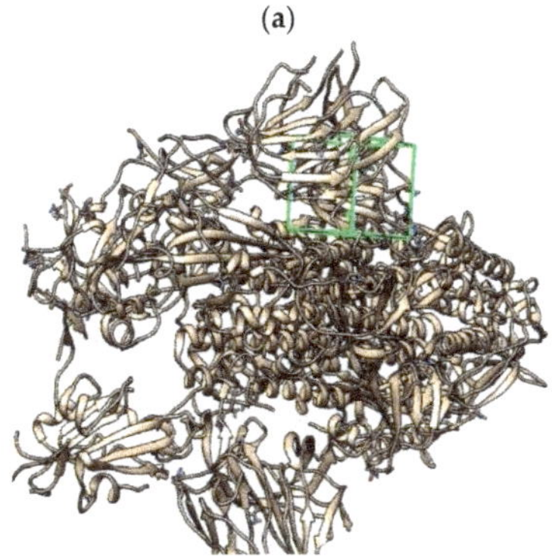

(**b**)

Figure 3. The stable pose of cubebol in receptor pocket using UCSF Chimera: (**a**) main protease Mpro; (**b**) spike glycoprotein targets.

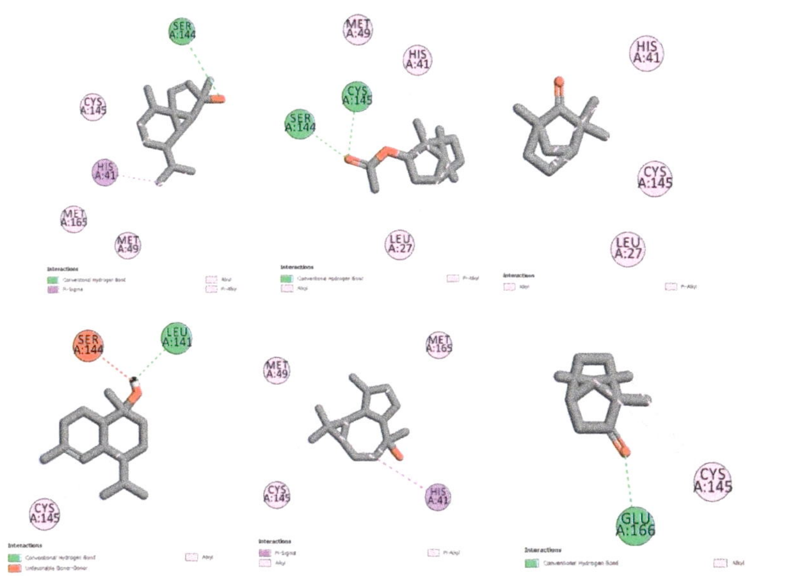

Figure 4. The molecular interactions between the studied compounds and main protease Mpro receptor (pdb:6lu7) using discovery studio visualizer.

The molecular docking of compounds with the spike glycoprotein (pdb:6vsb) receptor presented in Figure 5 shows pi-alkyl interactions between bornyl acetate and LEU C:293 and PHE C:58 and PHE C:59 residues, with a score of 3.55 (−5.53 with Chimera Vina). The compound cubebol is stabilized by two hydrogen bond interactions, with ASN C:606 and LYS C:300 residues. In addition, the presence of pi-alkyl interactions increases the stability of this compound in the receptor pocket, with a score of 3.37 (−5.6 with Chimera Vina). L-fenchone is stabilized by pi-alkyl interactions, with a score of 2.54 (−4.8 with UCSF Chimera). The three compounds tau-muurolol, viridiflorol, and camphor show a hydrogen bond interaction with the ARG B:1224 residue for the two first compounds and with the ARG B:1226 residue for camphor.

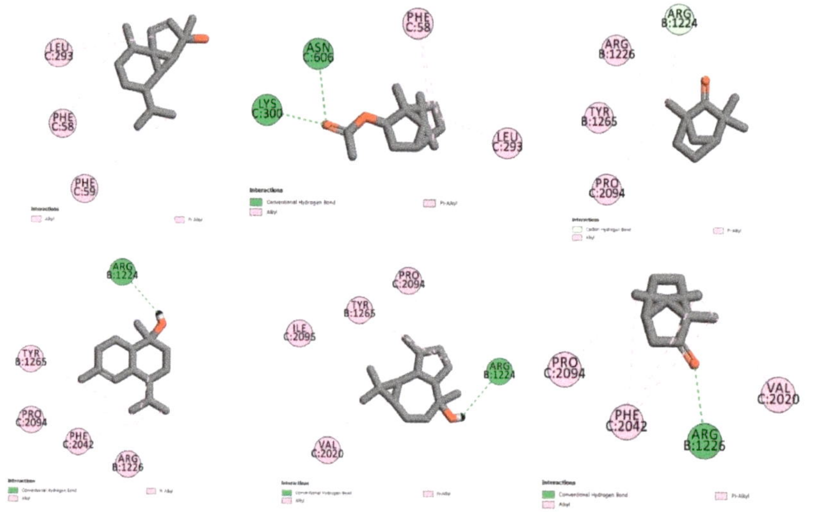

Figure 5. The molecular interactions between the studied compounds and spike glycoprotein receptor (pdb:6vsb) using discovery studio visualizer.

Based on the energy affinities presented in Table 5 and the molecular interactions described in Figures 4 and 5, cubebol and bornyl acetate are the compounds that show an excellent inhibition to both receptors, the main protease Mpro (pdb:6lu7) and spike glycoprotein (pdb:6vsb) targets. Moreover, the *L. stoechas* plants gathered from Khenifra and Beni Mellal show an interesting cubebol percentage. In addition, $LSEO_B$ presents bornyl acetate in its composition, indicating that the $LSEO_B$ plant could be a promising SARS-CoV-2 inhibitor.

The results of the current in silico molecular docking process, employing the binding affinity and interactions, support the use of LSEO compounds as possible candidate inhibitors in the treatment of COVID-19.

3.6. ADMET Predictions

The Lipinski rule is one of the best filters in the virtual screening of bioactive molecules to determine an effective drug in early preclinical development [76]. The values in Table 6, calculated using pkCSM, indicate that cubebol and bornyl acetate have molecular weights under 500, LogP and hydrogen bond donors less than 5, and rotatable bonds and hydrogen bond acceptors less than 10, with a polar surface under 140 Å^2, all indicating the drug permeability and ability of these two compounds.

Table 6. Physicochemical parameters (Lipinski Rule of Five) of cubebol and bornyl acetate compounds.

	MW	LogP	Rotatable Bonds	Donors	Acceptors	Surface
Cubebol	222.372	3.46	1	1	1	99.62
Bornyl acetate	196.29	2.76	1	2	0	86.01

Absorption, distribution, metabolism, excretion, and toxicity studies are essential for determining pharmacological properties to discover bioactive compounds with desirable pharmaceutical properties and therefore discuss their drug availability [77]. The calculation of intestinal absorption, and skin and $CaCO_2$ permeability indicate that cubebol and bornyl acetate have high $CaCO_2$ permeability ($CaCO_2 > 0.9$), high intestinal absorption (a compound with values less than 30% are poorly absorbed), and low skin permeability (a compound with values less than −2.5 has low skin permeability) (Table 7). Moreover, the distribution and metabolism results show that both cubebol and bornyl acetate present no inhibition for main cytochrome enzymes, while cubebol reveals that it can be a CYP3A4 substrate, which may be likely metabolized and present drug–drug interactions.

Table 7. Pharmacokinetic (ADMET) properties of cubebol and bornyl acetate compounds.

	Absorption			Distribution and Metabolism CYP450						Excretion and Toxicity		
	Skin Permeability	Intestinal Absorption	$CaCO_2$ Permeability	3A4 Substrate	3A4 Inhibitor	6D6 Substrate	6D6 Inhibitor	VDss	BBB	Total Clearance	AMES	Hepatotoxicity
Cubebol	−2.17	94.94	1.32	yes	no	no	no	0.45	0.66	0.88	no	no
Bornyl acetate	−2.23	95.36	1.85	no	no	no	no	0.30	0.55	1.03	no	no

Cubebol exhibited a high steady-state volume of distribution (VDss), >0.45, and was ready to cross the blood–brain barrier (BBB). Bornyl acetate showed medium VDss and was also ready to cross the BBB. Both compounds disclosed no AMES toxicity or hepatotoxicity, with total clearance of 0.88 and 1.03 for cubebol and bornyl acetate, respectively. These results indicate that bornyl acetate revealed the best pharmacokinetic properties compared to cubebol, and it can be considered in further experiments. Similar to our study, Wei et al., who found linalool and linalyl acetate (29.48 and 40.97%, respectively) in lavender, proved that these major LSEO components had no toxicity and were safe to be used as food or medication [78]. A recent study reported that linalyl acetate (39.7%), linalool (33.6%), and terpinen-4-ol (14.9%) were the most abundant lavender EOs and that they possessed antiviral activities against many DNA and RNA viruses [79].

In silico studies and ADMET prediction of the selected LSEO bioactive molecules demonstrated good pharmacokinetic properties. The phytochemical composition and some biological activity outcomes were slightly different compared to other studies and are a confirmation of the originality of our *Lavandula stoechas* research. The results are very promising and could encourage further in vitro and in vivo evaluations of this plant and its LSEO.

4. Conclusions

The present work is a detailed description of the chemical composition and biological effects of essential oils extracted from *Lavandula stoechas* harvested from three Moroccan sites. Our results showed that this plant synthesized various volatile compounds, such as L-fenchone, cubebol, camphor, bornyl acetate, and τ-muurolol, with qualitative and quantitative differences depending on their harvest site. The essential oils were in vitro analyzed for their antimicrobial, antioxidant, and anti-SARS-CoV-2 effects. The inhibition reached 81.1% for the antioxidant activity. The antimicrobial tests disclosed that the essential oils were effective against the growth of *B. subtilis*, *P. aeroginosa*, and *P. mirabilis*. In addition, $LSEO_K$, $LSEO_B$, and $LSEO_A$ inhibited the growth of *C. albicans*. In silico investigation of the volatile compounds of essential oils against SARS-CoV-2 revealed a strong affinity of these molecules with the targets of this virus. Future studies should focus on determining and/or validating the pharmacokinetic and pharmacodynamic parameters of the essential oils from *Lavandula*, as well as the toxic effects in clinical trials, before any application in the pharmaceutical, cosmetic, or food industries.

Author Contributions: Conceptualization, T.B. and A.L.; methodology, T.B., A.L., K.H. (Kaoutar Harboul), H.C., A.K., S.M.J. and A.G.; software, T.B., S.L. and A.B.; validation, T.B., S.L. and A.B.; investigation, T.B. and S.L.; writing—original draft preparation, T.B., A.L. and L.Q.-R.; writing—review and editing, T.B., L.-H.L., A.B., M.E.R. and M.A.; supervision, K.H. (Khalil Hammani) and M.A.; funding acquisition, L.-H.L. and M.E.R. All authors have read and agreed to the published version of the manuscript.

Funding: This research received no external funding.

Data Availability Statement: Not applicable.

Conflicts of Interest: The authors declare no conflict of interest.

References

1. Benali, T.; Habbadi, K.; Khabbach, A.; Marmouzi, I.; Zengin, G.; Bouyahya, A.; Chamkhi, I.; Chtibi, H.; Aanniz, T.; Achbani, E.H. GC–MS Analysis, Antioxidant and Antimicrobial Activities of *Achillea Odorata* Subsp. *Pectinata* and *Ruta Montana* Essential Oils and Their Potential Use as Food Preservatives. *Foods* **2020**, *9*, 668. [CrossRef] [PubMed]
2. Poudel, D.K.; Rokaya, A.; Ojha, P.K.; Timsina, S.; Satyal, R.; Dosoky, N.S.; Satyal, P.; Setzer, W.N. The Chemical Profiling of Essential Oils from Different Tissues of *Cinnamomum Camphora* L. and Their Antimicrobial Activities. *Molecules* **2021**, *26*, 5132. [CrossRef] [PubMed]
3. Aissi, O.; Boussaid, M.; Messaoud, C. Essential Oil Composition in Natural Populations of *Pistacia Lentiscus* L. from Tunisia: Effect of Ecological Factors and Incidence on Antioxidant and Antiacetylcholinesterase Activities. *Ind. Crops Prod.* **2016**, *91*, 56–65. [CrossRef]
4. Carvalho, S.; Macel, M.; Mulder, P.P.; Skidmore, A.; Van Der Putten, W.H. Chemical Variation in *Jacobaea Vulgaris* Is Influenced by the Interaction of Season and Vegetation Successional Stage. *Phytochemistry* **2014**, *99*, 86–94. [CrossRef]
5. Formisano, C.; Delfine, S.; Oliviero, F.; Tenore, G.C.; Rigano, D.; Senatore, F. Correlation among Environmental Factors, Chemical Composition and Antioxidative Properties of Essential Oil and Extracts of Chamomile (*Matricaria Chamomilla* L.) Collected in Molise (South-Central Italy). *Ind. Crops Prod.* **2015**, *63*, 256–263. [CrossRef]
6. Moghaddam, M.; Farhadi, N. Influence of Environmental and Genetic Factors on Resin Yield, Essential Oil Content and Chemical Composition of *Ferula Assa-Foetida* L. Populations. *J. Appl. Res. Med. Aromat. Plants* **2015**, *2*, 69–76. [CrossRef]
7. Moniodis, J.; Renton, M.; Jones, C.G.; Barbour, E.L.; Byrne, M. Genetic and Environmental Parameters Show Associations with Essential Oil Composition in West Australian Sandalwood (*Santalum Spicatum*). *Aust. J. Bot.* **2018**, *66*, 48–58. [CrossRef]
8. Sehaki, C.; Jullian, N.; Choque, E.; Dauwe, R.; Fontaine, J.X.; Molinie, R.; Ayati, F.; Fernane, F.; Gontier, E. Profiling of Essential Oils from the Leaves of *Pistacia Lentiscus* Collected in the Algerian Region of Tizi-Ouzou: Evidence of Chemical Variations Associated with Climatic Contrasts between Littoral and Mountain Samples. *Molecules* **2022**, *27*, 4148. [CrossRef]
9. Yosr, Z.; Imen, B.H.Y.; Rym, J.; Chokri, M.; Mohamed, B. Sex-Related Differences in Essential Oil Composition, Phenol Contents and Antioxidant Activity of Aerial Parts in *Pistacia Lentiscus* L. during Seasons. *Ind. Crops Prod.* **2018**, *121*, 151–159. [CrossRef]
10. Bousta, D.; Farah, A. A Phytopharmacological Review of a Mediterranean Plant: *Lavandula Stoechas* L. *Clin. Phytosci.* **2020**, *6*, 9.
11. Camejo-Rodrigues, J.; Ascensao, L.; Bonet, M.À.; Valles, J. An Ethnobotanical Study of Medicinal and Aromatic Plants in the Natural Park of "Serra de São Mamede" (Portugal). *J. Ethnopharmacol.* **2003**, *89*, 199–209. [CrossRef] [PubMed]
12. Novais, M.H.; Santos, I.; Mendes, S.; Pinto-Gomes, C. Studies on Pharmaceutical Ethnobotany in Arrábida Natural Park (Portugal). *J. Ethnopharmacol.* **2004**, *93*, 183–195. [CrossRef] [PubMed]
13. Polat, R.; Satıl, F. An Ethnobotanical Survey of Medicinal Plants in Edremit Gulf (Balıkesir–Turkey). *J. Ethnopharmacol.* **2012**, *139*, 626–641. [CrossRef]
14. Tardío, J.; Pardo-de-Santayana, M.; Morales, R. Ethnobotanical Review of Wild Edible Plants in Spain. *Bot. J. Linn. Soc.* **2006**, *152*, 27–71. [CrossRef]
15. Giray, E.S.; Kırıcı, S.; Kaya, D.A.; Türk, M.; Sönmez, Ö.; Inan, M. Comparing the Effect of Sub-Critical Water Extraction with Conventional Extraction Methods on the Chemical Composition of *Lavandula Stoechas*. *Talanta* **2008**, *74*, 930–935. [CrossRef]
16. Skoula, M.; Abidi, C.; Kokkalou, E. Essential Oil Variation of *Lavandula Stoechas* L. Ssp. Stoechas Growing Wild in Crete (Greece). *Biochem. Syst. Ecol.* **1996**, *24*, 255–260. [CrossRef]
17. Boukhatem, M.N.; Boumaiza, A.; Nada, H.G.; Rajabi, M.; Mousa, S.A. *Eucalyptus Globulus* Essential Oil as a Natural Food Preservative: Antioxidant, Antibacterial and Antifungal Properties In Vitro and in a Real Food Matrix (Orangina Fruit Juice). *Appl. Sci.* **2020**, *10*, 5581. [CrossRef]
18. Bouyahya, A.; Et-Touys, A.; Abrini, J.; Talbaoui, A.; Fellah, H.; Bakri, Y.; Dakka, N. *Lavandula Stoechas* Essential Oil from Morocco as Novel Source of Antileishmanial, Antibacterial and Antioxidant Activities. *Biocatal. Agric. Biotechnol.* **2017**, *12*, 179–184. [CrossRef]

19. Ez Zoubi, Y.; El Ouali Lalami, A.; Moschos, P.; Daferera, D.; Lachkar, M.; Abdessalam, E.; Farah, A. Chemical Composition, Antioxidant and Antimicrobial Activities of the Essential Oil and Its Fractions of *Lavandula Stoechas* L. From Morocco. *Int. J. Curr. Pharm. Rev. Res.* **2017**, *22*, 8. [CrossRef]
20. Ezzoubi, Y.; Bousta, D.; Lachkar, M.; Farah, A. Antioxidant and Anti-Inflammatory Properties of Ethanolic Extract of *Lavandula Stoechas* L. from Taounate Region in Morocco. *Int. J. Phytopharm.* **2014**, *5*, 21–26.
21. Insawang, S.; Pripdeevech, P.; Tanapichatsakul, C.; Khruengsai, S.; Monggoot, S.; Nakham, T.; Artrod, A.; D'Souza, P.E.; Panuwet, P. Essential Oil Compositions and Antibacterial and Antioxidant Activities of Five *Lavandula Stoechas* Cultivars Grown in Thailand. *Chem. Biodivers.* **2019**, *16*, e1900371. [CrossRef]
22. Messaoud, C.; Chograni, H.; Boussaid, M. Chemical Composition and Antioxidant Activities of Essential Oils and Methanol Extracts of Three Wild *Lavandula* L. Species. *Nat. Prod. Res.* **2012**, *26*, 1976–1984. [CrossRef]
23. Yassine, E.Z.; Dalila, B.; Latifa, E.M.; Smahan, B.; Lebtar, S.; Sanae, A.; Abdellah, F. Phytochemical Screening, Anti-Inflammatory Activity and Acute Toxicity of Hydro-Ethanolic, Flavonoid, Tannin and Mucilage Extracts of *Lavandula Stoechas* L. from Morocco. *Int. J. Pharm. Phytochem. Res.* **2016**, *8*, 31–37.
24. Lafraxo, S.; El Barnossi, A.; El Moussaoui, A.; Bourhia, M.; Salamatullah, A.M.; Alzahrani, A.; Ait Akka, A.; Choubbane, A.; Akhazzane, M.; Aboul-Soud, M.A. Essential Oils from Leaves of *Juniperus Thurifera* L., Exhibiting Antioxidant, Antifungal and Antibacterial Activities against Antibiotic-Resistant Microbes. *Horticulturae* **2022**, *8*, 321. [CrossRef]
25. Brahmi, F.; Guendouze, N.; Hauchard, D.; Okusa, P.; Kamagaju, L.; Madani, K.; Duez, P. Phenolic Profile and Biological Activities of *Micromeria Graeca* (L.) Benth. Ex Rchb. *Int. J. Food Prop.* **2017**, *20*, 2070–2083.
26. Benali, T.; Habbadi, K.; Bouyahya, A.; Khabbach, A.; Marmouzi, I.; Aanniz, T.; Chtibi, H.; Mrabti, H.N.; Achbani, E.H.; Hammani, K. Phytochemical Analysis and Study of Antioxidant, Anticandidal, and Antibacterial Activities of *Teucrium Polium* Subsp. *Polium and Micromeria Graeca (Lamiaceae) Essential Oils from Northern Morocco. Evid.-Based Complement. Altern. Med.* **2021**, *2021*, 6641720.
27. Gulluce, M.; Sahin, F.; Sokmen, M.; Ozer, H.; Daferera, D.; Sokmen, A.; Polissiou, M.; Adiguzel, A.; Ozkan, H. Antimicrobial and Antioxidant Properties of the Essential Oils and Methanol Extract from *Mentha Longifolia* L. Ssp. Longifolia. *Food Chem.* **2007**, *103*, 1449–1456. [CrossRef]
28. Rusu, M.E.; Fizesan, I.; Pop, A.; Mocan, A.; Gheldiu, A.M.; Babota, M.; Vodnar, D.C.; Jurj, A.; Berindan-Neagoe, I.; Vlase, L.; et al. Walnut (*Juglans regia* L.) Septum: Assessment of Bioactive Molecules and In Vitro Biological Effects. *Molecules* **2020**, *25*, 2187. [CrossRef] [PubMed]
29. Ćavar Zeljković, S.; Schadich, E.; Džubák, P.; Hajdúch, M.; Tarkowski, P. Antiviral Activity of Selected Lamiaceae Essential Oils and Their Monoterpenes Against SARS-Cov-2. *Front. Pharmacol.* **2022**, *13*, 1589. [CrossRef]
30. Elsebai, M.F.; Albalawi, M.A. Essential Oils and COVID-19. *Molecules* **2022**, *27*, 7893. [CrossRef] [PubMed]
31. Strub, D.J.; Talma, M.; Strub, M.; Rut, W.; Zmudzinski, M.; Brud, W.; Neyts, J.; Vangeel, L.; Zhang, L.; Sun, X. Evaluation of the Anti-SARS-CoV-2 Properties of Essential Oils and Aromatic Extracts. *Sci. Rep.* **2022**, *12*, 14230. [CrossRef] [PubMed]
32. Jain, A.N. Surflex: Fully Automatic Flexible Molecular Docking Using a Molecular Similarity-Based Search Engine. *J. Med. Chem.* **2003**, *46*, 499–511. [CrossRef] [PubMed]
33. Trott, O.; Olson, A.J. AutoDock Vina: Improving the Speed and Accuracy of Docking with a New Scoring Function, Efficient Optimization, and Multithreading. *J. Comput. Chem.* **2010**, *31*, 455–461. [CrossRef] [PubMed]
34. Jin, Z.; Du, X.; Xu, Y.; Deng, Y.; Liu, M.; Zhao, Y.; Zhang, B.; Li, X.; Zhang, L.; Peng, C. Electromechanical Coupling in the Hyperpolarization-Activated K+ Channel KAT1. *Nature* **2020**, *583*, 145–149.
35. Wrapp, D.; Wang, N.; Corbett, K.S.; Goldsmith, J.A.; Hsieh, C.-L.; Abiona, O.; Graham, B.S.; McLellan, J.S. Cryo-EM Structure of the 2019-NCoV Spike in the Prefusion Conformation. *Science* **2020**, *367*, 1260–1263. [CrossRef] [PubMed]
36. Kufareva, I.; Abagyan, R. Methods of Protein Structure Comparison. In *Homology Modeling*; Springer: Berlin/Heidelberg, Germany, 2011; pp. 231–257.
37. *Discovery Studio Visualizer*, version 17.2.0.16349; Accelrys Software Inc.: San Diego, CA, USA, 2016.
38. Ghaleb, A.; Aouidate, A.; Ayouchia, H.B.E.; Aarjane, M.; Anane, H.; Stiriba, S.-E. In Silico Molecular Investigations of Pyridine N-Oxide Compounds as Potential Inhibitors of SARS-CoV-2: 3D QSAR, Molecular Docking Modeling, and ADMET Screening. *J. Biomol. Struct. Dyn.* **2022**, *40*, 143–153. [CrossRef]
39. Pires, D.E.; Blundell, T.L.; Ascher, D.B. PkCSM: Predicting Small-Molecule Pharmacokinetic and Toxicity Properties Using Graph-Based Signatures. *J. Med. Chem.* **2015**, *58*, 4066–4072. [CrossRef]
40. Khaerunnisa, S.; Kurniawan, H.; Awaluddin, R.; Suhartati, S.; Soetjipto, S. Potential Inhibitor of COVID-19 Main Protease (Mpro) from Several Medicinal Plant Compounds by Molecular Docking Study. *Preprints* **2020**, *2020*, 2020030226.
41. Tahir ul Qamar, M.; Shahid, F.; Aslam, S.; Ashfaq, U.A.; Aslam, S.; Fatima, I.; Fareed, M.M.; Zohaib, A.; Chen, L.-L. Reverse Vaccinology Assisted Designing of Multiepitope-Based Subunit Vaccine against SARS-CoV-2. *Infect. Dis. Poverty* **2020**, *9*, 132. [CrossRef]
42. Xu, Y.; Li, X.; Zhu, B.; Liang, H.; Fang, C.; Gong, Y.; Guo, Q.; Sun, X.; Zhao, D.; Shen, J. Characteristics of Pediatric SARS-CoV-2 Infection and Potential Evidence for Persistent Fecal Viral Shedding. *Nat. Med.* **2020**, *26*, 502–505. [CrossRef]
43. Benabdelkader, T.; Zitouni, A.; Guitton, Y.; Jullien, F.; Maitre, D.; Casabianca, H.; Legendre, L.; Kameli, A. Essential Oils from Wild Populations of Algerian *Lavandula Stoechas* L.: Composition, Chemical Variability, and in Vitro Biological Properties. *Chem. Biodivers.* **2011**, *8*, 937–953. [CrossRef]

44. Biltekin, S.N.; Karadağ, A.E.; Demirci, B.; Demirci, F. ACE2 and LOX Enzyme Inhibitions of Different Lavender Essential Oils and Major Components Linalool and Camphor. *ACS Omega* **2022**, *7*, 36561–36566. [CrossRef]
45. Bozkurt, İ.A.; Soylu, S.; Merve, K.; Soylu, E.M. Chemical Composition and Antibacterial Activity of Essential Oils Isolated from Medicinal Plants against Gall Forming Plant Pathogenic Bacterial Disease Agents. *Kahramanmaraş Sütçü İmam Üniv. Tarım Ve Doğa Derg.* **2020**, *23*, 1474–1482.
46. Gören, A.C.; Topçu, G.; Bilsel, G.; Bilsel, M.; Aydoğmusç, Z.; Pezzuto, J.M. The Chemical Constituents and Biological Activity of Essential Oil of *Lavandula Stoechas* Ssp. *Stoechas*. *Z. Für Nat. C* **2002**, *57*, 797–800. [CrossRef]
47. Chamkhi, I.; Benali, T.; Aanniz, T.; El Menyiy, N.; Guaouguaou, F.-E.; El Omari, N.; El-Shazly, M.; Zengin, G.; Bouyahya, A. Plant-Microbial Interaction: The Mechanism and the Application of Microbial Elicitor Induced Secondary Metabolites Biosynthesis in Medicinal Plants. *Plant Physiol. Biochem.* **2021**, *167*, 269–295. [CrossRef]
48. Aboukhalid, K.; Al Faiz, C.; Douaik, A.; Bakha, M.; Kursa, K.; Agacka-Mo\ldoch, M.; Machon, N.; Tomi, F.; Lamiri, A. Influence of Environmental Factors on Essential Oil Variability in *Origanum Compactum* Benth. Growing Wild in Morocco. *Chem. Biodivers.* **2017**, *14*, e1700158. [CrossRef]
49. Aboukhalid, K.; Lamiri, A.; Agacka-Mołdoch, M.; Doroszewska, T.; Douaik, A.; Bakha, M.; Casanova, J.; Tomi, F.; Machon, N.; Faiz, C.A. Chemical Polymorphism of *Origanum Compactum* Grown in All Natural Habitats in Morocco. *Chem. Biodivers.* **2016**, *13*, 1126–1139. [CrossRef]
50. Angioni, A.; Barra, A.; Coroneo, V.; Dessi, S.; Cabras, P. Chemical Composition, Seasonal Variability, and Antifungal Activity of *Lavandula Stoechas* L. Ssp. *Stoechas Essential Oils from Stem/Leaves and Flowers*. *J. Agric. Food Chem.* **2006**, *54*, 4364–4370.
51. Carrasco, A.; Ortiz-Ruiz, V.; Martinez-Gutierrez, R.; Tomas, V.; Tudela, J. *Lavandula Stoechas* Essential Oil from Spain: Aromatic Profile Determined by Gas Chromatography–Mass Spectrometry, Antioxidant and Lipoxygenase Inhibitory Bioactivities. *Ind. Crops Prod.* **2015**, *73*, 16–27. [CrossRef]
52. Cherrat, L.; Espina, L.; Bakkali, M.; Pagán, R.; Laglaoui, A. Chemical Composition, Antioxidant and Antimicrobial Properties of *Mentha Pulegium*, *Lavandula Stoechas* and *Satureja Calamintha Scheele* Essential Oils and an Evaluation of Their Bactericidal Effect in Combined Processes. *Innov. Food Sci. Emerg. Technol.* **2014**, *22*, 221–229. [CrossRef]
53. Cosentino, S.; Tuberoso, C.I.G.; Pisano, B.; Satta, M.L.; Mascia, V.; Arzedi, E.; Palmas, F. In-Vitro Antimicrobial Activity and Chemical Composition of Sardinian *Thymus* Essential Oils. *Lett. Appl. Microbiol.* **1999**, *29*, 130–135. [CrossRef]
54. McGowan, J.E., Jr. Resistance in Nonfermenting Gram-Negative Bacteria: Multidrug Resistance to the Maximum. *Am. J. Infect. Control.* **2006**, *34*, S29–S37. [CrossRef]
55. Sokovic, M.; Marin, P.D.; Brkic, D.; van Griensven, L.J. Chemical Composition and Antibacterial Activity of Essential Oils against Human Pathogenic Bacteria. *Food* **2008**, *1*, 220–226.
56. Gill, A.O.; Delaquis, P.; Russo, P.; Holley, R.A. Evaluation of Antilisterial Action of Cilantro Oil on Vacuum Packed Ham. *Int. J. Food Microbiol.* **2002**, *73*, 83–92. [CrossRef]
57. Mourey, A.; Canillac, N. Anti-Listeria Monocytogenes Activity of Essential Oils Components of Conifers. *Food Control* **2002**, *13*, 289–292. [CrossRef]
58. Alminderej, F.; Bakari, S.; Almundarij, T.I.; Snoussi, M.; Aouadi, K.; Kadri, A. Antioxidant Activities of a New Chemotype of *Piper Cubeba* L. Fruit Essential Oil (Methyleugenol/Eugenol): In Silico Molecular Docking and ADMET Studies. *Plants* **2020**, *9*, 1534. [CrossRef]
59. Silva, A.M.d.O.e.; Machado, I.D.; Santin, J.R.; de Melo, I.L.P.; Pedrosa, G.V.; Genovese, M.I.; Farsky, S.H.P.; Mancini-Filho, J. Aqueous Extract of *Rosmarinus Officinalis* L. Inhibits Neutrophil Influx and Cytokine Secretion. *Phytother. Res.* **2015**, *29*, 125–133. [CrossRef]
60. Cetin, A. Some Flavolignans as Potent SARS-CoV-2 Inhibitors via Molecular Docking, Molecular Dynamic Simulations and ADME Analysis. *Curr. Comput.-Aided Drug Des.* **2022**, *18*, 337–346. [CrossRef]
61. Bahl, A.S.; Verma, V.K.; Bhatia, J.; Arya, D.S. Integrating In Silico and In Vivo Approach for Investigating the Role of Polyherbal Oil in Prevention and Treatment of COVID-19 Infection. *Chem.-Biol. Interact.* **2022**, *367*, 110179. [CrossRef]
62. Khan, M.T.; Ali, A.; Wei, X.; Nadeem, T.; Muhammad, S.; Al-Sehemi, A.G.; Wei, D. Efeito Inibitório da Timoquinona de *Nigella Sativa* Contra a Principal Protease Do SARS-CoV-2. Um Estudo In Silico. *Braz. J. Biol.* **2022**, *84*, e250667. [CrossRef]
63. Santos, E.S.; Silva, P.C.; Sousa, P.S.; Aquino, C.C.; Pacheco, G.; Teixeira, L.F.; Araujo, A.R.; Sousa, F.B.; Barros, R.O.; Ramos, R.M. Antiviral Potential of Diminazene Aceturate against SARS-CoV-2 Proteases Using Computational and in Vitro Approaches. *Chem.-Biol. Interact.* **2022**, *367*, 110161. [CrossRef]
64. da Silva, J.K.R.; Figueiredo, P.L.B.; Byler, K.G.; Setzer, W.N. Essential Oils as Antiviral Agents, Potential of Essential Oils to Treat SARS-CoV-2 Infection: An in-Silico Investigation. *Int. J. Mol. Sci.* **2020**, *21*, 3426. [CrossRef]
65. Malone, B.; Urakova, N.; Snijder, E.J.; Campbell, E.A. Structures and Functions of Coronavirus Replication–Transcription Complexes and Their Relevance for SARS-CoV-2 Drug Design. *Nat. Rev. Mol. Cell Biol.* **2022**, *23*, 21–39. [CrossRef]
66. Zhou, Y.; Huang, T.; Cheng, A.S.; Yu, J.; Kang, W.; To, K.F. The TEAD Family and Its Oncogenic Role in Promoting Tumorigenesis. *Int. J. Mol. Sci.* **2016**, *17*, 138. [CrossRef]
67. Gentile, D.; Patamia, V.; Scala, A.; Sciortino, M.T.; Piperno, A.; Rescifina, A. Inhibitors of SARS-CoV-2 Main Protease from a Library of Marine Natural Products: A Virtual Screening and Molecular Modeling Study. *Mar. Drugs* **2020**, *18*, 225. [CrossRef]

68. Joshi, R.S.; Jagdale, S.S.; Bansode, S.B.; Shankar, S.S.; Tellis, M.B.; Pandya, V.K.; Chugh, A.; Giri, A.P.; Kulkarni, M.J. Discovery of Potential Multi-Target-Directed Ligands by Targeting Host-Specific SARS-CoV-2 Structurally Conserved Main Protease. *J. Biomol. Struct. Dyn.* **2021**, *39*, 3099–3114. [CrossRef]
69. Manish, M. Studies on Computational Molecular Interaction between SARS-CoV-2 Main Protease and Natural Products. *ChemRxiv* 2020.
70. Thuy, B.T.P.; My, T.T.A.; Hai, N.T.T.; Hieu, L.T.; Hoa, T.T.; Thi Phuong Loan, H.; Triet, N.T.; Anh, T.T.V.; Quy, P.T.; Tat, P.V. Investigation into SARS-CoV-2 Resistance of Compounds in Garlic Essential Oil. *ACS Omega* **2020**, *5*, 8312–8320. [CrossRef]
71. Beck, H.C.; Petersen, J.; Nielsen, S.J.; Morszeck, C.; Jensen, P.B.; Sehested, M.; Grauslund, M. Proteomic Profiling of Human Colon Cancer Cells Treated with the Histone Deacetylase Inhibitor Belinostat. *Electrophoresis* **2010**, *31*, 2714–2721. [CrossRef]
72. Hofmarcher, M.; Mayr, A.; Rumetshofer, E.; Ruch, P.; Renz, P.; Schimunek, J.; Seidl, P.; Vall, A.; Widrich, M.; Hochreiter, S. Large-Scale Ligand-Based Virtual Screening for SARS-CoV-2 Inhibitors Using Deep Neural Networks. *arXiv* **2020**, arXiv:2004.00979. [CrossRef]
73. Altulea, D.; Maassen, S.; Baranov, M.V.; van den Bogaart, G. What Makes (Hydroxy) Chloroquine Ineffective against COVID-19: Insights from Cell Biology. *J. Mol. Cell Biol.* **2021**, *13*, 175–184. [CrossRef]
74. Kapuy, O.; Korcsmáros, T. Chloroquine and COVID-19—A Systems Biology Model Uncovers the Drug's Detrimental Effect on Autophagy and Explains Its Failure. *PLoS ONE* **2022**, *17*, e0266337. [CrossRef]
75. Law, W.Y.; Asaruddin, M.R.; Bhawani, S.A.; Mohamad, S. Pharmacophore Modelling of Vanillin Derivatives, Favipiravir, Chloroquine, Hydroxychloroquine, Monolaurin and Tetrodotoxin as MPro Inhibitors of Severe Acute Respiratory Syndrome Coronavirus-2 (SARS-CoV-2). *BMC Res. Notes* **2020**, *13*, 527. [CrossRef] [PubMed]
76. Foudah, A.I.; Alqarni, M.H.; Alam, A.; Salkini, M.A.; Alam, P.; Alkholifi, F.K.; Yusufoglu, H.S. Determination of Chemical Composition, In Vitro and In Silico Evaluation of Essential Oil from Leaves of *Apium Graveolens* Grown in Saudi Arabia. *Molecules* **2021**, *26*, 7372. [CrossRef]
77. Ghannay, S.; Kadri, A.; Aouadi, K. Synthesis, in Vitro Antimicrobial Assessment, and Computational Investigation of Pharmacokinetic and Bioactivity Properties of Novel Trifluoromethylated Compounds Using In Silico ADME and Toxicity Prediction Tools. *Mon. Für Chem.-Chem. Mon.* **2020**, *151*, 267–280. [CrossRef]
78. Wei, M.; Liu, F.; Raka, R.N.; Xiang, J.; Xiao, J.; Han, T.; Guo, F.; Yang, S.; Wu, H. In Vitro and In Silico Analysis of 'Taikong Blue'Lavender Essential Oil in LPS-Induced HaCaT Cells and RAW264. 7 Murine Macrophages. *BMC Complement. Med. Ther.* **2022**, *22*, 324. [CrossRef]
79. Abou Baker, D.H.; Amarowicz, R.; Kandeil, A.; Ali, M.A.; Ibrahim, E.A. Antiviral Activity of *Lavandula Angustifolia* L. and *Salvia Officinalis* L. Essential Oils against *Avian Influenza* H5N1 *Virus*. *J. Agric. Food Res.* **2021**, *4*, 100135. [CrossRef]

MDPI
St. Alban-Anlage 66
4052 Basel
Switzerland
www.mdpi.com

Plants Editorial Office
E-mail: plants@mdpi.com
www.mdpi.com/journal/plants

www.ingramcontent.com/pod-product-compliance
Lightning Source LLC
Chambersburg PA
CBHW061818210926
51850CB00006B/742